An Introduction to Involutive Structures

Detailing the main methods in the theory of involutive systems of complex vector fields, this book examines the major results from the last 25 years in the subject. One of the key tools of the subject – the Baouendi–Treves approximation theorem – is proved for many function spaces. This in turn is applied to questions in partial differential equations and several complex variables. Many basic problems such as regularity, unique continuation and boundary behavior of the solutions are explored. The local solvability of systems of partial differential equations is studied in some detail. The book provides a solid background for beginners in the field and also contains a treatment of many recent results which will be of interest to researchers in the subject.

SHIFERAW BERHANU is a Professor of Mathematics at Temple University in the US.

PAULO D. CORDARO is a Professor of Mathematics in the Institute of Mathematics and Statistics at the University of São Paulo in Brazil.

JORGE HOUNIE is a Professor of Mathematics at the Federal University of São Carlos in Brazil.

NEW MATHEMATICAL MONOGRAPHS

For information about Cambridge University Press mathematics publications
visit http://www.cambridge.org/mathematics

An Introduction to Involutive Structures

SHIFERAW BERHANU
Temple University

PAULO D. CORDARO
University of São Paulo

JORGE HOUNIE
Federal University of São Carlos

CAMBRIDGE
UNIVERSITY PRESS

CAMBRIDGE UNIVERSITY PRESS

Cambridge, New York, Melbourne, Madrid, Cape Town, Singapore, São Paulo

Cambridge University Press
The Edinburgh Building, Cambridge CB2 8RU, UK

Published in the United States of America by Cambridge University Press, New York

www.cambridge.org
Information on this title: www.cambridge.org/9780521878579

First published 2008

Printed in the United Kingdom at the University Press, Cambridge

A catalog record for this publication is available from the British Library

ISBN 978-0-521-87857-9 hardback

Contents

Preface

Since the first systematic exposition of the theory of involutive systems of
vector fields ([**T5**]) was published almost 15 years ago, the subject has under-
gone considerable development and many new applications have been found.
Systems of vector fields arise as a local basis of an involutive sub-bundle
$\mathcal{V}$ of the complexified tangent bundle $\mathbb{C}TM$. *Involutivity* of $\mathcal{V}$ means that
the commutation bracket of two smooth sections of $\mathcal{V}$ must also be a section
of $\mathcal{V}$. Examples of involutive structures $(M, \mathcal{V})$ include foliations, complex
structures, and CR structures. In these examples, $\mathcal{V} \cap \overline{\mathcal{V}}$ has constant rank.
However, in recent work on integral geometry, natural examples of involu-
tive structures have arisen for which the rank of $\mathcal{V} \cap \overline{\mathcal{V}}$ changes from point
to point ([**BE**], [**BEGM**], and [**EG1**]). In the works [**BE**] and [**BEGM**], the
cohomology of involutive structures is a key ingredient. Examples of invo-
lutive structures where the rank of $\mathcal{V} \cap \overline{\mathcal{V}}$ is not constant also arise naturally,
for instance, on the tangent bundle of symmetric spaces (see [**Sz**] and the
references therein) or in the study of the generalized similarity principle for
the equation

$$Lu = Au + B\overline{u}$$

where L is a planar complex vector field not necessarily elliptic, which is
intimately linked to the study of infinitesimal deformations of surfaces in $\mathbb{R}^3$
with non-negative curvature (see [**Me3**], [**Me4**], and the references therein).

This book introduces the reader to a number of results on systems of vector
fields with complex-valued coefficients defined on a smooth manifold M.
Most of the time, it will be assumed that the involutive structure $(M, \mathcal{V})$ is
locally integrable. The latter means that the orthogonal of $\mathcal{V}$, which is a sub-
bundle T' of the complexified cotangent bundle $\mathbb{C}T^*M$, is locally generated
by exact differentials. When $(M, \mathcal{V})$ is locally integrable, each point has a
neighborhood U such that if $\{L_1, \ldots, L_n\}$ are n smooth vector fields that form

a basis of V over U, then we can find $m = \dim \mathcal{M} - n$ smooth, complex-valued functions $Z_1, \ldots, Z_m$ which are solutions of the equations

$$L_j h = 0, \quad 1 \leq j \leq n \tag{1}$$

and whose differentials are linearly independent over $\mathbb{C}$ at each point of U. The m functions $Z = (Z_1, \ldots, Z_m)$ are sometimes referred to as a complete set of *first integrals* in the neighborhood U.

In 1981, in [**BT1**], Baouendi and Treves proved that in a locally integrable structure, each solution of (1) can be locally approximated by a sequence $P_k(Z)$ where the P_k are holomorphic polynomials of m variables and $Z = (Z_1, \ldots, Z_m)$ is a complete set of first integrals. This approximation theorem has enabled several researchers to use the methods of complex analysis, harmonic analysis, and partial differential equations to study many problems on locally integrable structures. These problems include: the local and microlocal regularity of the solutions of (1); the determination of sets of uniqueness for solutions of (1); the solvability of the differential complex associated with the structure $(\mathcal{M}, V)$; and many other properties of the solutions of (1).

This book attempts to present a systematic treatment of some of these results in a way that is accessible to graduate students with a background in real analysis, one complex variable, and basic introductions to several complex variables and linear PDEs including the theory of distributions.

Chapter I introduces the basic concepts in the theory of involutive and locally integrable structures. Special classes of involutive structures such as complex structures, CR structures, elliptic structures, and real analytic structures are identified and examples are provided. Useful local representations both for general involutive and locally integrable structures are also discussed. A proof of the Newlander–Nirenberg theorem is presented in the appendix to Chapter I. Chapter II is devoted to the approximation theorem of Baouendi and Treves. It is shown that the approximation is valid in many function spaces used in analysis: the Lebesgue spaces L^p, $1 \leq p < \infty$; Sobolev spaces; Hölder spaces; and localizable Hardy spaces h^p, $0 < p < \infty$. Applications to uniqueness in the Cauchy problem and extendability of CR functions are also included. Chapter III presents a variety of results on unique continuation for solutions and approximate solutions in a locally integrable structure $(\mathcal{M}, V)$. The orbits of Sussmann associated with the real parts $\Re L$ of the smooth sections of V play a crucial role in many problems, including the study of unique continuation and the chapter includes a discussion of some of the properties of these orbits. Chapter IV provides a detailed treatment of locally solvable vector fields. In the first part of the chapter, where the focus is on planar vector fields, the solvability condition ($\mathscr{P}$) of Nirenberg and Treves is discussed and a priori estimates are proved in L^p and in a mixed norm that involves the Hardy space $h^1(\mathbb{R})$. A duality argument is then used to derive local solvability results in L^p, $1 < p < \infty$ and in $L^\infty[\mathbb{R}; \text{bmo}(\mathbb{R})]$. The chapter also includes sections on the sufficiency and necessity of condition ($\mathscr{P}$) for local solvability in higher dimensions. The first part of Chapter V introduces certain submanifolds in an involutive structure $(\mathcal{M}, \mathcal{V})$ which are important in the study of solutions. These submanifolds are generalizations of the totally real and generic CR submanifolds encountered in CR manifolds. The second part of the chapter introduces the FBI transform first in $\mathbb{R}^n$ and then in a locally integrable structure. The FBI transform is then applied to derive edge-of-the-wedge type results. It is also applied to study the microlocal singularities of the solutions of a first-order nonlinear PDE and a generalization of the F. and M. Riesz theorem. Chapter VI studies some boundary properties of the solutions of locally integrable vector fields. These properties include the existence of a trace at the boundary, pointwise convergence of solutions to their boundary values, and the validity of Hardy space-like properties. Chapter VII describes the differential complex attached to a general involutive structure. An invariant definition of this complex is followed by a useful representation in appropriate coordinates. An approximate Poincaré Lemma for locally integrable structures is also proved in the chapter. Chapter VIII deals with the local solvability theory of the undetermined systems of partial differential equations naturally associated with a locally integrable structure, that is, the cohomology theory of its differential complex. Necessary and sufficient conditions are studied in some detail when the structure is analytic, or elliptic, or has corank one. Concerning the latter class, a thorough exposition of the geometric characterization of local solvability in degree one for real analytic structures is presented.

Finally we conclude with an epilogue which summarizes some of the results obtained in recent years on diverse areas such as the similarity principle, Mizohata structures, and hyperfunction solutions in hypoanalytic manifolds. Two applications of the similarity principle are described. The first application concerns uniqueness in the Cauchy problem for a class of semilinear equations. The second application involves the theory of bending of surfaces.

There are numerous interesting results on complex vector fields and involutive structures that have been obtained since the publication of [**BT1**] and which are not covered in this book. The authors have selected the material with which they have had first-hand experience. In the notes at the end of each chapter, we indicate some related works and provide additional references.

The reader is referred to [**BER**] for a further reference on CR manifolds and to [**T5**] for additional topics on involutive structures.

We are grateful to Elisandra Bär, Sagun Chanillo, Nicholas Hanges, Gustavo Hoepfner, and Gerson Petronilho for reading parts of the manuscript, pointing out errors and suggesting improvements.

We are also grateful to Peter Thompson of Cambridge University Press for the expedience with which our book has been handled.

I

Locally integrable structures

In this chapter we introduce the main concepts which will be studied through-out the book. In order to do so we recall some standard notions such as differentiable manifolds, vector fields, differential forms, etc., with the purpose mainly of laying down the basis for the presentation and to establish the notations.

Nevertheless, we assume from the reader some familiarity with these concepts. In particular, we freely use some standard results on complex vector fields and complex differential forms on $\mathbb{R}^N$.

I.1 Complex vector fields

Let Ω be a Hausdorff topological space, with a countable basis of open sets. A *differentiable structure over Ω of dimension N* is a collection of pairs $\mathcal{F} = \{(U, \mathbf{x})\}$, where $U \subset \Omega$ is a nonempty open set, $\mathbf{x} : U \longrightarrow \mathbb{R}^N$ is a homeomorphism onto an open subset $\mathbf{x}(U)$ of $\mathbb{R}^N$ and the following properties are satisfied:

(1) $\bigcup_{(U,\mathbf{x})\in\mathcal{F}} U = \Omega$;

(2) $\mathbf{x}(U \cap U') \xrightarrow{\mathbf{x}' \circ \mathbf{x}^{-1}} \mathbf{x}'(U \cap U')$ is C^∞ for each pair $(U, \mathbf{x})$, $(U', \mathbf{x}') \in \mathcal{F}$ with $U \cap U' \neq \emptyset$;

(3) $\mathcal{F}$ is maximal with respect to (1) and (2), that is, if $\emptyset \neq V \subset \Omega$ is open and $\mathbf{y} : V \longrightarrow \mathbf{y}(V)$ is a homeomorphism over an open subset of $\mathbb{R}^N$ such that, for any $(U, \mathbf{x}) \in \mathcal{F}$ with $U \cap V \neq \emptyset$, the composition $\mathbf{x}(U \cap V) \xrightarrow{\mathbf{y} \circ \mathbf{x}^{-1}} \mathbf{y}(U \cap V)$ is C^∞, then $(V, \mathbf{y}) \in \mathcal{F}$.

1

It is easy to see that given any family $\mathcal{F}^* = \{(U, \mathbf{x})\}$ as above satisfying (1) and (2) there is a unique differentiable structure $\mathcal{F}$ over Ω, of dimension N, such that $\mathcal{F}^* \subset \mathcal{F}$.

DEFINITION I.1.1. *A differentiable manifold (or smooth manifold) of dimension N is a Hausdorff topological space Ω, with a countable basis equipped with a differentiable structure of dimension N.*

If, in the above definitions, we replace C^∞ by real-analytic we obtain the concept of a *real-analytic manifold of dimension N*.

We give some examples:

(1) $\Omega = \mathbb{R}^N$, $\mathcal{F}^* = \{(\mathbb{R}^N, \text{identity map})\}$.
(2) Let Ω be a differentiable manifold of dimension N and let $W \subset \Omega$ be open. Then over W is defined a natural differentiable structure of dimension N, which is given by

$$\mathcal{F}_W = \{(W \cap U, \mathbf{x}|_{W \cap U}) : (U, \mathbf{x}) \in \mathcal{F}, W \cap U \neq \emptyset\}.$$

(3) Let $f : \mathbb{R}^{N+1} \to \mathbb{R}$ be a C^∞ function. Let

$$\Omega = \{x \in \mathbb{R}^{N+1} : f(x) = 0\}$$

and suppose that $df(x) \neq 0, \forall x \in \Omega$. Then a natural differentiable structure of dimension N is defined over Ω (as a consequence of the implicit function theorem).

NOTATION. An element $(U, \mathbf{x}) \in \mathcal{F}$ will be refered to as a *local chart* or as a *local system of coordinates*. If we write $\mathbf{x} = (x_1, \ldots, x_N)$ then for $p \in U$ its local coordinates (with respect to this given local chart) are given by $(x_1(p), \ldots, x_N(p))$.

From now on, unless otherwise stated, we shall fix a differentiable manifold Ω (of dimension N). We shall say that a function $f : \Omega \to \mathbb{C}$ is *smooth* if for every $(U, \mathbf{x}) \in \mathcal{F}$ the composition $f \circ \mathbf{x}^{-1}$ is C^∞ on $\mathbf{x}(U)$.[1] We shall denote by $C^\infty(\Omega)$ the set of all smooth functions on Ω. We observe that C^∞ is an algebra over $\mathbb{C}$ which contains, as an $\mathbb{R}$-subalgebra, the set $C^\infty(\Omega; \mathbb{R})$ of all smooth functions on Ω which are real-valued.

DEFINITION I.1.2. *A (smooth) complex vector field over Ω is a $\mathbb{C}$-linear map*

$$L : C^\infty(\Omega) \longrightarrow C^\infty(\Omega)$$

---

[1] More generally, we say that a function $f : \Omega \to \mathbb{C}$ is C^k ($k \geq 0$) if for every $(U, \mathbf{x}) \in \mathcal{F}$ the composition $f \circ \mathbf{x}^{-1}$ is C^k on $\mathbf{x}(U)$.

which satisfies the Leibniz rule

$$L(fg) = fL(g) + gL(f), \quad f, g \in C^\infty(\Omega). \tag{I.1}$$

We shall denote by $\mathfrak{X}(\Omega)$ the set of all complex vector fields over Ω.

PROPOSITION I.1.3. *If $L \in \mathfrak{X}(\Omega)$ and if f is constant then $Lf = 0$. We also have*

$$\operatorname{supp} Lf \subset \operatorname{supp} f, \quad \forall f \in C^\infty(\Omega), \ L \in \mathfrak{X}(\Omega). \tag{I.2}$$

PROOF. For the first statement it suffices to show that $L1 = 0$ and this follows from (I.1) together with the fact that $1^2 = 1$. We shall now prove (I.2); we must show that if f vanishes on an open set $V \subset \Omega$ then the same is true for Lf.

Let $p \in V$ be arbitrary. We select a local chart $(U, \mathbf{x})$ with $p \in U \subset V$ and take $\varphi \in C_c^\infty(\mathbf{x}(U))$ such that $\varphi(\mathbf{x}(p)) = 1$. Then the function $g : \Omega \to \mathbb{R}$ defined by the rule

$$g(q) = \begin{cases} \varphi(\mathbf{x}(q)) & \text{if } q \in U \\ 0 & \text{if } q \notin U \end{cases}$$

belongs to $C^\infty(\Omega; \mathbb{R})$ and vanishes on $\Omega \setminus V$. In particular,

$$f = (1 - g)f$$

and then

$$L(f)(p) = (1 - g(p))L(f)(p) + f(p)L(1 - g)(p) = 0,$$

since $g(p) = 1$. $\qquad\square$

A consequence of the preceding result is the possibility of defining the *restriction* of an element $L \in \mathfrak{X}(\Omega)$ to an open subset W of Ω. More precisely, there is a $\mathbb{C}$-linear map

$$\mathfrak{X}(\Omega) \ni L \longrightarrow L_W \in \mathfrak{X}(W)$$

which turns the diagram

$$\begin{array}{ccc} C^\infty(\Omega) & \overset{L}{\longrightarrow} & C^\infty(\Omega) \\ \downarrow & & \downarrow \\ C^\infty(W) & \overset{L_W}{\longrightarrow} & C^\infty(W) \end{array}$$

commutative (the vertical arrows denote the restriction map). Indeed, if $p \in W$ and $f \in C^\infty(W)$ we set

$$L_W(f)(p) = L(\tilde{f})(p),$$

where $\tilde{f}$ is any element in $C^\infty(\Omega)$ which coincides with f in a neighborhood of p. Such a definition is meaningful according to Proposition I.1.3 and it is very easy to check that L_W defines an element in $\mathfrak{X}(W)$. As usual we shall write L instead of L_W, since the meaning will always be clear from the context.

I.2 The algebraic structure of $\mathfrak{X}(\Omega)$

Given $g \in C^\infty(\Omega)$ and $L \in \mathfrak{X}(\Omega)$ we can define $gL \in \mathfrak{X}(\Omega)$ by

$$(gL)(f) = g \cdot L(f), \quad f \in C^\infty(\Omega).$$

Such external multiplication gives $\mathfrak{X}(\Omega)$ the structure of a $C^\infty(\Omega)$-*module*.

A very important (internal) operation in $\mathfrak{X}(\Omega)$ is the so-called *Lie bracket* (or *commutator*) between two vector fields. Given $L, M \in \mathfrak{X}(\Omega)$ we define

$$[L, M](f) = L(M(f)) - M(L(f)), \quad f \in C^\infty(\Omega). \tag{I.3}$$

It is a simple verification to check that $[L, M] \in \mathfrak{X}(\Omega)$. This bracket operation turns $\mathfrak{X}(\Omega)$ into a *Lie algebra*[2] *over* $\mathbb{C}$.

Let $(U, \mathbf{x})$ be a local chart in Ω and let also $L \in \mathfrak{X}(U)$. We fix $p \in U$ and write as before

$$\mathbf{x}(q) = (x_1(q), \ldots, x_N(q)), \quad q \in U.$$

Next we take $V \subset U$ open such that $\mathbf{x}(V)$ is an open ball centered at $\mathbf{x}(p) = a = (a_1, \ldots, a_N)$. Given $f \in C^\infty(U)$, write $f^* = f \circ \mathbf{x}^{-1}$. If $(x_1, \ldots, x_N) \in \mathbf{x}(V)$, the Fundamental Theorem of Calculus applied to the function $t \mapsto f^*(a_1 + t(x_1 - a_1), \ldots, a_N + t(x_N - a_N))$ gives

$$f^*(x_1, \ldots, x_N) = f^*(a_1, \ldots, a_N) + \sum_{j=1}^N h_j(x_1, \ldots, x_N)(x_j - a_j),$$

where $h_j \in C^\infty(\mathbf{x}(V))$ and $h_j(a) = (\partial f^*/\partial x_j)(a)$. If we further set $g_j = h_j \circ \mathbf{x} \in C^\infty(V)$, we obtain

$$f(q) = f(p) + \sum_{j=1}^N g_j(q)(x_j(q) - x_j(p)), \quad q \in V, \tag{I.4}$$

---

[2] Recall that a Lie algebra over $\mathbb{C}$ is a $\mathbb{C}$-vector space E over which is defined a bilinear form $E \times E \ni (v, w) \mapsto [v, w]$ which satisfies

$$[u, u] = 0, \quad [u, [v, w]] + [v, [w, u]] + [w, [u, v]] = 0, \quad u, v, w \in E.$$

and consequently the Leibniz rule gives

$$L(f)(p) = \sum_{j=1}^{N} g_j(p) \left(Lx_j \right)(p). \tag{I.5}$$

DEFINITION I.2.1. *The $\mathbb{C}$-linear map $C^\infty(U) \to C^\infty(U)$ given by*

$$f \mapsto \frac{\partial f^*}{\partial x_j} \circ \mathbf{x}$$

defines an element in $\mathfrak{X}(U)$, which will be denoted by $\frac{\partial}{\partial x_j}$.

Returning to the preceding argument and notation we can write

$$g_j(p) = h_j(\mathbf{x}(p)) = \frac{\partial f^*}{\partial x_j}(\mathbf{x}(p)) = \left(\frac{\partial}{\partial x_j} \right)(f)(p).$$

Inserting this in (I.5) gives

$$L(f)(p) = \sum_{j=1}^{N} \left(Lx_j \right)(p) \left(\frac{\partial}{\partial x_j} \right)(f)(p);$$

since p was an arbitrary point taken in U we obtain the *representation of L in the local coordinates* $(x_1, \ldots, x_N)$:

$$L = \sum_{j=1}^{N} \left(Lx_j \right) \frac{\partial}{\partial x_j}. \tag{I.6}$$

In particular this representation shows that the $C^\infty(U)$-module $\mathfrak{X}(U)$ is free, with basis $\{\partial/\partial x_1, \ldots, \partial/\partial x_N\}$.

Observe that if $M \in \mathfrak{X}(U)$ then the representation of $[L, M]$ in the local coordinates $(x_1, \ldots, x_N)$ is given by

$$[L, M] = \sum_{j=1}^{N} \left\{ L(Mx_j) - M(Lx_j) \right\} \frac{\partial}{\partial x_j}. \tag{I.7}$$

I.3 Formally integrable structures

Denote by $\mathcal{B}_p$ the set of all pairs (V, f), where V is an open neighborhood of p and $f \in C^\infty(V)$. In $\mathcal{B}_p$ we introduce the following equivalence relation: $(V_1, f_1) \sim (V_2, f_2)$ if there is an open neighborhood V of p, $V \subset V_1 \cap V_2$, such that f_1 and f_2 agree on V.

A *germ of a C^∞ function at p* is an element in the quotient space $C^\infty(p) \doteq \mathcal{B}_p/\sim$. We observe that $C^\infty(p)$ is also a $\mathbb{C}$-algebra. Given a C^∞ function f

defined in an open neighborhood of p, the germ at p defined by f will be denoted by $\underline{f}$. Notice that there is a natural $\mathbb{C}$-algebra homomorphism $C^\infty(p) \to \mathbb{C}$ defined by $\underline{f} \mapsto f(p)$.

DEFINITION I.3.1. *A complex tangent vector (to Ω) at p is a $\mathbb{C}$-linear map*

$$v : C^\infty(p) \longrightarrow \mathbb{C}$$

satisfying

$$v(\underline{f}\,\underline{g}) = f(p)v(\underline{g}) + g(p)v(\underline{f}), \quad \underline{f}, \underline{g} \in C^\infty(p). \tag{I.8}$$

The set of all complex tangent vectors at p, denoted by $\mathbb{C}T_p\Omega$, has a structure of a $\mathbb{C}$-vector space and is called the *complex tangent space to Ω at p*.

If $L \in \mathfrak{X}(\Omega)$ then $L_p : C^\infty(p) \to \mathbb{C}$ defined by

$$L_p(\underline{f}) = L(f)(p), \quad \underline{f} \in C^\infty(p)$$

belongs to $\mathbb{C}T_p\Omega$. Conversely, suppose that for each $p \in \Omega$ an element $v_p \in \mathbb{C}T_p\Omega$ is given such that

$$p \mapsto v_p(\underline{f}) \in C^\infty(\Omega), \quad \forall f \in C^\infty(\Omega).$$

Then there is $L \in \mathfrak{X}(\Omega)$ such that $L_p = v_p$ for all $p \in \Omega$.

Suppose now that $p \in U$ and that $(U, \mathbf{x})$ is a local chart. If $v \in \mathbb{C}T_p\Omega$ then, according to (I.5),

$$v(\underline{f}) = \sum_{j=1}^N g_j(p)v(\underline{x_j}) = \sum_{j=1}^N v(\underline{x_j}) \left(\frac{\partial}{\partial x_j}\right)_p (\underline{f}), \quad \underline{f} \in C^\infty(p).$$

In particular we conclude that $\{(\frac{\partial}{\partial x_j})_p : j = 1, \ldots, N\}$ is a basis of $\mathbb{C}T_p\Omega$.

The *complexified tangent bundle of Ω* is defined as the disjoint union

$$\mathbb{C}T\Omega = \bigcup_{p\in\Omega} \mathbb{C}T_p\Omega.$$

We shall also need the notion of a *complex vector sub-bundle of $\mathbb{C}T\Omega$ of rank n and corank $N - n$*. By this we mean a disjoint union

$$\mathcal{V} = \bigcup_{p\in\Omega} \mathcal{V}_p \subset \mathbb{C}T\Omega$$

satisfying the following conditions:

(a) For each $p \in \Omega$, $\mathcal{V}_p$ is a vector subspace of $\mathbb{C}T_p\Omega$ of dimension n.
(b) Given $p_0 \in \Omega$ there are an open set U_0 containing p_0 and vector fields $L_1, \ldots, L_n \in \mathfrak{X}(U_0)$ such that $L_{1p}, \ldots, L_{np}$ span $\mathcal{V}_p$ for every $p \in U_0$.

The vector space $\mathcal{V}_p$ is called the *fiber of $\mathcal{V}$ at p*.

Given a complex vector sub-bundle $\mathcal{V}$ of $\mathbb{C}T\Omega$ and an open subset W of Ω, a *section of $\mathcal{V}$ over W* is an element L of $\mathfrak{X}(W)$ such that $L_p \in \mathcal{V}_p$ for all $p \in W$. We are now in a position to introduce our main object of study:

DEFINITION I.3.2. *A formally integrable structure over Ω is a complex vector sub-bundle $\mathcal{V}$ of $\mathbb{C}T\Omega$ satisfying the involutive (or Frobenius) condition:*

- *If $W \subset \Omega$ is open and $L, M \in \mathfrak{X}(W)$ are sections of $\mathcal{V}$ over W then $[L, M]$ is also a section of $\mathcal{V}$ over W.*

The rank (resp. corank) of $\mathcal{V}$ will be referred to as the *rank* (resp. *corank*) of the formally integrable structure $\mathcal{V}$. Let $\mathcal{V}$ be a formally integrable structure over Ω and fix $p \in \Omega$. There is a local chart $(U, \mathbf{x})$ with $p \in U$ and vector fields $L_1, \ldots, L_n \in \mathfrak{X}(U)$ such that $\{L_{1q}, \ldots, L_{nq}\}$ is a basis of $\mathcal{V}_q$ for every $q \in U$. If we write $\mathbf{x} = (x_1, \ldots, x_N)$ and

$$L_j = \sum_{k=1}^{N} a_{jk}(x) \frac{\partial}{\partial x_k}$$

then the matrix (a_{jk}) has rank equal to n at every point; moreover, there are $c_{jk}^\nu \in C^\infty(U)$, $j, k, \nu = 1, \ldots, n$, such that

$$[L_j, L_k] = \sum_{\nu=1}^{n} c_{jk}^\nu L_\nu, \quad j, k = 1, \ldots, n.$$

DEFINITION I.3.3. *A (classical) solution for the formally integrable structure $\mathcal{V}$ over Ω is a C^1-function u on Ω such that $Lu = 0$ for every section L of $\mathcal{V}$ defined in an open subset of Ω.*

More generally, we can consider the concept of *(weak) solutions* for the formally integrable structure $\mathcal{V}$ over Ω: it suffices to consider u, in the preceding definition, belonging to the space of distributions on Ω (we refer to [**H2**] for the theory of distributions on manifolds).

I.4 Differential forms

We shall denote by $\mathfrak{N}(\Omega)$ the dual of the $C^\infty(\Omega)$-module $\mathfrak{X}(\Omega)$ and shall refer to its elements as *differential forms over Ω of degree one* (or *one-forms* for short). In other words, a one-form on Ω is a $C^\infty(\Omega)$-linear map

$$\omega : \mathfrak{X}(\Omega) \to C^\infty(\Omega).$$

Let $\omega \in \mathfrak{N}(\Omega)$, $L \in \mathfrak{X}(\Omega)$ and suppose that L vanishes on an open subset $V \subset \Omega$. Then $\omega(L)$ also vanishes on V. Indeed, let $p \in V$ and let $g \in C^\infty(\Omega, \mathbb{R})$ be equal to one at p and vanish on $\Omega \setminus V$. Then $L = (1 - g)L$ and consequently

$$\omega(L) = (1 - g)\omega(L)$$

vanishes at p. In fact, we have a more precise result:

LEMMA I.4.1. *Let $\omega \in \mathfrak{N}(\Omega)$, $L \in \mathfrak{X}(\Omega)$ and suppose that $L_p = 0$. Then $\omega(L)(p) = 0$.*

PROOF. By the preceding discussion it is clear that we can restrict a one-form on Ω to an open set $W \subset \Omega$, that is, given $\omega \in \mathfrak{N}(\Omega)$ there is $\omega|_W \in \mathfrak{N}(W)$ which makes the diagram

$$
\begin{array}{ccc}
\mathfrak{X}(\Omega) & \xrightarrow{\omega} & C^\infty(\Omega) \\
\downarrow & & \downarrow \\
\mathfrak{X}(W) & \xrightarrow{\omega|_W} & C^\infty(W)
\end{array}
$$

commutative (the vertical arrows denote restriction homomorphisms). Let then $(U, \mathbf{x})$ be a local chart with $p \in U$. Then, if $\mathbf{x} = (x_1, \ldots, x_N)$ we have by (I.6)

$$\omega(L)(p) = \omega_U(L_U)(p) = \sum_{j=1}^N (Lx_j)(p)\omega_U\left(\frac{\partial}{\partial x_j}\right)(p) = 0.$$

The proof of Lemma I.4.1 is complete. □

If we then define

$$\mathbb{C}T_p^*\Omega \doteq \text{dual of } \mathbb{C}T_p\Omega,$$

to each $\omega \in \mathfrak{N}(\Omega)$ we can associate an element $\omega_p \in \mathbb{C}T_p^*\Omega$ by the formula

$$\omega_p(v) = \omega(L)(p),$$

where $L \in \mathfrak{X}(\Omega)$ is such that $L_p = v$.

As in the case for vector fields, we have a converse: if for every $p \in \Omega$ an element $\eta_p \in \mathbb{C}T_p^*\Omega$ is given such that

$$p \mapsto \eta_p(L_p) \in C^\infty(\Omega), \quad \forall L \in \mathfrak{X}(\Omega),$$

then there is $\omega \in \mathfrak{N}(\Omega)$ such that $\omega_p = \eta_p$, for every $p \in \Omega$.

PROPOSITION I.4.2. $\mathbb{C}T_p^*\Omega = \{\omega_p : \omega \in \mathfrak{N}(\Omega)\}$.

PROOF. Let $(U, \mathbf{x})$ be a local chart with $p \in U$. Formula (I.6) allows one to define $\mathrm{d}x_j \in \mathfrak{N}(U)$, $j = 1, \ldots, N$, by the rule

$$\mathrm{d}x_j\left(\frac{\partial}{\partial x_k}\right) = \delta_{jk}, \quad j, k = 1, \dots, N.$$

Hence, if $\omega \in \mathfrak{N}(U)$ we have

$$\omega = \sum_{j=1}^{N} \omega\left(\frac{\partial}{\partial x_j}\right)\mathrm{d}x_j, \tag{I.9}$$

where $\omega(\partial/\partial x_j) \in C^\infty(U)$. If we now observe that $\{(\mathrm{d}x_j)_p\} \subset \mathbb{C}T_p^*\Omega$ is the dual basis of $\{(\partial/\partial x_j)_p\} \subset \mathbb{C}T_p\Omega$ then the conclusion will follow easily. $\quad\square$

DEFINITION I.4.3. *Given* $f \in C^\infty(\Omega)$ *we define* $\mathrm{d}f \in \mathfrak{N}(\Omega)$ *by the formula*

$$\mathrm{d}f(L) = L(f), \quad L \in \mathfrak{X}(\Omega). \tag{I.10}$$

From (I.9) we obtain the usual representation in local coordinates

$$\mathrm{d}f = \sum_{j=1}^{N} \mathrm{d}f\left(\frac{\partial}{\partial x_j}\right)\mathrm{d}x_j = \sum_{j=1}^{N} \frac{\partial f}{\partial x_j}\mathrm{d}x_j.$$

We now introduce the *complexified cotangent bundle of* Ω as being the disjoint union

$$\mathbb{C}T^*\Omega \doteq \bigcup_{p\in\Omega} \mathbb{C}T_p^*\Omega.$$

As before we can also introduce the notion of a *complex vector sub-bundle of* $\mathbb{C}T^*\Omega$ *of rank* m as being a disjoint union

$$W = \bigcup_{p\in\Omega} W_p,$$

where each W_p is a vector subspace of $\mathbb{C}T_p^*\Omega$ of dimension m, satisfying the following property:

- Given $p_0 \in \Omega$ there are an open set U_0 containing p_0 and one-forms $\omega_1, \dots, \omega_m \in \mathfrak{N}(U_0)$ such that $\omega_{1p}, \dots, \omega_{mp}$ span W_p for every $p \in U_0$.

As before we shall refer to the space W_p as the *fiber* of W at the point p.

PROPOSITION I.4.4. *Let* $\mathcal{V} = \cup_{p\in\Omega}\mathcal{V}_p$ *be a complex vector sub-bundle of* $\mathbb{C}T\Omega$ *and set, for each* $p \in \Omega$,

$$\mathcal{V}_p^\perp \doteq \{\lambda \in \mathbb{C}T_p^*\Omega : \lambda = 0 \text{ on } \mathcal{V}_p\}.$$

Then $\mathcal{V}^\perp \doteq \cup_{p\in\Omega}\mathcal{V}_p^\perp$ *is a complex vector sub-bundle of* $\mathbb{C}T^*\Omega$.

PROOF. Given $p_0 \in \Omega$ there is a local chart

$$(U_0, \mathbf{x}), \qquad \mathbf{x} = (x_1, \ldots, x_N),$$

with $p_0 \in U_0$, and vector fields on U_0

$$L_j = \sum_{k=1}^{N} a_{jk} \frac{\partial}{\partial x_k}, \quad j = 1, \ldots, n,$$

such that $\{L_{1p}, \ldots, L_{np}\}$ spans $\mathcal{V}_p$ for all $p \in U_0$. After a contraction of U_0 around p_0 and a relabeling of the indices we can assume that the matrix $(a_{jk})_{j,k=1,\ldots,n}$ is invertible in U_0. Let $(b_{jk})_{j,k=1,\ldots,n}$ be its inverse and set

$$L_j^{\#} = \sum_{\nu=1}^{n} b_{j\nu} L_\nu, \quad j = 1, \ldots, n.$$

Then $\{L_{1p}^{\#}, \ldots, L_{np}^{\#}\}$ also spans $\mathcal{V}_p$ for all $p \in U_0$. Moreover, we have

$$L_j^{\#} = \frac{\partial}{\partial x_j} + \sum_{k=1}^{m} c_{jk} \frac{\partial}{\partial x_{n+k}}, \quad j = 1, \ldots, n,$$

where c_{jk} are smooth in U_0 and $m = N - n$. Set

$$\omega_\ell = dx_{n+\ell} - \sum_{\gamma=1}^{n} c_{\gamma\ell} dx_\gamma, \quad \ell = 1, \ldots, m.$$

Then $\omega_{1p}, \ldots, \omega_{mp}$ are linearly independent for all $p \in U_0$ and furthermore

$$\omega_\ell(L_j^{\#}) = dx_{n+\ell}(L_j^{\#}) - c_{j\ell} = 0.$$

Hence $\{\omega_{1p}, \ldots, \omega_{mp}\}$ is a basis for $\mathcal{V}_p^{\perp}$ for each $p \in U_0$. $\qquad \square$

REMARK I.4.5. It is clear that the preceding argument can be reversed. If $\mathcal{V}^{\perp}$ is a vector sub-bundle of $\mathbb{C}T^*\Omega$ then it follows that $\mathcal{V}$ is a vector sub-bundle of $\mathcal{V}$.

When $\mathcal{V}$ is a formally integrable structure over Ω of dimension N we shall always denote the sub-bundle $\mathcal{V}^{\perp}$ by T'. We shall also always denote by n the rank of $\mathcal{V}$ and by m the rank of T'. In particular, $n + m = N$.

We shall also use the standard notation:

$$T_p\Omega \doteq \{v \in \mathbb{C}T_p\Omega : v \text{ is real}\};$$

$$T_p^*\Omega \doteq \{\xi \in \mathbb{C}T_p^*\Omega : \xi \text{ is real}\};$$

$$T\Omega \doteq \bigcup_{p\in\Omega} T_p\Omega;$$

$$T^*\Omega \doteq \bigcup_{p\in\Omega} T_p^*\Omega.$$

Given $L \in \mathfrak{X}(\Omega)$ its *(complex)-conjugate* is the vector field $\overline{L} \in \mathfrak{X}(\Omega)$ defined by

$$\overline{L}(f) = \overline{L(\overline{f})}, \quad f \in C^\infty(\Omega).$$

In particular we shall say that L is a *real vector field* if $L = \overline{L}$, that is, if $LC^\infty(\Omega, \mathbb{R}) \subset C^\infty(\Omega, \mathbb{R})$. In the same way we can define the (complex)-conjugate of an element in $\mathbb{C}T_p\Omega$. Given a subspace $\mathcal{V}_p \subset \mathbb{C}T_p\Omega$ we define

$$\overline{\mathcal{V}}_p \doteq \{\overline{v} : v \in \mathcal{V}_p\}.$$

It is clear from the definitions that if $\mathcal{V}$ is a complex vector sub-bundle of $\mathbb{C}T\Omega$ then the same is true for $\overline{\mathcal{V}} \doteq \cup_{p\in\Omega}\overline{\mathcal{V}}_p$. We shall refer to $\overline{\mathcal{V}}$ as the *(complex)-conjugate* of the sub-bundle $\mathcal{V}$. Analogous definitions and results can be introduced and obtained for $\mathbb{C}T^*\Omega$ and its fibers $\mathbb{C}T_p^*\Omega$. It is also important to mention the equality

$$\overline{\mathcal{V}}^\perp = \overline{\mathcal{V}^\perp},$$

which is valid for every complex vector sub-bundle $\mathcal{V}$ of $\mathbb{C}T\Omega$.

I.5 The Frobenius theorem

We start by considering a *real* vector field

$$L = \sum_{j=1}^{N} a_j(x)\frac{\partial}{\partial x_j}$$

defined in a neighborhood of the origin in $\mathbb{R}^N$. Assume that $L \neq 0$. Then it is possible to find local coordinates $y_1, y_2, \ldots, y_N$, defined near the origin, such that

$$L = \frac{\partial}{\partial y_1}.$$

The proof of this result is very simple and will be recalled here.

We assume that $a_1(0) \neq 0$ and solve, in some neighborhood of the origin, the following Cauchy problem:

$$\begin{cases} \partial x_j/\partial y_1 = a_j(x_1, \ldots, x_N) & j = 1, \ldots, N \\ x_1(0, y_2, \ldots, y_N) = 0 \\ x_j(0, y_2, \ldots, y_N) = y_j & j = 2, \ldots, N. \end{cases}$$

The fact that $a_1(0) \neq 0$ implies that $(y_1, \ldots, y_N) \mapsto (x_1, \ldots, x_N)$ is a smooth diffeomorphism at the origin and a simple computation shows our claim.

The generalization of this result to a larger number of vector fields is the classical *Frobenius theorem*:

THEOREM I.5.1. *Let $L_1, \ldots, L_n$ be linearly independent, real vector fields defined in a neighborhood V of the origin in $\mathbb{R}^N$. Assume that the sub-bundle $\mathcal{V}$ of $\mathbb{C}TV$ generated by $L_1, \ldots, L_n$ is a formally integrable structure. Then there are local coordinates $y_1, y_2, \ldots, y_N$, defined near the origin, such that $\mathcal{V}$ is generated by $\partial/\partial y_1, \ldots, \partial/\partial y_n$.*

PROOF. We shall proceed by induction on N. The case $N = 1$ is trivial. We then suppose that the result was proved for values $< N$. Applying the procedure described at the beginning of this section we can make a change of variables and assume that the given vector fields have the form:

$$L_1 = \frac{\partial}{\partial x_1}, \quad L_j = \sum_{k=1}^{N} a_{jk} \frac{\partial}{\partial x_k}, \quad j = 2, \ldots, n.$$

We then introduce a new set of generators for the bundle $\mathcal{V}$:

$$L_1^{\#} = L_1, \quad L_j^{\#} = L_j - a_{j1} L_1, \quad j = 2, \ldots, n.$$

Notice that when $j \geq 2$ the vector field $L_j^{\#}$ does not involve differentiation in the x_1-variable. Thus, in a neighborhood of the origin, we have

$$[L_j^{\#}, L_k^{\#}] = \sum_{\nu=2}^{n} C_{jk}^{\nu} L_{\nu}^{\#}, \quad j, k = 2, \ldots, n.$$

If we then consider, in a neighborhood W of the origin in $\mathbb{R}^{N-1}$, the vector fields

$$M_j = \sum_{k=2}^{N} a_{jk}(0, x_2, \ldots, x_N) \frac{\partial}{\partial x_k}, \quad j = 2, \ldots, n,$$

as well as the sub-bundle $\mathcal{V}'$ of $\mathbb{C}TW$ defined by them, we conclude the existence of a coordinate system $y_2, \ldots, y_N$ defined near the origin in $\mathbb{R}^{N-1}$ for which $\mathcal{V}'$ is spanned by

$$\partial/\partial y_2, \ldots, \partial/\partial y_n.$$

This argument has the following consequence: returning to the original coordinates $(x_1, \ldots, x_N)$, the induction hypothesis allows us to assume from the beginning that

$$a_{jk}(0, x_2, \ldots, x_N) = 0, \quad j = 2, \ldots, n, \; k > n.$$

Now, the coefficient of $\partial/\partial x_\ell$ in the commutator $[L_1^\#, L_j^\#]$ is equal to $\partial a_{j\ell}/\partial x_1$. On the other hand,

$$[L_1^\#, L_j^\#] = \sum_{\nu=1}^{n} C_{1j}^\nu L_\nu^\# = C_{1j}^1 \frac{\partial}{\partial x_1} + \sum_{\nu=2}^{n} C_{1j}^\nu \sum_{k=2}^{N} a_{\nu k} \frac{\partial}{\partial x_k},$$

and thus

$$\frac{\partial a_{j\ell}}{\partial x_1} = \sum_{\nu=2}^{n} C_{1j}^\nu a_{\nu\ell}, \quad \ell = 2, \ldots, N, \ j = 2, \ldots, n.$$

Hence for each fixed ℓ the vector $(a_{2\ell}, \ldots, a_{n\ell})$ satisfies a linear system of ordinary differential equations with trivial initial condition. By the uniqueness theorem for such systems we conclude that $a_{j\ell} = 0$ if $j = 2, \ldots, n$ and $\ell > n$. Thus we have

$$L_j^\# = \sum_{k=2}^{n} a_{jk} \frac{\partial}{\partial x_k}, \quad j = 2, \ldots, n,$$

which concludes the proof. $\qquad\qquad\qquad\qquad\qquad\qquad\qquad\qquad\square$

We now discuss the holomorphic version of the Frobenius theorem. Write the complex coordinates in $\mathbb{C}^\mu$ as $z_1, \ldots, z_\mu$, where $z_j = x_j + iy_j$, and identify $\mathbb{C}^\mu \simeq \mathbb{R}^{2\mu}$ by

$$z = (z_1, \ldots, z_\mu) \mapsto (x_1, y_1, \ldots, x_\mu, y_\mu).$$

Given an open set $\Omega \subset \mathbb{C}^\mu$ denote by $\mathcal{O}(\Omega)$ the algebra of holomorphic functions on Ω. An element $L \in \mathfrak{X}(\Omega)$ is said to be a *holomorphic vector field* if given any $f \in \mathcal{O}(\Omega)$ we have $Lf \in \mathcal{O}(\Omega)$ and $L\overline{f} = 0$. Introducing the standard notation

$$\frac{\partial}{\partial z_j} = \frac{1}{2}\left\{\frac{\partial}{\partial x_j} - i\frac{\partial}{\partial y_j}\right\}, \quad \frac{\partial}{\partial \overline{z}_j} = \frac{1}{2}\left\{\frac{\partial}{\partial x_j} + i\frac{\partial}{\partial y_j}\right\}$$

it is clear that every vector field $L \in \mathfrak{X}(\Omega)$ can be written as

$$L = \sum_{j=1}^{\mu}\left\{a_j \frac{\partial}{\partial z_j} + b_j \frac{\partial}{\partial \overline{z}_j}\right\}, \tag{I.11}$$

where $a_j, b_j \in C^\infty(\Omega)$; (I.11) is then a holomorphic vector field if and only if $b_j = 0$ and $a_j \in \mathcal{O}(\Omega)$, $j = 1, \ldots, \mu$.

We now state the holomorphic version of the Frobenius theorem, whose proof is the same as that of Theorem I.5.1, working now in the holomorphic category.

THEOREM I.5.2. *Let $L_1, \ldots, L_n$ be linearly independent, holomorphic vector fields defined in a neighborhood V of the origin in $\mathbb{C}^\mu$. Assume that the sub-bundle $\mathcal{V}$ of $\mathbb{C}TV$ generated by $L_1, \ldots, L_n$ is a formally integrable structure. Then there are local holomorphic coordinates $w_1, w_2, \ldots, w_\mu$, defined near the origin in $\mathbb{C}^\mu$ such that $\mathcal{V}$ is generated by $\partial/\partial w_1, \ldots, \partial/\partial w_n$.*

I.6 Analytic structures

Let Ω be a real-analytic manifold, defined by the differentiable (real-analytic) structure $\mathcal{F} = \{(V, \mathbf{x})\}$. A function $f: \Omega \to \mathbb{C}$ is *real-analytic* if for every $(V, \mathbf{x}) \in \mathcal{F}$ the composition $f \circ \mathbf{x}^{-1}$ is *real-analytic* on $\mathbf{x}(V)$. Given $U \subset \Omega$ an open set, we shall denote by $\mathcal{A}(U)$ the space of real-analytic functions on U. An element $L \in \mathfrak{X}(\Omega)$ is said to be a *real-analytic vector field on* Ω if

$$L\mathcal{A}(U) \subset \mathcal{A}(U), \quad \forall U \subset \Omega \text{ open.}$$

If L is given in local coordinates as in (I.6) then L is real-analytic if and only if its coefficients Lx_j, $j = 1, \ldots, N$, are real-analytic functions.

Analogously, we shall say that $\omega \in \mathfrak{N}(\Omega)$ is a *real-analytic one-form* on Ω if $\omega(L) \in \mathcal{A}(U)$ for every $U \subset \Omega$ open and every real-analytic vector field L.

From such definitions it is clear that one can introduce the notions of complex *analytic* vector sub-bundles of $\mathbb{C}T\Omega$ and of $\mathbb{C}T^*\Omega$; in particular we can refer to the notion of an *analytic* formally integrable structure over Ω.

REMARK I.6.1. Suppose that Ω is now an open subset of $\mathbb{R}^N$ and let $L \in \mathfrak{X}(\Omega)$ be real-analytic. Write

$$L = \sum_{j=1}^N a_j(x) \frac{\partial}{\partial x_j}.$$

Let also $u \in \mathcal{A}(\Omega)$ and take an open set $\Omega^\mathbb{C} \subset \mathbb{C}^N$, where the holomorphic coordinates are written as $(z_1, \ldots, z_N)$, such that

- $\Omega^\mathbb{C} \cap \mathbb{R}^N = \Omega$;
- u, a_j extend as holomorphic functions $\tilde{u}, \tilde{a}_j$ on $\Omega^\mathbb{C}$.

Then

$$Lu = (\tilde{L}\tilde{u})|_\Omega, \tag{I.12}$$

where $\tilde{L}$ is the holomorphic vector field

$$\tilde{L} = \sum_{j=1}^N \tilde{a}_j(z) \frac{\partial}{\partial z_j}.$$

I.7 The characteristic set

Let $\mathcal{V} \subset \mathbb{C}T\Omega$ be a formally integrable structure over Ω. The *characteristic set of* $\mathcal{V}$ is the subset of $T^*\Omega$ defined by

$$T^0 \doteq T' \cap T^*\Omega. \tag{I.13}$$

We shall also write $T^0_p = T'_p \cap T^*_p\Omega$ if $p \in \Omega$. If we recall that the *symbol* of a vector field $L \in \mathfrak{X}(\Omega)$ is the function

$$\sigma(L) : T^*\Omega \to \mathbb{C}, \quad \sigma(L)(\xi) = \xi(L_p) \quad \text{if } \xi \in T^*_p\Omega,$$

then we see that $\xi \in T^0_p$ if and only if $\sigma(L)(\xi) = 0$ for every section L of $\mathcal{V}$.

Let $(U, \mathbf{x})$, $\mathbf{x} = (x_1, \ldots, x_N)$ be a local chart on Ω. Take $p \in U$ and $\xi \in T^*_p\Omega$. If we write $\xi = \sum_{j=1}^N \xi_j \mathrm{d}x_{jp}$ ($\xi_j \in \mathbb{R}$) and $L = \sum_{j=1}^N a_j(\partial/\partial x_j)$ then

$$\sigma(L)(\xi) = \sum_{j=1}^N a_j(p)\xi_j.$$

Thus, if $L_j = \sum_{k=1}^N a_{jk}(\partial/\partial x_k)$ are n linearly independent sections of $\mathcal{V}$ over U we can describe $T^0 \cap T^*U$ by the system of equations

$$\sum_{k=1}^N a_{jk}(p)\xi_k = 0, \quad p \in U, \ \xi_k \in \mathbb{R}, \ j = 1, \ldots, n.$$

EXAMPLE I.7.1 (The Mizohata operator). If we write the coordinates in $\Omega = \mathbb{R}^2$ as (x, t) then

$$M \doteq \frac{\partial}{\partial t} - it\frac{\partial}{\partial x} \in \mathfrak{X}(\mathbb{R}^2) \tag{I.14}$$

is called the *Mizohata vector field* or *Mizohata operator*. We now describe the characteristic set of the formally integrable structure defined by M. From the equation $\tau - it\xi = 0$ we get

$$T^0_{(x,t)} = \begin{cases} 0 & \text{if } t \neq 0 \\ \{\xi \mathrm{d}x_p : \xi \in \mathbb{R}\} & \text{if } p = (x, 0). \end{cases}$$

This example in particular shows that T^0 is not, in general, a vector sub-bundle of $T^*\Omega$.

I.8 Some special structures

Let $\mathcal{V}$ be a formally integrable structure over Ω. We shall say that $\mathcal{V}$ defines

- an *elliptic structure* if $T^0_p = 0$, $\forall p \in \Omega$;
- a *complex structure* if $\mathcal{V}_p \oplus \overline{\mathcal{V}}_p = \mathbb{C}T_p\Omega$, $\forall p \in \Omega$;

- a *Cauchy–Riemann (CR) structure* if $\mathcal{V}_p \cap \overline{\mathcal{V}}_p = 0$, $\forall p \in \Omega$;
- an *essentially real structure* if $\mathcal{V}_p = \overline{\mathcal{V}}_p$, $\forall p \in \Omega$.

Before we proceed further we state some easy consequences of the preceding definitions.

PROPOSITION I.8.1. *Every essentially real structure is locally generated by real vector fields.*

PROOF. Given $p_0 \in \Omega$ we take vector fields $L_1, \ldots, L_n$ which generate $\mathcal{V}$ in a neighborhood of p_0. By hypothesis the real vector fields $\Re L_j$, $\Im L_j$ are also sections of $\mathcal{V}$. Moreover,

$$\text{span}\left\{\left(\Re L_j\right)_{p_0}, \left(\Im L_j\right)_{p_0} : j = 1, \ldots, n\right\} = \mathcal{V}_{p_0}$$

and consequently n of the tangent vectors $\left(\Re L_j\right)_{p_0}$, $\left(\Im L_j\right)_{p_0}$ are linearly independent. Since this remains true in a neighborhood of p_0 the result is proved.   □

Next we recall a very elementary but useful result.

LEMMA I.8.2. *If V is a vector subspace of $\mathbb{C}^N = \mathbb{R}^N + i\mathbb{R}^N$ and if $V^0 = V \cap \mathbb{R}^N$ then $V^0 \otimes_{\mathbb{R}} \mathbb{C} \simeq V^0 + iV^0 = V \cap \overline{V}$.*

PROOF. We only verify the equality. If $x, y \in V^0$ then $x \pm iy \in V$ and so $V^0 + iV^0 \subset V \cap \overline{V}$. For the reverse inclusion take $z \in V \cap \overline{V}$. Then

$$z = \frac{1}{2}(z + \overline{z}) - \frac{i}{2}(iz - i\overline{z}) \in V^0 + iV^0.\qquad \square$$

As a consequence, given any formally integrable structure $\mathcal{V}$ over Ω we have

$$T^0_p \otimes_{\mathbb{R}} \mathbb{C} \simeq T'_p \cap \overline{T}'_p, \quad \forall p \in \Omega. \tag{I.15}$$

Since for a complex structure we also have

$$T'_p \oplus \overline{T}'_p = \mathbb{C}T^*_p\Omega, \quad \forall p \in \Omega$$

we obtain:

COROLLARY I.8.3. *Every complex structure is elliptic.*

Unlike what happens with Mizohata structures we have:

PROPOSITION I.8.4. *If V defines a CR structure over Ω then T^0 is a vector sub-bundle of $T^*\Omega$ of rank $d \doteq N - 2n$.*

PROOF. If $V_p \cap \overline{V}_p = 0$, for all $p \in \Omega$, then

$$W = V \oplus \overline{V} \doteq \bigcup_{p \in \Omega} (V_p \oplus \overline{V}_p)$$

is a vector sub-bundle of $\mathbb{C}T\Omega$ (of rank $2n$) which defines an essentially real structure over Ω. By Proposition I.4.4, $W^\perp$ is a vector sub-bundle of $\mathbb{C}T^*\Omega$ of rank d which of course satisfies $W_p^\perp = \overline{W}_p^\perp$ for all $p \in \Omega$. The same argument used in the proof of Proposition I.8.1 shows that $W^\perp$ has local *real* generators. Since these generators span T^0 the proof is complete. $\square$

In order to obtain appropriate local generators for a formally integrable structure we shall need an elementary result:

LEMMA I.8.5. *Let V be a complex subspace of $\mathbb{C}^N$ of dimension m. Let $V_0 = V \cap \mathbb{R}^N$, $d \doteq \dim_\mathbb{R} V_0$, $\nu \doteq m - d$. Let also $V_1 \subset \mathbb{C}^N$ be a subspace such that $(V_0 \oplus iV_0) \oplus V_1 = V$ and take:*

$$\{\zeta_1, \ldots, \zeta_\nu\} : \text{ basis for } V_1; \quad \{\xi_{\nu+1}, \ldots, \xi_m\} : \text{ real basis for } V_0.$$

If we write $\zeta_j = \xi_j + i\eta_j$, $j = 1, \ldots, \nu$, then:

$$\{\zeta_1, \ldots, \zeta_\nu, \xi_{\nu+1}, \ldots, \xi_m\} \quad \text{is a basis for } V; \tag{I.16}$$

$$\{\xi_1, \ldots, \xi_m, \eta_1, \ldots, \eta_\nu\} \quad \text{is linearly independent over } \mathbb{R}; \tag{I.17}$$

$$\nu + m \leq N. \tag{I.18}$$

PROOF. Notice that (I.16) is trivial since $\{\xi_{\nu+1}, \ldots, \xi_m\}$ is also a basis for $V_0 \oplus iV_0$.

Next we notice that $V \cap \overline{V}_1 = 0$. Indeed, let $z \in V \cap \overline{V}_1$. Then $\overline{z} \in V_1 \subset V$ and consequently $\Re z, \Im z \in V_0$, which gives $\overline{z} \in (V_0 \oplus iV_0) \cap V_1 = 0$. Hence

$$\{\zeta_1, \ldots, \zeta_\nu, \overline{\zeta}_1, \ldots, \overline{\zeta}_\nu, \xi_{\nu+1}, \ldots, \xi_m\}$$

is linearly independent. In particular, $2\nu + d = \nu + m \leq N$ and (I.17) holds. $\square$

Given a formally integrable structure V over Ω and fixing $p \in \Omega$ we shall apply Lemma I.8.5 with the choices

$$V = T'_p, \quad V_0 = T^0_p.$$

If $\{\zeta_1, \ldots, \zeta_\nu, \xi_{\nu+1}, \ldots, \xi_m\}$ is the basis given in (I.16) we first take a system of local coordinates

$$x_1, \ldots, x_\nu, y_1, \ldots, y_\nu, s_1, \ldots, s_d, t_1, \ldots, t_{n'}$$

vanishing at p such that, writing $z_j = x_j + iy_j$ we have

$$dz_j|_p = \zeta_j, \quad ds_k|_p = \xi_{\nu+k}, \quad j = 1, \ldots, \nu, \ k = 1, \ldots, d.$$

Afterwards we take one-forms $\omega_1, \ldots, \omega_\nu, \theta_1, \ldots, \theta_d$ which span T' in a neighborhood of p and such that

$$\omega_j|_p = dz_j|_p, \quad \theta_k|_p = ds_k|_p, \quad j = 1, \ldots, \nu, \ k = 1, \ldots, d.$$

If L is a complex vector field on Ω defined near p we can write it in the form

$$L = \sum_j A_j \frac{\partial}{\partial z_j} + \sum_j B_j \frac{\partial}{\partial \bar{z}_j} + \sum_k C_k \frac{\partial}{\partial s_k} + \sum_\ell D_\ell \frac{\partial}{\partial t_\ell}.$$

If, furthermore, L is a section of V we necessarily must have $A_j = C_k = 0$ at p for all j and k. Since $\nu + n' = n$, it follows that after a linear substitution we can find a set of local generators of the sub-bundle V in a neighborhood of p of the form

$$L_j = \frac{\partial}{\partial \bar{z}_j} + \sum_{j'=1}^{\nu} a_{jj'} \frac{\partial}{\partial z_{j'}} + \sum_{k=1}^{d} b_{jk} \frac{\partial}{\partial s_k}, \quad j = 1, \ldots, \nu, \tag{I.19}$$

$$\tilde{L}_\ell = \frac{\partial}{\partial t_\ell} + \sum_{j'=1}^{\nu} \tilde{a}_{\ell j'} \frac{\partial}{\partial z_{j'}} + \sum_{k=1}^{d} \tilde{b}_{\ell k} \frac{\partial}{\partial s_k}, \quad \ell = 1, \ldots, n', \tag{I.20}$$

where the coefficients $a_{jj'}, \tilde{a}_{\ell j'}, b_{jk}, \tilde{b}_{\ell k}$ all vanish at p.

We notice that the elliptic case corresponds to the situation when $d = 0$, the complex case to the one when $d = n' = 0$, and the CR case to the one when $n' = 0$.

Next we introduce a generalization of the structure defined by the Mizohata operator (*cf.* Example I.7.1).

DEFINITION I.8.6. *We shall say that a formally integrable structure V over Ω is a generalized Mizohata structure at $p_0 \in \Omega$ if $V_{p_0} = \overline{V}_{p_0}$.*

Thus in the case of generalized Mizohata structures the coordinates vanishing at p_0 can be taken as $(s_1, \ldots, s_m, t_1, \ldots, t_n)$ $[d = m, n = n'$ in this case$]$ and V is spanned by the vector fields

$$L_\ell = \frac{\partial}{\partial t_\ell} + \sum_{k=1}^{d} \tilde{b}_{\ell k}(s, t) \frac{\partial}{\partial s_k}, \quad \ell = 1, \ldots, n,$$

where $b_{\ell k} = 0$ at the origin for every ℓ, k.

Finally we recall the classical notion of the so-called *CR functions*:

DEFINITION I.8.7. *Given a CR formally integrable structure* $\mathcal{V}$ *over* Ω, *any classical solution (for the formally integrable structure* $\mathcal{V}$*) is called a CR function.*

Needless to add, we can also introduce the concept of CR distributions, etc.

I.9 Locally integrable structures

A complex vector sub-bundle $\mathcal{V}$ of $\mathbb{C}T\Omega$, of rank n, is said to define a *locally integrable structure* if given an arbitrary point $p_0 \in \Omega$ there are an open neighborhood U_0 of p_0 and functions $Z_1, \dots, Z_m \in C^\infty(U_0)$, with $m = N - n$, such that

$$\text{span}\{dZ_{1p}, \dots, dZ_{mp}\} = \mathcal{V}_p^\perp, \quad \forall p \in U_0. \tag{I.21}$$

If one observes that the differential of a smooth function g is a section of $\mathcal{V}^\perp$ if and only if $Lg = 0$ for every section of $\mathcal{V}$, it follows easily that every locally integrable structure satisfies the Frobenius condition. Hence, *every locally integrable structure defines a formally integrable structure.*

We have:

- *The formally integrable structure* $\mathcal{V}$ *is locally integrable if and only if, given* $p_0 \in \Omega$ *and vector fields* $L_1, \dots, L_n$ *which span* $\mathcal{V}$ *in an open neighborhood* U_0 *of* p_0, *there are an open neighborhood* $V_0 \subset U_0$ *of* p_0 *and smooth functions* $Z_1, \dots, Z_m \in C^\infty(V_0)$ *such that:*

$$dZ_1 \wedge \dots \wedge dZ_m \neq 0 \quad \text{in} \quad V_0;$$

$$L_j Z_k = 0, \quad j = 1, \dots, n, \ k = 1, \dots, m.$$

Thus, checking local integrability is equivalent to looking for a maximal number of nontrivial solutions to the (in general overdetermined) homogeneous system defined by a fixed set of independent sections of $\mathcal{V}$.

THEOREM I.9.1. *Every essentially real structure is locally integrable.*

PROOF. By Frobenius Theorem I.5.1, in conjunction with Proposition I.8.1, given $p \in \Omega$ we can find a local chart $(U, \mathbf{x})$, $\mathbf{x} = (x_1, \dots, x_N)$, with $p \in U$, such that

$$\frac{\partial}{\partial x_j}, \quad j = 1, \dots, n$$

are sections of $\mathcal{V}$ over U. It suffices to take

$$Z_k = x_{k+n}, \quad k = 1, \ldots, m.$$

$\square$

THEOREM I.9.2. *Every analytic formally integrable structure is locally integrable.*

PROOF. We shall prove that if $L_1, \ldots, L_n$ are linearly independent, real-analytic vector fields in an open ball B centered at the origin in $\mathbb{R}^N$ such that

$$[L_\alpha, L_\beta] = \sum_{\gamma=1}^{n} C_{\alpha\beta}^\gamma L_\gamma,$$

where $C_{\alpha\beta}^\gamma \in \mathcal{A}(B)$, then we can find real-analytic functions $Z_1, \ldots, Z_m$ defined in a neighborhood of the origin and satisfying

$$L_j Z_\ell = 0, \quad j = 1, \ldots, n, \ \ell = 1, \ldots, m;$$

$$dZ_1 \wedge \ldots \wedge dZ_m \neq 0.$$

We write

$$L_j = \sum_{k=1}^{N} a_{jk} \frac{\partial}{\partial x_k}$$

and take an open, connected set $U \subset \mathbb{C}^N$ such that $U \cap \mathbb{R}^N = B$ and such that there are $\tilde{a}_{jk}, \tilde{C}_{\alpha\beta}^\gamma \in \mathcal{O}(U)$ satisfying

$$\tilde{a}_{jk} = a_{jk}, \ \tilde{C}_{\alpha\beta}^\gamma = C_{\alpha\beta}^\gamma \quad \text{in } B.$$

Consider then the holomorphic vector fields in U:

$$\tilde{L}_j = \sum_{k=1}^{N} \tilde{a}_{jk} \frac{\partial}{\partial z_k}.$$

By analytic continuation the coefficients of the holomorphic vector fields

$$[\tilde{L}_\alpha, \tilde{L}_\beta] - \sum_{\gamma=1}^{n} \tilde{C}_{\alpha\beta}^\gamma \tilde{L}_\gamma$$

must vanish identically in U since they vanish on B and the former is connected. By the holomorphic version of the Frobenius theorem we can find holomorphic functions $W_1, \ldots, W_m$ defined in an open neighborhood $V \subset U$ of the origin in $\mathbb{C}^N$ such that

$$\tilde{L}_j W_\ell = 0, \quad j = 1, \ldots, n, \ \ell = 1, \ldots, m;$$

$$dW_1 \wedge \ldots \wedge dW_m \neq 0.$$

It suffices then to set $Z_k \doteq W_k|_{V \cap B}$ in order to obtain the desired solutions (*cf.* (I.12)). □

EXAMPLE I.9.3. For the Mizohata vector field (I.14) we have $MZ = 0$ in $\mathbb{R}^2$, where $Z(x, t) = x + it^2/2$. Notice that $dZ \neq 0$ everywhere.

I.10 Local generators

In this section we shall construct appropriate local coordinates and local generators of the sub-bundle T' when the structure $\mathcal{V}$ is locally integrable. Once more we shall apply Lemma I.8.5.

Let $p \in \Omega$ and let also $G_1, \ldots, G_m$ be smooth functions defined in a neighborhood of p such that $dG_1, \ldots, dG_m$ span T'. As in Section I.8 we make the choices: $V = T'_p$, $V_0 = T^0_p$. If $\{\zeta_1, \ldots, \zeta_\nu, \xi_{\nu+1}, \ldots, \xi_m\}$ is the basis given in (I.16) then we can find $(c_{jk}) \in \mathrm{GL}(m, \mathbb{C})$ such that

$$\sum_{k=1}^m c_{jk} dG_k(p) = \zeta_j, \quad j = 1, \ldots, \nu,$$

$$\sum_{k=1}^m c_{jk} dG_k(p) = \xi_j, \quad j = \nu+1, \ldots, m.$$

We then set

$$Z_j = \sum_{k=1}^m c_{jk} \{G_k - G_k(p)\}, \quad j = 1, \ldots, \nu,$$

$$W_\ell = \sum_{k=1}^m c_{\nu+\ell,k} \{G_k - G_k(p)\}, \quad \ell = 1, \ldots, d.$$

It is clear that $dZ_1, \ldots, dZ_\nu, dW_1, \ldots, dW_d$ also span T' in a neighborhood of p. If we further set

$$x_j = \Re Z_j, \quad y_j = \Im Z_j, \quad s_\ell = \Re W_\ell$$

then (I.17) gives that

$$dx_1, \ldots, dx_\nu, dy_1, \ldots, dy_\nu, ds_1, \ldots, ds_d$$

are linearly independent at p. We are now ready to state and prove the following important result:

THEOREM I.10.1. *Let $\mathcal{V}$ be a locally integrable structure defined on a manifold Ω. Let $p \in \Omega$ and d be the real dimension of T_p^0. Then there is a coordinate system vanishing at p,*

$$\{x_1, \ldots, x_\nu, y_1, \ldots, y_\nu, s_1, \ldots, s_d, t_1, \ldots, t_{n'}\}$$

and smooth, real-valued functions $\phi_1, \ldots, \phi_d$ defined in a neighborhood of the origin and satisfying

$$\phi_k(0) = 0, \ \mathrm{d}\phi_k(0) = 0, \quad k = 1, \ldots, d,$$

such that the differentials of the functions

$$Z_j(x, y) = z_j \doteq x_j + iy_j, \quad j = 1, \ldots, \nu; \tag{I.22}$$

$$W_k(x, y, s, t) = s_k + i\phi_k(z, s, t), \quad k = 1, \ldots, d, \tag{I.23}$$

span T' in a neighborhood of the origin. In particular, we have $\nu + d = m$, $\nu + n' = n$ and also

$$T_p^0 = \mathrm{span}\,\{\mathrm{d}s_1|_0, \ldots, \mathrm{d}s_d|_0\}. \tag{I.24}$$

PROOF. The proof follows almost immediately from the preceding discussion: it suffices to take smooth, real-valued functions $t_1, \ldots, t_{n'}$ defined near p and vanishing at p such that

$$\mathrm{d}x_1, \ldots, \mathrm{d}x_\nu, \mathrm{d}y_1, \ldots, \mathrm{d}y_\nu, \mathrm{d}s_1, \ldots, \mathrm{d}s_d, \mathrm{d}t_1, \ldots \mathrm{d}t_{n'}$$

are linearly independent. Notice that $\mathrm{d}W_k(p) = \xi_{\nu+k}$ is real, from which we derive that $\mathrm{d}\phi_k = 0$ at the origin. $\qquad\square$

Since we have

$$\frac{\partial W_k}{\partial s_{k'}}(0, 0, 0) = \delta_{kk'}, \quad k, k' = 1, \ldots, d,$$

we can introduce, in a neighborhood of the origin in $\mathbb{R}^{2\nu+d+n'}$, the vector fields

$$M_k = \sum_{k'=1}^{d} \mu_{kk'}(z, s, t) \frac{\partial}{\partial s_{k'}}, \quad k = 1, \ldots, d \tag{I.25}$$

characterized by the relations

$$M_k W_{k'} = \delta_{kk'}. \tag{I.26}$$

Consequently the vector fields

$$L_j = \frac{\partial}{\partial \overline{z}_j} - i \sum_{k=1}^{d} \frac{\partial \phi_k}{\partial \overline{z}_j}(z, s, t) M_k, \quad j = 1, \ldots, \nu, \tag{I.27}$$

$$\tilde{L}_\ell = \frac{\partial}{\partial t_\ell} - i \sum_{k=1}^{d} \frac{\partial \phi_k}{\partial t_\ell}(z, s, t) M_k, \quad \ell = 1, \ldots, n' \tag{I.28}$$

are linearly independent and satisfy

$$L_j Z_{j'} = \tilde{L}_\ell Z_{j'} = L_j W_k = \tilde{L}_\ell W_k = 0$$

for all $j, j' = 1, \ldots, \nu$, $\ell = 1, \ldots, n'$, and $k = 1, \ldots, d$. Hence

$$L_1, \ldots, L_\nu, \tilde{L}_1, \ldots, \tilde{L}_{n'} \text{ span } \mathcal{V} \text{ in a neighborhood of the origin.} \tag{I.29}$$

Notice that the one-forms

$$dz_1, \ldots, dz_\nu, d\bar{z}_1, \ldots, d\bar{z}_\nu, dW_1, \ldots, dW_d, dt_1, \ldots, dt_{n'} \tag{I.30}$$

span $\mathbb{C}T^*\Omega$ near the origin. Moreover, the dual basis of (I.30) is given by

$$L_1^\flat, \ldots, L_\nu^\flat, L_1, \ldots, L_\nu, M_1, \ldots, M_d, \tilde{L}_1, \ldots, \tilde{L}_{n'}, \tag{I.31}$$

where

$$L_j^\flat = \frac{\partial}{\partial z_j} - i \sum_{k=1}^{d} \frac{\partial \phi_k}{\partial z_j}(z, s, t) M_k, \quad j = 1, \ldots, \nu. \tag{I.32}$$

Finally we observe that

$$\text{the vector fields (I.31) are pairwise commuting.} \tag{I.33}$$

Indeed it suffices to notice that if P, Q are any two of the vector fields (I.31) and if F is any one of the functions $\{Z_j, \overline{Z}_j, W_k, t_\ell\}$, the fact that (I.30) is dual to (I.31) gives

$$dF([P, Q]) = [P, Q](F) = 0,$$

from which we obtain that $[P, Q] = 0$.

In many cases we do not need the precise information provided by Theorem I.10.1 and the following particular case is enough:

COROLLARY I.10.2. *Same hypotheses as in Theorem I.10.1. Then there is a coordinate system vanishing at p,*

$$\{x_1, \ldots, x_m, t_1, \ldots, t_n\}$$

and smooth, real-valued $\phi_1, \ldots, \phi_m$ defined in a neighborhood of the origin and satisfying

$$\phi_k(0, 0) = 0, \quad d_x \phi_k(0, 0) = 0, \quad k = 1, \ldots, m,$$

such that the differentials of the functions

$$Z_k(x, t) = x_k + i\phi_k(x, t), \quad k = 1, \ldots, m, \tag{I.34}$$

span T' in a neighborhood of the origin. □

If we write $Z(x, t) = (Z_1(x, t), \ldots, Z_m(x, t))$ then $Z_x(0, 0)$ equals the identity $m \times m$ matrix. Hence we can introduce, in a neighborhood of the origin in $\mathbb{R}^N$, the vector fields

$$M_k = \sum_{\ell=1}^m \mu_{k\ell}(x, t) \frac{\partial}{\partial x_\ell}, \quad k = 1, \ldots, m \qquad (\text{I.35})$$

characterized by the relations

$$M_k Z_\ell = \delta_{k\ell}. \qquad (\text{I.36})$$

Consequently the vector fields

$$L_j = \frac{\partial}{\partial t_j} - i \sum_{k=1}^m \frac{\partial \phi_k}{\partial t_j}(x, t) M_k, \quad j = 1, \ldots, n \qquad (\text{I.37})$$

are linearly independent and satisfy $L_j Z_k = 0$, for $j = 1, \ldots, n$, $k = 1, \ldots, m$. The same argument as before gives:

$$L_1, \ldots, L_n \text{ span } \mathcal{V} \text{ in a neighborhood of the origin;} \qquad (\text{I.38})$$
$$L_1, \ldots, L_n, M_1, \ldots, M_m \text{ are pairwise commuting and} \qquad (\text{I.39})$$
$$\text{span } \mathbb{C}T\mathbb{R}^N \text{ in a neighborhood of the origin in } \mathbb{R}^N.$$

Let U be an open set of $\mathbb{R}^n$ and assume, given a smooth function $\Phi : U \to \mathbb{R}^m$, $\Phi(t) = (\phi_1(t), \ldots, \phi_m(t))$. We shall call a *tube structure* on $\mathbb{R}^m \times U$ the locally integrable structure $\mathcal{V}$ on $\mathbb{R}^m \times U$ for which T' is spanned by the differentials of the functions

$$Z_k = x_k + i\phi_k(t), \quad k = 1, \ldots, m.$$

A tube structure $\mathcal{V}$ has remarkably simple global generators. Indeed if we set, as usual, $Z = (Z_1, \ldots, Z_m)$ we have $Z_x(x, t) = I$, the identity $m \times m$ matrix, for *every* $(x, t) \in \mathbb{R}^m \times U$. This gives $M_k = \partial/\partial x_k$ and consequently the vector fields (I.37) take the form

$$L_j = \frac{\partial}{\partial t_j} - i \sum_{k=1}^m \frac{\partial \phi_k}{\partial t_j}(t) \frac{\partial}{\partial x_k} \quad j = 1, \ldots, n. \qquad (\text{I.37}')$$

Observe that these vector fields span $\mathcal{V}$ on $\mathbb{R}^m \times U$.

I.11 Local generators in analytic structures

When $\mathcal{V}$ is real-analytic then the functions ϕ_k in Corollary I.10.2 can be taken real-analytic. We keep the notation established in the preceding section and consider the equation

$$Z(x, t) - z = 0$$

for $(x, t, z) \in \mathbb{C}^m \times \mathbb{C}^n \times \mathbb{C}^m$ in a neighborhood of the origin. Since

$$\frac{\partial Z}{\partial x}(0, 0) = I$$

we can find, by the implicit function theorem, a holomorphic function $x = H(z, t) = (H_1(z, t), \dots, H_m(z, t))$ defined in a neighborhood of the origin in $\mathbb{C}^m \times \mathbb{C}^n$ satisfying

$$H(0, 0) = 0, \quad H(Z(x, t), t) = x.$$

We set

$$Z_k^\#(x, t) \doteq H_k(Z(x, t), 0), \quad k = 1, \dots, m.$$

Then we also have

$$L_j Z_k^\# = 0, \quad j = 1, \dots, n, k = 1, \dots, m,$$
$$dZ_1^\# \wedge \dots \wedge dZ_m^\# \neq 0.$$

Moreover, $Z_k^\#(x, 0) = x_k$ for every k. Hence, if we consider the real-analytic diffeomorphism

$$(x, t) \mapsto (X, T) = (\Re Z^\#(x, t), t)$$

in these new variables we can write $Z_k^\#(X, T) = X_k + i\Phi_k^\#(X, T)$ where now we have $\Phi_k^\#(X, 0) = 0$ for every k. Summing up we can state:

COROLLARY I.11.1. *Let $\mathcal{V}$ be a locally integrable real-analytic structure defined on a real-analytic manifold Ω. Let $p \in \Omega$. Then there is a real-analytic coordinate system vanishing at p,*

$$\{x_1, \dots, x_m, t_1, \dots, t_n\}$$

and real-analytic, real-valued $\phi_1, \dots, \phi_m$ defined in a neighborhood of the origin and satisfying

$$\phi_k(x, 0) = 0, \quad k = 1, \dots, m$$

such that the differentials of the functions

$$Z_k = x_k + i\phi_k(x, t), \quad k = 1, \dots, m \tag{I.40}$$

span T' in a neighborhood of the origin.

REMARK I.11.2. We point out that in the coordinates (x, t) given by Corollary I.11.1 it is elementary to find the unique analytic solution u to the Cauchy problem:

$$\begin{cases} L_j u = 0 & j = 1, \dots, n, \\ u(x, 0) = h(x) \end{cases} \tag{I.41}$$

where h is real-analytic. Indeed,

$$u(x, t) = h(Z_1^{\#}(x, t), \dots, Z_m^{\#}(x, t))$$

solves (I.41) and in order to see that this is the unique analytic solution it suffices to notice that if v is analytic, if $v(x, 0) = 0$, and if $L_j v = 0$ for every j then v must vanish identically since all its derivatives vanish at the origin.

Uniqueness for the distribution solutions of (I.41) holds when the structure is only C^∞. This, though, is a much deeper result and its discussion will be postponed to Chapter II.

I.12 Integrability of complex and elliptic structures

The celebrated theorem of Newlander and Nirenberg ([NN]) states that *every complex structure is locally integrable*. We shall postpone the proof of this result to the appendix of this chapter and now we will apply it to prove the more general statement that in fact every elliptic structure is locally integrable. This result is due to L. Nirenberg.

THEOREM I.12.1. *Let V be an elliptic structure over a smooth manifold Ω. Then V is locally integrable.*

PROOF. By (I.15) we have $T'_p \cap \overline{T}'_p = 0$ for every $p \in \Omega$ and then

$$T' \oplus \overline{T}' \doteq \bigcup_{p \in \Omega} \left(T'_p \oplus \overline{T}'_p \right)$$

is a vector sub-bundle of $\mathbb{C}T^*\Omega$ of rank $2m$. In particular, if n is the dimension of Ω, we obtain that $2m \leq n$. Thus

$$V \cap \overline{V} \doteq \bigcup_{p \in \Omega} (V_p \cap \overline{V}_p) = \bigcup_{p \in \Omega} \left(T'_p \oplus \overline{T}'_p \right)^{\perp}$$

is a vector sub-bundle of $\mathbb{C}T\Omega$. By the argument that led to the proof of Proposition I.8.1 we see then that

$$V \cap T\Omega \doteq \bigcup_{p \in \Omega} (V_p \cap T_p\Omega)$$

is a vector sub-bundle of $T\Omega$. Notice that

$$n' \doteq \dim_{\mathbb{R}} \left(\mathcal{V}_p \cap T_p\Omega \right) = n - 2m, \quad p \in \Omega.$$

Let $p_0 \in \Omega$ be fixed. By the Frobenius Theorem I.5.1 we can find a coordinate system $(x_1, \ldots, x_{2m}, t_1, \ldots, t_{n'})$ around p_0 such that $\mathcal{V} \cap T\Omega$ is generated near p_0 by the vector fields

$$\frac{\partial}{\partial t_j}, \quad j = 1, \ldots, n'.$$

Next we select m complex vector fields

$$L_k = \sum_{\ell=1}^{2m} a_{k\ell}(x, t) \frac{\partial}{\partial x_\ell}$$

in such a way that $L_1, \ldots, L_m, \partial/\partial t_1, \ldots, \partial/\partial t_{n'}$ span $\mathcal{V}$ in a neighborhood of p_0. After a linear substitution (as in the proof of Proposition I.4.4) we can assume that the vector fields L_k take the form

$$L_k = \frac{\partial}{\partial x_k} + \sum_{\ell=m+1}^{2m} b_{k\ell}(x, t) \frac{\partial}{\partial x_\ell}, \quad k = 1, \ldots, m.$$

Since $\mathcal{V}$ is a formally integrable structure, we know $[\partial/\partial t_\alpha, L_k]$ must be a linear combination of $L_1, \ldots, L_m, \partial/\partial t_1, \ldots, \partial/\partial t_{n'}$. Due to the special form of the vector fields L_k these brackets must vanish identically, that is:

$$\sum_{\ell=m+1}^{2m} \frac{\partial b_{k\ell}}{\partial t_\alpha}(x, t) \frac{\partial}{\partial x_\ell} = 0, \quad \forall \alpha, k.$$

Consequently, the functions $b_{k\ell}$ do not depend on $t_1, \ldots, t_{n'}$ in a full neighborhood of p_0. Since, moreover,

$$L_1, \ldots, L_m, \overline{L}_1, \ldots, \overline{L}_m, \frac{\partial}{\partial t_1}, \ldots, \frac{\partial}{\partial t_{n'}}$$

span $\mathbb{C}T\Omega$ it follows that $L_1, \ldots, L_m, \overline{L}_1, \ldots, \overline{L}_{n'}$ are linearly independent. We conclude then that $L_1, \ldots, L_m$ define a complex structure (in the x-space) in a neighborhood of p_0. By the Newlander–Nirenberg theorem there are $Z_1(x), \ldots, Z_m(x)$ with linearly independent differentials such that

$$L_k Z_\ell = 0, \quad k, \ell = 1, \ldots, m.$$

Since, moreover,

$$\frac{\partial Z_\ell}{\partial t_j} = 0, \quad \ell = 1, \ldots, m, \ j = 1, \ldots, n'$$

the proof is complete. □

Theorem I.10.1 gives a particularly simple local representation for an elliptic structure. Let $\mathcal{V}$ and Ω be as in Theorem I.12.1 and fix $p \in \Omega$. With the notation as in Theorem I.10.1 we have $d = 0$, $\nu = m$ and thus there is a coordinate system

$$(x_1, \ldots, x_m, y_1, \ldots, y_m, t_1, \ldots t_{n'})$$

vanishing at p such that, setting $z_j = x_j + iy_j$, the differentials dz_j span T' near p, and the vector fields $\partial/\partial \overline{z}_k$, $\partial/\partial t_j$ span $\mathcal{V}$ near p. Notice also that $n' = 0$ corresponds to the case when $\mathcal{V}$ defines a complex structure.

I.13 Elliptic structures in the real plane

In this section we depart a bit from the spirit we have adopted in the exposition up to now and make use of some standard results on Fourier analysis and pseudo-differential operators in order to study elliptic structures in two-dimensional manifolds. The results contained here are not necessary for the comprehension of the remaining parts of the chapter and the section can be avoided in a first reading.

If Ω is an open subset of $\mathbb{R}^2$ any sub-bundle $\mathcal{V}$ of $\mathbb{C}T\Omega$ of rank one defines a formally integrable structure over Ω, for the involutive condition is automatically satisfied. Suppose that Ω contains the origin and let L be a complex vector field that spans $\mathcal{V}$ in a neighborhoord of 0. After division by a nonvanishing smooth factor it can be assumed that, in suitable coordinates (x_1, x_2), we can write

$$L = \frac{\partial}{\partial x_2} + a(x_1, x_2) \frac{\partial}{\partial x_1}.$$

As at the beginning of Section I.5 we can find a smooth diffeomorphism $(x, t) \mapsto (x_1, x_2)$, $x_2 = t$, which reduces $\Re L$ to $\partial/\partial t$. Since also $\partial/\partial x_1$ is a multiple of $\partial/\partial x$ in these new variables, L can be written as a nonvanishing multiple of

$$L_\bullet = \frac{\partial}{\partial t} + ib(x, t) \frac{\partial}{\partial x}, \tag{I.42}$$

where b is smooth and *real-valued*. Since both L and $L_\bullet$ span $\mathcal{V}$ in a neighborhood of the origin of $\mathbb{R}^2$, there is no loss of generality in assuming that our original L takes the form (I.42).

The structure $\mathcal{V}$ is elliptic if and only if L and $\overline{L}$ are linearly independent at every point. This is equivalent to saying that the function b in (I.42) never vanishes (in the p.d.e. terminology, L is an *elliptic* operator). We shall now recall the standard elliptic estimates satisfied by L and its transpose ${}^t L$ in a neighborhood of the origin. Let

$$L_0 = \frac{\partial}{\partial t} + ib(0,0)\frac{\partial}{\partial x}. \tag{I.43}$$

If $\varphi \in C_c^\infty(\mathbb{R}^2)$ then taking Fourier transforms gives

$$\mathcal{F}(L_0\varphi)(\xi,\tau) = (i\tau - b(0,0)\xi)\mathcal{F}(\varphi)(\xi,\tau).$$

Since $b(0,0) \neq 0$ we have

$$\tau^2 + \xi^2 \leq \max\left\{1, \frac{1}{b(0,0)^2}\right\} |i\tau - b(0,0)\xi|^2$$

and thus by Parseval's formula we obtain, in Sobolev norms,

$$\|\varphi\|_1 \leq C\left(\|L_0\varphi\|_0 + \|\varphi\|_0\right), \quad \varphi \in C_c^\infty(\mathbb{R}^2), \tag{I.44}$$

where for any real s we denote by $\|\varphi\|_s$ the norm in the Sobolev space $L_s^2(\mathbb{R}^2)$ (see Section II.3.2 for the definition of Sobolev norms). We select an open neighborhood of the origin $U \subset \Omega$ such that $|b(x,t) - b(0,0)| \leq 1/(2C)$ for $(x,t) \in U$. If $\varphi \in C_c^\infty(U)$ then by (I.44)

$$\begin{aligned}
\|\varphi\|_1 &\leq C\left(\|L\varphi\|_0 + \|(L - L_0)\varphi\|_0 + \|\varphi\|_0\right) \\
&= C\left(\|L\varphi\|_0 + \|(b(x,t) - b(0,0))\varphi_x\|_0 + \|\varphi\|_0\right) \\
&\leq C\left(\|L\varphi\|_0 + \|\varphi\|_0\right) + \frac{1}{2}\|\varphi\|_1,
\end{aligned}$$

and thus

$$\|\varphi\|_1 \leq 2C\left(\|L\varphi\|_0 + \|\varphi\|_0\right), \quad \varphi \in C_c^\infty(U). \tag{I.45}$$

Let now $V \subset\subset U$ be an open set and let also $\theta \in C_c^\infty(U)$ be identically equal to one in V. We denote by Θ the operator 'multiplication by θ' and by Λ the operator $(1 - \Delta)^{1/2}$. For a real number s and for $\varphi \in C_c^\infty(V)$, we obtain

$$\|\varphi\|_{s+1} = \|\Lambda^s\Theta(\varphi)\|_1 \leq \|\Theta\Lambda^s\varphi\|_1 + C_1\|\varphi\|_s$$

since the commutator between Λ^s and Θ has order $s-1$. If we now apply (I.45) we obtain

$$\begin{aligned}
\|\varphi\|_{s+1} &\leq C_2\left\{\|L\Theta\Lambda^s\varphi\|_0 + \|\Theta\Lambda^s\varphi\|_0 + \|\varphi\|_s\right\} \\
&\leq C_3\left\{\|(\Theta\Lambda^s)L\varphi\|_0 + \|\varphi\|_s\right\}
\end{aligned}$$

since both $\Theta\Lambda^s$ and its commutator with L have order s. We then obtain:

- For every $V \subset\subset U$ open and every $s \in \mathbb{R}$ there is $C^\bullet > 0$ such that

$$\|\varphi\|_{s+1} \leq C^\bullet\left\{\|L\varphi\|_s + \|\varphi\|_s\right\}, \quad \varphi \in C_c^\infty(V). \tag{I.46}$$

PROPOSITION I.13.1. *If $u \in \mathcal{D}'(U)$ and $Lu \in L^{2,s}_{loc}(U)$ then $u \in L^{2,s+1}_{loc}(U)$. In particular, if $u \in \mathcal{D}'(U)$ and $Lu \in C^{\infty}(U)$ then $u \in C^{\infty}(U)$.*

PROOF. Let $W \subset\subset V \subset\subset U$ be open sets and let $\theta \in C^{\infty}_c(V)$ be identically equal to one in W. Since there is $\sigma \leq s$ such that $u \in L^{2,\sigma}_{loc}(V)$ it will suffice to show that $\theta u \in L^2_{\sigma+1}$, for iteration of the argument will give the result.

Let $B_\epsilon = \rho_\epsilon *\, \cdot\,$, where $\{\rho_\epsilon\}$ is the usual family of mollifiers in $\mathbb{R}^2$. We have $B_\epsilon(\theta u) \to \theta u$ in L^2_σ as $\epsilon \to 0$ and also

$$LB_\epsilon(\theta u) = B_\epsilon L(\theta u) + [L, B_\epsilon](\theta u) \xrightarrow{\epsilon \to 0} L(\theta u) \text{ in } L^2_\sigma$$

by Friedrich's lemma, since $L(\theta u) \in L^2_\sigma$. Thus, if take $\epsilon_n \to 0$ and if we apply (I.46) for $s = \sigma$ and $\varphi = B_{\epsilon_m}(\theta u) - B_{\epsilon_n}(\theta u)$ we conclude that $\{B_{\epsilon_n}(\theta u)\}$ is a Cauchy sequence in $L^2_{\sigma+1}$. Hence $\theta u \in L^2_{\sigma+1}$ and the proof is complete. □

We shall now derive from (I.45) an estimate for the transpose of L which will lead us to a solvability result. If we notice that $^tL = -L - ib_x(x, t)$ then from (I.45) we obtain, for some constant $C' > 0$,

$$\|\varphi\|_1 \leq C' \left(\|^tL\varphi\|_0 + \|\varphi\|_0 \right), \quad \varphi \in C^{\infty}_c(U). \tag{I.47}$$

Now, it is elementary that

$$\|\varphi\|_0 \leq 2\delta\|\varphi_t\|_0, \quad \varphi \in C^{\infty}_c(U),$$

where $\delta = \sup\{|t| : (x, t) \in U\}$. Consequently, if we further contract U about the origin in order to achieve $2\delta C' \leq 1/2$, from (I.47) we finally obtain

$$\|\varphi\|_0 \leq 2C'\|^tL\varphi\|_0, \quad \varphi \in C^{\infty}_c(U). \tag{I.48}$$

PROPOSITION I.13.2. *For every $f \in L^2(U)$ there is $u \in L^2(U)$ such that $Lu = f$ in U.*

PROOF. Given $f \in L^2(U)$ consider the functional

$$^tL\varphi \mapsto \int f(x, t)\varphi(x, t)\,dxdt \tag{I.49}$$

defined on $\{^tL\varphi : \varphi \in C^{\infty}_c(U)\}$, where the latter is considered as a subspace of $L^2(U)$. By (I.48) it follows that (I.49) is well-defined and continuous. By the Hahn–Banach theorem we extend (I.49) to a continuous functional λ on $L^2(U)$ and by the Riesz representation theorem we find $u \in L^2(U)$ such that

$$\lambda(g) = \int g(x, t)u(x, t)\,dxdt, \quad g \in L^2(U).$$

In particular, if $\varphi \in C_c^\infty(U)$

$$\lambda({}^t L\varphi) = \int f(x,t)\varphi(x,t)\mathrm{d}x\mathrm{d}t,$$

which is precisely the meaning of the equality $Lu = f$ in the weak sense. $\qquad\square$

COROLLARY I.13.3. *Let $D \subset\subset U$ be an open disk centered at the origin. Then*

$$L C^\infty(\overline{D}) = C^\infty(\overline{D}). \tag{I.50}$$

PROOF. Given $f \in C^\infty(\overline{D})$ we extend it to an element $\tilde{f} \in C_c^\infty(U)$ and by Proposition I.13.2 we find $u \in L^2(U)$ solving $Lu = \tilde{f}$ in U. Finally, by Proposition I.13.1, we have $u \in C^\infty(U)$ and thus its restriction to D belongs to $C^\infty(\overline{D})$. $\qquad\square$

Still under the assumption that L is elliptic we apply (I.50) in order to find $v \in C^\infty(\overline{D})$ such that

$$Lv = -ib_x. \tag{I.51}$$

If we set

$$u(x,t) = \int_0^x e^{v(x',t)}\mathrm{d}x'$$

we get

$$\begin{aligned}
Lu(x,t) &= \int_0^x v_t(x',t)e^{v(x',t)}\mathrm{d}x' + ib(x,t)e^{v(x,t)} \\
&= \int_0^x (-ibv_x - ib_x)(x',t)e^{v(x',t)}\mathrm{d}x' + ib(x,t)e^{v(x,t)} \\
&= -i\int_0^x \partial_x\{be^v\}(x',t)\mathrm{d}x' + ib(x,t)e^{v(x,t)} \\
&= ib(0,t)e^{v(0,t)}.
\end{aligned}$$

Then if we set

$$Z(x,t) = u(x,t) - i\int_0^t b(0,t')e^{v(0,t')}\mathrm{d}t' \tag{I.52}$$

we obtain

$$LZ = 0, \quad Z_x = e^v \neq 0, \tag{I.53}$$

that is, our original elliptic structure $\mathcal{V}$ is locally integrable. We have thus obtained a proof of the *Newlander–Nirenberg theorem in the particular case when $N = 2$*. We emphasize for this situation the conclusion that we have reached at the end of Section I.12:

COROLLARY I.13.4. *If L is an elliptic operator in an open subset $\Omega \subset \mathbb{R}^2$ and if $p \in \Omega$ then we can find local coordinates (x, y) vanishing at p such that L can be written, in a neighborhood of p, as*

$$L = g(x, y) \left\{ \frac{\partial}{\partial x} + i \frac{\partial}{\partial y} \right\},$$

where g never vanishes. $\square$

REMARK I.13.5. Our discussion indeed leads to a general criterion that characterizes when a rank one formally integrable structure $\mathcal{V} \subset \mathbb{C}T\Omega$, $\Omega \subset \mathbb{R}^2$ open, is locally integrable. Suppose that $\mathcal{V}$ is spanned, in a neighborhood of the origin, by the vector field (I.42).

PROPOSITION I.13.6. *The following properties are equivalent:*

($\dagger$) *there is $Z \in C^\infty$ near the origin solving $LZ = 0$, $Z_x \neq 0$;*
($\ddagger$) *there is $v \in C^\infty$ near the origin solving (I.51).*

PROOF. We have already presented the argument that ($\ddagger$) $\Rightarrow$ ($\dagger$). For the reverse implication we notice that

$$0 = (LZ)_x = L(Z_x) + i b_x Z_x$$

and consequently

$$L\left(\log(Z_x)\right) = Z_x^{-1} L(Z_x) = -i b_x.$$ $\square$

I.14 Compatible submanifolds

Let Ω be a smooth manifold. A subset $\mathcal{M}$ of Ω is called an *embedded submanifold* (or *submanifold* for short) *of* Ω if there is $r \in \{0, 1, \ldots, N\}$ for which the following is true:

• Given $p_0 \in \mathcal{M}$ arbitrary there is a local chart $(U_0, \mathbf{x})$, with $p_0 \in U_0$ and $\mathbf{x} = (x_1, \ldots, x_N)$, such that

$$U_0 \cap \mathcal{M} = \{q \in U_0 : x_{r+1}(q) = x_{r+1}(p_0), \ldots, x_N(q) = x_N(p_0)\}.$$

When p_0 runs over $\mathcal{M}$ the pairs $(U_0, \mathbf{x}_0)$, where

$$\mathbf{x}_0 = (x_1|_{U_0 \cap \mathcal{M}}, \ldots, x_r|_{U_0 \cap \mathcal{M}}),$$

make up a family $\mathcal{F}^*$ that satisfies properties (1) and (2) of Section I.1 Hence $\mathcal{M}$ is a smooth manifold of dimension r. We shall refer to the number $N - r$ as the *codimension of $\mathcal{M}$ (in Ω).*

Let $p \in \mathcal{M}$ and denote by $C_\mathcal{M}^\infty(p)$ the space of germs of smooth functions on $\mathcal{M}$ at p. It is clear that the *restriction to $\mathcal{M}$* defines a surjective homomorphism o f $\mathbb{C}$-algebras $C^\infty(p) \to C_\mathcal{M}^\infty(p)$ which gives us then a natural injection

$$\iota_p : \mathbb{C}T_p\mathcal{M} \hookrightarrow \mathbb{C}T_p\Omega. \tag{I.54}$$

By transposition we thus obtain a surjection

$$(\iota_p)^* : \mathbb{C}T_p^*\Omega \longrightarrow \mathbb{C}T_p^*\mathcal{M}, \tag{I.55}$$

whose kernel will be denoted by $\mathbb{C}N_p^*\mathcal{M}$. We shall sometimes refer to the disjoint union

$$\mathbb{C}N^*\mathcal{M} \doteq \bigcup_{p \in \mathcal{M}} \mathbb{C}N_p^*\mathcal{M} \tag{I.56}$$

as the *complex conormal bundle of $\mathcal{M}$ in Ω*.

Let now $U \subset \Omega$ be open and let $\omega \in \mathfrak{N}(U)$. Given $L \in \mathfrak{X}(U \cap \mathcal{M})$ the map

$$p \mapsto \left(\iota_p^*(\omega_p)\right)(L_p)$$

is easily seen to be smooth on $U \cap \mathcal{M}$. By the discussion that precedes Proposition I.4.2, there is a form $\omega^\bullet \in \mathfrak{N}(U \cap \mathcal{M})$ such that

$$\omega_p^\bullet = (\iota_p)^*(\omega_p)$$

for every $p \in U \cap \mathcal{M}$. We shall denote $\omega^\bullet$ by $\iota^*\omega$ and shall refer to it as the *pullback of ω to $U \cap \mathcal{M}$*. It is clear that ι^* is a homomorphism which is moreover surjective when $U \cap \mathcal{M}$ is closed in U. Observe also that

$$\iota^*(\mathrm{d}f) = \mathrm{d}(f|_{U \cap \mathcal{M}}), \quad f \in C^\infty(U). \tag{I.57}$$

Let now $\mathcal{V}$ be a formally integrable structure over Ω, with $T' = \mathcal{V}^\perp$, and let $\mathcal{M} \subset \Omega$ be a submanifold. If $p \in \mathcal{M}$ we set

$$\mathcal{V}(\mathcal{M})_p \doteq \mathcal{V}_p \cap \mathbb{C}T_p\mathcal{M}, \quad \mathcal{V}(\mathcal{M}) \doteq \bigcup_{p \in \mathcal{M}} \mathcal{V}(\mathcal{M})_p. \tag{I.58}$$

With orthogonal now taken in the duality $(\mathbb{C}T_p\mathcal{M}, \mathbb{C}T_p^*\mathcal{M})$ we have

$$(\iota_p)^*(T_p') = \mathcal{V}(\mathcal{M})_p^\perp, \tag{I.59}$$

since the left-hand side is the image of the composition

$$T_p' \hookrightarrow \mathbb{C}T_p^*\Omega \xrightarrow{(\iota_p)^*} \mathbb{C}T_p^*\mathcal{M}$$

and consequently is equal to the orthogonal to the kernel of the composition

$$\mathbb{C}T_p\mathcal{M} \hookrightarrow \mathbb{C}T_p\Omega \longrightarrow \mathbb{C}T_p\Omega/\mathcal{V}_p.$$

DEFINITION I.14.1. *We shall say that $\mathcal{M}$ is compatible with the formally integrable structure $\mathcal{V}$ if $\mathcal{V}(\mathcal{M})$ defines a formally integrable structure over $\mathcal{M}$.*

When $\mathcal{M}$ is compatible with $\mathcal{V}$ then, according to our previous notation,

$$T'(\mathcal{M})_p \doteq \mathcal{V}(\mathcal{M})_p^\perp = (\iota_p)^*(T'_p)$$

(*cf.* (I.59)). The next result gives a very useful criterion:

PROPOSITION I.14.2. *The submanifold $\mathcal{M}$ is compatible with $\mathcal{V}$ if (and only if)*

$$p \mapsto \dim \mathcal{V}(\mathcal{M})_p \text{ is constant on } \mathcal{M}. \tag{I.60}$$

PROOF. We must prove that (I.60) implies that $\mathcal{V}(\mathcal{M})$ is a vector sub-bundle of $\mathcal{V}$ which satisfies the Frobenius condition.

First we observe that (I.60) and (I.59) give the existence of α such that

$$\dim(\iota_p)^*(T'_p) = \alpha, \ \forall p \in \mathcal{M}. \tag{I.61}$$

Let $p_0 \in \mathcal{M}$ and take $\omega_1, \ldots, \omega_m \in \mathfrak{N}(U_0)$, where U_0 is an open subset of Ω that contains p_0, such that $(\omega_1)_q, \ldots, (\omega_m)_q$ span T'_q for every $q \in U_0$. Select $j_1, \ldots, j_\alpha$ such that

$$\left\{ (\iota^* \omega_{j_1})_{p_0}, \ldots, (\iota^* \omega_{j_\alpha})_{p_0} \right\}$$

form a basis for $(\iota_{p_0})^*(T'_{p_0})$. Then

$$\left\{ (\iota^* \omega_{j_1})_p, \ldots, (\iota^* \omega_{j_\alpha})_p \right\}$$

will still be linearly independent when p belongs to an open neighborhood V_0 of p_0 in $\mathcal{M}$ and consequently, thanks to (I.61), will form a basis to $(\iota_p)^*(T'_p)$ for all such p. By the remark that follows Proposition I.4.4 we conclude that $\mathcal{V}(\mathcal{M})$ is a vector sub-bundle of $\mathcal{V}$.

To conclude the argument it suffices to observe that if U is an open subset of Ω and if $L, M \in \mathfrak{X}(U)$ are such that $L_p, M_p \in \mathbb{C}T_p\mathcal{M}$ for every $p \in U \cap \mathcal{M}$ then $[L, M]_p \in \mathbb{C}T_p\mathcal{M}$ also for every $p \in U \cap \mathcal{M}$. This property will easily imply that $\mathcal{V}(\mathcal{M})$ satisfies the Frobenius condition. $\square$

PROPOSITION I.14.3. *If $\mathcal{V}$ is a locally integrable structure over Ω and if $\mathcal{M}$ is a submanifold of Ω which is compatible with $\mathcal{V}$ then $\mathcal{V}(\mathcal{M})$ is a locally integrable structure over $\mathcal{M}$.*

PROOF. It follows from the proof of Proposition I.14.2 in conjunction with (I.57). $\square$

EXAMPLE I.14.4. *Generic submanifolds of complex space.* As in Section I.5 we shall write the complex coordinates in $\mathbb{C}^\mu$ as $z_1, \ldots, z_\mu$, where $z_j = x_j + iy_j$. If f is a smooth function on an open subset of $\mathbb{C}^\mu$ we shall write, as usual,

$$\partial f = \sum_{j=1}^{\mu} \frac{\partial f}{\partial z_j}\, dz_j, \tag{I.62}$$

$$\overline{\partial} f = \sum_{j=1}^{\mu} \frac{\partial f}{\partial \overline{z}_j}\, d\overline{z}_j. \tag{I.63}$$

DEFINITION I.14.5. *Let M be a submanifold of $\mathbb{C}^\mu$ of codimension d. We shall say that M is generic if given $p_0 \in M$ there are an open neighborhood U_0 of p_0 in $\mathbb{C}^\mu$ and real-valued functions $\rho_1, \ldots, \rho_d \in C^\infty(U_0)$ such that*

$$M \cap U_0 = \{z \in U : \rho_k(z) = 0, \ k = 1, \ldots, d\}$$

and

$$\overline{\partial}\rho_1, \ldots, \overline{\partial}\rho_d \text{ are linearly independent at each point of } M \cap U_0.$$

Notice that every one-codimensional submanifold of $\mathbb{C}^\mu$ is automatically generic. Denote by $\mathcal{V}^{0,1}$ the sub-bundle of $\mathbb{C}T\mathbb{C}^\mu$ which defines the complex structure on $\mathbb{C}^\mu$, that is, the sub-bundle spanned by the vector fields $\partial/\partial\overline{z}_j$, $j = 1, \ldots, \mu$.

PROPOSITION I.14.6. *If M is a generic submanifold of $\mathbb{C}^\mu$ of codimension d then M is compatible with $\mathcal{V}^{0,1}$. Moreover, $\mathcal{V}^{0,1}(M)$ is a locally integrable, CR structure for which n and m satisfy:*

$$\dim M = 2n + d, \quad m = \mu = n + d.$$

The sub-bundle $T'(M)$ is spanned by the differentials of the restriction to M of the complex coordinate functions on $\mathbb{C}^\mu$.

PROOF. Let $p \in M$. A vector $\sum_{j=1}^{\mu} a_j (\partial/\partial\overline{z}_j)_p$ belongs to $\mathbb{C}T_p M \cap \mathcal{V}_p^{0,1}$ if and only if

$$\sum_{j=1}^{\mu} a_j \frac{\partial \rho_k}{\partial \overline{z}_j}(p) = 0, \quad k = 1, \ldots, d.$$

Since M is generic it follows that

$$\dim_{\mathbb{C}}\left(\mathbb{C}T_p M \cap \mathcal{V}_p^{0,1}\right) = \mu - d, \quad \forall p \in M.$$

By Propositions I.14.2 and I.14.3 we conclude that M is compatible with $\mathcal{V}^{0,1}$ and that $\mathcal{V}^{0,1}(M)$ is locally integrable. Moreover, since $(\mathcal{V}^{0,1})_p \cap (\overline{\mathcal{V}^{0,1}})_p = 0$ for every $p \in \mathbb{C}^\mu$ we obtain

$$\mathcal{V}^{0,1}(\mathcal{M})_p \cap \overline{\mathcal{V}^{0,1}(\mathcal{M})}_p = 0, \quad \forall p \in \mathcal{M},$$

which shows that $\mathcal{V}^{0,1}(\mathcal{M})$ defines a CR structure over $\mathcal{M}$. Finally, we have $n \doteq \operatorname{rank} \mathcal{V}^{0,1}(\mathcal{M}) = \mu - d$ and thus $\dim \mathcal{M} = 2\mu - d = 2n + d$ and $m = \dim \mathcal{M} - n = n + d$.

The last statement follows immediately from the proof of Proposition I.14.2.

$\square$

I.15 Locally integrable CR structures

When $\mathcal{V}$ defines a locally integrable CR structure over Ω then, according to Proposition I.8.4, $d \doteq \dim T_p^0 = N - 2n$, for all $p \in \Omega$. Using Theorem I.10.1 we obtain $m = N - n = n + d$, $\nu = m - d = n$ and $n' = N - 2\nu - d = 0$. We summarize:

- Given $p \in \Omega$ there is a coordinate system vanishing at p,

$$\{x_1, \ldots, x_n, y_1, \ldots, y_n, s_1, \ldots, s_d\}$$

and smooth, real-valued functions $\phi_1, \ldots, \phi_d$ defined in a neighborhood of the origin and satisfying

$$\phi_k(0) = 0, \ \mathrm{d}\phi_k(0) = 0, \quad k = 1, \ldots, d, \tag{I.64}$$

such that the differentials of the functions

$$Z_j = x_j + iy_j, \quad j = 1, \ldots, n; \tag{I.65}$$

$$W_k = s_k + i\phi_k(z, s), \quad k = 1, \ldots, d \tag{I.66}$$

span T' in a neighborhood of the origin.

Notice that $\mathcal{V}$ is spanned, in a neighborhood of the origin, by the pairwise commuting vector fields (I.27), where $\nu = n$ and there is no t-variable.

Suppose that $\phi = (\phi_1, \ldots, \phi_d)$ is defined in a neighborhood U of the origin in $\mathbb{C}^n \times \mathbb{R}^d$. Then the map

$$\mathfrak{F} : U \to \mathbb{C}^{n+d}, \quad \mathfrak{F}(z, s) = (z, s + i\phi(z, s)) \tag{I.67}$$

has rank $2n + d$ and consequently $\mathfrak{F}(U)$ is an embedded submanifold of $\mathbb{C}^{n+d}$ of dimension $2n + d$ (and of codimension d).

Now we write the coordinates in $\mathbb{C}^{n+d}$ as

$$(z_1, \ldots, z_n, w_1, \ldots, w_d),$$

where $w = s + it$, $w_j = s_j + it_j$. Then $\mathfrak{F}(U)$ is defined by the equations

$$\rho_k(z, w) \doteq \phi_k(z, s) - t_k = 0, \quad k = 1, \ldots, d.$$

Since

$$\frac{\partial \rho_k}{\partial w_\ell} = \frac{1}{2} \left\{ \frac{\partial \phi_k}{\partial s_\ell} + i\delta_{k\ell} \right\}, k, \ell = 1, \dots, d,$$

we conclude, taking into account (I.64), that $\mathfrak{F}(U)$ is generic if U is taken small enough so that

$$\left\| \frac{\partial \phi}{\partial s}(z, s) \right\| \le \frac{1}{2}, \quad (z, s) \in U. \tag{I.68}$$

By Proposition I.14.6 the complex structure $\mathcal{V}^{0,1}$ on $\mathbb{C}^{n+d}$ defines a locally integrable CR structure $\mathcal{V}(\mathfrak{F}(U))$ on $\mathfrak{F}(U)$ for which the sub-bundle $T'(\mathfrak{F}(U)$ is spanned by the differentials of the restrictions of the functions $z_1, \dots, z_n$, $w_1, \dots, w_d$ to $\mathfrak{F}(U)$. Since in the local coordinates (z, s) we have

$$z_j|_{\mathfrak{F}(U)} = Z_j(z, s), \quad w_k|_{\mathfrak{F}(U)} = W_k(z, s)$$

(*cf.* (I.65), (I.66)), we can state:

PROPOSITION I.15.1. *Every locally integrable CR structure can be locally realized as the CR structure induced by the complex structure on a generic submanifold of the complex space.* $\qquad\square$

REMARK I.15.2. Let $\mathcal{V}$ be a tube structure on $\mathbb{R}^m \times U$ (*cf.* Section I.10). Thus U is an open subset of $\mathbb{R}^n$ and we assume given smooth, real-valued functions $\phi_1, \dots, \phi_m$ on U such that T' is spanned by the differential of the functions $Z_k = x_k + i\phi_k(t)$, $k = 1, \dots, m$. Recall that $\mathcal{V}$ is then spanned on $\mathbb{R}^m \times U$ by the vector fields (I.37'). Let us now assume that $\mathcal{V}$ is also a CR structure. Let $d = m - n$ be the rank of the characteristic set T^0 (*cf.* Proposition I.8.4). Since $\mathcal{V}$ being CR demands that $T'_{(x,t)} + \overline{T'}_{(x,t)} = \mathbb{C}T^*_{(x,t)}(\mathbb{R}^m \times U)$ for every $(x, t) \in \mathbb{R}^m \times U$, we must then have

$$\text{rank } \Phi'(t) = n, \quad \forall t \in U,$$

where $\Phi = (\phi_1, \dots, \phi_m)$. This implies that $\mathcal{M} \doteq \Phi(U)$ is an embedded submanifold of $\mathbb{R}^m$ of dimension n and it is clear that $\mathcal{V}$ can be realized as the CR structure induced by the complex structure on the generic submanifold $\mathbb{R}^m + i\mathcal{M}$ of $\mathbb{R}^m + i\mathbb{R}^m = \mathbb{C}^m$.

One very important model of a CR structure is the *Hans Lewy* structure. We take as Ω the space $\mathbb{C} \times \mathbb{R}$, where the coordinates are written as $z = x + iy$ and s, and consider the formally integrable structure $\mathcal{V}$ spanned by the Hans Lewy vector field (or operator)

$$L = \frac{\partial}{\partial \overline{z}} - iz\frac{\partial}{\partial s}. \tag{I.69}$$

Since L and $\overline{L}$ are linearly independent at every point it follows that $\mathcal{V}$ defines a CR structure which is furthermore locally integrable, since the differential of the functions z and $W = s + i|z|^2$ span T' on $\mathbb{C} \times \mathbb{R}$. Notice also that the Hans Lewy structure can be globally realized as the CR structure induced on the hyperquadric

$$\mathfrak{Q} \doteq \{(z, w) \in \mathbb{C}^2 : w = s + it, \, t = |z|^2\} \tag{I.70}$$

by the complex structure on $\mathbb{C}^2$.

More generally, given $\epsilon_j \in \{-1, 1\}$, $j = 1, \ldots, n$, we can consider the CR structure $\mathcal{V}$ on $\mathbb{C}^n \times \mathbb{R}$ spanned by the pairwise commuting vector fields

$$L_j = \frac{\partial}{\partial \overline{z}_j} - i\epsilon_j z_j \frac{\partial}{\partial s}, \quad j = 1, \ldots, n. \tag{I.71}$$

Such a structure is also locally integrable for the differential of the functions $z_1, \ldots, z_n$ and $W = s + i\phi(z)$, with

$$\phi(z) = \sum_{j=1}^{n} \epsilon_j |z_j|^2,$$

span T' on $\mathbb{C}^n \times \mathbb{R}$.

I.16 A CR structure that is not locally integrable

In this section we shall prove the following quite involved result:

PROPOSITION I.16.1. *Let*

$$\epsilon_1 = 1, \quad \epsilon_j = -1, \, j = 2, \ldots, n. \tag{I.72}$$

There is a smooth function $g(z, s)$ defined in an open neighborhood $\mathcal{O}$ of the origin in $\mathbb{C}^n \times \mathbb{R}$ and vanishing to infinite order at $z_1 = 0$, such that if we set

$$L_j^\# = \frac{\partial}{\partial \overline{z}_j} - i\epsilon_j z_j (1 + g(z, s)) \frac{\partial}{\partial s}, \quad j = 1, \ldots, n, \tag{I.73}$$

then the following is true:

(a) *the vector fields $L_j^\#$ are pairwise commuting;*
(b) *if h is a C^1 function near the origin satisfying $L_j^\# h = 0$ $(j = 1, \ldots, n)$ then $(\partial h / \partial s)(0, 0) = 0$.*

Before we embark on the proof we shall state and prove the important consequence of this result:

COROLLARY I.16.2. *The vector fields* (I.73) *span a CR structure which is not locally integrable in any neighborhood of the origin.*

Indeed, first we notice that $L_1^\#, \ldots, L_n^\#, \overline{L_1^\#}, \ldots, \overline{L_n^\#}$ are linearly independent over $\mathcal{O}$ which together with property (a) shows that (I.73) define a CR structure over $\mathcal{O}$.

Now, given any smooth solution h to the system

$$L_j^\# h = 0, \quad j = 1, \ldots, n, \tag{I.74}$$

we necessarily have $(\partial h / \partial \overline{z}_j)(0, 0) = 0$ for all $j = 1, \ldots, n$. By property (b) we then obtain $dh = \sum_{j=1}^n a_j dz_j$ at the origin and hence any set $h_1, \ldots, h_{n+1}$ of smooth solutions to (I.74) must have linearly dependent differentials at the origin. In particular, the CR structure defined by the vector fields (I.73) cannot be locally integrable.

PROOF OF PROPOSITION I.16.1. The first step in the proof is the construction of the function g. In the complex plane we denote the variable by $w = s + it$ and consider a sequence of closed, disjoint disks $\{D_j\}$, all of them contained in the sector $\{w : |s| < t\}$ and such that $D_j \to \{0\}$ as $j \to \infty$.

Let $F \in C^\infty(\mathbb{C}, \mathbb{R})$ have support contained in the union of the disks D_j and satisfy

$$F(w) > 0, \quad \forall w \in \text{int}(D_j), \quad \forall j. \tag{I.75}$$

As before we shall write $W(z, s) = s + i\phi(z)$, with

$$\phi(z) = |z_1|^2 - |z_2|^2 - \ldots - |z_n|^2. \tag{I.76}$$

LEMMA I.16.3. *The function $F \circ W$ vanishes to infinite order at $z_1 = 0$.*

PROOF. Denote by H the Heaviside function. For every $\ell \in \mathbb{Z}_+$ there is $C_\ell > 0$ such that

$$|F(w)| \le C_\ell (tH(t))^\ell.$$

Then

$$|F(W(z, s))| \le C_\ell (\phi(z)H(\phi(z)))^\ell.$$

Since moreover $\phi(z)H(\phi(z)) \le |z_1|^2$, the lemma is proved. □

We then set

$$g(z, s) \doteq \frac{F(W(z, s))}{z_1 - F(W(z, s))}. \tag{I.77}$$

Since

$$g(z, s) = \frac{F(W(z, s))}{z_1} \frac{1}{1 - F(W(z, s))/z_1}$$

it follows from Lemma I.16.3 that g is smooth in an open neighborhood of the origin in $\mathbb{C}^n \times \mathbb{R}$ and that g vanishes to infinite order at $z_1 = 0$.

We shall now proceed to the proof of (a). We shall write

$$L_j^{\#} = L_j - i\epsilon_j z_j g(z, s) \frac{\partial}{\partial s},$$

(*cf.* (I.76), (I.73)). Since $[L_j, L_k] = 0$ and $L_j z_k = 0$ for all j and k we obtain

$$[L_j^{\#}, L_k^{\#}] = -i \left\{ \epsilon_k z_k L_j g - \epsilon_j z_j L_k g \right\} \frac{\partial}{\partial s}. \tag{I.78}$$

Now

$$L_j g = \frac{z_1}{(z_1 - F \circ W)^2} L_j (F \circ W)$$

and an easy computation making use of the chain rule gives

$$L_j \{ F(W(z, s)) \} = -2i\epsilon_j z_j \frac{\partial F}{\partial w} (W(z, s)).$$

Hence from (I.78) we obtain

$$\begin{aligned} [L_j^{\#}, L_k^{\#}] &= \frac{-iz_1}{(z_1 - F \circ W)^2} \left\{ \epsilon_k z_k L_j (F \circ W) - \epsilon_j z_j L_k (F \circ W) \right\} \frac{\partial}{\partial s} \\ &= \frac{-2z_1}{(z_1 - F \circ W)^2} \left\{ \epsilon_k z_k \epsilon_j z_j - \epsilon_j z_j \epsilon_k z_k \right\} \frac{\partial}{\partial s} = 0. \end{aligned}$$

We now start to prove (b). For this we set

$$\chi(z, s) = h(z, 0, \ldots, 0, s)$$

and will show that $(\partial \chi / \partial s)(0, 0) = 0$. We assume that χ is C^1 in a set of the form

$$V = \{ (z, s) \in \mathbb{C} \times \mathbb{R} : |z| < r, |s| < \delta \}$$

and observe that

$$L\chi - iz f(z, s) \frac{\partial \chi}{\partial s} = 0, \tag{I.79}$$

where L is the Hans Lewy operator given in (I.79) and

$$f(z, s) = \frac{F(s + i|z|^2)}{z - F(s + i|z|^2)}$$

is smooth in V (contracting V if necessary).

Let $U \doteq \{w = s + it \in \mathbb{C} : |s| < \delta, \, 0 < t < r^2\}$ and assume that $D_j \subset U$ for all j. Define

$$I(w) = \int_{|z|=\sqrt{t}} \chi(z, s)dz, \quad w \in U. \tag{I.80}$$

By Stokes' theorem we have

$$I(w) = \int_{|z| \le \sqrt{t}} \frac{\partial \chi}{\partial \bar{z}}(z, s)d\bar{z} \wedge dz = 2i \int_0^{2\pi} \int_0^{\sqrt{t}} \frac{\partial \chi}{\partial \bar{z}}(\rho e^{i\theta}, s)\rho d\rho d\theta$$

from where we obtain

$$\frac{\partial I}{\partial t}(w) = i \int_0^{2\pi} \frac{\partial \chi}{\partial \bar{z}}(\sqrt{t}e^{i\theta}, s)d\theta = \int_{|z|=\sqrt{t}} \frac{1}{z}\frac{\partial \chi}{\partial \bar{z}}(z, s)dz.$$

Consequently,

$$\frac{\partial I}{\partial \bar{w}}(w) = \frac{i}{2} \int_{|z|=\sqrt{t}} \frac{1}{z}(L\chi)(z, s)dz \tag{I.81}$$

(*cf.* (I.79)). From (I.79), (I.81) and from the fact that F is supported in the union of the disks D_j we conclude that I is a holomorphic function of w in the connected open set $U \setminus \bigcup_j D_j$. Since, moreover, $I(w) \to 0$ when $t \to 0^+$ the Schwarz reflection principle implies that I vanishes identically in $U \setminus \bigcup_j D_j$. In particular,

$$I \equiv 0 \quad \text{on} \quad \partial D_j, \quad \forall j. \tag{I.82}$$

Next we consider, for each j, the map

$$\partial D_j \times S^1 \longrightarrow \mathbb{R}^3, \quad (w, \theta) \mapsto (\sqrt{t}e^{i\theta}, s) \tag{I.83}$$

whose image defines a torus $T_j \subset V$. If we set

$$u \doteq \chi \, dz \wedge dW^\flat, \tag{I.84}$$

where $W^\flat(z, s) = s + i|z|^2$, we have $\int_{T_j} u = 0$ for all j, as a consequence of (I.82). Consequently,

$$\int_{S_j} du = 0, \quad \forall j, \tag{I.85}$$

where S_j is the solid torus whose boundary is equal to T_j.

We shall now exploit property (I.85). Since $dz, d\bar{z}, dW^\flat$ are linearly independent we can write

$$d\chi = A dz + B d\bar{z} + C dW^\flat, \tag{I.86}$$

where A, B and C are continuous functions. If we apply both sides of (I.86) to L we obtain that $B = L\chi$, since $Lz = LW^\flat = 0$. Hence, from (I.84) we obtain

$$
\begin{aligned}
\mathrm{d}u &= (L\chi)\mathrm{d}\bar{z} \wedge \mathrm{d}z \wedge \mathrm{d}W^\flat \\
&= izf(z,s)\frac{\partial\chi}{\partial s}\,\mathrm{d}\bar{z} \wedge \mathrm{d}z \wedge \mathrm{d}s \\
&= -2zf(z,s)\frac{\partial\chi}{\partial s}\,\mathrm{d}x \wedge \mathrm{d}y \wedge \mathrm{d}s,
\end{aligned}
$$

which in conjunction with (I.85) gives

$$
\int_{S_j} zf(z,s)\frac{\partial\chi}{\partial s}\mathrm{d}x\mathrm{d}y\mathrm{d}s = 0, \quad \forall j. \tag{I.87}
$$

Now we observe that $zf(z,s) = F(s+i|z|^2)\psi(z,s)$, where ψ is smooth and satisfies $\psi(0,0) = 1$. From (I.87) we conclude the existence of points $P_j, Q_j \in S_j$ such that

$$
\Re\left\{\psi(P_j)\frac{\partial\chi}{\partial s}(P_j)\right\} = \Im\left\{\psi(Q_j)\frac{\partial\chi}{\partial s}(Q_j)\right\} = 0
$$

for all j. It suffices to let $j \to \infty$ to obtain that $(\partial\chi/\partial s)(0,0) = 0$ and hence to conclude the proof of the proposition. $\qquad\square$

I.17 The Levi form on a formally integrable structure

Let $\mathcal{V}$ be a formally integrable structure over a smooth manifold Ω and let $\xi \in T_p^0$, $\xi \neq 0$ be fixed (recall that in particular $\xi \in T_p^*\Omega \subset \mathbb{C}T_p^*\Omega$). We start with the following result:

LEMMA I.17.1. *Let L and M be sections of $\mathcal{V}$ in a neighborhood of p. If either $L_p = 0$ or $M_p = 0$ then $\xi\left([L,\overline{M}]_p\right) = 0$.*

PROOF. We take complex vector fields $L_1, \ldots, L_n$ which span $\mathcal{V}$ at each point in a neighborhood of p.

Assume for instance that $M_p = 0$ (for the other case the argument is analogous). Then we can write

$$
M = \sum_{j=1}^{n} g_j L_j,
$$

where g_j are smooth functions and $g_j(p) = 0$ for all $j = 1, \ldots, n$. We have

$$[L, \overline{M}] = \sum_{j=1}^{n} \left\{ (L\overline{g_j}) \overline{L}_j + \overline{g_j}[L, \overline{L}_j] \right\}$$

and thus $\xi([L, \overline{M}]_p) = 0$ since $\xi(\overline{L}_j|_p) = 0$ (because ξ is real) and $\overline{g_j}(p) = 0$. $\qquad\qquad\square$

From Lemma I.17.1 it follows that the following definition is meaningful:

DEFINITION I.17.2. *The Levi form of the formally integrable structure $\mathcal{V}$ at the characteristic point $\xi \in T_p^0$, $\xi \neq 0$ is the hermitian form on $\mathcal{V}_p$ defined by*

$$\mathfrak{L}_{(p,\xi)}(v, w) = \frac{1}{2i} \xi([L, \overline{M}]_p), \qquad (I.88)$$

where L and M are smooth sections of $\mathcal{V}$ defined in a neighborhood of p and satisfying $L_p = v$, $M_p = w$.

Given a hermitian form $\mathfrak{H}$ on a finite-dimensional complex vector space V, its main invariants are the subspaces V^+, V^- and $V^\perp$ of V, which give a decomposition

$$V = V^+ \oplus V^- \oplus V^\perp$$

and are characterized by:

- $v \mapsto \mathfrak{H}(v, v)$ is positive definite on V^+;
- $v \mapsto \mathfrak{H}(v, v)$ is negative definite on V^-;
- $V^\perp = \{v \in V : \mathfrak{H}(v, w) = 0, \forall w \in V\}$.

Thus $\mathfrak{H}$ is itself positive definite (resp. positive negative) if $V = V^+$ (resp. $V = V^-$). More generally, $\mathfrak{H}$ is said to be *positive* (resp. *negative*) if $V^- = \{0\}$ (resp. $V^+ = \{0\}$). Also, $\mathfrak{H}$ is said to be *nondegenerate* if $V^\perp = \{0\}$. Finally, we recall that it is common to call the positive integer $|\dim V^+ - \dim V^-|$ the *signature* of $\mathfrak{H}$. Notice that the signature does not change after multiplication of $\mathfrak{H}$ by a nonzero real number.

A formally integrable structure $\mathcal{V}$ over Ω is *nondegenerate* if given any $\xi \in T_p^0$, $\xi \neq 0$ the Levi form $\mathfrak{L}_{(p,\xi)}$ is a nondegenerate hermitian form.

We now describe the Levi form for a formally integrable CR structure over Ω. Let $p \in \Omega$, $\xi \in T_p^0$, $\xi \neq 0$. According to the results described in Section I.8 we can find a system of coordinates

$$(x_1, \ldots, x_n, y_1, \ldots, y_n, s_1, \ldots, s_d)$$

vanishing at p and vector fields of the form

$$L_j = \frac{\partial}{\partial \bar{z}_j} + \sum_{j'=1}^d a_{jj'}(z, s) \frac{\partial}{\partial z_{j'}} + \sum_{k=1}^d b_{jk}(z, s) \frac{\partial}{\partial s_k}, \quad j = 1, \dots, n,$$

with $a_{jj'}(0, 0) = b_{jk}(0, 0) = 0$ for all j, j', k, which span $\mathcal{V}$ in a neighborhood of the origin in $\mathbb{R}^{2n+d}$. Notice, moreover, that T_p^0 is equal to the span of $\{ds_1|_0, \dots, ds_d|_0\}$.

Write $\xi = \xi_1 ds_1|_0 + \dots + \xi_d ds_d|_0$ and denote by $(A_{jj'})$ the matrix of the Levi form $\mathfrak{L}_{(p,\xi)}$ with respect to the basis $\{(\partial/\partial \bar{z}_1)_p, \dots, (\partial/\partial \bar{z}_n)_p\}$ of $\mathcal{V}_p$. Thus, by definition, $A_{jj'} = \mathfrak{L}_{(p,\xi)}\big((\partial/\partial \bar{z}_j)_p, (\partial/\partial \bar{z}_{j'})_p\big)$ and then

$$\begin{aligned} A_{jj'} &= \frac{1}{2i}(\xi_1 ds_1|_0 + \dots + \xi_d ds_d|_0)([L_j, \bar{L}_{j'}]_p) \\ &= \frac{1}{2i}\sum_{k=1}^d \xi_k \{L_j \overline{b_{j'k}} - \bar{L}_{j'} b_{jk}\}(0, 0), \end{aligned}$$

that is

$$A_{jj'} = \frac{1}{2i}\sum_{k=1}^d \xi_k \left\{ \frac{\overline{\partial b_{j'k}}}{\partial z_j}(0, 0) - \frac{\partial b_{jk}}{\partial z_{j'}}(0, 0) \right\}. \tag{I.89}$$

As an example, let us consider the CR structure defined by the vector fields $L_j^\#$ given by (I.73). In this case $d = 1$ and we take $\xi = ds|_0$. We also have $b_j = -i\epsilon_j z_j(1 + g(z, s))$, where g vanishes to infinite order at $z_1 = 0$. Then

$$\frac{\partial b_{j'}}{\partial z_j}(0, 0) = -i\epsilon_j \delta_{jj'}$$

and (I.89) gives

$$(A_{jj'}) = \text{diag}\{\epsilon_1, \dots, \epsilon_n\}.$$

Thus, Corollary I.16.2 has provided an example of a nondegenerate CR structure, defined in a neighborhood of the origin in $\mathbb{C}^n \times \mathbb{R}$, for which the signature of the Levi form at $\lambda ds|_0 \in T_0^0$, $\lambda \neq 0$ is equal to $n - 1$.

In connection with this example we mention the following deep result which gives a positive answer to the problem of local integrability (or local realizability, as we have seen in Proposition I.15.1) for certain classes of CR structures. It shows that the value of the signature of the Levi form plays a crucial role.

Recall that by Proposition I.8.4 the characteristic set of a CR structure is a sub-bundle of the cotangent bundle.

THEOREM I.17.3. *Let $\mathcal{V}$ be a nondegenerate CR structure over a smooth manifold Ω and assume that its characteristic set has rank equal to one. Let* *n* denote the rank of $\mathcal{V}$ (and thus the dimension of Ω is equal to $2n+1$). Suppose that for some $p \in \Omega$ the signature of the Levi form at $\xi \in T_p^0$, $\xi \neq 0$, is equal to n. If $n \geq 3$ then $\mathcal{V}$ is locally integrable in a neighborhood of p.

Finally, we shall compute the expression of the matrix $(A_{jj'})$ of the Levi form when $\mathcal{V}$ is locally integrable and CR. Invoking the local coordinates described at the beginning of Section I.15, and in particular the functions (I.66) satisfying (I.64), we see that we can take the vector fields L_j in the form (*cf.* (I.27))

$$L_j = \frac{\partial}{\partial \bar{z}_j} - i \sum_{k=1}^{d} \frac{\partial \phi_k}{\partial \bar{z}_j}(z,s) M_k, \quad j = 1, \dots, n,$$

where

$$M_k = \sum_{k'=1}^{d} \mu_{kk'}(z,s) \frac{\partial}{\partial s_{k'}}, \quad k = 1, \dots, d,$$

characterized by the relations $M_k\{s_{k'} + i\phi_{k'}\} = \delta_{kk'}$. In particular, (I.64) gives

$$\mu_{kk'}(0,0) = \delta_{kk'}. \tag{I.90}$$

According to our previous notation, we have $a_{jj'} \equiv 0$ for all j, j' and

$$b_{jk} = -i \sum_{k'=1}^{d} \frac{\partial \phi_{k'}}{\partial \bar{z}_j} \mu_{k'k}.$$

Again by (I.64) and by (I.90) we have

$$\frac{\partial b_{jk}}{\partial z_{j'}}(0,0) = -i \frac{\partial^2 \phi_k}{\partial z_{j'} \partial \bar{z}_j}(0,0)$$

and then by (I.89) we obtain

$$A_{jj'} = \sum_{k=1}^{d} \xi_k \frac{\partial^2 \phi_k}{\partial z_{j'} \partial \bar{z}_j}(0,0). \tag{I.91}$$

EXAMPLE I.17.4. The following discussion justifies our terminology and makes a connection with the theory of several complex variables.

Let U be an open subset of $\mathbb{C}^{n+1}$ with a smooth boundary. Let $\rho \in C^\infty(\mathbb{C}^{n+1}, \mathbb{R})$ be such that $U = \{z : \rho(z) < 0\}$ and that $d\rho \neq 0$ on $\partial U = \{z : \rho(z) = 0\}$. We say that U satisfies the *Levi condition* at the point $p \in \partial U$ if the restriction of the hermitian form

$$\zeta \mapsto \sum_{j,k=1}^{n+1} \frac{\partial^2 \rho}{\partial z_j \partial \bar{z}_k}(p) \zeta_j \bar{\zeta}_k$$

to the space $\mathcal{T}_p = \{\zeta \in \mathbb{C}^{n+1} : \sum_{j=1}^{n+1}(\partial \rho/\partial z_j)(p)\zeta_j = 0\}$ is positive.

The Levi condition is independent of the choice of the defining function ρ: it is also a holomorphic invariant. After a translation and a $\mathbb{C}$-linear tranformation we can assume that $0 \in \partial\Omega$ and that the tangent space to $\partial\Omega$ at the origin is given by the real-hyperplane $\Im w = 0$, where now we are writing the complex coordinates as $(z_1, \ldots, z_n, w)$. We can also assume that the exterior normal to Ω at the origin is the vector $(0, \ldots, 0, -i) \in \mathbb{C}^{n+1}$.

By the implicit function theorem we conclude the existence of a smooth, real-valued function ϕ satisfying $\phi(0, 0) = 0$, $d\phi(0, 0) = 0$ such that ρ can be written, near the origin and in these new complex variables, as

$$\rho(z, w) = \phi(z, \Re w) - \Im w. \tag{I.92}$$

Since then $\mathfrak{T}_0 = \{\zeta_{n+1} = 0\}$, the Levi condition at the origin can be written as:

$$\sum_{j,k=1}^n \frac{\partial^2 \phi}{\partial z_j \partial \bar{z}_k}(0, 0)\zeta_j \bar{\zeta_k} \geq 0, \quad \forall \zeta \in \mathbb{C}^n. \tag{I.93}$$

The boundary of U is a one-codimensional submanifold of $\mathbb{C}^{n+1}$ and consequently it is generic. The complex structure $\mathcal{V}^{0,1}$ of $\mathbb{C}^{n+1}$ induces on ∂U a CR structure $\mathcal{V}^{0,1}(\partial U)$ and, according to the discussion in Section I.15, the differentials of the functions

$$Z_j = z_j, \quad j = 1, \ldots, n, \quad W(z, s) = s + i\phi(z, s)$$

span $T'(\partial U)$ near the origin [we are writing $s = \Re w$ and considering (z, s) as local coordinates in ∂U]. From (I.91) we obtain the following equivalent statement to (I.93): *the Levi form of the CR structure $\mathcal{V}^{0,1}(\partial U)$ at the characteristic point $ds|_0$ is positive.*

To obtain an invariant statement let us first denote by $T^{1,0}$ the orthogonal sub-bundle $\left(\mathcal{V}^{0,1}\right)^{\perp}$. Given an open set U with a smooth boundary ∂U as above, and given $p \in \partial U$, the map $\iota_p^* : \mathbb{C}T_p^*\mathbb{C}^{n+1} \to \mathbb{C}T_p^*\partial U$ induces an isomorphism

$$\gamma_p : T_p^{1,0} \xrightarrow{\sim} T_p'(\partial U).$$

Let $\xi \in T_p^0(\partial U)$, $\xi \neq 0$. We shall say that ξ is *inward pointing* if

$$\Im\left(\gamma_p^{-1}(\xi)\right)(\mathsf{v}) > 0$$

for every $\mathsf{v} \in T_p\mathbb{C}^{n+1}$ which is inward pointing toward U. In the preceding set-up, when $p = 0$ and ρ is given by (I.92), then $\gamma_0^{-1}(\lambda ds|_0) = \lambda dw|_0$ and then $\xi = \lambda ds|_0$ is inward pointing if and only if $\lambda > 0$. Summing up we can state:

PROPOSITION I.17.5. *Let $U \subset \mathbb{C}^{n+1}$ be an open set with a smooth boundary. Then U satisfies the Levi condition at $p \in \partial U$ if and only if the Levi form associated with the CR structure $\mathcal{V}(\partial U)$ is positive at every $\xi \in T_p^0(\partial U)$, $\xi \neq 0$ which is inward pointing.*

Appendix: Proof of the Newlander–Nirenberg theorem

In this appendix we shall present an argument due to B. Malgrange ([**Mal**]) which leads to the proof of the Newlander–Nirenberg theorem. We start by recalling some of the results we need from the theory of nonlinear elliptic equations.

Let us consider then an *overdetermined* system of nonlinear partial differential equations

$$\Phi\left[x, \vec{u}, \partial_{x_1}\vec{u}, \ldots, \partial^\alpha \vec{u}, \ldots\right] = 0, \quad |\alpha| \le M, \tag{I.94}$$

where x varies in an open subset Ω of $\mathbb{R}^N$,

$$\vec{u} = (u_1, \ldots, u_q) \in C^M(\Omega; \mathbb{R}^q),$$

$\Phi = (\phi_1, \ldots, \phi_p)$ is smooth and real-valued and $q \le p$. The system (I.94) is elliptic at $\vec{u}_0 \in C^M(\Omega; \mathbb{R}^q)$ in Ω if the *linear* differential operator

$$\vec{v} \mapsto \frac{\mathrm{d}}{\mathrm{d}\lambda}\Phi\left[x, \vec{u}_0 + \lambda\vec{v}, \partial_{x_1}(\vec{u}_0 + \lambda\vec{v}), \ldots, \partial^\alpha(\vec{u}_0 + \lambda\vec{v}), \ldots\right]\big|_{\lambda=0} \tag{I.95}$$

is elliptic in the following sense: if

$$\sigma : \Omega \times (\mathbb{R}^N \setminus \{0\}) \to L(\mathbb{R}^q, \mathbb{R}^p)$$

denotes the principal symbol of (I.95) then

$$\operatorname{rank} \sigma(x, \xi) = q, \quad \forall (x, \xi) \in \Omega \times \mathbb{R}^N \setminus \{0\}.$$

We call (*I.95*) the *linearization of (I.94) at* $\vec{u}_0$.

Here is an important remark that will be quite important in what follows: if $x_0 \in \Omega$ and if

$$\vec{v} \mapsto \frac{\mathrm{d}}{\mathrm{d}\lambda}\Phi\left[x_0, \vec{u}_0(x_0) + \lambda\vec{v}, \partial_{x_1}(\vec{u}_0)(x_0) + \lambda\partial_{x_1}(\vec{v}), \ldots\right]\big|_{\lambda=0} \tag{I.96}$$

is an elliptic linear system (with constant coefficients!) then (I.94) is elliptic at $\vec{u}_0$ in a neighborhood of x_0. Accordingly, we shall call (I.96) the *linearization of (I.94) at* $\vec{u}_0$ *at the point* x_0.

The two main results that are essential for Malgrange's argument are:

- If $\vec{u}$ is a C^M-solution of (I.94), if (I.94) is elliptic at $\vec{u}$ in the sense just defined, and if the function Φ is real-analytic then $\vec{u}$ is real-analytic.
- Now assume that $q = p$ and that (I.94) is elliptic at $\vec{u}_0 \in C^M(\Omega; \mathbb{R}^q)$. Let $x_0 \in \Omega$ be such that

$$\Phi\left[x_0, \vec{u}_0(x_0), \partial_{x_1}\vec{u}_0(x_0), \ldots, \partial^\alpha \vec{u}_0(x_0), \ldots\right] = 0.$$

Then there are $\epsilon_0 > 0$, $C > 0$ and $0 < \mu < 1$ such that for every $0 < \epsilon \le \epsilon_0$ there is a smooth solution $\vec{u}_\epsilon$ to (I.94) on $|x - x_0| < \epsilon$ satisfying the bounds

$$\left|\partial^\alpha\left(\vec{u}_\epsilon(x) - \vec{u}_0(x)\right)\right| \le C\epsilon^{M-|\alpha|+\mu}, \quad |x - x_0| < \epsilon, \ |\alpha| \le M.$$

We now embark on the proof of the Newlander–Nirenberg theorem. The starting point is the description of the special generators presented after Lemma I.8.5, particularly the vector fields given by (I.19), taking into account that when the structure is complex then $d = n' = 0$. In other words, we can assume that our (complex) formally integrable structure is defined, in an open neighborhood of the origin in $\mathbb{C}^m$, by the pairwise commuting vector fields

$$L_j = \frac{\partial}{\partial \bar{z}_j} + \sum_{k=1}^m a_{jk}(z)\frac{\partial}{\partial z_k}, \quad j = 1, \ldots, m, \tag{I.97}$$

where $a_{jk} = 0$ at the origin. For technical reasons, which are going to be clear in the argument, it is convenient to assume that $a_{jk}(z) = O(|z|^2)$, and this property can be achieved after performing a local diffeomorphism of the form $z' = z + Q(z, \bar{z})$, where Q is a homogeneous polynomial of degree two in $(z_1, \ldots, z_m, \bar{z}_1, \ldots, \bar{z}_m)$ chosen suitably. We leave the details of this (simple) computation to the reader.

Malgrange's key idea is to show the existence of a local diffeomorphism $w = H(z)$, defined near the origin in $\mathbb{C}^m$, such that, in the new variables $w_1, \ldots, w_m$, the structure has a set of generators which have *real-analytic* coefficients. This implies the sought-for conclusion thanks to Theorem I.9.2.

In order to shorten the notation and make the computations more apparent, we shall describe all the systems involved in vector and matrix notation. Thus we set

$$\vec{L} = \begin{bmatrix} L_1 \\ \vdots \\ L_m \end{bmatrix}, \quad \frac{\partial}{\partial z} = \begin{bmatrix} \frac{\partial}{\partial z_1} \\ \vdots \\ \frac{\partial}{\partial z_m} \end{bmatrix}, \quad \frac{\partial}{\partial \bar{z}} = \begin{bmatrix} \frac{\partial}{\partial \bar{z}_1} \\ \vdots \\ \frac{\partial}{\partial \bar{z}_m} \end{bmatrix}$$

and rewrite the system (I.97) as

$$\vec{L} = \frac{\partial}{\partial \bar{z}} + A(z)\frac{\partial}{\partial z},$$

where $A(z)$ denotes the matrix $\{a_{jk}(z)\}$.

Let $w = H(z)$ be a local diffeomorphism near the origin in $\mathbb{C}^m$ satisfying

$$H_z(0) \text{ is invertible.} \tag{I.98}$$

Since

$$\frac{\partial}{\partial \bar{z}} = {}^t H_{\bar{z}} \frac{\partial}{\partial w} + {}^t \overline{H}_{\bar{z}} \frac{\partial}{\partial \overline{w}}, \quad \frac{\partial}{\partial z} = {}^t H_z \frac{\partial}{\partial w} + {}^t \overline{H}_z \frac{\partial}{\partial \overline{w}},$$

a new set of generators for the structure is defined, in the new variables $w_1, \ldots, w_m$, by the system

$$\vec{L}^\bullet = \frac{\partial}{\partial \overline{w}} + B(w)\frac{\partial}{\partial w}, \tag{I.99}$$

where

$$B(w) = \left({}^t \overline{H}_{\bar{z}} + A\ {}^t \overline{H}_z \right)^{-1} \left({}^t H_{\bar{z}} + A\ {}^t H_z \right)\big|_{z=H^{-1}(w)}. \tag{I.100}$$

If $L_1^\bullet, \ldots, L_m^\bullet$ denote the components of $\vec{L}^\bullet$ then a fortiori we must have

$$[L_j^\bullet, L_k^\bullet] = 0, \quad \forall j, k = 1, \ldots, m, \ j < k.$$

Writing $B = \{b_{jk}\}$ this property is equivalent to

$$\frac{\partial b_{k\ell}}{\partial \overline{w}_j} - \frac{\partial b_{j\ell}}{\partial \overline{w}_k} - \sum_{r=1}^{m} \left\{ b_{kr}(w)\frac{\partial b_{j\ell}}{\partial w_r} - b_{jr}(w)\frac{\partial b_{k\ell}}{\partial w_r} \right\} = 0, \ \forall j, k, \ell, \ j < k. \tag{I.101}$$

We emphasize: given any local diffeomorphism H satisfying (I.98) then equations (I.101) are satisfied by $B = \{b_{jk}\}$ defined by (I.100).

The system (I.101) together with the additional equations

$$\sum_{j=1}^{m} \frac{\partial b_{jk}}{\partial w_j} = 0, \quad k = 1, \ldots, m \tag{I.102}$$

make up a system of quasi-linear partial differential equations in the unknowns $\{b_{jk}\}$. Let us write $\vec{V} = (\Re b_{1,1}, \Re b_{1,2}, \ldots, \Im b_{m,m-1}, \Im b_{m,m}) \in \mathbb{R}^{2m^2}$. Then systems (I.101) and (I.102) can be written as

$$\mathcal{P}\vec{V} + \Gamma(\vec{V}, \vec{\nabla}\vec{V}) = 0, \tag{I.103}$$

where $\mathcal{P}$ is an elliptic linear operator with constant coefficients and Γ is a bilinear form in its arguments. It then follows that there is a small number $\sigma > 0$ such that if $|B(0)| \leq \sigma$ then (I.101), (I.102) is elliptic at B in an open neighborhood of the origin. Hence any such B is a real-analytic function of w and the argument will be complete if we can show that a diffeomorphism H satisfying (I.98) can be chosen in such a way that B, defined by (I.100), is a solution of (I.102) satisfying $|B(0)| \leq \sigma$.

We are left to solve the determined system

$$\sum_{j=1}^{m} \frac{\partial}{\partial w_j} \left[({}^t\overline{H}_{\bar{z}} + A \, {}^t\overline{H}_z)^{-1} ({}^tH_{\bar{z}} + A \, {}^tH_z) \right]_{jk} = 0, \quad k = 1, \ldots, m, \quad \text{(I.104)}$$

whose unknown is $(\Re H(z), \Im H(z))$ (we look at (I.104) as a determined system of $2m$ *real* equations). It is important to emphasize that these equations are now being considered in the $z_1, \ldots, z_m$ variables.

Since $A(z) = O(|z|^2)$ it is easily seen that $H_0(z) = z$ satisfies (I.104) at the origin. Furthermore, taking

$$H(z) = H_0(z) + \lambda G(z)$$

then for $\lambda \in \mathbb{R}$, $|\lambda|$ small we have

$$({}^t\overline{H}_{\bar{z}} + A \, {}^t\overline{H}_z)^{-1} ({}^tH_{\bar{z}} + A \, {}^tH_z) = A + \lambda \left[{}^tG_{\bar{z}} + AF \right] + O(\lambda^2)$$

for some F smooth. Furthermore, since

$$\frac{\partial}{\partial w} = {}^tH_z^{-1} \frac{\partial}{\partial z} - {}^tH_z^{-1} \, {}^t\overline{H}_z \frac{\partial}{\partial \overline{w}}$$

we obtain

$$\frac{\partial}{\partial w_j} = \frac{\partial}{\partial z_j} + O(\lambda).$$

Hence, using once more the fact that $A(z) = O(|z|^2)$, we can easily conclude that the linearization of (I.104) at H_0 at the origin can be identified, in a natural way, with the *complex* operator

$$\begin{aligned}
G &\mapsto \left(\sum_{j=1}^{m} \frac{\partial}{\partial z_j} \left[{}^tG_{\bar{z}} \right]_{j1}, \ldots, \sum_{j=1}^{m} \frac{\partial}{\partial z_j} \left[{}^tG_{\bar{z}} \right]_{jm} \right) \\
&= \left(\sum_{j=1}^{m} \frac{\partial^2 G_1}{\partial z_j \partial \bar{z}_j}, \ldots, \sum_{j=1}^{m} \frac{\partial^2 G_m}{\partial z_j \partial \bar{z}_j} \right),
\end{aligned}$$

which is clearly elliptic (in the usual sense). We conclude that there are $\epsilon_0 > 0$, $C > 0$ and $\mu < 1$ such that for every $0 < \epsilon \leq \epsilon_0$ there is a smooth solution H_ϵ to (I.104) satisfying

$$\|H_\epsilon - H_0\|_{C^2\{z:|z|\leq\epsilon\}} \leq C\epsilon^{2+\mu}, \quad \epsilon \leq \epsilon_0. \quad \text{(I.105)}$$

In particular, if $\epsilon > 0$ is small enough we can ensure that:

- H_ϵ is a local diffeomorphism near the origin satisfying (I.98);
- B defined by (I.100) satisfies $|B(0)| \le \sigma$.

The proof is complete. ☐

Notes

The first treatment of formally and locally integrable structures as presented here appeared in [**T4**], the main point for this being the discovery of the Approximation Formula by M. S. Baouendi and F. Treves in 1981 ([**BT1**]); such structures were then studied extensively in [**T5**]. The pioneering work though seems to be the article by Andreotti-Hill ([**AH1**]), where the concept of what we now call a real-analytic locally integrable structure was introduced in its full generality.

This introductory chapter contains mainly results that have already been presented in standard textbooks. We mention, for instance, the Frobenius theorem, whose proof was taken from L. Hörmander's book [**H4**] and the integrability of elliptic vector fields in the plane, of which we give an almost self-contained proof, depending only on very simple facts concerning commutators of certain pseudo-differential operators that can be found, for instance, in [**Fo**].

As mentioned in the text, Theorem I.12.1 is due to L. Nirenberg ([**N2**]) and the proof we present was taken from [**T5**].

Proposition I.16.1 is a particular case of a more general result due to H. Jacobowitz and F. Treves ([**JT1**]). We also refer to [**JT2**] where the same authors study, via a category argument, the set of all formally integrable CR structures of rank n on an open subset of $\mathbb{R}^{2n+1}$ whose Levi form has, at each nonzero characteristic point, signature $n-1$.

Theorem I.17.3 was originally due to M. Kuranishi ([**Ku1**], [**Ku2**]) in the case $n \ge 4$. Later, T. Akahori ([**Ak**]) presented an improvement to Kuranishi's argument which allowed him to prove Theorem I.17.3 also for the case $n = 3$. The case $n = 2$ is still an open problem, whereas when $n = 1$ the conclusion is false, according to [**N3**] (see also Theorem I.12.1). A proof of Theorem I.17.3 can also be found in [**W3**].

Finally, Malgrange's proof of the Newlander–Nirenberg theorem that we presented in the appendix was taken from [**N1**], where the use of a solvability result on elliptic determined systems of nonlinear partial differential equations makes the argument a bit simpler.

II

The Baouendi–Treves approximation formula

In this chapter we prove what is probably the most important single result in the theory of locally integrable structures. It states that in a small neighborhood of a given point of the domain of a locally integrable structure $\mathcal{L}$, any solution of the equation $\mathcal{L}u = 0$ may be approximated by polynomials in a set of a finite number of homogeneous solutions as soon as the solutions in that set are chosen with linearly independent differentials and the number of them is equal to the corank of $\mathcal{L}$. Such a set is called a complete set of first integrals of the locally integrable structure.

The proof is relatively simple for classical solutions and depends on the construction of a suitable approximation of the identity modeled on the kernel of the heat equation as shown in Section II.1. The extension to distribution solutions is carried out in Section II.2. Section II.3 studies the convergence of the formula in some of the standard spaces used in analysis: Lebesgue spaces L^p, $1 \leq p < \infty$; Sobolev spaces; Hölder spaces; and (localizable) Hardy spaces h^p, $0 < p < \infty$. The last section is devoted to applications.

II.1 The approximation theorem

Since the approximation formula is of a local nature it will be enough to restrict our attention to a locally integrable structure $\mathcal{L}$ defined in an open subset Ω of $\mathbb{R}^N$ over which $\mathcal{L}^\perp$ is spanned by the differentials $dZ_1, \ldots, dZ_m$ of m smooth functions $Z_j \in C^\infty(\Omega)$, $j = 1, \ldots, m$, at every point of Ω. Thus, if n is the rank of $\mathcal{L}$, we recall that $N = n + m$.

Given a distribution $u \in \mathcal{D}'(\Omega)$ we say that u is a homogeneous solution of $\mathcal{L}$ and write $\mathcal{L}u = 0$ if

$$Lu = 0 \quad \text{on } U$$

52

for every local section L of $\mathcal{L}$ defined on an open subset $U \subset \Omega$. Simple examples of homogeneous solutions of $\mathcal{L}$ are the constant functions and also the functions $Z_1, \ldots, Z_m$, since $LZ_j = \langle dZ_j, L \rangle = 0$ because $dZ_j \in \mathcal{L}^{\perp}$, $j = 1, \ldots, m$. By the Leibniz rule, any product of smooth homogeneous solutions is again a homogeneous solution, so a polynomial with constant coefficients in the m functions Z_j, i.e., a function of the form

$$P(Z) = \sum_{|\alpha| \leq d} c_\alpha Z^\alpha, \quad \alpha = (\alpha_1, \ldots, \alpha_m) \in \mathbb{Z}^m, \quad c_\alpha \in \mathbb{C}, \qquad \text{(II.1)}$$

is also a homogeneous solution. The approximation theorem states that any distribution solution u of $\mathcal{L}u = 0$ is the weak limit of polynomial solutions such as (II.1).

THEOREM II.1.1. *Let $\mathcal{L}$ be a locally integrable structure on Ω and assume that $dZ_1, \ldots, dZ_m$ span $\mathcal{L}^{\perp}$ at every point of Ω. Then, for any $p \in \Omega$, there exist two open sets U and W, with $p \in U \subset \overline{U} \subset W \subset \Omega$, such that*

(i) *every $u \in \mathcal{D}'(W)$ that satisfies $\mathcal{L}u = 0$ on W is the limit in $\mathcal{D}'(U)$ of a sequence of polynomial solutions $P_j(Z_1, \ldots, Z_m)$:*

$$u = \lim_{j \to \infty} P_j \circ Z \quad \text{in } \mathcal{D}'(U);$$

(ii) *if $u \in C^k(W)$ the convergence holds in the topology of $C^k(U)$, $k = 0, 1, 2, \ldots, \infty$.*

Some well-known approximation results in analysis are particular cases of Theorem II.1.1.

EXAMPLE II.1.2. Let $\mathcal{L}$ be the locally integrable structure generated over an open set $\Omega \subset \mathbb{C}$ by the Cauchy–Riemann vector field

$$\bar{\partial} = \frac{1}{2} \left(\frac{\partial}{\partial x} + i \frac{\partial}{\partial y} \right), \quad z = x + iy.$$

Then a distribution solution of $\bar{\partial}u = 0$ is just a holomorphic function and the theorem simply states that any holomorphic function can be locally approximated by polynomials in the complex variable z.

Later we will give several applications of the approximation theorem but we wish to point out already one interesting consequence. Assume that two points $p, q \in U$ are such that $Z(p) = Z(q)$ and let $u \in C^0(\Omega)$ satisfy $\mathcal{L}u = 0$. Then $P \circ Z(p) = P \circ Z(q)$ for any polynomial P in m variables and, by the uniform approximation of u on U by polynomials in Z, it follows that $u(p) = u(q)$. The *fibers of Z in U* are, by definition, the equivalence classes of the equivalence relation defined by '$p \sim q$ if and only if $Z(p) = Z(q)$'. Thus, every solution $u \in C^0(\Omega)$ of $\mathcal{L}u = 0$ is constant on the fibers of Z. In particular, if the differentials of $Z_1^\#, \ldots, Z_m^\#$ span $\mathcal{L}^\perp$ over Ω it follows that $Z^\# = (Z_1^\#, \ldots, Z_m^\#)$ is constant on the fibers of Z in U. Applying the theorem with $Z^\#$ in the place of Z we may as well find a neighborhood $U^\# \subset U$ of p such that Z is constant on the fibers of $Z^\#$ in $U^\#$, which shows that the fibers of Z and the fibers of $Z^\#$ on $U^\#$ are identical. Thus, in the sense of germs of sets at p, the equivalence classes defined by Z and those defined by any other $Z^\# = (Z_1^\#, \ldots, Z_m^\#)$ such that $dZ_1^\#, \ldots, dZ_m^\#$ generates $\mathcal{L}^\perp$ coincide. This independence of the particular choice of Z allows us to talk about the *germs at p of the fibers of $\mathcal{L}$* which are invariants of the structure.

The fact that u is constant on the fibers of Z in U when $\mathcal{L}u = 0$, $u \in C^0(\Omega)$, may be expressed by saying that there exists a function $\widehat{u} \in C^0(Z(U))$ such that $u = \widehat{u} \circ Z$. Thus, any continuous solution of $\mathcal{L}u = 0$ can be factored as the composition with Z of a continuous function defined on a subset of $\mathbb{C}^m$. In general, the set $Z(U)$ may be irregular but if it happens to be a submanifold of $\mathbb{C}^m$, then $\widehat{u}$ will satisfy in the weak sense the induced Cauchy–Riemann equations on $Z(U)$. Hence, at a conceptual level, the theorem links the study of solutions of $\mathcal{L}u = 0$ to solutions of the induced Cauchy–Riemann equations on certain sets of $\mathbb{C}^m$.

We will prove Theorem II.1.1 in several steps. The first step consists of taking convenient local coordinates in a neighborhood of p. Applying Corollary I.10.2, there exists a local coordinate system vanishing at p,

$$\{x_1, \ldots, x_m, t_1, \ldots, t_n\}$$

and smooth, real-valued functions $\phi_1, \ldots, \phi_m$ defined in a neighborhood of the origin and satisfying

$$\phi_k(0,0) = 0, \quad d_x \phi_k(0,0) = 0, \quad k = 1, \ldots, m,$$

such that the functions Z_k, $k = 1, \ldots, m$, may be written as

$$Z_k(x,t) = x_k + i\phi_k(x,t), \quad k = 1, \ldots, m, \tag{II.2}$$

on a neighborhood of the origin. To do so we need to assume that the real parts of $dZ_1, \ldots, dZ_m$ are linearly independent, for which we might have to replace Z_j by iZ_j for some of the indexes $j \in \{1, \ldots, m\}$. Notice that this will not change the conclusion of the theorem. Thus, we may choose a number R such that if

$$V = \{q : \quad |x(q)| < R, \ |t(q)| < R\}$$

then (II.2) holds in a neighborhood of $\overline{V}$ and we may assume that

$$\left\|\left(\frac{\partial \phi_j(x, t)}{\partial x_k}\right)\right\| < \frac{1}{2}, \quad (x, t) \in \overline{V}, \tag{II.3}$$

where the double bar indicates the norm of the matrix $\phi_x(x, t) = (\partial \phi_j(x, t)/\partial x_k)$ as a linear operator in $\mathbb{R}^m$. Modifying the functions ϕ_k's off a neighborhood of $\overline{V}$ may assume without loss of generality that the functions $\phi_k(x, t)$, $k = 1, \ldots, m$, are defined throughout $\mathbb{R}^N$, have compact support and satisfy (II.3) everywhere, that is

$$\left\|\left(\frac{\partial \phi_j(x, t)}{\partial x_k}\right)\right\| < \frac{1}{2}, \quad (x, t) \in \mathbb{R}^N. \tag{II.3$'$}$$

Modifying also $\mathcal{L}$ off a neighborhood of $\overline{V}$ we may assume as well that the differentials dZ_j, $j = 1, \ldots, m$, given by (II.2), span $\mathcal{L}^\perp$ over $\mathbb{R}^N$. Of course, the new structure $\mathcal{L}$ and the old one coincide on V so any conclusion we draw about the new $\mathcal{L}$ on V will hold as well for the original $\mathcal{L}$. We will make use of the vector fields L_j, $j = 1, \ldots, n$ and M_k, $k = 1, \ldots, m$ entirely analogous to those introduced in Chapter I after Corollary I.10.2, with the only difference that here they are defined throughout $\mathbb{R}^N$. We recall from Chapter I that the vector fields

$$M_k = \sum_{\ell=1}^{m} \mu_{k\ell}(x, t)\frac{\partial}{\partial x_\ell}, \quad k = 1, \ldots, m$$

are characterized by the relations

$$M_k Z_\ell = \delta_{k\ell}, \quad k, \ell = 1, \ldots, m$$

and that the vector fields

$$L_j = \frac{\partial}{\partial t_j} - i\sum_{k=1}^{m} \frac{\partial \phi_k}{\partial t_j}(x, t)M_k, \quad j = 1, \ldots, n$$

are linearly independent and satisfy $L_j Z_k = 0$, for $j = 1, \ldots, n$, $k = 1, \ldots, m$. Hence, $L_1, \ldots, L_n$ span $\mathcal{L}$ at every point while the $N = n + m$ vector fields

$$L_1, \ldots, L_n, M_1 \ldots, M_m$$

are pairwise commuting and span $\mathbb{C}T_p(\mathbb{R}^N)$, $p \in \mathbb{R}^N$. Since

$$dZ_1, \ldots, dZ_m, dt_1, \ldots, dt_n \quad \text{span} \quad \mathbb{C}T^*\mathbb{R}^N$$

the differential dw of a C^1 function $w(x, t)$ may be expressed in this basis. In fact, we have

$$dw = \sum_{j=1}^{n} L_j w\, dt_j + \sum_{k=1}^{m} M_k w\, dZ_k \tag{II.4}$$

which may be checked by observing that $L_j Z_k = 0$ and $M_k t_j = 0$ for $1 \leq j \leq n$ and $1 \leq k \leq m$, while $L_j t_k = \delta_{jk}$ for $1 \leq j, k \leq n$ and $M_k Z_j = \delta_{jk}$ for $1 \leq j, k \leq m$ ($\delta_{jk} = $ Kronecker delta).

We now choose the open set W as any fixed neighborhood of $\overline{V}$ in Ω. In proving the theorem we will assume initially that u is a smooth homogeneous solution of $\mathcal{L}u = 0$ defined in W with continuous derivatives of all orders, i.e., $u \in C^\infty(W)$ satisfies on W the overdetermined system of equations

$$\begin{cases} L_1 u = 0, \\ L_2 u = 0, \\ \cdots\cdots\cdots \\ L_n u = 0. \end{cases} \tag{II.5}$$

Given such u we define a family of functions $\{E_\tau u\}$ that depend on a real parameter τ, $0 < \tau < \infty$, by means of the formula

$$E_\tau u(x, t) = (\tau/\pi)^{m/2} \int_{\mathbb{R}^m} e^{-\tau[Z(x,t)-Z(x',0)]^2} u(x', 0) h(x') \det Z_x(x', 0) \, dx'$$

which we now discuss. For $\zeta = (\zeta_1, \ldots, \zeta_m) \in \mathbb{C}^m$ we will use the notation $[\zeta]^2 = \zeta_1^2 + \cdots + \zeta_m^2$, which explains the meaning of $[Z(x, t) - Z(x', 0)]^2$ in the formula. The function $h(x) \in C_c^\infty(\mathbb{R}^m)$ satisfies $h(x) = 0$ for $|x| \geq R$ and $h(x) = 1$ in a neighborhood of $|x| \leq R/2$ (recall that R was introduced right before (II.3) in the definition of the set V). Note that since u is assumed to be defined in a neighborhood of $\overline{V}$, the product $u(x', 0)h(x')$ is well-defined on $\mathbb{R}^m$, compactly supported, and of class C^∞. Since Z has m components we may regard Z_x as the $m \times m$ matrix $(\partial Z_j / \partial x_k)$ and denote by $\det Z_x$ its determinant. Furthermore, since the exponential in the integrand is an entire function of $(Z_1, \ldots, Z_m)$, the chain rule shows that it satisfies the homogenous system of equations (II.5) and the same holds for $E_\tau u(x, t)$ by differentiation under the integral sign. The second step of the proof will be to show that $E_\tau u(x, t) \to u(x, t)$ as $\tau \to \infty$ uniformly for $|x| < R/4$ and $|t| < T < R$ if T is conveniently small. Once this is proved we may approximate in the C^∞ topology the exponential $e^{-\tau[\zeta]^2}$ (for fixed large τ) by the partial sum of degree k, $P_k(\zeta)$, of its Taylor series on a fixed polydisk that contains the set $\{\sqrt{\tau}(Z(x, t) - Z(x', 0)) : |x|, |x'| < R, |t| < R\}$, so replacing the exponential in the definition of E_τ by $P_k(Z(x, t) - Z(x', 0))$ we will find polynomials in $Z(x, t)$ that approximate $E_\tau u(x, t)$ in the C^∞ topology for $|x| < R/4$ and $|t| < T$ when k is large. Hence, from now on we fix our attention on the convergence of $E_\tau u \to u$. We consider the following modification of the operator E_τ:

$$G_\tau u(x, t) = (\tau/\pi)^{m/2} \int_{\mathbb{R}^m} e^{-\tau[Z(x,t)-Z(x',t)]^2} u(x', t) h(x') \det Z_x(x', t) \, dx'.$$

Notice that in the trivial case in which the functions ϕ_k, $k = 1, \dots, m$, vanish identically so $Z(x, t) = x$ and $\det Z_x = 1$, G_τ is just the convolution of $u(x, 0) h(x)$ with a Gaussian in $\mathbb{R}^m$, which is a well-known approximation of the identity as $\tau \to \infty$. In general, the functions ϕ_k do not vanish but they are relatively small because they vanish at the origin and (II.3′) holds, so G_τ is still an approximation of the identity. The idea is then to prove that $G_\tau u \to u$ and then estimate the difference $R_\tau u = G_\tau u - E_\tau u$ using the fact that $\mathcal{L}u = 0$.

LEMMA II.1.3. *Let B be an $m \times m$ matrix with real coefficients and norm $\|B\| < 1$ and set $A = I + iB$ where I is the identity matrix. Then*

$$\det A \int_{\mathbb{R}^m} e^{-[Ax]^2} \, dx = \pi^{m/2}.$$

PROOF. We may write $[Ax]^2 = {}^t A A x \cdot x$ (the dot indicates the standard inner product in $\mathbb{R}^m$ and also its extension as a $\mathbb{C}$-bilinear form to $\mathbb{C}^m$) so $e^{-[Ax]^2} = e^{-Cx \cdot x}$ where the matrix $C = {}^t A A$ has positive definite real part $\Re C = I - {}^t B B$ because $\|B\| < 1$. It is then known that (see, e.g., [**H2,** page 85])

$$\int_{\mathbb{R}^m} e^{-Cx \cdot x} \, dx = \pi^{m/2} (\det C)^{-1/2}$$

where the branch of the square root is chosen so $(\det C)^{1/2} > 0$ when C is real. Since $\det C = (\det A)^2$ the proof is complete. $\square$

Set $h(x) u(x, t) \det Z_x(x, t) = v(x, t)$. For (x, t) fixed, the matrix $Z_x(x, t) = I + i\phi_x(x, t)$ satisfies the hypotheses of the lemma in view of (II.3′). Thus, we may write

$$h(x) u(x, t) = \pi^{-m/2} \int_{\mathbb{R}^m} e^{-[Z_x(x,t)x']^2} v(x, t) \, dx'.$$

Introducing the change of variables $x' \mapsto x + \tau^{-1/2} x'$ in the integral that defines $G_\tau u$ we get

$$G_\tau u(x, t) = \pi^{-m/2} \int_{\mathbb{R}^m} e^{-\tau[Z(x,t)-Z(x+\tau^{-1/2}x',t)]^2} v(x + \tau^{-1/2} x', t) \, dx'.$$

Then

$$G_\tau u(x, t) - h(x) u(x, t) = I_\tau + J_\tau,$$

where

$$I_\tau(x, t) = \pi^{-m/2} \int_{\mathbb{R}^m} e^{-[Z_x(x,t)x']^2} (v(x + \tau^{-1/2}x', t) - v(x, t)) \, dx'$$

and

$$J_\tau(x, t) = \pi^{-m/2}$$

$$\int_{\mathbb{R}^m} \left(e^{-\tau[Z(x,t)-Z(x+\tau^{-1/2}x',t)]^2} - e^{-[Z_x(x,t)x']^2} \right) v(x + \tau^{-1/2}x', t) \, dx'.$$

To estimate I_τ we observe that $|e^{-[Z_x(x,t)x']^2}| = e^{-|x'|^2 + |\phi_x(x,t)x'|^2} \leq e^{-3|x'|^2/4}$ in view of (II.3′). We also observe that $|\nabla_x v(x, t)|$ is bounded in $\mathbb{R}^m \times \{|t| \leq R\}$ because v vanishes for large x, so the mean value theorem gives

$$|I_\tau(x, t)| \leq C\tau^{-1/2} \int_{\mathbb{R}^m} e^{-3|x'|^2/4} |x'| \, dx' \leq C'\tau^{-1/2},$$

showing that $|I_\tau(x, t)| \to 0$ as $\tau \to \infty$ uniformly on $\mathbb{R}^m \times \{|t| \leq R\}$. To estimate J_τ we first observe that $|e^{-\tau[Z(x,t)-Z(x+\tau^{-1/2}x',t)]^2} - e^{-[Z_x(x,t)x']^2}| \leq 2e^{-3|x'|^2/4}$, so

$$|J_\tau(x, t)| \leq C \int_{|x'|<K} |e^{-\tau[Z(x,t)-Z(x+\tau^{-1/2}x',t)]^2} - e^{-[Z_x(x,t)x']^2}| \, dx'$$

$$+ C \exp(-K^2/2).$$

Thus, to show that $|J_\tau(x, t)| \to 0$ uniformly we need only estimate the integral on $|x'| < K$ for any large K. When $|x'| \leq K$ and $|t| \leq R$, the Leibniz quotient $\zeta_1 = (Z(x, t) - Z(x + \tau^{-1/2}x', t))/\tau^{-1/2}$ converges to $\zeta_2 = -Z_x(x, t)x'$ uniformly in x as $\tau \to \infty$ in view of (II.3′), which also implies that $\Re[\zeta_1]^2 \geq 0$ and $\Re[\zeta_2]^2 \geq 0$. Since $e^{-\zeta}$ is a Lipschitz function on $\Re\zeta \geq 0$ and $|[\zeta_1]^2 - [\zeta_2]^2| \leq C\tau^{-1/2}$ (note that ζ_2 remains bounded as $(x, t) \in \mathbb{R}^N$ and $|x'| \leq K$), we have

$$|J_\tau(x, t)| \leq CK^m \tau^{-1/2} + C \exp(-K^2/2),$$

which shows that $J_\tau(x, t) \to 0$ uniformly for $x \in \mathbb{R}^m$ and $|t| \leq R$ as $\tau \to \infty$. Thus, $G_\tau u(x, t) \to h(x)u(x, t)$ uniformly and the limit $h(x)u(x, t) = u(x, t)$ for $|x| < R/2$.

We will now estimate the remainder $R_\tau = G_\tau - E_\tau$ by means of Stokes' theorem. The fact that u satisfies the system (II.5)—which was not used to prove that $G_\tau u \to hu$—is essential at this point. For $(x, t) \in \mathbb{R}^N$ fixed consider the m-form on $\mathbb{R}^N$ given by

$$\omega(x', t') = (\tau/\pi)^{m/2} e^{-\tau[Z(x,t)-Z(x',t')]^2} u(x', t') h(x') \, dZ(x', t')$$

$$= v(x', t') \, dZ(x', t'),$$

where $dZ = dZ_1 \wedge \cdots \wedge dZ_m$. Hence, we may write

$$G_\tau u(x, t) = \int_{\mathbb{R}^m \times \{t\}} \omega \quad \text{and} \quad E_\tau u(x, t) = \int_{\mathbb{R}^m \times \{0\}} \omega,$$

observing that the pullback of $dZ(x', t')$ to a slice $\{t = c = \text{const.}\}$ is given by $\det Z_x(x', c) \, dx_1 \wedge \cdots \wedge dx_m$. Keeping in mind that ω vanishes identically for $|x'| > R$ and invoking Stokes' theorem, we have

$$G_\tau u(x, t) - E_\tau u(x, t) = \int_{\mathbb{R}^m \times [0, t]} d\omega$$

where $[0, t]$ denotes the segment joining the origin of $\mathbb{R}^n$ to the point $t \in \mathbb{R}^n$. To compute $d\omega$ we will take advantage of expression (II.4). We have $d\omega = dv \wedge dZ$ so the only terms in (II.4) that matter here are those that do not contain dZ_j, $j = 1, \ldots, m$, i.e., $d\omega = \sum_{j=1}^n L_j v \, dt_j \wedge dZ$. Since the exponential factor in v is an entire function of $Z_1, \ldots, Z_n$, and thus satisfies (II.5) as well as u, we obtain

$$R_\tau u(x, t) = (\tau/\pi)^{m/2} \sum_{j=1}^n \int_{\mathbb{R}^m \times [0, t]} e^{-\tau[Z(x,t) - Z(x', t')]^2} u(x', t') L_j h(x') \, dt_j \wedge dZ(x', t').$$

Assume now that $|x| \le R/4$ and $|t| \le T$, where T will be chosen momentarily. We wish to estimate the exponential factor

$$\left| e^{-\tau[Z(x,t) - Z(x', t')]^2} \right| = e^{\tau(|\phi(x,t) - \phi(x', t')|^2 - |x - x'|^2)}.$$

We have

$$|\phi(x, t) - \phi(x', t')| \le |\phi(x, t) - \phi(x', t)| + |\phi(x', t) - \phi(x', t')|$$

$$\le \frac{1}{2}|x - x'| + C|t - t'|$$

$$\le \frac{1}{2}|x - x'| + CT$$

because $t' \in [0, t]$ and $|t| \le T$. Hence,

$$|\phi(x, t) - \phi(x', t')|^2 \le \frac{1}{2}|x - x'|^2 + 2\lambda T^2$$

and

$$\left| e^{-\tau[Z(x,t) - Z(x', t')]^2} \right| = e^{\tau(2\lambda T^2 - |x - x'|^2/2)},$$

where λ is a bound that depends only on ϕ and does not depend on u. Since $L_j h$ vanishes for $|x'| \le R/2$ we have that $|x'| \ge R/2$ in all integrands in the expression of R_τ, so $|x - x'| \ge R/4$ and

$$|R_\tau u(x, t)| \le C e^{\tau(2\lambda T^2 - R^2/32)}.$$

We may now choose T small enough so as to achieve $|R_\tau u(x, t)| \leq C e^{-\tau R^2/33}$. This proves that $|R_\tau u(x, t)| \to 0$ uniformly on $U = \{|x| \leq R/4\} \times \{|t| \leq T\}$. Summing up, we have found a neighborhood of the origin U such that for any C^∞-solution u of (II.5) defined in W, $E_\tau u \to u$ uniformly on U, which partially proves part (i) of the theorem for very regular distributions.

The third step is to prove part (ii) of the theorem for $k = \infty$ (the cases $1 \leq k < \infty$ will be proved later). The main tool is the use of commutation formulas for the vector fields M_k with G_τ.

LEMMA II.1.4. *For $u \in C^1(W)$ and $k = 1, \ldots, m$, the following identity holds:*

$$M_k G_\tau u(x, t) - G_\tau M_k u(x, t) = [M_k, G_\tau] u(x, t)$$

$$= (\tau/\pi)^{m/2} \int_{\mathbb{R}^m} e^{-\tau[Z(x,t)-Z(x',t)]^2} u(x', t) M_k' h(x') \det Z_x(x', t) \, dx'. \qquad (II.6)$$

PROOF. By the symmetry in the variables x and x' of the expression

$$Z_j(x, t) - Z_j(x', t), \quad j = 1, \ldots, m$$

we have

$$\delta_{jk} = M_k(x, t, D_x)(Z_j(x, t) - Z_j(x', t)))$$
$$= -M_k(x', t, D_{x'})(Z_j(x, t) - Z_j(x', t))).$$

Thus, if $F(\zeta)$ is an entire holomorphic function and we set

$$f(x, x', t) = F(Z(x, t) - Z(x', t))$$

we also have, by the chain rule,

$$M_k(x, t, D_x) f(x, t, t') = -M_k(x', t, D_{x'}) f(x, t, t').$$

Applying this to $F(\zeta) = e^{-\tau[\zeta]^2}$ we get, after differentiation under the integral sign that

$$M_k G_\tau u(x, t) = -(\tau/\pi)^{m/2}$$

$$\int_{\mathbb{R}^m} M_k(x', t, D_{x'})(e^{-\tau[Z(x,t)-Z(x',t)]^2}) u(x', t) h(x') \, dZ(x', t)$$

where we have used the fact that the pullback to any slice $t' = \text{const.}$ of the m-form $dZ_1 \wedge \cdots \wedge dZ_n$ is given by $\det Z_x(x', t) \, dx'$. Next, using the 'integration by parts' formula

$$\int_{\mathbb{R}^m} M_k v \, w \, dZ = -\int_{\mathbb{R}^m} v \, M_k w \, dZ \qquad (II.7)$$

which is valid if v and w are of class C^1 and one of them has compact support, we get

$$M_k G_\tau u(x, t) = (\tau/\pi)^{m/2}$$

$$\int_{\mathbb{R}^m} e^{-\tau[Z(x,t)-Z(x',t)]^2} \left(M_k u(x', t) h(x') + u(x', t) M_k h(x') \right) dZ(x', t)$$

which proves (II.6). To complete the proof we show that (II.7) holds. Consider the exact m-form defined by

$$\omega_k = d(uv\, dZ_1 \wedge \cdots \wedge \widehat{dZ_k} \wedge \cdots \wedge dZ_m)$$

$$= d(uv) \wedge dZ_1 \wedge \cdots \wedge \widehat{dZ_k} \wedge \cdots \wedge dZ_m$$

where the hat indicates that the factor dZ_k has been omitted. The pullback of ω_k to the slice $\{t\} \times \mathbb{R}^m$ is exact, so

$$\int_{\{t\} \times \mathbb{R}^m} \omega_k = 0. \tag{II.8}$$

Using (II.4) to compute $d(uv)$ and observing that the pullback to the slice of terms that contain a factor dt_j vanish, we get

$$\omega_k|_{\{t\} \times \mathbb{R}^m} = (-1)^{k+1} (v M_k u + u M_k v) \, dZ|_{\{t\} \times \mathbb{R}^m},$$

so (II.8) implies (II.7). $\qquad\qquad\qquad\qquad\qquad\qquad\qquad\square$

Next we prove for the L_j commutation formulas analogous to (II.6). We write

$$L_j = \frac{\partial}{\partial t_j} - i \sum_{k=1}^{m} \frac{\partial \phi_k}{\partial t_j}(x, t) M_k,$$

$$= \frac{\partial}{\partial t_j} + \sum_{k=1}^{m} \lambda_{jk} \frac{\partial}{\partial x_k}, \quad j = 1, \ldots, n.$$

We start with a technical lemma.

LEMMA II.1.5.

$$\frac{\partial \det Z_x}{\partial t_j} + \sum_{k=1}^{m} \frac{\partial (\lambda_{jk} \det Z_x)}{\partial x_k} \equiv 0, \quad j = 1, \ldots, n. \tag{II.9}$$

PROOF. Note that (II.9) says that the vector field $(\det Z_x) L_j$ is divergence free, i.e., $\operatorname{div}((\det Z_x) L_j) = 0$, or that ${}^t L_j (\det Z_x) = 0$ where ${}^t L_j$ is the transpose of L_j. Take a test function $v(x, t)$ and consider the compactly supported exact form

$$\omega_j = d\left(v\,dZ \wedge dt_1 \wedge \cdots \wedge \widehat{dt_j} \wedge \cdots \wedge dt_n\right)$$
$$= dv \wedge dZ \wedge dt_1 \wedge \cdots \wedge \widehat{dt_j} \wedge \cdots \wedge dt_n$$
$$= (-1)^{m+j-1} L_j v\,dZ \wedge dt$$
$$= (-1)^{m+j-1} L_j v\,(\det Z_x)\,dx \wedge dt$$

whose integral over $\mathbb{R}^N$ vanishes, that is,

$$\int_{\mathbb{R}^N} L_j v(\det Z_x)\,dxdt = \int_{\mathbb{R}^N} v\,{}^t L_j (\det Z_x)\,dxdt = 0.$$

Since v is arbitrary, ${}^t L_j(\det Z_x) \equiv 0$ and (II.9) is proved. $\qquad\square$

If $\tilde{g}(\zeta, t)$ is a smooth function on $\mathbb{C}^m \times \mathbb{R}^n$ that is holomorphic with respect to ζ and we set $g(x, t) = \tilde{g}(Z(x, t), t)$ we have, by the chain rule, that

$$L_j g(x, t) = \frac{\partial \tilde{g}}{\partial t_j}(Z(x, t), t)$$

because $L_j Z_k = 0$, $k = 1, \ldots, m$. To take advantage of this fact we may write $G_\tau u(x, t) = (\tau/\pi)^{m/2}(\tilde{G}_\tau u)(Z(x, t), t)$, where

$$\tilde{G}_\tau u(\zeta, t) = \int_{\mathbb{R}^m} e^{-\tau[\zeta - Z(x', t)]^2} u(x', t) h(x') \det Z_x(x', t)\,dx',$$

so

$$L_j G_\tau u(x, t) = (\tau/\pi)^{m/2} \frac{\partial \tilde{G}_\tau u}{\partial t_j}(Z(x, t), t).$$

To compute the right-hand side of the last identity we write $e_\tau(\zeta, x', t) = e^{-\tau[\zeta - Z(x', t)]^2}$, differentiate with respect to t_j under the integral sign, and observe that

$$\frac{\partial(e_\tau uh \det Z_x)}{\partial t_j} = \frac{\partial(e_\tau uh)}{\partial t_j} \det Z_x + e_\tau uh \frac{\partial(\det Z_x)}{\partial t_j}$$

$$= \det Z_x L_j(e_\tau uh) - \det Z_x \sum_{k=1}^{m} \lambda_{jk} \frac{\partial}{\partial x_k}(e_\tau uh)$$

$$+ e_\tau uh \frac{\partial(\det Z_x)}{\partial t_j}.$$

Note that the integral over $\mathbb{R}^m$ of the second term of the right-hand side may be written, after integration by parts, as

$$\int e_\tau uh \sum_{k=1}^{m} \frac{\partial}{\partial x_k}(\lambda_{jk} \det Z_x)\,dx$$

so the integral of the second and third terms together yields

$$\int e_\tau u h \left(\frac{\partial(\det Z_x)}{\partial t_j} + \sum_{k=1}^{m} \frac{\partial}{\partial x_k} (\lambda_{jk} \det Z_x) \right) dx = 0,$$

in view of (II.9). Since $L_j(e_\tau) = 0$, we also have that $\det Z_x L_j(e_\tau u h) = \det Z_x e_\tau (L_j u)h + \det Z_x e_\tau u L_j h$. This shows that

$$\frac{\partial}{\partial t_j} \tilde{G}_\tau u(\zeta, t) = \tilde{G}_\tau L_j u(\zeta, t) + \int e_\tau(\zeta, x', t)(u(L_j h) \det Z_x)(x', t) \, dx'.$$

When $\zeta = Z(x, t)$ we obtain

LEMMA II.1.6. *For $u \in C^1(W)$ and $j = 1, \ldots, m$ the following identity holds:*

$$L_j G_\tau u(x, t) - G_\tau L_j u(x, t) = [L_j, G_\tau] u(x, t)$$

$$= (\tau/\pi)^{m/2} \int_{\mathbb{R}^m} e^{-\tau[Z(x,t) - Z(x',t)]^2} u(x', t) L_j h(x') \det Z_x(x', t) \, dx'. \quad \text{(II.10)}$$
□

Let us assume now that $u \in C^\infty(W)$ satisfies $\mathcal{L}u = 0$ and we wish to prove that $E_\tau u(x, t) \to u(x, t)$ in $C^\infty(U)$. We have already proved that $G_\tau u \to hu$ uniformly in $\{|t| \le T\} \times \mathbb{R}^m$. Since $L_j M_k u = M_k L_j u = 0$, $1 \le j \le n$, $1 \le k \le m$, $M_k u$ is a smooth solution of the system, so we also have that $G_\tau M_k u \to h M_k u$ uniformly on $\{|t| \le T\} \times \mathbb{R}^m$. Now, the expression (II.6) of $[M_k, G_\tau]u$ is almost identical to that of G_τ, the only difference being that h has been replaced by $M_k h$, so $[M_k, G_\tau]u \to (M_k h)u$. Restricting our attention to U where $h = 1$ and $M_k h = 0$, we conclude that $M_k G_\tau u = G_\tau M_k u + [M_k, G_\tau]u \to M_k u$ uniformly on U as $\tau \to \infty$. A similar conclusion can be obtained for $L_j G_\tau u$ using (II.10) instead of (II.6), that is, $L_j G_\tau u \to L_j u$ uniformly on U. Since any first-order derivative D may expressed as a linear combination with smooth coefficients of the M_k's and the L_j's, we see that $DG_\tau u \to Du$ uniformly on U. This shows that $G_\tau u \to u$ in $C^1(U)$. Of course, the argument can be iterated for higher-order derivatives to conclude that $G_\tau u \to u$ in $C^\infty(U)$.

II.2 Distribution solutions

We continue the proof of Theorem II.1.1, keeping the notations of Section II.1. In order to extend the arguments of the previous section to a distribution $u \in \mathcal{D}'(W)$ such that $\mathcal{L}u = 0$—which is the fourth step of the proof of Theorem II.1.1—it is enough to check the following facts:

(a) $E_\tau u$ is well-defined for $u \in \mathcal{D}'(W)$;
(b) $G_\tau u$ is well-defined for $u \in \mathcal{D}'(W)$;

(c) $G_\tau u \to u$ in $\mathcal{D}'(U)$ as $\tau \to \infty$ for $u \in \mathcal{D}'(W)$;

(d) $R_\tau u = G_\tau u - E_\tau u \to 0$ in $\mathcal{D}'(U)$ as $\tau \to \infty$ for $u \in \mathcal{D}'(W)$.

We start by observing that since u satisfies the system of equations (II.5) on a neighborhood of $\overline{V}$, the wave front set $WF(u)$ of u is contained in the characteristic set of $\mathcal{L}$ and therefore does not intersect the set

$$\{(x, t, 0, \tau) \in \mathbb{R}^N \times \mathbb{R}^N, \quad |x|, |t| < R', \tau \neq 0\},$$

for some $R' > R$. Thus, $WF(hu)$ is contained in the same set and, in particular, the restriction of u to W belongs to

$$C^\infty(\{|t| \leq R\}; \mathcal{D}'(\{|x| < R\})).$$

On the connection between wave front sets and restrictions of distributions, we refer to [**H2**, chapter VIII]. Moreover, since $V = \{|x| < R\} \times \{|t| < R\}$ is relatively compact in W we may assume that $t \mapsto u(\cdot, t)$ is a continuous function with values in the L^2 based local Sobolev space $L_{\text{loc}}^{2,s}(B_R)$ of order s, for all $|t| \leq R$ and some real s, where B_R denotes the ball of radius R centered at the origin of $\mathbb{R}^m$ (for the definition of local Sobolev spaces see Section II.3.2 below). Thus, for any $|t| \leq R$, the trace $u(\cdot, t)$ is well-defined and belongs to $L_{\text{loc}}^{2,s}(B_R)$. Then, $E_\tau u(x, t)$ (resp. $G_\tau u(x, t)$) is well-defined if we interpret the integral as duality between the distribution $u(\cdot, 0)$ and the test function $(\tau/\pi)^{m/2} e^{-\tau[Z(x,t)-Z(x',0)]^2} h(x') \det Z_x(x', 0)$ (resp. $u(\cdot, t)$ and the test function $(\tau/\pi)^{m/2} e^{-\tau[Z(x,t)-Z(x',t)]^2} h(x') \det Z_x(x', t)$). This takes care of (a) and (b). To prove (d), it is convenient to express $R_\tau u$ by a reinterpretation of the formula obtained for smooth u using Stokes' theorem. We point out that the formula could also have been written as

$$R_\tau u(x, t) = \int_{[0,t]} \sum_{j=1}^n r_j(x, t, t', \tau) \, dt'_j, \tag{II.11}$$

where

$$r_j(x, t, t', \tau) = (\tau/\pi)^{m/2} \int_{\mathbb{R}^m} e^{-\tau[Z(x,t)-Z(x',t')]^2} u(x', t') L_j h(x') \det Z_x(x', t') \, dx' \tag{II.12}$$

and $[0, t]$ denotes the straight segment joining 0 to t. In other words, by integrating first in x we may express the integral of an $m + 1$-form over the cell $\mathbb{R}^m \times [0, t]$ as the integral of a 1-form over the segment $[0, t]$. In this form, Stokes' theorem is just a restatement of the fundamental theorem of calculus for a 1-form. To prove this claim, write for fixed ζ and τ

$$g(t') = \tilde{G}_\tau u(\zeta, t') = \int_{\mathbb{R}^m} e^{-\tau[\zeta-Z(x',t')]^2} u(x', t') h(x') \det Z_x(x', t') \, dx'.$$

Then,

$$g(t) - g(0) = \int_{[0,t]} \sum_{j=1}^{n} \frac{\partial g}{\partial t'_j}(t') \, dt'_j. \tag{*}$$

To compute the derivatives of g we write $e_\tau(\zeta, x', t) = e^{-\tau[\zeta - Z(x',t)]^2}$, differentiate with respect to t'_j under the integral sign, and recall that

$$\frac{\partial(e_\tau uh \det Z_x)}{\partial t'_j} = \frac{\partial(e_\tau uh)}{\partial t'_j} \det Z_x + e_\tau uh \frac{\partial(\det Z_x)}{\partial t'_j}$$

$$= \det Z_x L_j(e_\tau uh) - \det Z_x \sum_{k=1}^{m} \lambda_{jk} \frac{\partial}{\partial x_k}(e_\tau uh)$$

$$+ e_\tau uh \frac{\partial(\det Z_x)}{\partial t'_j},$$

a fact we already used in the proof of (II.10). Once again, the integral over $\mathbb{R}^m$ of the second term of the right-hand side may be written, after integrating by parts, as

$$\int e_\tau uh \sum_{k=1}^{m} \frac{\partial}{\partial x_k}(\lambda_{jk} \det Z_x) \, dx$$

so the integral of the second and third terms together yields

$$\int e_\tau uh \left(\frac{\partial(\det Z_x)}{\partial t_j} + \sum_{k=1}^{m} \frac{\partial}{\partial x_k}(\lambda_{jk} \det Z_x) \right) dx = 0,$$

in view of (II.9). Since $L_j(e_\tau u) = 0$, we also have that $\det Z_x L_j(e_\tau uh) = \det Z_x e_\tau u L_j h$. This shows that

$$\frac{\partial g}{\partial t'_j}(t') = \tilde{r}_j(\zeta, t', \tau) \tag{**}$$

where

$$\tilde{r}_j(\zeta, t', \tau) = \int_{\mathbb{R}^m} e^{-\tau[\zeta - Z(x',t')]^2} u(x', t') L_j h(x') \det Z_x(x', t') \, dx'.$$

Hence, (*) for $\zeta = Z(x, t)$ gives an alternative proof of the fact that $R_\tau u = G_\tau u - E_\tau u$ as given by (II.11) and (II.12). Notice that (II.12) makes sense if $u \in C^\infty(\{|t| \le R\}; \mathcal{D}'(\{|x| < R\}))$ as soon as we change the integral symbol by the duality pairing between the distribution $u(\cdot, t')$ and the appropriate test function; furthermore, $R_\tau u = G_\tau u - E_\tau u$ is still given by (II.11) and (II.12) in the case of distribution solutions since (**) is easily seen to remain valid in this case. Note also that $R_\tau u(x, t)$ is a smooth function of (x, t). We will prove a stronger form of (d).

PROPOSITION II.2.1. *Let $u \in \mathcal{D}'(W)$ satisfy the system (II.5). Then,*

$$R_\tau u(x, t) \to 0 \quad \text{in } C^\infty(U). \tag{II.13}$$

PROOF. We already saw that the exponential in (II.12) may be majorized by $e^{-c\tau}$ for some positive constant $c > 0$ when $|x| < R/4$, $|x'| \geq R/2$, $|t| < T$ and $t' \in [0, T]$. Let Δ_x denote the Laplacian in $\mathbb{R}^m$. For $k \in \mathbb{Z}_+$ we may write

$$L_j h(x') u(x', t') \det Z_x(x', t') = \chi(x')(1 - \Delta_{x'})^k (1 - \Delta_{x'})^{-k} \big(L_j h(x') \\ u(x', t') \det Z_x(x', t') \big),$$

where $\chi(x')$ is a cut-off function that vanishes for $|x'| \leq R/4$ such that $\chi(x') L_j h(x') = L_j h(x')$. Let us write

$$v_j(x', t') = (1 - \Delta_{x'})^{-k} [(L_j h(x')) u(x', t') \det Z_x(x', t')].$$

It follows that $v_j \in C^0(V)$ for an appropriate choice of k and we may write, after an integration by parts,

$$r_j(x, t, t', \tau) = (\tau/\pi)^{m/2} \int v_j(x', t')(1 - \Delta_{x'})^k [\chi(x') e^{-\tau[Z(x,t) - Z(x',t')]^2}] \, dx'.$$

Indeed, the convolution operator

$$(1 - \Delta)^{-k} f(x) = \frac{1}{(2\pi)^m} \int e^{ix \cdot \xi} (1 + |\xi|^2)^{-k} \widehat{f}(\xi) \, d\xi, \quad f \in \mathcal{S}(\mathbb{R}^m)$$

maps continuously $L^{2,s}(\mathbb{R}^m)$ onto $L^{2,s+2k}(\mathbb{R}^m)$ and the latter is contained in $L^\infty(\mathbb{R}^m) \cap C^0(\mathbb{R}^m)$ if $s + 2k > m/2$ by Sobolev's embedding theorem. Hence, $r_j(x, t, t', \tau)$ is continuous with respect to t' and converges to 0 uniformly for $|x| \leq R/2$, $|t'| \leq |t| \leq T$, as $\tau \to \infty$, since the derivatives in $(1 - \Delta_{x'})^k$ produce powers of τ that are dominated by the exponential $e^{-c\tau}$. Hence, $R_\tau u(x, t) \to 0$ uniformly as $\tau \to \infty$ and it is easy to see, by differentiating (II.11), that the same holds for the derivatives of any order with respect to x and t of $R_\tau u(x, t)$, as we wished to prove. $\qquad \square$

Finally, it is enough to prove that (c) holds assuming that $u \in C^0(\{|t| \leq R\}, L^{2,k}_{\text{loc}}(B_R))$ for some integer k. Let us start with the case $k = 0$. We assume that $u \in C^0(\{|t| \leq R\}, L^2_{\text{loc}}(B_{R'}))$ (with R' slightly larger that R) and we wish to prove that

$$\int_{|x| \leq R/4} |G_\tau u(x, t) - u(x, t)|^2 \, dx \to 0 \quad \text{uniformly in } |t| \leq T,$$

which certainly implies (c) in this case. Redefining u by zero off $B_R \times \mathbb{R}^n$ we may assume that $u(x, t) \in L^2(\mathbb{R}^m)$ for each fixed t, $|t| \leq T$. Using once more (II.3′), we see that for any $x, x' \in \mathbb{R}^m$ and $t \in \mathbb{R}^n$

$$\mathcal{R}[Z(x, t) - Z(x', t)]^2 = |x - x'|^2 - |\phi(x, t) - \phi(x', t)|^2$$
$$\geq (3/4)|x - x'|^2,$$

so the exponential inside the integral that defines $G_\tau u$ has a bound $|e^{-\tau[Z(x,t)-Z(x',t')]^2}| \leq e^{-3\tau|x-x'|^2/4}$. If we set

$$F_\tau(x) = \tau^{m/2} e^{-3\tau|x|^2/4}, \quad 0 < \tau < \infty,$$

we easily conclude for fixed $|t| \leq R$ that

$$|G_\tau u(x, t)| \leq C \left(F_\tau * |u|\right)(x, t)$$

where the convolution is performed in the x variable and t plays the role of a parameter. Since $\|F_\tau\|_{L^1} = \|F_1\|_{L^1} = C$, Young's inequality for convolution implies

$$\sup_{|t| \leq T} \|G_\tau u(\cdot, t)\|_{L^2(\mathbb{R}^m)} \leq C \sup_{|t| \leq T} \|u(\cdot, t)\|_{L^2(\mathbb{R}^m)}. \tag{II.14}$$

On the other hand, we proved in Section II.1 that if $u \in C_c^\infty(V)$ then $G_\tau u \to u$ uniformly in $U = B_{R/4} \times \{|t| < T\}$, which implies convergence in the mixed norm space $C^0(\{|t| \leq T\}; L^2(B_{R/4}))$. So the operator $G_\tau|_U$ converges to the restriction operator $u \mapsto u|_U$, as $\tau \to \infty$, on a dense subset of $C^0(\{|t| \leq T\}, L^2(B_R))$ and the family of operators $\{G_\tau|_U\}$ is equicontinuous because of (II.14). Thus, $G_\tau u|_U \to u|_U$ in the whole space $C^0(\{|t| \leq T\}, L^2(B_R))$.

Assume now that $u \in C^0(\{|t| \leq T\}, L^{2,1}(B_{R'}))$, $R' > R$. Introducing a cut-off function we may assume that $u \in C^0(\{|t| \leq T\}, L^{2,1}(\mathbb{R}^m))$ without modifying u for $|x| < R$. Thus, for $|t| \leq T$ fixed, we see that u, $(\partial u/\partial x_k)$ and $(\partial u/\partial t_j)$ are in $L^2(\mathbb{R}^m)$ for $1 \leq k \leq m$, $1 \leq j \leq n$. Since we are assuming that $\phi(x, t)$ is compactly supported, the coefficients of L_j and M_k are bounded, with bounded derivatives. In particular, $L_j u$ and $M_k u$ are in $L^2(\mathbb{R}^m)$ for $1 \leq k \leq m$, $1 \leq j \leq n$, uniformly in $|t| \leq T$. To obtain the convergence result for $k = 1$ we will be able to reason as with the case $k = 0$ as soon as we prove an estimate analogous to (II.14) for the $L^{2,1}$ norm, i.e.,

$$\sup_{|t| \leq T} \|G_\tau u(\cdot, t)\|_{L^{2,1}(\mathbb{R}^m)} \leq C \sup_{|t| \leq T} \|u(\cdot, t)\|_{L^{2,1}(\mathbb{R}^m)}. \tag{II.15}$$

Any first-order derivative with respect to x is a linear combination with bounded coefficients of the M_k's, so it is enough to prove for $|t| \leq T$, $1 \leq k \leq m$, $1 \leq j \leq n$, that

$$\|M_k G_\tau u(\cdot, t)\|_{L^2(\mathbb{R}^m)} \le C \sup_{|t| \le T} \|u(\cdot, t)\|_{L^{2,1}(\mathbb{R}^m)}. \tag{II.16}$$

Writing $M_k G_\tau = [M_k, G_\tau] + G_\tau M_k$ we are led to estimate $\|G_\tau M_k u\|_{L^2}$ and $\|[M_k, G_\tau]u\|_{L^2}$. By (II.14) we have $\|G_\tau M_k u\|_{L^2} \le C\|M_k u\|_{L^2} \le C'\|u\|_{L^{2,1}}$. Notice that an estimate like (II.14) holds as well with $[M_k, G_\tau]$ in the place of G_τ because G_τ and $[M_k, G_\tau]$ have very similar kernels, as (II.10) shows. Thus, $\|[M_k, G_\tau]u\|_{L^2} \le C\|u\|_{L^{2,1}}$, which proves (II.16) and gives (II.15). This process can be continued to prove

$$\sup_{|t| \le T} \|G_\tau u(\cdot, t)\|_{L^{2,k}(\mathbb{R}^m)} \le C_k \sup_{|t| \le T} \|u(\cdot, t)\|_{L^{2,k}(\mathbb{R}^m)}, \quad k = 1, 2 \ldots \tag{II.17}$$

To deal with the case in which k' is a negative integer, i.e., $k' = -|k'| = -k$, we consider a slight modification of G_τ, namely, $G'_\tau u(x) = h(x) G_\tau u(x)$. Of course, $\{G_\tau u|_U\} = \{G'_\tau u|_U\}$ because $h(x) = 1$ for $|x| \le R/2$, so this change will not affect our conclusions for $|x| \le R/4$. The advantage of considering G'_τ is that for fixed t it becomes a formally symmetric operator in the x-variables, as soon as we use the pairing given by the complex measure $dZ(x, t) = \det Z_x(x, t)\,dx$. More precisely, for fixed t and $v, w \in C_c^\infty(\mathbb{R}^m)$ we have $\langle G'_\tau v, w \rangle = \langle v, G'_\tau w \rangle$ where we are using the notation $\langle a, b \rangle = \int a(x)b(x) \det Z_x(x, t)\,dx$, when $a, b \in C^\infty(\mathbb{R}^m)$ and one of them has compact support. Thus,

$$\begin{aligned}
\|G'_\tau u(\cdot, t)\|_{L^{2,k'}(\mathbb{R}^m)} &\le C \sup_{\substack{w \in C_c^\infty(\mathbb{R}^m) \\ \|w\|_{L^{2,k}} \le 1}} |\langle G'_\tau u(\cdot, t), w \rangle| \\
&= C \sup_{\substack{w \in C_c^\infty(\mathbb{R}^m) \\ \|w\|_{L^{2,k}} \le 1}} |\langle u(\cdot, t), G'_\tau w \rangle| \\
&\le C \sup_{\substack{w \in C_c^\infty(\mathbb{R}^m) \\ \|w\|_{L^{2,k}} \le 1}} \|u(\cdot, t)\|_{L^{2,k'}} \|G'_\tau w\|_{L^{2,k}} \\
&\le C\|u(\cdot, t)\|_{L^{2,k'}},
\end{aligned} \tag{II.18}$$

where we have used (II.17) for the positive integer k in the last inequality. This extends (II.17) to all integers $k \in \mathbb{Z}$, proving the equicontinuity of G'_τ in all spaces $C^0(\{|t| \le T\}, L^{2,k}(B_{R'}))$, $k \in \mathbb{Z}$, which together with the convergence of $G_\tau u|_U$ to $u|_U$ for the space of test functions $C_c^\infty(\overline{B_{R'}} \times \{|t| \le T\})$ which is dense in any $C^0(\{|t| \le T\}, L^{2,k}(B_{R'}))$ proves that $G_\tau u \to u$ in $C^0(\{|t| \le T\}, L^{2,k}(B_{R/4}))$ for any $u \in C^0(\{|t| \le T\}, L^{2,k}(B_{R'}))$. This proves (c) and concludes the proof of part (i) of Theorem II.1.1.

To prove part (ii) of the theorem—this is the fifth and final step of the proof—using the same method of proof, it will be enough to prove the equicontinuity of G_τ on the spaces

$$C^j\big(\{|t| \le T\}, C_b^k(\mathbb{R}^m)\big), \quad j, k = 0, 1, 2 \ldots,$$

where $C_b^k(\mathbb{R}^m)$ is the space of functions on $\mathbb{R}^m$ possessing continuous bounded derivatives of order $\le k$. For $j, k = 0$ this is easily achieved by noting that

$$|G_\tau u(x, t)| \le C\,(F_\tau * |u|)\,(x, t) \le C'\|u\|_{C^0\big(\{|t| \le T\}, C_b^0(\mathbb{R}^m)\big)}.$$

For $j, k \le 1$ one expresses the derivatives in terms of the vector fields L_j and M_k and reduces the equicontinuity for the norms of $C^j\big(\{|t| \le T\}, C_b^k(\mathbb{R}^m)\big)$ to the case $j = k = 0$ by introduction of the commutators $[G_\tau, L_j]$ and $[G_\tau, M_k]$, as was done before for Sobolev norms; iteration of this process gives the result for $k = 2, 3, \ldots$ This concludes the proof of Theorem II.1.1.

II.3 Convergence in standard functional spaces

As proved in Proposition II.2.1, $R_\tau u = G_\tau u - E_\tau u \to 0$ in $C^\infty(U)$, for any distribution u satisfying $\mathcal{L}u = 0$ in a larger open set V. This reduces the problem of the convergence $E_\tau u \to u$ in any space with coarser topology than C^∞-topology to the convergence of $G_\tau u \to u$ in the same space. Now, as the reader probably noticed in the proof of Theorem II.1.1, the operator G_τ is very close to convolution with a Gaussian in the x-variables with t playing the role of a parameter, and as such it is a very well-behaved approximation of the identity. Hence, loosely speaking, we may expect that the convergence $G_\tau u \to u$ on U holds in the topology of many functional spaces used in analysis, provided that u belongs to that space over the larger set V. In this section we deal with this question and the approach will always be the same: to prove convergence in a given space of distributions $X(U)$ we will first prove the equicontinuity of $\{G_\tau\}$ in the space $X(\mathbb{R}^N)$ and then try to apply the standard fact that under the hypotheses of equicontinuity it is enough to check the convergence on a convenient dense subset of $X(V)$. Usually the dense subset will be the space of test functions $\psi \in C_c^\infty(V)$, for which we know that $G_\tau \psi \to \psi$ in $C^\infty(\overline{U})$. Thus, this approach works if (i) $X(V)$ is a normal space of distributions (i.e., $C_c^\infty(V)$ is dense in $X(V)$), and (ii) $C^\infty(\overline{U}) \subset X(U)$ with continuous inclusion. We have already applied this principle in the proof of Theorem II.1.1 with $X(V) = C^0\big(\{|t| \le R\}, L^{2,k}(B_R)\big)$.

II.3.1 Convergence in L^p

The main result of this subsection is:

THEOREM II.3.1. *Let $\mathcal{L}$ be a locally integrable structure on Ω and assume that $dZ_1, \ldots, dZ_m$ span $\mathcal{L}^\perp$ at every point of Ω. Then, for any $z \in \Omega$, there exist two open sets U and W, with $z \in U \subset \overline{U} \subset W \subset \Omega$, such that for any $u \in L^p_{\mathrm{loc}}(W)$, $1 \leq p \leq \infty$, satisfying $\mathcal{L}u = 0$,*

$$E_\tau u(x, t) \longrightarrow u(x, t) \quad a.e. \ in \ U \ as \ \tau \to \infty. \tag{II.19}$$

In case p is finite, i.e., $1 \leq p < \infty$, we also have

$$E_\tau u(x, t) \longrightarrow u(x, t) \quad in \ L^p(U) \ as \ \tau \to \infty. \tag{II.20}$$

In (II.19) and (II.20) we may replace the operator E_τ by a convenient sequence of polynomials in Z, $P_\ell(Z_1, \ldots, Z_m)$.

In the proof of Theorem II.3.1 we may assume from the start by shrinking W that $u \in L^p(W)$ and we will do so. We are also tacitly assuming that we are using special coordinates (x, t) adapted to a given set of local generators $dZ_1, \ldots, dZ_m$ of $\mathcal{L}^\perp$ with linearly independent real parts so that $Z = x + i\phi(x, t)$, where $\phi(x, t)$ is smooth, real, has compact support and satisfies (II.3′). Once the special coordinates (x, t) are fixed, the operator E_τ referred to in (II.19) and (II.20) is defined precisely as in the proof of Theorem II.1.1.

We will also prove below theorems similar to Theorem II.3.1 for different norms and in all of them the first step will be to choose special local coordinates where Z has this special form where the operators E_τ and G_τ are defined and have good convergence properties. To avoid repetitions we will always assume that this step has already been carried out, even if not mentioned explicitly.

According to the considerations made at the beginning of the section, we need only prove that

$$G_\tau u \longrightarrow hu \quad in \quad L^p(W), \quad \tau \longrightarrow \infty, \quad u \in L^p(W). \tag{II.21}$$

For $1 \leq p < \infty$, the space $C^0_c(W)$ is dense in $L^p(W)$ and (II.20) will be a consequence of

$$G_\tau u \longrightarrow hu \quad uniformly, \quad \tau \longrightarrow \infty, \quad u \in C^0_c(W)$$

(which we already know by Theorem II.1.1) and the uniform bound that we will prove later:

$$\|G_\tau u\|_p \leq C\|u\|_p, \quad u \in L^p(\mathbb{R}^N), \quad \tau > 0, \tag{II.22}$$

where $\| \ \|_p$ denotes the L^p-norm.

Let us set $W = B_x \times B_t$, where $B_x = \{|x| < R\}$ and $B_t = \{|t| < R\}$. Let $u \in L^p(W)$ and set $u^t(x) = u(x, t)$. Fubini's theorem guarantees that u^t is defined for a.e. t, it is measurable, and it belongs to $L^p(B_x)$. If, moreover, u satisfies $\mathcal{L}u = 0$, we know that u has a trace $T_t u$ and $B_t \ni t \mapsto T_t u \in \mathcal{D}'(B_x)$ is a smooth function. It will be useful to compare both types of restrictions of u to the slices $t = \mathrm{const}$.

LEMMA II.3.2. *If $u \in L^p(W)$, $1 \leq p \leq \infty$, and u is a solution of the system* (II.5) *then $T_t u = u^t$ for a.e. $t \in B_t$. In particular, $T_t u \in L^p(B_x)$ for a.e. $t \in B_t$.*

PROOF. We take functions $\phi \in C_c^\infty(B_x)$ and $\psi \in C_c^\infty(B_t)$. We know that $t \mapsto \langle T_t u, \phi \rangle$ is a C^∞-function defined in B_t, $t \mapsto \langle u^t, \phi \rangle$ belongs to $L^p(B_t)$ and

$$
\int \langle T_t u, \phi \rangle \psi(t)\, dt = \int \left(\int u(x, t)\phi(x)\, dx \right) \psi(t)\, dt \tag{II.23}
$$
$$
= \int \langle u^t, \phi \rangle \psi(t)\, dt.
$$

If we take $\psi(t) = \chi_j(t - t_0)$, $\chi_j(t) = j^n \chi(jt)$, $0 \leq \chi \in C_c^\infty(\{|t| \leq 1\})$, $\int \chi\, dt = 1$, and let $j \to \infty$, the left-hand side of (II.23) converges for every $t \in B_t$ to $\langle T_t u, \phi \rangle$ while the right-hand side converges a.e. to $\langle u^t, \phi \rangle$. Hence, there is a null set $N(\phi) \subset B_t$ such that

$$
\langle T_t u, \phi \rangle = \langle u^t, \phi \rangle, \qquad \phi \in C_c^\infty, \quad t \notin N(\phi).
$$

If we apply the last identity to a dense sequence $\{\phi_k\} \subset C_c^\infty(B_x)$ and set $N = \bigcup N(\phi_n)$ we obtain that $T_t u = u^t$ as elements of $\mathcal{D}'(B_x)$ when t is not in the null set N. $\qquad \square$

REMARK II.3.3. One cannot expect in general that, under the conditions of Lemma II.3.2, $T_t u \in L^p$ for all t. For instance, if $\Omega = (-1, 1) \times (-1, 1) \subset \mathbb{R}^2$, $Z = x + it^2/2$, $L = \partial_t - it\partial_x$ is the Mizohata operator and $u(x, t) = 1/Z(x, t)$, it is simple to verify that $u \in L^p(\Omega)$ for $1 \leq p < 3/2$, $Lu = 0$ in the sense of distributions and $T_t u \in C^\infty([-1, 1]) \subset L^\infty(-1, 1) \subset L^p(-1, 1)$ for $t \neq 0$ but for $t = 0$ we have $T_0 u = \mathrm{pv}(1/x) - i\pi\delta(x) \notin L^p(-1, 1)$.

We now prove Theorem II.3.1. Consider the maximal operator associated with $G_\tau u$:

$$
G_\tau^* u(x, t) = \sup_{\tau \geq 1} |G_\tau u(x, t)|.
$$

We claim that, for $u \in L^1(\Omega)$, there exists a constant $C > 0$ such that

$$
G_\tau^* u(x, t) \leq C M(h(x) T_t u(x)), \tag{II.24}
$$

for any t such that $T_t u \in L^1(B_x)$. Here

$$Mf(x, t) = \sup_{r>0} \frac{1}{|B(x, r)|} \int_{B(x,r)} |f(x', t)| \, dx'$$

is the Hardy–Littlewood maximal operator acting in the x-variable, $B(x, r)$ is the ball of radius r centered at x, and $|B(x, r)|$ denotes its Lebesgue measure. In fact, $|G_\tau u(x, t)|$ can be estimated by

$$(\tau/\pi)^{m/2} \int_{\mathbb{R}^m} e^{-\tau(|x-x'|^2 - |\phi(x,t) - \phi(x',t)|^2)} |T_t u(x')| |h(x')| |\det Z_x(x', t)| \, dx'$$

and this expression can be dominated by the maximal operator

$$\sup_{\tau \geq 1} F_\tau * |h \, T_t u \det Z_x| = C \sup_{\tau \geq 1} \tau^{m/2} \int_{\mathbb{R}^m} e^{-3\tau|x-x'|^2/4}$$
$$|T_t u(x')| |h(x')| |\det Z_x(x', t)| \, dx'$$

where

$$F_\tau(x) = C\tau^{m/2} e^{-3\tau|x|^2/4}$$

and C is a constant. Hence,

$$G_\tau^* u(x, t) \leq \sup_{\tau \geq 1} F_\tau * |h \, T_t u \det Z_x| \leq C M(h(\cdot) T_t u(\cdot))(x).$$

The last inequality follows from the fact that $F_1(x) = Ce^{-3|x|^2/4}$ is radial decreasing and belongs to $L^1(\mathbb{R}^m)$ (see, for instance, [**S1**, page 62]). Thus, (II.24) is proved.

If $u \in C_c^0(W)$, we know that $G_\tau u(x, t) \to h(x)u(x, t)$, $\tau \to \infty$ uniformly. The standard properties of the maximal operator allow us to conclude that for any $t \in B_t$ such that $T_t u(x) \in L^1(B_x)$ there exists a subset $N_t \subset B_x$ with $|N_t| = 0$ such that

$$G_\tau u(x, t) \to h(x)u(x, t), \quad x \notin N_t.$$

Hence, if we choose $(x, t) \in U$ such that $T_t u \in L^1(B_x)$ and $x \notin N_t$, we get (recalling that $R_\tau u \to 0$ uniformly in U)

$$E_\tau u(x, t) \to h(x)u(x, t) = u(x, t) \quad \text{a.e. in } U$$

and therefore $E_\tau u(x, t) \to u(x, t)$ a.e. in U as we wished to prove.

We now prove (II.22). We observe that

$$|G_\tau u(x, t)| \leq F_\tau * |h \, T_t u \det Z_x|$$

and then Young's inequality for convolution implies

$$\|G_\tau u(\cdot, t)\|_{L^p(dx)} \leq \|F_\tau\|_1 \|h \, T_t u \det Z_x\|_{L^p(dx)} \leq C\|T_t u\|_{L^p(dx)},$$

since the L^1 norm of F_τ does not depend on τ and $h \det Z_x$ is bounded. Raising this inequality to the pth power and integrating with respect to t we obtain (II.22). Since $G_\tau u \to hu$ uniformly in W as $\tau \to \infty$ when u is continuous, the usual density argument shows that (II.21) holds for $1 \leq p < \infty$. Thus, (II.19) and (II.20) have been proved. Finally, since $E_\tau u$ can be approximated in $C^\infty(U)$ by polynomials in Z for fixed τ, the proof is complete. $\qquad\square$

It is obvious that (II.20) is, in general, false for $p = \infty$ because the uniform limit of a sequence of continuous functions, such as $E_\tau u(x, t)$, is continuous.

A simple consequence of Theorem II.3.1 is:

COROLLARY II.3.4. *Let $\mathcal{L}$ be a locally integrable structure over a C^∞ manifold U and let $u \in L^p_{\mathrm{loc}}(U)$, $1 \leq p \leq \infty$, $v \in L^q_{\mathrm{loc}}(U)$, $1/p + 1/q = 1$, be solutions of the system (II.5). Then the product $w = uv \in L^1_{\mathrm{loc}}(U)$ also satisfies (II.5).*

PROOF. By localization we may assume that U is the neighborhood where the conclusions of Theorem II.3.1 hold. Set $u_\tau = E_\tau u$, $w_\tau = u_\tau v$. Leibniz's rule shows that $\mathcal{L}w_\tau = 0$, as $u_\tau \in C^\infty(U)$. By Theorem II.3.1 and Hölder's inequality $w_\tau \to w$ in $L^1_{\mathrm{loc}}(U)$, $\tau \to \infty$, showing that $\mathcal{L}w = 0$ in the sense of distributions. $\qquad\square$

II.3.2 Convergence in Sobolev spaces

In this subsection we prove

THEOREM II.3.5. *Let $\mathcal{L}$ be a locally integrable structure with first integrals $Z_1, \ldots, Z_m$, defined in a neighborhood of the closure of $W = B_x \times B_t$. There exists a neighborhood $U \subset W$ of the origin such that for any $u \in L^{p,s}_{\mathrm{loc}}(W)$, $1 < p < \infty$, $s \in \mathbb{R}$, satisfying $\mathcal{L}u = 0$,*

$$E_\tau u(x, t) \longrightarrow u(x, t) \quad \text{in } L^{p,s}_{\mathrm{loc}}(U), \quad \tau \longrightarrow \infty. \tag{II.25}$$

As usual, we may replace the operator E_τ in (II.25) by a convenient sequence of polynomials in Z, $P_\ell(Z_1, \ldots, Z_m)$.

We recall that for $1 \leq p \leq \infty$, $s \in \mathbb{R}$,

$$L^p_s(\mathbb{R}^N) = \{f \in \mathcal{S}'(\mathbb{R}^N) : \|f\|_{p,s} \doteq \|\Lambda^s f\|_p < \infty\}$$

where $\Lambda^s f(x) = \mathcal{F}^{-1}[(1 + |\xi|^2)^{s/2} \mathcal{F} f(\xi)](x)$ and $\mathcal{F}$ denotes the Fourier transform in $\mathbb{R}^N$ (Λ^s is the Bessel potential and $\mathcal{S}'$ denotes the space of tempered distributions). For $k \in \mathbb{Z}_+$ and p in the range $1 < p < \infty$ the space $L^p_k(\mathbb{R}^N)$ is exactly the subspace of the functions in $L^p(\mathbb{R}^N)$ whose derivatives of order $\leq k$ in the sense of distributions belong to $L^p(\mathbb{R}^N)$. This space is equivalently normed by ([S1])

$$\|u\|_{L^p_k} = \sum_{|\alpha| \leq k} \|D^\alpha u\|_p. \tag{II.26}$$

The space $L^{p,s}_{\text{loc}}(\Omega)$ is the subspace of $\mathcal{D}'(\Omega)$ of the distributions u such that $\psi u \in L^p_s(\mathbb{R}^N)$ for all test functions $\psi \in C^\infty_c(\Omega)$, equipped with the locally convex topology given by the seminorms $u \mapsto \|\psi u\|_{p,s}$, $\psi \in C^\infty_c(\Omega)$. Fix $p \in (1, \infty)$, $s \in \mathbb{R}$ and choose the open sets U and W as in Theorem II.1.1. The theorem will be proved if we show that

$$\lim_{\tau \to \infty} G_\tau v = h v \quad \text{in} \quad L^p_s(W), \quad \forall v \in C^\infty_c(W), \tag{II.27}$$

and there exists a positive constant C such that

$$\|G_\tau w\|_{p,s} \leq C\|w\|_{p,s} \quad \forall w \in L^p_s(\mathbb{R}^N). \tag{II.28}$$

Indeed, (II.27) and (II.28) imply as usual, by density and triangular approximation, that $\|G_\tau w - hw\|_{p,s} \to 0$ as $\tau \to \infty$ for any $w \in L^p_s(\mathbb{R}^N) \cap \mathcal{E}'(W)$—where $\mathcal{E}'(W)$ denotes the space of distributions compactly supported in W—which implies that $G_\tau w \to w$ in the topology of $L^{p,s}_{\text{loc}}(U)$. We know that for $u \in C^\infty_c(U)$, $G_\tau u \to u$ in $C^\infty(U)$, thus (II.27) is clearly true and we need only worry about proving (II.28), which we prove first for a positive integer $s = k \in \mathbb{Z}_+$. The vector fields L_j and M_k form a basis of $\mathbb{C}T\mathbb{R}^n$ and we may express the derivatives D^α in (II.26) in terms of the vector fields L_j, $j = 1, \ldots, n$, M_k, $k = 1, \ldots, m$. This gives

$$\|G_\tau w\|_{L^p_k} \leq C \sum_{|\alpha_1| + |\alpha_2| \leq k} \|M^{\alpha_1} L^{\alpha_2} G_\tau w\|_p. \tag{II.29}$$

We write

$$L_j G_\tau w = G_\tau L_j w + [L_j, G_\tau] w,$$
$$M_k G_\tau w = G_\tau M_k w + [M_k, G_\tau] w.$$

As shown in Lemmas II.1.4 and II.1.6, the operators $[L_j, G_\tau]$ and $[M_k, G_\tau]$ are given by the same expression as G_τ with $h(x)$ replaced respectively by $L_j h(x)$ and $M_k h(x)$. Hence, the proof of Theorem II.3.1 gives bounds in L^p for the commutators that may be written as

$$\|[L_j, G_\tau]v\|_p + \|[M_k, G_\tau]v\|_p \leq C\|v\|_p, \quad v \in L^p(\mathbb{R}^N). \tag{II.30}$$

Thus, for $1 \leq j \leq n$, $1 \leq k \leq m$,

$$\|L_j G_\tau w\|_p + \|M_k G_\tau w\|_p \leq C(\|L_j w\|_p + \|M_k w\|_p + \|w\|_p)$$
$$\leq C(\|w\|_{p,1} + \|w\|_p)$$
$$\leq C\|w\|_{p,1}, \tag{II.31}$$

where we have used (II.22) to estimate $G_\tau L_j w$ and $G_\tau M_k w$ in the first inequality. Thus, combining (II.26) for $u = G_\tau w$ and $k = 1$ with (II.31) we get (II.28) for $k = 1$. This reasoning can be iterated for any $s = k \in \mathbb{Z}_+$ and the theorem is proved for $s \in \mathbb{Z}_+$.

To prove (II.28) for nonintegral $s > 0$, we use interpolation of Sobolev spaces (on the subject of interpolation see, for instance, [**C1**] and [**C2**]). First we take $k \in \mathbb{Z}_+$ such that $0 < s < k$. The operator G_τ is of type $(p, p, 0, 0)$ and also of type (p, p, k, k), $k \in \mathbb{Z}_+$, that is, it verifies

$$\|G_\tau w\|_p \leq C\|w\|_p, \quad w \in C_c^\infty(\mathbb{R}^N)$$

and

$$\|G_\tau w\|_{p,k} \leq C\|w\|_{p,k}, \quad w \in C_c^\infty(\mathbb{R}^N).$$

By complex interpolation we obtain that G_τ is of type (p, p, s, s); that is, (II.28) holds for $0 < s < k$ and $w \in C_c^\infty(\mathbb{R}^N)$ and by density it also holds for $w \in L_s^p(\mathbb{R}^N)$. Finally, to prove (II.28) for $s < 0$, we invoke a slight variation of the duality argument that was used to extend (II.18) from positive integers to negative integers: we consider the modification of G_τ, $G_\tau' u(x) = h(x) G_\tau u(x)$ which is formally symmetric in the x-variables for fixed t for the pairing given by integration with respect to $dZ(x, t) = \det Z_x(x, t) \, dx$ and thus also symmetric in both variables x and t for the pairing given by integration with respect to $dZ(x, t) \wedge dt = \det Z_x(x, t) \, dx dt$. Since this is a nonsingular continuous pairing for the spaces $L_s^p(\mathbb{R}^N)$ and $L_{-s}^q(\mathbb{R}^N)$, $1/p + 1/q = 1$, it extends (II.28) to $s < 0$ as follows:

$$\|G_\tau' w\|_{L_s^p(\mathbb{R}^N)} \leq C \sup_{\substack{\psi \in C_c^\infty(\mathbb{R}^N) \\ \|\psi\|_{L_{-s}^q} \leq 1}} |\langle G_\tau' w(\cdot, t), \psi \rangle|$$

$$\leq C \sup_{\substack{\psi \in C_c^\infty(\mathbb{R}^N) \\ \|\psi\|_{L_{-s}^q} \leq 1}} |\langle w, G_\tau' \psi \rangle|$$

$$\leq C \sup_{\substack{\psi \in C_c^\infty(\mathbb{R}^N) \\ \|\psi\|_{L_{-s}^q} \leq 1}} \|w\|_{L_s^p} \|G_\tau' \psi\|_{L_{-s}^q}$$

$$\leq C_s \|w\|_{L_s^p(\mathbb{R}^N)},$$

where in the last inequality we used (II.28) with q in the place of p and $-s > 0$ in the place of s. Thus, (II.28) is completely proved and the proof of Theorem II.3.5 is complete. □

II.3.3 Convergence in Hölder spaces

Let $\Omega \subset \mathbb{R}^N$ be an open, bounded, convex set. The Hölder space $C^\alpha(\Omega)$ is defined as

$$C^\alpha(\Omega) = \{u \in C^k(\overline{\Omega}), \|u\|_\alpha < \infty\}$$

where

$$\|u\|_\alpha = |u|_\alpha + |u|_0,$$

$$|u|_0 = \sup_{x \in \overline{\Omega}} |u(x)|,$$

$$|u|_\alpha = \sup_{\substack{x,y \in \overline{\Omega}, \\ x \neq y}} \frac{|u(x) - u(y)|}{|x - y|^\alpha}, \quad 0 < \alpha \leq 1,$$

$$|u|_\alpha = \sum_{\sigma \leq k} |D^\sigma u|_{\alpha - k}, \quad k < \alpha \leq k+1, \quad k \in \mathbb{Z}+, \quad u \in C^k(\Omega).$$

The spaces $C^\alpha(\mathbb{R}^N)$ are defined similarly. The approximation theorem is:

THEOREM II.3.6. *Let $\mathcal{L}$ be a locally integrable structure with first integrals $Z_1, \ldots, Z_m$, defined in a neighborhood of the closure of $W = B_x \times B_t$. There exists a convex neighborhood $U \subset \Omega$ of the origin such that for any $u \in C^\beta(W)$, $\beta > 0$ satisfying $\mathcal{L}u = 0$ in a neighborhood of $\overline{W}$ and any $0 \leq \alpha < \beta$*

$$E_\tau u(x, t) \longrightarrow u(x, t) \text{ in } C^\alpha(U), \quad \tau \longrightarrow \infty. \tag{II.32}$$

As usual, we may replace the operator E_τ in (II.32) by a convenient sequence of polynomials in Z, $P_\ell(Z_1, \ldots, Z_m)$.

PROOF. As always, since $C_c^\infty(W)$ is dense in $C_c^\beta(W)$ for the C^α norm, we need only prove

$$G_\tau u \longrightarrow u \quad \text{in} \quad C^\alpha(W), \quad u \in C_c^\infty(W),$$

and the inequality

$$\|G_\tau u\|_\alpha \leq C\|u\|_\alpha, \quad u \in C_c^\alpha(W).$$

It is obvious that $\|G_\tau u - u\|_\alpha \to 0$ when $\tau \to \infty$, $u \in C_c^\infty(W)$, because by Theorem II.1.1 $G_\tau u \to u$, $\tau \to \infty$ in $C^k(W)$ for every positive integer k. We may assume without loss of generality, as we always do, that $Z(x, t) = x + i\phi(x, t)$ is defined and satisfies (II.3′) throughout $\mathbb{R}^N$ and reduces to

$Z(x, t) \equiv x$ for (x, t) outside a compact set. We shall then prove

$$\|G_\tau u\|_\alpha \le C\|u\|_\alpha, \quad u \in C_c^\alpha(\mathbb{R}^N). \tag{II.33}$$

We assume first that $0 < \alpha < 1$. It will be useful to use the following well-known characterization of $C^\alpha(\mathbb{R}^N)$ ([**S2**, page 256]):

LEMMA II.3.7. *A function u belongs to $C^\alpha(\mathbb{R}^N)$, $0 < \alpha < 1$, if and only if there exist a sequence of functions $(u_k) \in C^1(\mathbb{R}^N)$, bounded and with bounded gradients, such that*

(i) $\|u_k\|_{L^\infty} \le K 2^{-\alpha k}$, $k = 0, 1, \ldots$;
(ii) $\|\nabla u_k\|_{L^\infty} \le K 2^{(1-\alpha)k}$, $k = 0, 1, \ldots$;
(iii) $u(z) = \sum_{k=0}^\infty u_k(z)$, $z \in \mathbb{R}^N$.

It also follows that the best constant K in (i) and (ii) above is proportional to $\|u\|_\alpha$. Such a sequence is usually called a sequence of best approximation for u. We start by writing $u = \sum u_k$ with (u_k) a sequence of best approximation for u. Then, $G_\tau u = \sum G_\tau u_k$ and we need to estimate the essential supremum of $G_\tau u_k$ and $\nabla G_\tau u_k$. Taking account of (II.22) with $p = \infty$ and (i) of Lemma II.3.7 we derive

$$\|G_\tau u_k\|_{L^\infty} \le C\|u_k\|_{L^\infty} \le CK2^{-\alpha k}, \quad k \in \mathbb{Z}_+. \tag{II.34}$$

In order to estimate $\nabla G_\tau u_k$ it is convenient to express any partial derivative in terms of the vector fields L_j and M_ℓ, $1 \le j \le n$, $1 \le \ell \le m$. Then, we are led to estimate $L_j G_\tau u_k$, $j = 1, \ldots, n$ and $M_\ell G_\tau u_k$, $\ell = 1, \ldots, m$. We may write $L_j G_\tau u_k = G_\tau L_j u_k + [L_j, G_\tau] u_k$ and recall that

$$\|[L_j, G_\tau] u_k\|_{L^\infty} \le C\|u_k\|_{L^\infty},$$

which follows from (II.30) with $p = \infty$. We get

$$\|L_j G_\tau u_k\|_{L^\infty} \le C(\|L_j u_k\|_{L^\infty} + \|u_k\|_{L^\infty})$$
$$\le C(\|\nabla u_k\|_{L^\infty} + \|u_k\|_{L^\infty}), \quad j = 1, \ldots, n, \quad k = 1, 2, \ldots$$

Similar estimates are true for $M_\ell G_\tau u_k$, $\ell = 1, \ldots, m$, $k \in \mathbb{Z}_+$ and we obtain

$$\|\nabla G_\tau u_k\|_{L^\infty} \le C(\|u_k\|_{L^\infty} + \|\nabla u_k\|_{L^\infty}) \le C'K2^{(1-\alpha)k}, \quad k \in \mathbb{Z}_+. \tag{II.35}$$

Thus, (II.34), (II.35) and Lemma II.3.7 imply that (II.33) holds for $0 < \alpha < 1$.

Let us assume next that there is a positive integer k such that $\alpha = k + \eta$, $0 < \eta < 1$ and we wish to estimate

$$\|G_\tau u\|_\alpha \sim \sum_{|\sigma| \le k} \|D^\sigma G_\tau u\|_\eta \le C \sum_{|\sigma_1| + |\sigma_2| \le k} \|M^{\sigma_1} L^{\sigma_2} G_\tau u\|_\eta.$$

Using the commutation formulas of Lemmas II.1.4 and II.1.6 it is easy to prove (II.33) by induction on k, adapting the reasonings we used to deal with Sobolev norms of integral order in Section II.3.2; we leave the details to the reader. Finally, to prove (II.33) for $\alpha = k = 1, 2, \ldots$, we observe that in this case $\|u\|_\alpha = \|u\|_k \sim \|u\|_{L_k^\infty}$ so (II.33) is a variation of the estimates already considered for Sobolev norms. This completes the proof of Theorem II.3.6.

$\square$

It is not possible to take $\alpha = \beta$ in Theorem II.3.6, as we will see next.

EXAMPLE II.3.8. Consider in $\mathbb{R}^2$, where we denote the coordinates by (x, t), the structure $\mathcal{L}$ spanned by ∂_t with first integral $Z(x, t) = x$ and let $0 < \beta \le 1$. Consider a function $u(x) \in C_c^\beta(\mathbb{R}^2)$ independent of t (so it satisfies $\mathcal{L}u = 0$) such that $u(x) = |x|^\beta$ for $|x| \le 1$. If $w(x, t)$ is of class C^1 in a neighborhood of the origin, we have for $0 < \varepsilon < 1$ sufficiently small,

$$|u - w|_\beta \ge \frac{|u(\varepsilon) - w(\varepsilon, 0) - (u(0) - w(0, 0))|}{\varepsilon^\beta} \ge 1 - C\varepsilon^{1-\beta}$$

and the left-hand side is $\ge 1/2$ for ε small, showing that u cannot be approximated by continuously differentiable functions in the C^β topology.

II.3.4 Convergence in Hardy spaces

We recall that the real Hardy space $H^p(\mathbb{R}^N)$, $0 < p < \infty$, introduced by Stein and Weiss ([**SW**]), is equal to $L^p(\mathbb{R}^N)$ for $p > 1$, is properly contained in $L^1(\mathbb{R}^N)$ for $p = 1$, and is a space of not necessarily locally integrable distributions for $0 < p < 1$. For $p \le 1$, $H^p(\mathbb{R}^N)$ is a substitute for $L^p(\mathbb{R}^N)$ ([**S2**]), as the latter is not a space of distributions and has trivial dual if $p < 1$; even for $p = 1$, $L^1(\mathbb{R}^N)$ does not behave as well as $L^p(\mathbb{R}^N)$, $1 < p < \infty$, for example on questions concerning the continuity of pseudo-differential operators. Let us choose a function $\Phi \in \mathcal{S}(\mathbb{R}^N)$, with $\int \Phi dz \ne 0$ and write $\Phi_\varepsilon(z) = \varepsilon^{-N}\Phi(z/\varepsilon)$, $z \in \mathbb{R}^N$, and

$$M_\Phi f(z) = \sup_{0 < \varepsilon < \infty} |(\Phi_\varepsilon * f)(z)|.$$

Then ([**S2**])

$$H^p(\mathbb{R}^N) = \{f \in \mathcal{S}'(\mathbb{R}^N) : \quad M_\Phi f \in L^p(\mathbb{R}^N)\}.$$

An obstacle to the localization of the elements of $H^p(\mathbb{R}^N)$, $0 < p \le 1$, is that ψu may not belong to $H^p(\mathbb{R}^N)$ for $\psi \in C_c^\infty(\mathbb{R}^N)$ and $u \in H^p(\mathbb{R}^N)$. A way around this is the definition of localizable Hardy spaces $h^p(\mathbb{R}^N)$ ([**G**],[**S2**]) by means of the truncated maximal function

$$m_\Phi f(z) = \sup_{0<\varepsilon\leq 1} |(\Phi_\varepsilon * f)(z)|,$$

$$h^p(\mathbb{R}^N) = \{f \in \mathcal{S}'(\mathbb{R}^N) : \quad m_\Phi f \in L^p(\mathbb{R}^N)\}.$$

It turns out that if Φ is replaced in the definition of $h^p(\mathbb{R}^N)$ by any other function $\Phi \in \mathcal{S}(\mathbb{R})$ only required to satisfy $\int \Phi \neq 0$, this will not change the space $h^p(\mathbb{R}^N)$. It is also known that the space $h^p(\mathbb{R}^N)$ is stable under multiplication by test functions and also that $h^p(\mathbb{R}^N) = L^p(\mathbb{R}^N)$ for $1 < p < \infty$. For $0 < p \leq 1$, which we henceforth assume, $h^p(\mathbb{R}^N)$ is a metric space with the distance $d(f, g) = \int (m_\Phi(f - g)(z))^p \, dz$. If $\Omega \subset \mathbb{R}^N$ is an open set, the space $H^p_{\mathrm{loc}}(\Omega)$ is the subspace of $\mathcal{D}'(\Omega)$ of the distributions u such that $\psi u \in h^p(\mathbb{R}^N)$ for all test functions $\psi \in C^\infty_c(\Omega)$. A sequence u_n converges to zero in $H^p_{\mathrm{loc}}(\Omega)$ if $\psi u_n \to 0$ in $h^p(\mathbb{R}^N)$ for every $\psi \in C^\infty_c(\Omega)$. We have

THEOREM II.3.9. *Let $\mathcal{L}$ be a locally integrable structure with first integrals $Z_1, \ldots, Z_m$, defined in a neighborhood of the closure of $W = B_x \times B_t$. There exists a neighborhood $U \subset W$ of the origin such that for any $u \in H^p_{\mathrm{loc}}(W)$, $0 < p < \infty$, satisfying $\mathcal{L}u = 0$,*

$$E_\tau u(x, t) \longrightarrow u(x, t) \text{ in } H^p_{\mathrm{loc}}(U), \quad \tau \longrightarrow \infty. \tag{II.36}$$

As usual, we may replace the operator E_τ in (II.36) by a convenient sequence of polynomials in Z, $P_\ell(Z_1, \ldots, Z_m)$.

PROOF. Since $H^p_{\mathrm{loc}}(W) = L^p_{\mathrm{loc}}(W)$ for $p > 1$, Theorem II.3.9 follows from Theorem II.3.1 for these values of p and it is enough to assume that $0 < p \leq 1$. The space $C^\infty_c(W)$ is continuously included in $H^p_{\mathrm{loc}}(W)$ and the theorem may be proved by showing once again that

$$\lim_{\tau\to\infty} G_\tau v = h v \text{ in } h^p(\mathbb{R}^N), \quad \forall v \in C^\infty_c(W) \tag{II.37}$$

$$\|G_\tau w\|_{h^p} \leq C \|w\|_{h^p} \quad \forall w \in h^p(\mathbb{R}^N) \tag{II.38}$$

with the notation $\|w\|_{h^p} = (\int (m_\Phi w(z))^p \, dz)^{1/p}$, in spite of the fact that $w \to \|w\|_{h^p}$ is not a norm for $p < 1$. To prove (II.37) and (II.38) we use the atomic decomposition of h^p ([**G**],[**S2**]). An h^p atom, $p \leq 1$, is a bounded, compactly supported function $a(z)$ satisfying the following property: there exists a cube Q with sides parallel to the coordinate axes that contains the support of a and furthermore

(i) $|a(z)| \leq |Q|^{-1/p}$, a.e., with $|Q|$ denoting the Lebesgue measure of Q;
(ii) $\int z^\alpha a(z)\,dz = 0$, $|\alpha| \leq N(1/p - 1)$, if the side length of Q happens to be less than 1.

Notice that if the support of a is contained in a cube Q such that (i) holds and the side of Q has length ≥ 1, then a is an atom, as condition (ii) is vacuous and only (i) is required in this case.

As always, (II.37) follows from the convergence $G_\tau v \to v$ in $C_c^\infty(\Omega)$, $v \in C_c^\infty(\Omega)$. So, to prove Theorem II.3.9, we need only show (II.38) and the density of $C_c^\infty(\mathbb{R}^N)$ in $h^p(\mathbb{R}^N)$. To prove the density, it is enough to approximate h^p atoms by smooth h^p atoms in the h^p norm. This is simply approximating a rough atom a by the convolution $a_\varepsilon = a * \psi_\varepsilon$, where $\psi_\varepsilon(z) = \varepsilon^{-N}\psi(z/\varepsilon)$, and $\psi \in C_c^\infty(\mathbb{R}^N)$ has integral equal to 1. Then, a_ε satisfies the vanishing moments condition (ii) because a does and satisfies (i) for a cube Q slightly larger than the one that worked for a, if $\varepsilon > 0$ is sufficiently small. Moreover, $a_\varepsilon \to a$ in the h^p 'norm' as $\varepsilon \to 0$. To check the last fact use Hölder's inequality to write

$$\int (m_\Phi(a - a_\varepsilon)(z))^p\,dz \leq |Q|^{1-p/2}\|m_\Phi(a - a_\varepsilon)\|_{L^2}$$
$$\leq C|Q|^{1-p/2}\|M(a - a_\varepsilon)\|_{L^2}$$
$$\leq C|Q|^{1-p/2}\|a - a_\varepsilon\|_{L^2}$$

where we have majorized the maximal function $m_\Phi(a - a_\varepsilon)$ by the Hardy–Littlewood maximal function $M(a - a_\varepsilon)$ which is continuous in L^2.

Any $w \in h^p$ can be written as a convergent series in h^p, $w = \sum_k \lambda_k a_k$, where the a_k are atoms and λ_k are complex numbers such that $\sum_k |\lambda_k|^p \sim \|w\|_{h^p}$ ([S2]) (since atoms may be approximated by smooth atoms we may even assume that $a_k \in C_c^\infty(\mathbb{R}^N)$ for all k). Then, to prove (II.38) it is enough to verify that there is a constant $C > 0$ such that for all h^p atoms $a(z)$

$$\|G_\tau a\|_{h^p}^p = \int (m_\Phi G_\tau a(z))^p\,dz \leq C, \quad \tau \geq 1. \tag{II.39}$$

Indeed,

$$\int \left(m_\Phi G_\tau \sum_k \lambda_k a_k\right)^p dz \leq \int \left(\sum_k |\lambda_k| m_\Phi G_\tau a_k\right)^p dz$$
$$\leq \sum_k |\lambda_k|^p \int (m_\Phi G_\tau a_k)^p\,dz$$

because $p \leq 1$. We assume without loss of generality that $\Phi \geq 0$ is supported in the unit ball (in fact, changing the function Φ by any other function in $\mathcal{S}(\mathbb{R}^N)$

with nonvanishing integral will produce an equivalent 'norm' in $H^p(\mathbb{R}^N)$).
We set $F(x) = e^{-3|x|^2/4}$, $x \in \mathbb{R}^m$, $F_\sigma(x) = \sigma^{-m} F(s/\sigma)$ and we check that by
the estimates of Section II.3.1 (see (II.24)):

$$|(\Phi_\varepsilon * G_\tau a)(x, t)| \leq C|\Phi_\varepsilon * (F_\sigma \overset{(x)}{*} a)(x, t)|$$

$$= C|(\Phi_\varepsilon * a) \overset{(x)}{*} F_\sigma(x, t)|, \quad \sigma = \tau^{-1/2},$$

where the symbol $\overset{(x)}{*}$ denotes convolution in the x-variable. Let $Q = Q_1 \times Q_2$,
$Q_1 \subset \mathbb{C}^m$, $Q_2 \subset \mathbb{C}^n$, be a cube containing the support of a. Thus, invoking (i),
we get

$$m_\Phi(G_\tau a)(x, t) \leq C|Q|^{-1/p} \chi_{Q_2}(t). \tag{II.40}$$

Here and in the sequel, χ_A will denote the characteristic function of a measurable set A. Let Q_1^* (resp. Q_1^{**}) be the cube in $\mathbb{R}^m$ concentric with Q_1 having
twice (resp. four times) the side length. Then (II.40) shows that

$$\int_{Q_1^{**} \times \mathbb{R}^n} |m_\Phi(G_\tau a)(x, t)|^p \, dx \, dt \leq C, \tag{II.41}$$

with $C > 0$ independent of $0 < \varepsilon \leq 1$, $\tau \geq 1$, $a(z)$ an atom. Thus, (II.39) will
be proved as soon as we obtain

$$\int_{(\mathbb{R}^m \setminus Q_1^{**}) \times \mathbb{R}^n} \sup_{0 < \varepsilon \leq 1} |\Phi_\varepsilon * (F_\sigma \overset{(x)}{*} a)(x, t)|^p \, dx \, dt \leq C, \quad 0 < \sigma \leq 1. \tag{II.42}$$

Assuming that $\Phi(x, t) = \Phi^1(x)\Phi^2(t)$, Φ_1 and Φ_2 supported in the unit ball of
$\mathbb{R}^m$ and $\mathbb{R}^n$ respectively, we are led to consider the convolution $\Phi_\varepsilon^1 \overset{(x)}{*} a \overset{(x)}{*} F_\sigma$.
In order to simplify the notation we simply write $\Phi_\varepsilon^1 * a * F_\sigma$, letting t play
the role of a parameter. Let us assume first that the side r of the cube Q
is ≥ 1. Since Φ^1 is supported in the unit ball, $\Phi_\varepsilon^1 * a \doteq a^\varepsilon$, $0 < \varepsilon \leq 1$, is
supported in Q_1^*. Therefore, if $x \notin Q_1^{**}$, letting x_0 be the center of Q_1 and
$C_L = \sup_{x \in \mathbb{R}^n} |x|^L F(x)$, we have

$$|(\Phi_\varepsilon^1 * a * F_\sigma)(x, t)| \leq \chi_{Q_2}(t) \left| \int a^\varepsilon(y, t) F_\sigma(x - y) \, dy \right|$$

$$\leq CC_L \chi_{Q_2}(t) |Q|^{-1/p} |Q_1^*| \sigma^{-m} \left[\frac{|x - x_0|}{\sigma} \right]^{-L}$$

where we have used that $|x - x_0| \sim |x - y|$ for $y \in Q_1^*$ and $x \notin Q_1^{**}$. Since
$|Q_1^*| = (2r)^m \leq (2|x - x_0|)^m$ and $\sigma^{L-m} \leq 1$ if we take $L > m$, we obtain for a
large integer $d = L - m$

$$|(\Phi_\varepsilon^1 * a * F_\sigma)(x, t)| \leq C\chi_{Q_2}(t) |Q|^{-1/p} |x - x_0|^{-d}.$$

Convolving with $\Phi_\varepsilon^2(t)$ gives, for $x \notin Q_1^{**}$ and $t \in \mathbb{R}^n$,

$$|\Phi_\varepsilon * (F_\sigma \overset{(x)}{*} a)(x, t)| \leq C|Q|^{-1/p}|x - x_0|^{-d}(\Phi_\varepsilon^2 \overset{(t)}{*} \chi_{Q_2})(t)$$
$$\leq C|Q|^{-1/p}|x - x_0|^{-d}\chi_{Q_2^*}(t).$$

Choose $d = m + 1$. If we take the supremum in $0 < \varepsilon \leq 1$, raise both sides to the pth power and integrate in $(\mathbb{R}^m \backslash Q_1^{**}) \times \mathbb{R}^n$, we obtain (II.42), under the assumption $r \geq 1$.

Let us assume now that $r < 1$, so $a(z)$ satisfies the moment conditions (ii). It is clear that these properties are inherited by $a^\varepsilon(z)$, i.e., $\int z^\alpha a^\varepsilon(z)\,dz = 0$, $|\alpha| \leq N(1/p - 1)$. We start by writing $F(x)$ as a convergent series in $\mathcal{S}(\mathbb{R}^m)$, $F(x) = \sum_k F^{(k)}(x)$ with $F^{(0)}$ supported in the unit ball $B = B(0, 1)$ and each $F^{(k)}$ supported in some ball of radius 1. We aim at proving (II.42) with $F^{(k)}$ in the place of F. Using the vanishing of the moments of a

$$(a^\varepsilon \overset{(x)}{*} F_\sigma^{(k)})(x, t) = \chi_{Q_2^*}(t) \int a(y, t)\, G_{\sigma,\varepsilon}^{(k)}(x - y)\,dy$$

$$= \int a(y, t)[G_{\sigma,\varepsilon}^{(k)}(x - y) - q_{x,\varepsilon}(y)]\,dy \qquad (\text{II.43})$$

where $G_{\sigma,\varepsilon}^{(k)} = \Phi_\varepsilon^1 * F_\sigma^{(k)}$ and $q_{x,\varepsilon}(y)$ is the Taylor polynomial of degree d of the function $y \to G_{\sigma,\varepsilon}^{(k)}(x - y)$ expanded about x_0 and d is the integral part of $N(1/p - 1)$. The usual estimates for the remainder of the Taylor expansion imply that the integrand in (II.43) is $\leq C|Q|^{-1/p}\sigma^{-(d+1+m)}\,r^{d+1}$. We assume first that $k = 0$ so $F^{(0)}$ is supported in the unit ball. Since $|x - x_0| \leq C|x - y|$ when $y \in Q_1^*$ and $x \notin Q_1^{**}$, $|x - y| \leq \sigma$ on the support of $F_\sigma^{(0)}(x - y)$, and a is supported in the cube Q_1^* of measure $(2r)^m$ it follows that for any $0 < \epsilon \leq 1$ and $0 < \sigma \leq 1$

$$|(a^\varepsilon * F_\sigma^{(0)})(x, t)|^p \leq C_0\chi_{Q_2}(t)\left(\frac{r}{|x - x_0|}\right)^{(d+m+1)p}, \qquad x \notin Q_1^{**},$$

which after integration gives

$$\int_{(\mathbb{R}^m \backslash Q_1^{**}) \times \mathbb{R}^n} \sup_{0 < \varepsilon \leq 1} |\Phi_\varepsilon * (F_\sigma^{(0)} * a)(x, t)|^p\,dx\,dt \leq C_0. \qquad (\text{II.44})$$

On the other hand, the proof of (II.41) shows that

$$\int_{Q_1^{**} \times \mathbb{R}^n} \sup_{0 < \varepsilon \leq 1} |\Phi_\varepsilon * (F_\sigma^{(0)} * a)(x, t)|^p\,dx\,dt \leq C_0,$$

which combined with (II.44) gives

$$\int_{\mathbb{R}^m \times \mathbb{R}^n} \sup_{0 < \varepsilon \leq 1} |\Phi_\varepsilon * (F_\sigma^{(0)} * a)(x, t)|^p\,dx\,dt \leq 2C_0. \qquad (\text{II.45})$$

For other values of k we consider an appropriate translate $\tilde{F}^{(k)}$ of $F^{(k)}$ so that $\tilde{F}^{(k)}$ is supported in $B(0, 1)$. If for any given σ we replace the atom a by a convenient translate $\tilde{a}$, which of course is also an atom, we may write $a^\varepsilon * F_\sigma^{(k)} = \tilde{a}^\varepsilon * \tilde{F}_\sigma^{(k)}$. Reasoning as before we get the analogue of (II.45):

$$\int_{\mathbb{R}^m \times \mathbb{R}^n} \sup_{0 < \varepsilon \leq 1} |\Phi_\varepsilon * (F_\sigma^{(k)} * a)(x, t)|^p \, dx \, dt \leq C_k. \tag{II.46}$$

The proof also shows that there is a continuous seminorm p in $\mathcal{S}$ involving derivatives of order $\leq d + 1$ such that $C_k \leq p(F^{(k)})$ and since the series $F = \sum_k F^{(k)}$ converges absolutely in $\mathcal{S}$ we see that $\sum_k C_k < \infty$. Estimates (II.46) imply (II.41) by subadditivity and the theorem is proved. $\qquad\square$

II.4 Applications

In this section we discuss two typical applications of the Baouendi–Treves approximation formula. The first one deals with extensions of CR functions and the second with uniqueness of solutions of the equation $\mathcal{L}u = 0$ where $\mathcal{L}$ is a locally integrable structure. The principle that governs the first application is conceptually very simple: suppose that we know that a sequence of polynomials $P_\ell(\zeta)$, $\zeta \in \mathbb{C}^m$, converges uniformly in a compact set $K \subset \mathbb{C}^m$, then it converges uniformly in the holomorphic convex hull $\widehat{K}$ of K in $\mathbb{C}^m$. We recall that

$$\widehat{K} = \bigcap_{P \in \mathcal{P}} \{\zeta \in \mathbb{C}^m : \quad |P(\zeta)| \leq \sup_K |P|\},$$

where $\mathcal{P}$ denotes the space of polynomials in m complex variables. Since on a ball that contains K any entire function, that is any holomorphic function defined throughout $\mathbb{C}^m$, can be uniformly approximated by the partial sums of its Taylor series, we also have

$$\widehat{K} = \{\zeta \in \mathbb{C}^m : \quad |f(\zeta)| \leq \sup_K |f| \quad \text{for all entire functions } f\}.$$

Let $u \in C^0(W)$ satisfy $\mathcal{L}u = 0$ on W and let $K = Z(\overline{V})$ where $\overline{V} \subset U$ and U, W are the neighborhoods in the statement of Theorem II.1.1. We already noticed that we may write $u = \hat{u} \circ Z$ on $\overline{V}$ where $\hat{u} \in C^0(K)$ because u is constant on the fibers of Z in U. Now, we have a function $\widehat{U}(\zeta)$ defined on $\widehat{K}$ by $\widehat{U}(\zeta) = \lim_{\ell \to \infty} P_\ell(\zeta)$, $\zeta \in \widehat{K}$, which clearly extends $\hat{u}$. Depending on the geometry of $Z(\overline{V})$, $\widehat{K}$ may have nonempty interior and on this open set the extension $\widehat{U}$ will be holomorphic because it is the uniform limits of polynomials in ζ. Composition with Z gives the required extension. When u

is not continuous but, say, belongs to L^p, things are technically more involved but essentially the same principle works.

This type of approach may also be seen at work in the following simple example. Consider the operator in $\mathbb{R}^2$

$$L = \frac{\partial}{\partial t} - 3it^2 \frac{\partial}{\partial x} \tag{a}$$

with first integral $Z(x, t) = x + it^3$. Indeed, it is easily verified that $LZ = 0$ and clearly dZ never vanishes. The operator L has real analytic coefficients and is elliptic off the x-axis but is not elliptic at $t = 0$, nevertheless it shares with elliptic vector fields with real analytic coefficients the following regularity property: if u is a C^1 solution of $Lu = 0$, then u is real-analytic ([M]). This is also true for distribution solutions (thus, (a) is analytic hypoelliptic) but to keep matters simple let us restrict ourselves to classical solutions. To prove the claim, it will be enough to prove that u is real-analytic at any point $(x, 0)$ of its domain, since for points (x, t) with $t \neq 0$ this follows from ellipticity. Let us prove, for instance, that u is real-analytic at the origin in case it is defined in a neighborhood of the origin. By Theorem II.1.1 we may find $\delta > 0$ such that for $|x| \leq \delta$ and $|t| \leq \delta$ the uniform limit $u(x, t) = \lim_{\ell \to \infty} P_\ell(x + it^3)$ holds for a certain sequence of polynomials P_ℓ, $\ell \in \mathbb{Z}_+$. This implies that the sequence $P_\ell(z) = P_\ell(x + iy)$ is a Cauchy sequence in the space $C^0(K)$ where $K = [-\delta, \delta] \times [-\delta^3, \delta^3]$. Hence, $\lim_{\ell \to \infty} P_\ell(z) \doteq \widehat{u}(z)$ is a continuous function on K which is a holomorphic function on $(-\delta, \delta) \times (-\delta^3, \delta^3)$ and we have that $u(x, t) = \widehat{u}(x + it^3)$ for $|x|, |t| \leq \delta$. Since $\widehat{u}$ is real-analytic in a neighborhood of the origin and so is $Z(x, t) = x + it^3$, it follows that u is real-analytic in a neighborhood of the origin as we wished to prove.

II.4.1 Extendability of CR functions

Consider the Heisenberg group

$$\mathbb{H}^n \simeq \mathbb{C}^n \times \mathbb{R} = \{(z, s) = (z_1, \ldots, z_n, s) : \quad z \in \mathbb{C}^n, \ s \in \mathbb{R}\}$$

with the group law

$$(z, s) \cdot (w, s') = \left(z + w, s + s' + \Im \sum_{j=1}^n z_j \bar{w}_j\right).$$

Then $\mathbb{H}^n$ can be topologically identified with the boundary of the Siegel upper half-space

$$\mathbb{D}^{n+1} = \{(z_1, \ldots, z_{n+1}) \in \mathbb{C}^{n+1} : \quad \Im z_{n+1} > \sum_{j=1}^n |z_j|^2\}$$

via the map

$$Z : (z_1, \ldots, z_n, t) \longmapsto (z_1, \ldots, z_n, t + i|z|^2). \qquad (\text{II}.47)$$

This identification endows $\mathbb{H}^n$ with the CR structure transported from the boundary $\partial\mathbb{D}^{n+1}$ which possesses a standard CR structure as a smooth boundary of an open subset of $\mathbb{C}^{n+1}$ induced by the anti-holomorphic differentiations. A function $f \in C^1(\mathbb{H}^n)$ (or more generally a distribution) is a CR function (resp. CR distribution) if and only if it satisfies the overdetermined first-order linear system of equations

$$\tilde{L}_j f = \frac{\partial f}{\partial \bar{z}_j} - i z_j \frac{\partial f}{\partial s} = 0, \quad j = 1, \ldots, n. \qquad (\text{II}.48)$$

Observe that the vector fields $\tilde{L}_j$ are left-invariant under the action of $\mathbb{H}^n$. The components of the map (II.47), that is, the functions $Z_1(z, s) = z_1, \ldots, Z_n(z, s) = z_n$, $W(z, s) = s + i|z|^2$ satisfy (II.48) and it is of interest to determine which solutions of (II.48) may be expressed as the composition of the map (II.47) with a holomorphic function defined in $\mathbb{D}^{n+1}$ and having a suitable trace in $\partial\mathbb{D}^{n+1}$. It is known ([**FS**]) that a function $f \in C^1(\mathbb{H}^n)$ is a CR function if and only if there exists a function $F \in C^1(\overline{\mathbb{D}}^{n+1})$ which is holomorphic in $\mathbb{D}^{n+1}$ and whose composition with the map (II.47) is equal to f. There is also a similar local result due to Hans Lewy ([**L1**]) which holds in the general set-up of CR structures of hypersurface type with nondegenerate Levi form which we now describe. Consider a hypersurface Ω in $\mathbb{C}^{n+1}$ with the CR structure $\mathcal{L}$ induced by the standard anti-holomorphic differentiations of $\mathbb{C}^{n+1}$. We may assume that, in a suitable neighborhood of the origin in $\mathbb{C}^{n+1}$, Ω is given by

$$t = \Phi(z_1, z_2, \ldots, z_n, s), \qquad z_i \in \mathbb{C}, \quad s \in \mathbb{R}, \quad i = 1, \ldots, n$$

where

$$\Phi(z, s) = \sum_{i,j=1}^{n} \frac{\partial^2 \Phi}{\partial z_j \partial \bar{z}_k}(0, 0) z_j \bar{z}_k + O(|z|^3 + |s||z| + s^2).$$

Then $\mathcal{L}$ is orthogonal to the differential of the functions

$$Z_j(z, s) = z_j, \quad j = 1, \ldots, n, \quad z = (z_1, \ldots, z_n)$$
$$W(z, s) = s + i\Phi(z, s),$$

and generated by the vector fields

$$L_j = \frac{\partial}{\partial \bar{z}_j} - i\Phi_{\bar{z}_j}(z, s)[1 + i\Phi_s(z, s)]^{-1}\frac{\partial}{\partial s}, \quad j = 1, \ldots, n. \qquad (\text{II}.49)$$

Using z_j and $w = s + it$ as a system of coordinates, the Levi form at $(0, ds)$ is represented by the matrix

$$\frac{\partial^2 \Phi}{\partial z_k \partial \bar{z}_j}(0, 0).$$

The aforementioned result of Hans Lewy asserts that, when the Levi form of $\mathcal{L}$ at $(0, ds)$ has a positive eigenvalue, there exists a neighborhood V of the origin in $\mathbb{C}^{n+1}$ such that every continuous function satisfying

$$L_j u = 0 \qquad\qquad (\text{II}.50)$$

in $Z^{-1}(\Omega \cap V)$, $Z = (z_1, ..., z_n, s + i\Phi(z, s))$, can written as

$$u = F \circ Z$$

where F is a continuous function defined in $\{(z, w) \in V, \, t \geq \Phi(z, s)\}$ and holomorphic in $V^+ = \{(z, w) \in V, \, t > \Phi(z, s)\}$.

We now return to the Heisenberg group $\mathbb{H}^n$ and recall that the (global) holomorphic Hardy space $\mathcal{H}^p(\mathbb{D}^{n+1})$, $0 < p < \infty$, is the set of functions F, holomorphic in $\mathbb{D}^{n+1}$, which satisfy

$$\sup_{0<\rho<\infty} \int_{\mathbb{C}\times\mathbb{R}} |F(z, s + i(|z|^2 + \rho))|^p \, dm(z) \, ds < \infty.$$

Here dm is the Lebesgue measure on $\mathbb{C}^n$, ds is the Lebesgue measure on the real line and it turns out that the pullback of the product measure $dm \times ds$ is the Haar measure on $\mathbb{H}^n$. If $F \in \mathcal{H}^p(\mathbb{D}^{n+1})$, F has a pointwise boundary value f at almost every point of $\partial\mathbb{D}^{n+1}$ given by the normal limit which exists also in L^p norm and, of course, f is a CR distribution. We now prove an analogue of Lewy's local extension result within the framework of local L^p spaces, $1 \leq p < \infty$.

THEOREM II.4.1. *Let Ω be a smooth hypersurface of $\mathbb{C}^{n+1}$ passing through the origin and assume that the Levi form has a nonzero eigenvalue. Then, for any $1 \leq p < \infty$ and $f \in L^p_{loc}(\Omega)$ which is a CR distribution in a neighborhood of the origin, there exists an open set $V \ni 0$ of $\mathbb{C}^{n+1}$ and a holomorphic function F in $L^p(V^+)$ (V^+ denotes the portion of V lying on the 'convex' side of Ω) such that f is the trace of F.*

PROOF. In view of the hypothesis we may assume V^+ is given by $t = \Im z_{n+1} > \Phi(z, s)$ with

$$\Phi(z, s) = |z_1|^2 + \sum_{j=2}^{n} \epsilon_j |z_j|^2 + O(|z|^3 + |s||z| + s^2) \qquad (\text{II}.51)$$

where each ϵ_j may assume the values $+1$, -1, or 0. We will assume initially that the remainder terms vanish identically because the proof is very simple in this case. Hence, we assume that

$$\Phi(z, s) = \Phi(z) = |z_1|^2 + \sum_{j=2}^{n} \epsilon_j |z_j|^2. \qquad \text{(model case)}$$

Since f is a CR function, it follows that $f \circ Z$ satisfies the overdetermined system (II.50) where the vector fields L_j are given by (II.49). By Theorem II.3.1 there is a sequence of polynomials $P_\ell(Z)$, $Z = (z_1, \ldots, z_n, s + i\Phi(z, s))$ that converges to $f \circ Z$ in L^p norm in a neighborhood of the origin in $\mathbb{C}_z^n \times \mathbb{R}_s$. We may assume that the closure of the Cartesian product of the polydisk $\Delta(0, 2\sqrt{a})$ of radius $0 < a \leq 1$ times the interval $(-a, a)$ is contained in that neighborhood. Let us write $z' = (z_2, \ldots, z_n)$. Then, for each z' and t fixed, the set

$$\{z_1 : \quad (z_1, z', s + it) \in V^+\}$$

is a disk centered at the origin of radius $R(z', t) = (t - \sum_{j=2}^{n} \epsilon_j |z_j|^2)^{1/2}$ if $(t - \sum_{j=2}^{n} \epsilon_j |z_j|^2) \geq 0$ and empty if the latter quantity is negative. We will denote this (possibly empty) disk by $D(z', t)$. Given an entire function u defined on $\mathbb{C}^{n+1}$ (actually we will only use that u is harmonic in the first variable), we wish to estimate the L^p norm of u on

$$V_a^+ = \{(z_1, z', s + it) \in V^+ : \quad |z_j| \leq a, \ j = 2, \ldots, n, \ |s|, |t| \leq a\}$$

in terms of the L^p norm of the restriction of u to the boundary of V^+. As the disks $D(z', r)$ sweep V^+, their boundaries sweep the boundary of V^+, which suggests the use of Poisson's formula. A change to polar coordinates $re^{i\theta}$ in the variable (x_1, y_1) allows us to express the integral

$$I = \int_{-a}^{a} ds \int_{-a}^{a} dt \int_{\Delta'(0,a)} dx' dy' \int_{D(z',t)} |u(x_1 + iy_1, z', s, t)|^p dx_1 dy_1$$

as

$$I = \int_{-a}^{a} ds \int_{-a}^{a} dt \int_{\Delta'(0,a)} dx' dy' \int_{0}^{2\pi} d\theta \int_{0}^{R(z',t)} |u(re^{i\theta}, z', s, t)|^p r dr.$$

It is a well-known consequence of Poisson's formula and Young's inequality for convolution that

$$\int_{0}^{R(z',t)} \int_{0}^{2\pi} |u(re^{i\theta}, z', s, t)|^p d\theta \, r dr \leq \frac{R(z', t)^2}{2} \int_{0}^{2\pi} |u(R(z', t)e^{i\theta}, z', s, t)|^p d\theta.$$

A more geometric way of writing this inequality for any disk D is

$$\int_{D} |u|^p dA \leq \frac{\text{diam}(D)}{4} \int_{\partial D} |u|^p d\sigma \qquad \text{(II.52)}$$

where dA is the element of area and dσ indicates arc length. Hence,

$$I \leq \int_{-a}^{a} ds \int_{\Delta'(0,a)} dx' \, dy' \int_{\alpha(z')}^{a} dt \, \frac{R(z',t)^2}{2} \int_{0}^{2\pi} |u(R(z',t)e^{i\theta}, z', s, t)|^p \, d\theta,$$

where, for a given z', $\alpha(z')$ indicates the value of t below which the disk $D(z', t)$ becomes empty (if this ever happens) or $-a$, whichever is larger. Now the substitution $\tau = R(z', t)$ in the integral with respect to t (so that $t = \Phi(\tau, z')$) yields, assuming a is sufficiently small,

$$I \leq \int_{-a}^{a} ds \int_{\Delta'(0,a)} dx' \, dy' \int_{-2\sqrt{a}}^{2\sqrt{a}} |\tau|^3 d\tau \int_{0}^{2\pi} |u(\tau e^{i\theta}, z', s, \Phi(\tau, z'))|^p \, d\theta$$

$$\leq \int_{-a}^{a} ds \int_{\Delta'(0,a)} dx' \, dy' \int_{-2\sqrt{a}}^{2\sqrt{a}} |\tau| d\tau \int_{0}^{2\pi} |u(\tau e^{i\theta}, z', s, \Phi(\tau, z'))|^p \, d\theta$$

$$\leq 2 \int_{\Delta(0,2\sqrt{a}) \times (-a,a)} |u \circ Z|^p \, dx dy ds.$$

Thus, we have proved that

$$\int_{V_a^+} |u|^p \, dx dy ds dt \leq 2 \int_{\Delta(0,2\sqrt{a}) \times (-a,a)} |(u \circ Z)(z, s)|^p \, dx dy ds \qquad \text{(II.53)}$$

and applying this to $u = P_\ell - P_{\ell'}$ we conclude that the sequence P_ℓ converges in $L^p(V_a^+)$ to a holomorphic function F that has a trace $F/\partial V_a^+$ such that $F/\partial V_a^+ \circ Z = f \circ Z$ and this implies that $F/\partial V_a^+ = f$, as we wished to prove (it follows from Cauchy's formula that L^p-convergence implies local uniform convergence). To deal with a general Φ given by (II.51) we may reason exactly in the same way, except that now the domains of $\mathbb{C}$

$$\tilde{D}(z', s, t) = \{z_1 : \quad (z_1, z', s + it) \in V^+\}$$

will no longer be round disks centered at the origin. However, they are simply connected and may be regarded as smooth perturbations of a disk $D(z', t)$ of radius $R(z', t)$ which can be mapped by a Riemann map $z_1 \mapsto \Psi(z_1; z', s, t)$ onto $D(z', t)$. Thus, we will be able to reason as in the proof of (II.52) as soon as we prove the following substitute for (II.52):

$$\int_{\tilde{D}(z',s,t)} |u|^p \, dA \leq C \operatorname{diam}(\tilde{D}(z', s, t)) \int_{\partial \tilde{D}(z',s,t)} |u|^p \, d\sigma$$

where $C > 0$ is independent of (z', s, t) in a neighborhood of the origin and u is any harmonic function defined in $\tilde{D}(z', s, t)$ and continuous in its closure. To simplify the notation we omit any reference to the variables (z', s) that play the role of parameters and write $z = x + iy$ instead of $z_1 = x_1 + y_1$. Thus, we are led to consider the class $\mathcal{F}_\epsilon$ of smooth functions $\phi(x, y)$ in $\mathbb{R}^2$ whose Taylor series at the origin is $\phi(x, y) \sim a + bx + cy + x^2 + y^2 + O(|z|^3)$, when

$z \to 0$, where $|a| + |b| + |c| < \epsilon$ and such that $|D^\alpha \phi(x, y)| \le C_\alpha$ (here $\epsilon > 0$ is a conveniently chosen small number and (C_α) is a given fixed sequence of positive constants). We will need to study the sublevel sets in a fixed small neighborhood of the origin,

$$\tilde{D}(t) = \{z = x + iy : \ |z| < r, \ \phi(x, y) < t\},$$

for an arbitrary $\phi \in \mathcal{F}_\epsilon$. Observe that any $\phi \in \mathcal{F}_\epsilon$ has a small local minimum m at a point $z_0 = (x_0, y_0)$ located close to the origin for small ϵ. It follows that

$$m + 2^{-1}|z - z_0|^2 \le \phi(x, y) \le m + 2|z - z_0|^2,$$

in a neighborhood of the origin and thus

$$D(z_0, \sqrt{t/2}) \subset \tilde{D}(m + t) \subset D(z_0, \sqrt{2t}).$$

We see that $\tilde{D}(m + t)$ is empty for $t \le 0$ and contained between concentric disks of radius comparable to $\sqrt{t}$ if t is positive and small. Furthermore, the implicit function theorem shows that, in the latter case, $\tilde{D}(m + t)$ has a smooth boundary made up of a simple closed curve contained in the annulus $t/2 < |z - z_0|^2 < 2t$.

LEMMA II.4.2. *There exist* $t_0, r_0 > 0$ *such that for all* $0 < t \le t_0$ *and* $\phi \in \mathcal{F}_\epsilon$, $\tilde{D}(m + t)$ *is a relatively compact simply connected open subset of the disk* $D(0, r_0)$. *Furthermore, there exists* $C > 0$ *such that for every harmonic function* u *defined in a neighborhood of the closure of* $\tilde{D}(m + t)$ *and any* $1 \le p < \infty$, *the following a priori inequality holds:*

$$\int_{\tilde{D}(m+t)} |u|^p \, dA \le C \operatorname{diam}(\tilde{D}(m+t)) \int_{\partial \tilde{D}(m+t)} |u|^p \, d\sigma. \tag{II.54}$$

PROOF. After a translation, we may assume that $z_0 = 0$. For small $t > 0$, the level curve $\phi(x, y) = m + t$, which is implicitly given in polar coordinates by $r^2(A(\theta) + rB(r, \theta)) = t$ where $A(\theta) = \alpha \cos^2 \theta + 2\beta \sin \theta \cos \theta + \gamma \sin^2 \theta$ and all derivatives of B with respect to x and y are bounded, may also be explicitly expressed by $r = r(\theta, t)$. Observe that if ϵ is small, α and γ are close to 1 and β is close to zero. Implicit differentiation shows that

$$r' = \frac{\partial r}{\partial \theta} = -\frac{rA_\theta + r^2 B_\theta}{2A + 3rB + r^2 B_r} = O(\sqrt{t}), \quad t \to 0.$$

Differentiating further the expression above we conclude that the higher-order derivatives $r^{(n)}$, $n = 1, 2, \ldots$, are also $O(\sqrt{t})$ as $t \to 0$. Consider a dilation of $\tilde{D}(m+t)$, $\mathcal{D}_t = (1/\sqrt{t})\tilde{D}(m+t)$, whose boundary is given by $R_t(\theta) = r(\theta, t)/\sqrt{t} = A^{-1/2}(\theta) + O(\sqrt{t})$. Observe that we also have

$$d^n R_t/d\theta^n = d^n A^{-1/2}/d\theta^n + O(\sqrt{t}) \quad \text{for } n \geq 1.$$

Since (II.54) is invariant under dilations of the domain, it will be enough to prove it for the dilate $\mathcal{D}_t$ that converges in C^∞ to the domain $\mathcal{D}_0$ with equation $R < A^{-1/2}(\theta)$ as $t \to 0$. To do so it is enough to show that, for small t, the derivative F_t' of the Riemann map F_t from $\mathcal{D}_t$ to the unit disk satisfies $1/C \leq |F_t'| \leq C$. Indeed, if u is harmonic in $\mathcal{D}_t$ and continuous up to the boundary, so will be $v = u \circ F_t^{-1}$ on the unit disk, and starting from (II.52) applied to v, the change of variables $w = F_t(z)$ will give

$$\int_{\mathcal{D}_t} |u|^p \, dA \leq C \int_{\partial \mathcal{D}_t} |u|^p \, d\sigma. \tag{II.55}$$

Notice that if we introduce the factor $\operatorname{diam}(\mathcal{D}_t)$ on the right-hand side of (II.55) the inequality remains valid because $2/\sqrt{2} \leq \operatorname{diam}(\mathcal{D}_t) \leq 2\sqrt{2}$. Hence, the proof of (II.54) will be finished as soon as we prove

LEMMA II.4.3. *There exist $t_0 > 0$ and $C > 0$ such that for $0 \leq t \leq t_0$ the Riemann map F_t from $\mathcal{D}_t$ to the unit disk D satisfies $1/C \leq |F_t'| \leq C$.*

PROOF. Let u be the solution of the Dirichlet problem

$$\begin{cases} \Delta u = \dfrac{\partial^2 u}{\partial x^2} + \dfrac{\partial^2 u}{\partial y^2} = 0, & \text{on } \mathcal{D}_t, \\[2mm] u|_{\partial \mathcal{D}_t} = u(R_t(\theta)e^{i\theta}) = \log(R_t(\theta)), & 0 \leq \theta \leq 2\pi. \end{cases} \tag{II.56}$$

Let v be the harmonic conjugate of u in $\mathcal{D}_t$ (say, normalized by $v(0) = 0$) and set $f_t = u + iv$. Then a Riemann map from $\mathcal{D}_t$ onto the unit disk $D = D(0, 1)$ is (*cf.* the proof of theorem 3.3 in [**F**])

$$F_t(z) = z e^{-f_t(z)}.$$

Thus, $F_t' = e^{-f_t(z)}(1 - z f_t'(z))$ and $|F_t'| = e^{-u(z)}|1 - z f_t'(z)|$ which implies, by the maximum principle, that

$$C^{-1} \inf_{\mathcal{D}_t} |1 - z f_t'(z)| \leq |F_t'| \leq C \sup_{\mathcal{D}_t} |1 - z f_t'(z)|,$$

with $C > 0$ independent of t, for small t. Indeed, $\log(R_t(\theta))$ converges to $-(1/2)\log A(\theta) = -(1/2)\log[\alpha\cos^2\theta + 2\beta\sin\theta\cos\theta + \gamma\sin^2\theta]$ as $t \to 0$ and the domain $\mathcal{D}_0$ is close to the unit disk for small ϵ. Therefore, to conclude the proof, we need only show that $|f_t'| \leq 1/2$ for small t and ϵ. Since $f_t' = u_x - iu_y$ we must show that the derivatives of u are uniformly small in $\mathcal{D}_t$. The domains $\mathcal{D}_t$ change with t and the analysis may be simplified by mapping $\mathcal{D}_t \cup \partial \mathcal{D}_t$

onto the fixed domain $\mathcal{D}_0 \cup \partial\mathcal{D}_o$ by a diffeomorphism (of manifolds with boundaries) Φ_t such that all derivatives of Φ_t and Φ_t^{-1} are bounded uniformly with bounds that do not depend on $t \in [0, t_0]$. Such Φ_t are easily constructed. Then, $U_t = u \circ \Phi_t^{-1}$ is the solution of a Dirichlet problem on $\mathcal{D}_0$ with respect to an elliptic second-order differential operator $P_t(x, y, D_x, D_y)U_t = 0$ and in particular satisfies the boundary condition $U_t|_{\partial\mathcal{D}_0} = (\log|\Phi_t^{-1}|)|_{\partial\mathcal{D}_0}$. The coefficients of $P_t(x, y, D_x, D_y)$ depend continuously on $t \in [0, t_0]$ as well as their derivatives. The usual regularity theory of smooth elliptic boundary value problems implies that there exists a positive integer $N > 0$ with the following property: given $\rho > 0$ there exists $\delta > 0$ such that any function U that satisfies the equation $P_t(x, y, D_x.D_y)U = 0$ for some $t \in [0, t_0]$, and has in addition all tangential derivatives at the boundary bounded up to order N by δ, will satisfy the estimate $|\nabla U(x, y)| \leq \rho$. Since $\mathcal{D}_0$ is close to the unit disk for small ϵ, it follows that $(\log|\Phi_t^{-1}|)|_{\partial\mathcal{D}_0}$ will have small tangential derivatives up to any fixed order, and thus $U_t = u \circ \Phi_t^{-1}$ will have uniformly small gradient. The chain rule now implies that $u = U \circ \Phi_t$ has small gradient, uniformly in $(x, y) \in \mathcal{D}_t$ and $t \in [0, t_0)$, proving that $|f_t'| = |\nabla u| \leq 1/2$ for small t and ϵ. $\square$

Since Lemma II.4.3 implies (II.54), Lemma II.4.2 is proved. $\square$

As we pointed out, the control of the L^p norm of u on the sublevel sets $\tilde{D}(m + t)$ in terms of the L^p norm of u on their boundaries $\partial\tilde{D}(m + t)$ given by Lemma II.4.2 is all that is needed in order to extend the proof carried out in the model case to the general case. The proof of Theorem II.4.1 is then complete. $\square$

REMARK II.4.4. Stronger results than Theorem II.4.1 are known. In fact, it is possible to sweep V^+ with suitable translates of Ω so that the L^p norm of the restriction of F to those translates is uniformly bounded ([**Ro**]). Theorem II.4.1 then follows from an application of Fubini's theorem.

II.4.2 Propagation of zeros of homogeneous solutions

Given a locally integrable structure $\mathcal{L}$ in a manifold Ω, and a solution u of $\mathcal{L}u = 0$ a natural question is: what additional conditions must the solution u satisfy in order to conclude that u vanishes identically? The local version of the question is: given $p \in \Omega$, and a neighborhood V of p, what conditions guarantee that there exists a neighborhood $p \in U \subset V$ on which u vanishes identically? A natural additional condition would be to require that u vanish in some subset of V. In a small neighborhood of p, $\mathcal{L}u = 0$ may be expressed as an overdetermined system of equations (II.5). To get some insight, let us consider the simplest case of a single vector field $L = A\partial_x + B\partial_y$, $|A| + |B| > 0$, defined in an open set $\Omega \subset \mathbb{R}^2$. Since the constant functions $u = C$ always satisfy $Lu = 0$ it is apparent that some additional condition is needed; for instance, requiring that u vanishes at p certainly rules out the nonzero constants, but for most vector fields this is not enough (there exist, however, vector fields whose only homogeneous solutions are the constant functions [**N1**]). If $L = \bar{\partial}$ is the Cauchy–Riemann operator of Example II.1.2, one could require that u vanishes at p to *infinite order* which would imply that u vanishes throughout any connected open set U that contains p. However, this condition will not be enough for the vector field $L = \partial_x$ since a smooth function $u(y)$ independent of x could vanish to infinite order at p and yet not vanish identically in any neighborhood U of p. A better condition for $L = \partial_x$ would be then to require that u vanishes on the curve $\Sigma = \{(p_1, y)\}$, $p = (p_1, p_2)$. So requiring that u vanishes on Σ, that is a submanifold of Ω of codimension one, works for both $\bar{\partial}$ and ∂_x but it does not work for ∂_y (show this). The main point is that ∂_y is tangent to $\{(p_1, y)\}$ while the two previous vector fields are transversal to any vertical line (for a complex vector field $L = X + iY$ with real part X and imaginary part Y, L transversal to Σ means that at least one of the two vectors X and Y is transversal). This suggests that we should look at the case where u vanishes on a submanifold Σ of codimension one to which L is transversal. Note that if the structure $\mathcal{L}$ of rank $n = 1$ generated by L is locally integrable, the corank m of $L^\perp$ must be one, so we have $N = 2$, $m = 1$, and $n = 1$. Elaboration of this type of consideration for the case of a locally integrable structure $\mathcal{L}$ of rank n and corank m defined in a manifold of dimension $N = n + m$ leads to the following definition:

DEFINITION II.4.5. *Let $\Sigma \subset \Omega$ be an embedded submanifold. We say that Σ is maximally real with respect to $\mathcal{L}$ if*

(i) *the dimension of Σ is equal to m;*
(ii) *for every $p \in \Sigma$, any nonvanishing section L of $\mathcal{L}$ defined in a neighborhood of p is transversal to Σ at p.*

If local coordinates $\{x_1, \ldots, x_m, t_1, \ldots, t_n\}$ vanishing at p are chosen according to Corollary I.10.2, then $\mathcal{L}^\perp$ is generated in a neighborhood of $x = 0$, $t = 0$, by $dZ_1, \ldots, dZ_m$, where $Z_j(x, t) = x_j + i\phi_j(x, t)$, $\phi_j(0, 0) = 0$, $(\partial\phi_j/\partial x_k(0, 0)) = 0$, $1 \le j, k, \le m$, and the vectors $L_1, \ldots, L_n$ become $L_j = \partial/\partial t_j$, $j = 1, \ldots, n$ at the origin. If Σ is maximally real, the vectors $\partial_{t_j}|_0$ are transversal to Σ at the origin, so by the implicit function theorem we may find locally defined functions $\tau_j(x)$ such that $\Sigma = \{(x, \tau(x)\}$, where $\tau(x) = (\tau_1(x), \ldots, \tau_n(x))$. If we perform the change of coordinates $x' = x$, $t' = t - \tau(x)$ the expression of Z in the new coordinates is $Z'(x', t') = x' + i\phi(x', t' + \tau(x')) = x' + i\phi'(x', t')$ and now Σ is given by $t' = 0$. In other words, if Σ is maximally real, we may always assume that the set of coordinates (x, t) of Corollary I.10.2 are such that not only Z has the form $Z(x, t) = x + i\phi$ with ϕ real, $\phi(0, 0) = 0$, $d_x\phi(0, 0) = 0$ but also that Σ is given locally by $\Sigma = \{(x, 0)\}$. In particular, if u is a distribution solution of $\mathcal{L}u = 0$ we may always consider its restriction to Σ, $u|_\Sigma$, which is just the trace $u(x, 0)$ which we have seen to exist from considerations on the wave front set of u.

THEOREM II.4.6. *Let $\mathcal{L}$ be a locally integrable structure on the manifold Ω and let $\Sigma \subset \Omega$ be an embedded submanifold maximally real with respect to $\mathcal{L}$. If $u \in \mathcal{D}'(\Omega)$ satisfies*

(i) *$\mathcal{L}u = 0$ in Ω;*
(ii) *$u|_\Sigma = 0$;*

then u vanishes identically in a neighborhood V of Σ.

PROOF. It is enough to see that any point $p \in \Sigma$ is contained in a neighborhood U on which u vanishes identically. According to our previous remarks, given $p \in \Sigma$ we may assume that the special coordinates of Corollary I.10.2 that were used to prove Theorem II.1.1 are such that Σ is given by $\Sigma = \{(x, 0)\}$ and $p = (0, 0)$. We may find open sets $0 \in U \subset W$ as in Theorem II.1.1 so that W is contained in the coordinate neighborhood and u is approximated in U by $E_\tau u$ in the sense of $\mathcal{D}'(U)$. However, the formula that defines $E_\tau u$ right after (II.5) shows that $E_\tau u(x, t) = 0$ because $u(x, 0)$ vanishes on $\Sigma \cap W$. Thus, $u \equiv 0$ on U. $\square$

COROLLARY II.4.7. *Let $\mathcal{L}$ be a locally integrable structure on a manifold Ω and let $u \in \mathcal{D}'(\Omega)$ satisfy $\mathcal{L}u = 0$ in Ω. Let L be a local section of $\mathcal{L}$, let $X = \Re L$. Assume that γ is an integral curve of X joining the points p and $q \in \Omega$. Then $p \in \operatorname{supp} u \implies q \in \operatorname{supp} u$.*

PROOF. If X vanishes at p then $p = q$ and there is nothing to prove. We may assume that $\gamma : [0, 1] \to \Omega$ is a nonconstant solution of $\gamma'(s) = X \circ \gamma(s)$, $0 \le s \le 1$, with $\gamma(0) = q$ and $\gamma(1) = p$, so X does not vanish in a neighborhood of γ. Denote by $K = \operatorname{supp} u$ the support of u and let us assume for the sake of a contradiction that $p \in K$ and $q \notin K$. Replacing p by the first point $\gamma(s)$ such that $\gamma(s) \in K$ we may assume that p and q are as close as we wish and all points in γ between q and p are not in K. We may find a local set of generators of $\mathcal{L}$, $L \doteq L_1, L_2, \ldots, L_n$ such that in appropriate coordinates (x, t), $|x| < 1$, $|t| < 2$, that rectify the flow of $X_1 \doteq X$ we have

(i)
$$X_1 = \Re L_1 = \frac{\partial}{\partial t_1}$$

$$X_j = \Re L_j = \frac{\partial}{\partial t_j} + \sum_{k=1}^{m} \lambda_{jk} \frac{\partial}{\partial x_k}, \quad j = 2, \ldots, n$$

and $p = (0, 0)$;

(ii) $\gamma(s) = (s - 1, 0, \ldots, 0)$, $q = \gamma(0) = (-1, 0, \ldots, 0)$;

(iii) for some $a > 0$ the embedded closed m-ball given by $|x| \le a$, $t' = 0$, $t_1 = -1$ does not meet K (here $t' = (t_2, \ldots, t_n)$). Since it is an embedded submanifold with boundary we may denote this m-ball as $\Sigma_0 \cup \partial \Sigma_0$, where Σ_0 is the corresponding open m-ball.

Consider now the one-parameter family of embedded submanifolds Σ_σ (without their boundaries) given by the equations

$$t_1 = \sigma - 1 - \sigma \frac{|x|^2}{a^2}, \quad t_2 = \cdots = t_n = 0, \quad |x| < a, \quad 0 \le \sigma \le 1.$$

Since $\Sigma_0 \cap K = \emptyset$ and $(0, 0) \in \Sigma_1 \cap K$ there is a largest $\sigma_0 \in (0, 1]$ such that $\Sigma_\sigma \cap K = \emptyset$ for $0 \le \sigma < \sigma_0$. Note that the submanifolds Σ_σ are all maximally real with respect to $\mathcal{L}$. Indeed, the vector fields X_j, $1 \le j \le n$, are transversal to any Σ_σ. This is clear for $j \ge 2$ because Σ_σ is contained in the slice $t_2 = \cdots = t_n = 0$ and it is also obvious for $j = 1$ because $(\partial / \partial t_1)$ is never tangent to Σ_σ. Hence, the trace $u|_{\Sigma_\sigma}$ is well-defined and furthermore $u|_{\Sigma_\sigma} = 0$ for $0 < \sigma < \sigma_0$ and, since $\sigma \mapsto u|_{\Sigma_\sigma}$ depends continuously on σ, we conclude that

$$u|_{\Sigma_{\sigma_0}} = 0. \tag{A}$$

We claim that

$$\Sigma_{\sigma_0} \cap K \neq \emptyset. \tag{B}$$

Indeed, since $\operatorname{dist}(\Sigma_{\sigma_0}, K) = 0$, this is certainly true if we replace Σ_{σ_0} by its closure $\overline{\Sigma_{\sigma_0}}$ which amounts to adding to Σ_{σ_0} its boundary points $\partial \Sigma_{\sigma_0}$. But, for any $\sigma \in [0, 1]$, $\partial \Sigma_\sigma$ is given by $|x| = a$, $t_1 = -1$, $t_2 = \cdots = t_n = 0$, so (iii) shows that $\partial \Sigma_\sigma \cap K = \emptyset$. Hence, $\Sigma_{\sigma_0} \cap K = \overline{\Sigma_{\sigma_0}} \cap K \neq \emptyset$. However, applying Theorem II.4.6 to Σ_{σ_0}, (A) implies that u vanishes in a neighborhood of Σ_{σ_0} in $|x| < a$, $|t| < 2$. This contradicts (B). $\qquad \square$

Let Ω be a manifold and consider a collection $D = \{X\}$ of locally defined, smooth, real vector fields X. In Chapter III, the notion of *orbit* of D is defined. Suppose now that $\mathcal{L}$ is a locally integrable structure and we consider the collection $D_{\mathcal{L}} = \{\Re L\}$ of all vector fields that are real parts of local sections of $\mathcal{L}$. In this case the orbits of $D_{\mathcal{L}}$ are simply called the orbits of $\mathcal{L}$. In the language of orbits, Corollary II.4.7 implies that if an orbit of $\mathcal{L}$ intersects the support K of a solution u of the equation $\mathcal{L}u = 0$ it must be entirely contained in K. This is equivalent to saying that K is a union of orbits of $\mathcal{L}$. Thus, Corollary II.4.7 gives an alternative proof of Theorem III.2.1. The proof presented in Chapter III follows in a remarkable simple way—thanks to the use of a criterion of Bony about flow-invariant sets—from a related uniqueness result that we now describe. An embedded submanifold of Ω of codimension 1 will be called a hypersurface. A hypersurface $\Sigma \subset \Omega$ is noncharacteristic with respect to $\mathcal{L}$ at $p \in \Sigma$ if there exists a local section L of $\mathcal{L}$ defined in a neighborhood of p that is transversal to Σ at p (which means, changing L by iL if necessary, that $X = \Re L$ is transversal to Σ at p). Notice that if u is a solution of $\mathcal{L}u = 0$ defined in a neighborhood U of p, the trace $u|_{\Sigma \cap U}$ is defined because u satisfies the equation $Lu = 0$ for any local section of $\mathcal{L}$, so choosing L transversal to Σ we see that the wave front set of u does not contain Σ's conormal directions.

DEFINITION II.4.8. *Let $\mathcal{L}$ be a formally integrable structure in the manifold Ω. We say that $\mathcal{L}$ has the Uniqueness in the Cauchy Problem property for noncharacteristic hypersurfaces if and only if the following holds: for every hypersurface Σ, every point $p \in \Sigma$ such that Σ is noncharacteristic at p and every distribution solution u of $\mathcal{L}u = 0$ defined in a neighborhood U of p,*

$$u|_{U \cap \Sigma} = 0 \implies u \text{ vanishes in a neighborhood of } p.$$

COROLLARY II.4.9. *The Uniqueness in the Cauchy Problem property for noncharacteristic hypersurfaces holds for every locally integrable structure $\mathcal{L}$.*

PROOF. Let Σ be a noncharacteristic hypersurface at p. As usual, we denote by N the dimension of the manifold Ω, by n the rank of $\mathcal{L}$ and set $m = N - n$. In appropriate local coordinates (x, t) we may assume that $\mathcal{L}^{\perp}$ is generated by $dZ_1, \ldots, dZ_m$, $Z = x + i\phi(x, t)$, $\phi(0, 0) = 0$, $d_x\phi(0, 0) = 0$, $p = (0, 0)$. Hence, $\mathcal{L}$ is spanned at $(0, 0)$ by

$$\frac{\partial}{\partial t_1}, \ldots, \frac{\partial}{\partial t_n},$$

and since $\mathcal{L}$ is transversal to Σ at $p = (0, 0)$, the implicit function theorem gives a local representation of Σ as $t_1 = t_1(t', x)$, $t' = (t_2, \ldots, t_n)$, after renumbering the t-coordinates if necessary. Let Σ_1 be given by $t_1 = t_1(0, x)$, $t' = 0$. Then, Σ_1 is a maximally real submanifold contained in Σ that contains

$p = (0, 0)$. Consider now a neighborhood U of $p = (0, 0)$ and $u \in \mathcal{D}'(U)$ such that $\mathcal{L}u = 0$ and $u|_{U \cap \Sigma} = 0$. Since $\Sigma_1 \subset \Sigma$ we also have that $u|_{U \cap \Sigma_1} = 0$ and it follows from Theorem II.4.6 that u vanishes in a neighborhood of p. □

EXAMPLE II.4.10. P. Cohen ([**Co**]) (see also [**Zu**] and the references therein) constructed smooth functions $u(x, y)$ and $a(x, y)$ defined on $\mathbb{R}^2$ such that

(1) $Lu(x, y) = \dfrac{\partial u}{\partial y} + a(x, y) \dfrac{\partial u}{\partial x} = 0;$

(2) $u(x, y) = a(x, y) = 0$ for all $y \le 0;$

(3) $\operatorname{supp} u = \operatorname{supp} a = \{(x, y) : \quad y \ge 0\}.$

Thus, the formally integrable structure $\mathcal{L}$ spanned by the vector field L fails to have the Uniqueness in the Cauchy Problem property for the noncharacteristic curve $\Sigma = \{t = 0\}$ and, by Corollary II.4.9, cannot be locally integrable in any open set that intersects the x-axis. The construction of $a(x, y)$ shows that $a(x, y)$ is real-analytic for $y \ne 0$, so for any point $p = (x, y)$ with $y \ne 0$ we may find a function Z defined in a neighborhood of p such that $LZ = 0$ and $dZ(p) \ne 0$. On the other hand, if Z is a smooth function defined in a neighborhood of $p = (x, 0)$ such that $LZ = 0$ we must have that $dZ(p) = 0$, otherwise $\mathcal{L}$ would be locally integrable in some open set that intersects the x-axis, a contradiction. A nonlocally integrable vector field was first exhibited by Nirenberg ([**N1**]) who used a completely different method to construct a vector field whose only homogeneous solutions are contant.

II.4.3 An extension

In the applications to uniqueness we have seen so far, the 'initial' maximally real manifold $t = 0$ is in the interior of the domain where the solution u of $\mathcal{L}u = 0$ is defined. This is quite convenient because in this case the trace $u(\cdot, t)$ exists and $t \mapsto u(\cdot, t)$ is a continuous function of t valued in the space of distributions. However, in the study of one-sided Cauchy problems or boundary values of solutions, it is desirable to consider the case where the solution is not defined in a neighborhood of the 'initial' manifold. We will say that a set $\Gamma \subset \mathbb{R}^n \setminus \{0\}$ is a cone (or a cone with vertex at the origin to be explicit) if $t \in \Gamma \Longleftrightarrow \rho t \in \Gamma \ \forall \ 0 < \rho < \infty$. A set $\Gamma_T \subset \mathbb{R}^n \setminus \{0\}, 0 < T$, will be called a truncated cone if there exists a cone Γ such that $\Gamma_T = \Gamma \cap \{|t| < T\}$. An open truncated cone is a truncated cone which is an open set. Notice that the origin is in the closure of Γ and Γ_T but it does not belong to them. A cone Γ' is said to be a proper subcone of Γ if $\overline{\Gamma'} \cap \{|x| = 1\}$ is a compact subset of Γ. This is, for instance, the case if Γ and Γ' are circular cones with the same axis and Γ' has a smaller aperture than Γ. If Γ' is a proper subcone of Γ and $T' < T$ we say that $\Gamma'_{T'}$ is a proper truncated subcone of the truncated cone Γ_T. When $n = 1$, a truncated cone is an interval of the form $(0, T)$ or $(-T, 0)$ or the union of both. If $W \subset \mathbb{R}^m$ is an open set and $\Gamma_T \subset \mathbb{R}^n$ is an open truncated cone, the set $W \times \Gamma_T \subset \mathbb{R}^m \times \mathbb{R}^n$ is usually called a wedge with edge W.

Consider a locally integrable structure $\mathcal{L}$ of rank n in an N-manifold and assume that the standard coordinates (x, t) used in the proof of Theorem II.1.1 had been chosen in a neighborhood of the origin. Let $B_x \subset \mathbb{R}^m$, $m = N - n$, be a ball centered at the origin, $\Gamma_T \subset \mathbb{R}^n$ a truncated open cone, and assume that u is a distribution satisfying the system (II.5) in $B_x \times \Gamma_T$. Under this circumstances we can assert that the trace $u|_{B_x \times \{t\}} = T_t u(x) = u(x, t)$ is defined and depends smoothly on $t \in \Gamma_T$ as a map valued in $\mathcal{D}'(B_x)$, but $u(x, 0)$ might not be defined. On the other hand, we may *assume* that $\lim_{t \to 0} T_t u \doteq bu$ exists in $\mathcal{D}'(B_x)$ as $t \to 0$.

If $n = N - m = 1$, $B_x \times \{0\}$ divides $\Omega = B_x \times (-T, T)$ into two components $\Omega^+ = \{(x, t) \in \Omega : t > 0\}$ and $\Omega^- = \{(x, t) \in \Omega : t < 0\}$ and in this case we consider distributions u that satisfy the system (II.5) in Ω^+ and such that $\lim_{t \searrow 0} T_t u = bu$ exists. In other words, we assume that $u \in C^0(\Gamma_T \cup \{0\}, \mathcal{D}'(B_x))$ (resp. $u \in C^0([0, T), \mathcal{D}'(B_x))$ for $n = 1$). We see that $E_\tau u$ can still be defined by

$$E_\tau u(x, t) = (\tau/\pi)^{m/2} \int_{\mathbb{R}^m} e^{-\tau[Z(x,t)-Z(x',0)]^2} u(x', 0) h(x') \det Z_x(x', 0) \, dx'$$

as soon as we interpret $u(x', 0)$ as $bu(x')$. For a given $t \in \Gamma_T$ and $0 < \epsilon < 1$ consider

$$R_\tau^\epsilon u(x, t) = G_\tau u(x, t) - E_\tau^\epsilon u(x, t)$$

where $E_\tau^\epsilon u$ is given by

$$E_\tau^\epsilon u(x, t) = (\tau/\pi)^{m/2} \int_{\mathbb{R}^m} e^{-\tau[Z(x,t)-Z(x',0)]^2} u(x', \epsilon t) h(x') \det Z_x(x', 0) \, dx'$$

and

$$G_\tau u(x, t) = (\tau/\pi)^{m/2} \int_{\mathbb{R}^m} e^{-\tau[Z(x,t)-Z(x',t)]^2} u(x', t) h(x') \det Z_x(x', t) \, dx'.$$

As in the proof of Theorem II.1.1, the remainder $R_\tau^\epsilon u$ is given by

$$R_\tau^\epsilon u(x, t) = \int_{[\epsilon t, t]} \sum_{j=1}^m r_j(x, t, t', \tau) \, dt'_j,$$

where

$$r_j(x, t, t', \tau) = (\tau/\pi)^{m/2} \int_{\mathbb{R}^m} e^{-\tau[Z(x,t)-Z(x',t')]^2} u(x', t') L_j h(x') \det Z_x(x', t') \, dx'.$$

Letting $\epsilon \to 0$ we obtain

$$R_\tau u = G_\tau u - E_\tau u,$$

with R_τ given by

$$R_\tau u(x, t) = \int_{[0,t]} \sum_{j=1}^{m} r_j(x, t, t', \tau) \, dt'_j,$$

$$r_j(x, t, t', \tau) = (\tau/\pi)^{m/2} \int_{\mathbb{R}^m} e^{-\tau[Z(x,t)-Z(x',t')]^2} u(x', t') L_j h(x') \det Z_x(x', t') \, dx'.$$

The proof of Theorem II.1.1 now shows that there is a ball $B'_x = B'_x(0, \delta)$ and proper subcone $\Gamma'_\rho \subset \Gamma_T$ such that $R_\tau u \to 0$ uniformly in $B'_x \times \Gamma'_\rho$ as $\tau \to \infty$. Indeed, we can find a fixed k such that $v_j(x, t) = (1 - \Delta_x)^{-k}[(L_j h(x)) u(x, t) \det Z_x(x, t))]$ is continuous in $B_x \times \Gamma'_\rho$, since the distributions $x \to L_j h(x) u(x, t) \det Z_x(x, t)$ lie in a bounded set of some Sobolev space when t ranges over a compact subset of $\Gamma_T \cup \{0\}$ because $u \in C^0(\Gamma_T \cup \{0\}, \mathcal{D}'(B_x))$. Since the continuity of $\Gamma_T \cup \{0\} \ni t \to T_t u(x) \in \mathcal{D}'(B_x)$ implies the continuity of $\Gamma_T \cup \{0\} \ni t \to D_x^\alpha T_t u(x) = T_t D_x^\alpha u(x) \in \mathcal{D}'(B_x)$ and equation (II.5) allows us to express the derivatives of u with respect to t as a linear combination with smooth coefficients of derivatives of u with respect to x for $t \neq 0$, we conclude that actually $u \in C^\infty(\Gamma_T \cup \{0\}, \mathcal{D}'(B_x))$. The derivatives of $R_\tau u$ can be estimated in the same fashion and we obtain

COROLLARY II.4.11. *Let $u \in C^0(\Gamma_T \cup \{0\}, \mathcal{D}'(B_x))$ (resp. $u \in C^0([0, T), \mathcal{D}'(B_x))$ for $n = 1$) be a distribution satisfying the system (II.5) in $\Omega = B_x \times \Gamma_T$ (resp. in $\Omega^+ = B_x \times (0, T)$ for $n = 1$). There exist $\delta > 0$, and a proper subcone $\Gamma'_\rho \subset \Gamma_T$ (resp. a number $\rho > 0$ for $n = 1$) such that for all multi-indexes $\alpha \in \mathbb{Z}_+^m$ and $\beta \in \mathbb{Z}_+^n$*

$$D_x^\alpha D_t^\beta R_\tau u(x, t) \longrightarrow 0 \quad \text{uniformly on } B_x(0, \delta) \times (\Gamma_\rho \cup \{0\})$$

(resp. on $B_x(0, \delta) \times [0, \rho)$ for $n = 1$).

Corollary II.4.11 reduces the study of the approximation of u by E_τ to the problem of approximating u by $G_\tau u$. As an illustration, we sketch the proof of a version of the approximation for wedges. Consider a wedge $W = B_x \times \Gamma_T$—where $B_x \subset \mathbb{R}^m$ is a ball centered at the origin and $\Gamma_T \subset \mathbb{R}^n$ is an open truncated cone—and a locally integrable structure $\mathcal{L}$ with first integrals $Z_1 = x_1 + i\phi_1(x, t), \ldots, Z_m = x_m + i\phi_m(x, t)$, $\phi(0, 0) = d_x\phi(0, 0) = 0$, defined in a neighborhood of the closure of W. Let $u \in C^0(\Gamma_T \cup \{0\}, L_{loc}^p(B_x))$, $1 \leq p < \infty$ satisfy $\mathcal{L}u = 0$ and we wish to approximate u by polynomials in Z in the topology of $C^0(\Gamma_\rho \cup \{0\}, L_{loc}^p(B_x(\delta)))$, where Γ'_ρ is a proper subcone of Γ of height ρ, $B_x(\delta) \subset B_x$ is a ball of radius δ and $\rho, \delta > 0$ are small.

Shrinking B_x we may assume that $u(\cdot, t) \in L^p$ and by Corollary II.4.11 it will be enough to approximate u by $G_\tau u$ in the norm

$$\sup_{t \in \Gamma_T} \| u(\cdot, t) - G_\tau u(\cdot, t) \|_{L^p(\mathbb{R}^m)}.$$

By the proof of Theorem II.3.1 we know that the norm of G_τ as an operator on $L^p(\mathbb{R}^m)$ (depending on t as a parameter) may be bounded by a constant independent of $t \in \Gamma_T$. Thus, it is enough to check that G_τ converges strongly to the identity on a dense subset of $C^0(\Gamma_T \cup \{0\}, L^p(B_x))$. This is indeed the case, because if $v(x, t)$ is continuous and supported in $(\Gamma_T \cup \{0\}) \times B'_x$ where B'_x is a ball concentric with B_x and of smaller radius, we know by the proof of Theorem II.1.1 that $G_\tau v(x, t) \to v(x, t)$ uniformly on $\Gamma_T \times B_x$ and this implies convergence in the norm of $C^0(\Gamma_T \cup \{0\}, L^p(B_x))$. This proves

THEOREM II.4.12. *Let $\mathcal{L}$ be a locally integrable structure with first integrals $Z_1, \ldots, Z_m$, defined in a neighborhood of the closure of $W = B_x \times \Gamma_T$. There exist a ball $B'_x \subset B_x$ and a proper truncated subcone Γ'_ρ of Γ_T such that for any $u \in C^0(\Gamma_T \cup \{0\}, L^p(B_x))$, $1 \le p < \infty$, satisfying $\mathcal{L}u = 0$*

$$E_\tau u(x, t) \longrightarrow u(x, t) \text{ in } C^0(\Gamma'_\rho \cup \{0\}, L^p(B'_x)), \quad \tau \longrightarrow \infty. \tag{II.57}$$

As usual, we may replace the operator E_τ in (II.57) by a convenient sequence of polynomials in Z, $P_\ell(Z_1, \ldots, Z_m)$. $\qquad\square$

Notes

The approximation formula of Section II.1 for classical solutions was first proved by Baouendi and Treves in [**BT1**], building upon their previous work ([**BT2**]) that dealt with a corank one system of real-analytic vector fields. For distribution solutions, the proof in [**BT1**] relied on a local representation formula proved under a supplementary hypothesis on the locally integrable structure. This representation formula, which is of independent interest and states that any distribution solution u of $\mathcal{L}u = 0$ may be written as $u = P(x, D)v$, where v is a classical solution of $\mathcal{L}v = 0$ and $P(x, D)$ is a differential operator that commutes with the local generators L_j, $1 \le j \le n$, of $\mathcal{L}$. This representation formula was proved in general by Treves in [**T4**], who also stated and proved the approximation formula for distribution solutions in all generality. Metivier studied the case of a nonlinear first-order analytic single equation and proved an approximation formula for solutions of class C^2, obtaining as a consequence uniqueness in the Cauchy problem ([**Met**]).

The convergence in L^p of the approximation formula for solutions in L^p is an unpublished observation of S. Chanillo and S. Berhanu; the proofs presented here for L^p as well as for other functional spaces follow [**HMa1**].

It was soon realized by researchers in several complex variables theory that the approximation formula, although formulated in the rather general context of locally integrable structures, could be applied with success to deal with classical questions and it was used early as a tool in the problem of extending CR functions ([**BP**],[**W1**], [**BT3**]) and other matters like the study of the Radó property for CR functions ([**RS**]) (see also [**HT1**] for the Radó property for solutions of locally solvable vector fields).

Because the approximation is obtained through the operator E_τ that depends linearly on the trace of the solution on a maximally real submanifold, it is hardly surprising that it would have consequences for uniqueness questions. One remarkable feature is that it applies directly to distribution solutions in sharp contrast with other methods, like Carleman's estimates, which were devised to deal with functions rather than with less regular distributions. Before the definition of orbits by Sussmann in 1973 ([**Su**]), the propagation of zeros had been observed for some operators with real-analytic coefficients ([**Z**]) using as propagators Nagano's leaves ([**Na**]), which coincide with Sussmann's orbits in the real-analytic set-up. The theorem stating that the support of a solution is a union of Sussmann's orbits was initially stated and proved in [**T4**]. Another early application to uniqueness is [**BT4**]. Nowadays, the use of the approximation formula is so standard that probably there is no point in keeping track of its use in the literature. Anyway, we mention [**BH3**] as a recent uniqueness result that takes advantage of the approximation formula. Another application outside the scope of the theory of holomorphic functions is its use in the study of removable singularities for solutions of locally solvable vector fields ([**HT2**], [**HT3**], [**HT4**]).

III

Sussmann's orbits and unique continuation

In this chapter we will present various results on unique continuation for solutions and approximate solutions of locally integrable structures. Our main focus will be on those results where Sussmann's orbits have played a decisive role. We will begin with some general discussion of these orbits, taken mainly from [**Su**] and [**BM**].

III.1 Sussmann's orbits

Let $\mathcal{M}$ be a C^∞, paracompact manifold. Let D be a set of locally defined, smooth real vector fields. That is, each X in D is defined on some open subset of $\mathcal{M}$ and it is smooth there. Assume that the union of the domains of the elements of D equals $\mathcal{M}$. We define an equivalence relation on $\mathcal{M}$ as follows: two points p and q are related if there is a curve $\gamma : [0, T] \longrightarrow \mathcal{M}$ such that

(1) $\gamma(0) = p$, $\gamma(T) = q$;
(2) there exist $t_0 = 0 < t_1 < \cdots < t_n = T$ and vector fields $X_i \in D$ ($i = 1, \ldots, n$) such that for each i, the restriction $\gamma : [t_{i-1}, t_i] \longrightarrow \mathcal{M}$ is an integral curve of X_i or $-X_i$.

The equivalence classes of this relation will be called the *orbits* of D. In [**Su**], Sussmann showed that these orbits can be equipped with a natural topology and differentiable structure which makes them immersed submanifolds of $\mathcal{M}$. We will next briefly describe the orbit topology and C^∞ structure (the reader is referred to [**Su**] and [**BER**] for more details). If $X \in D$ is defined near p in $\mathcal{M}$, let $\Phi_t^X(p)$ denote the integral curve of X which at $t = 0$ equals p and is defined on a maximal interval. If $Y = (X_1, \ldots, X_m) \in D^m$ (i.e., each $X_i \in D$),

$s = (t_1, \ldots, t_m) \in \mathbb{R}^m$, and $p \in \mathcal{M}$, we write

$$\Phi_s^Y(p) = \Phi_{t_1}^{X_1}(\Phi_{t_2}^{X_2}(\ldots \Phi_{t_m}^{X_m}(p) \ldots)),$$

and let $\Omega(Y)$ denote the open subset of $\mathbb{R}^m \times \mathcal{M}$ consisting of the points (s, p) where $\Phi_s^Y(p)$ is defined. For $p \in \mathcal{M}$ and $Y \in D^n$, let $\Phi^Y(p)$ denote the map $s \longmapsto \Phi_s^Y(p)$, and let $\Omega(Y, p)$ be its domain. Note that $\Omega(Y, p)$ is a subset of $\mathbb{R}^n$. Suppose that $\mathcal{L} = \mathcal{L}_x$ is an orbit of D through a point x. Observe that $\mathcal{L}$ is the union of the sets $\Phi^Y(x)(\Omega(Y, x))$, where $Y \in D^n$ for $n = 1, 2, \ldots$ The orbit $\mathcal{L}$ is topologized by giving it the strongest topology that makes all the $\Phi^Y(x)$ continuous (for all n, and for all $Y \in D^n$). Note that since each $\Phi^Y(x) : \Omega(Y, x) \longrightarrow \mathcal{M}$ is continuous, it follows that the topology of $\mathcal{L}$ is finer than the subspace topology. Equivalently, the inclusion map from $\mathcal{L}$ into $\mathcal{M}$ is continuous. As the examples below will show, in general, this inclusion won't be a homeomorphism. For the independence of the topology of $\mathcal{L}$ on the point x, we refer the reader to [**Su**, page 176]. We will briefly recall the differentiable structure on $\mathcal{L}$ by describing the coordinate charts. Let $\Gamma(D)$ be the smallest set of locally defined C^∞ vector fields on $\mathcal{M}$ satisfying:

(1) $D \subseteq \Gamma(D)$, and
(2) for any $p \in \mathcal{M}$, $\{X_p : X \in \Gamma(D)\}$ is a subspace of $T_p\mathcal{M}$.

We will use $\widehat{\Gamma(D)}$ to denote the smallest set of locally defined, smooth vector fields which contains $\Gamma(D)$ and is invariant under the group of local diffeomorphisms generated by $\Gamma(D)$. It is not hard to see that the dimension of the fibers $\widehat{\Gamma(D)}_x$ is constant as x varies in the orbit $\mathcal{L}$. Suppose now $q \in \mathcal{L}$. By lemmas 5.1 and 5.2 in [**Su**], there exist $Y \in D^n$ for some n, $q' \in \mathcal{L}$ and $s \in \Omega(Y, q')$ such that

$$\Phi^Y(q')(s) = q,$$

and the rank k of the differential of

$$\Phi^Y(q') : \Omega(Y, q') \longrightarrow \mathcal{M}$$

at the point s is maximal, and that in fact, this rank equals $\dim \widehat{\Gamma(D)}_x$ for any $x \in \mathcal{L}$. By the rank theorem, we can find neighborhoods U of s in $\mathbb{R}^n$, V of q in $\mathcal{M}$, diffeomorphisms F from U onto C^n, G from V onto C^N ($N = $ dimension of $\mathcal{M}$) such that

$$G \circ \Phi^Y(q') \circ F^{-1}(x_1, \ldots, x_n) = (x_1, \ldots, x_k, 0, \ldots, 0).$$

Here C^l denotes the cube

$$\{(x_1, \ldots, x_l) \in \mathbb{R}^l : |x_i| < 1 \quad \forall i\}.$$

Let $\Lambda = \Phi^Y(q')(U)$. Λ is an open subset of $\mathcal{L}$ (see [**Su**]). Moreover, Λ is a submanifold of $\mathcal{M}$ since

$$G(\Lambda) = \{(x_1, \ldots, x_k, 0, \ldots, 0)\}.$$

The differentiable structure on the orbit $\mathcal{L}$ is defined by taking the pairs $\{(\Lambda, G|_\Lambda)\}$ as charts. One of the main results proved by Sussmann may be stated as follows:

THEOREM III.1.1 (Theorem 4.1 in [**Su**]). *Let $\mathcal{M}$ be a C^∞ manifold, and let D be a set of locally defined, smooth vector fields such that the union of the domains of the elements of D is $\mathcal{M}$. Then*

(1) *If $\mathcal{L}$ is an orbit of D then $\mathcal{L}$ (with the topology described above) admits a unique differentiable structure such that $\mathcal{L}$ is a submanifold of $\mathcal{M}$.*

(2) *With the topology and differentiable structure as above, every orbit of D is a maximal integral submanifold of $\widehat{\Gamma(D)}$.*

We will next present several examples.

EXAMPLE III.1.2. Let $\mathcal{M}$ be a manifold and suppose P is a sub-bundle of the tangent bundle $T\mathcal{M}$ of dimension k. That is, for each $x \in \mathcal{M}$, the fiber P_x is a k-dimensional subspace of $T_x\mathcal{M}$, and for each $y \in \mathcal{M}$, there exists a neighborhood U of y and smooth vector fields $X^1, X^2, \ldots, X^k$ on U such that $\{X^j(x) : 1 \le j \le k\}$ is a basis of P_x for each $x \in U$. We assume that P is closed under Lie brackets. Then by the Frobenius theorem, the manifold $\mathcal{M}$ is foliated by leaves each of which is an integral manifold of P. If we set D to be equal to the set of smooth local sections of P, then these leaves are precisely the orbits of D. Note that in this example, the orbits have the same dimension. Thus the concept of Sussmann's orbits may be viewed as a generalization of Frobenius foliations. For a concrete example of this kind, consider the 2-torus $\mathbb{T}^2 = S^1 \times S^1$. Use the angles (θ_1, θ_2) as coordinates for points in $\mathbb{T}^2$, so θ_1 and θ_2 are determined modulo integral multiples of 2π. Pick two real numbers a and b, not both equal to zero, and consider the sub-bundle of the tangent bundle of $\mathbb{T}^2$ generated by the vector field

$$L = a\frac{\partial}{\partial\theta_1} + b\frac{\partial}{\partial\theta_2}.$$

The orbits are the integral curves of L. If a and b are linearly dependent over the rational numbers, then each orbit is diffeomorphic to S^1. In this case, an orbit is an embedded submanifold of $\mathbb{T}^2$ and so its orbit topology agrees with the induced subspace topology. If a and b are linearly independent over the rational numbers, each orbit is diffeomorphic to the real line. In this case the orbits are dense in $\mathbb{T}^2$, and hence are not embedded submanifolds.

EXAMPLE III.1.3. Let $\mathcal{M}_1 = \mathbb{R}^2$ and $\mathcal{M}_2 = \{x \in \mathcal{M}_1 : \|x\| < 1\}$. Let $g(t) \in C^\infty(\mathbb{R})$, $g > 0$ on $(1, 2)$ and $g \equiv 0$ outside $(1, 2)$. Let $D = \left\{ \frac{\partial}{\partial x_1}, g(x_1) \frac{\partial}{\partial x_2} \right\}$. $\mathcal{M}_1$ is the only orbit for D. However, if we consider D on $\mathcal{M}_2$, the orbits are the horizontal segments in $\mathcal{M}_2$. Notice also that the tangent space of the orbit $\mathcal{M}_1$ at points (x_1, x_2) with $x_1 \notin (1, 2)$ does not coincide with the fiber of the Lie algebra generated by $\dfrac{\partial}{\partial x_1}$ and $g(x_1) \dfrac{\partial}{\partial x_2}$.

EXAMPLE III.1.4. Consider the orbits of $\left\{ \frac{\partial}{\partial x_2}, x_1 \frac{\partial}{\partial x_1} \right\}$ in $\mathbb{R}^2$. There are three orbits: $\{x_1 > 0\}$, $\{x_1 < 0\}$, and $\{x_1 = 0\}$. Thus the dimension of orbits is not locally constant. In general, if $d(x) =$ the dimension of the orbit through x, then $d(x)$ is a lower semicontinuous function.

EXAMPLE III.1.5. The analytic case: suppose $\mathcal{M}$ is a real-analytic manifold and D is a set of real-analytic vector fields on $\mathcal{M}$. Let D^* be the smallest Lie algebra (under brackets) of real-analytic vector fields that contains D. It is well known (see [**Su**], for example) that if $p \in \mathcal{M}$, then there are a finite number of elements $X_1, \ldots, X_k$ of D^* such that every $X \in D^*$ can be expressed in a neighborhood of p as

$$\sum_{j=1}^k f_j X_j$$

for some real-analytic functions f_j. Moreover, in this case, if $\mathcal{L}$ is an orbit of D and $p \in \mathcal{L}$, then its tangent space

$$T_p \mathcal{L} = D_p^* \quad \text{where} \quad D_p^* = \{X(p) : X \in D^*\}.$$

This makes it easier to compute the dimensions of orbits in the analytic case. The concept of orbits in the analytic case dates back to Nagano's paper ([**Na**]). Orbits arise in a locally integrable structure $(\mathcal{M}, \mathcal{V})$ by taking D as the collection of the real parts of smooth, local sections of $\mathcal{V}$. Below we will give an example of orbits arising from the CR structure of a hypersurface in $\mathbb{C}^2$. More examples will be given in the rest of the sections.

EXAMPLE III.1.6. Let $z = x + iy$, $w = s + it$ denote the variables in $\mathbb{C}^2$ and suppose $g = g(x, y)$ is a real-valued, real-analytic function defined on the plane such that

(1) $g(0,0) = 0$, $g(x, y) > 0$ for $(x, y) \neq (0, 0)$; and

(2) $\Delta g < 2 \dfrac{|\nabla g|^2}{g}$.

Define

$$\rho(z, w) = s^2 + (t - g(x, y))^2 - g(x, y)^2$$

and let

$$\mathcal{M} = \{(z, w) \in \mathbb{C}^2 \backslash \{0\} : \rho(z, w) = 0\}.$$

Notice that since $d\rho \neq 0$ on $\mathcal{M}$, $\mathcal{M}$ is a real-analytic hypersurface. We consider the orbits arising from the CR structure of $\mathcal{M}$. Observe first that the complex line $\Sigma = \mathbb{C} \backslash \{0\} \times \{0\} \subset \mathcal{M}$. Since the bundle $\mathcal{V}$ is tangent to Σ, the bracket $[X, Y]$ of any two smooth sections X and Y of $\mathfrak{R}\mathcal{V}$ is also tangent to Σ. Hence by the remarks in Example III.1.5, Σ is an orbit. We will next show that $\mathcal{M} \backslash \Sigma$ is strictly pseudo-convex. For any $a = (a_1, a_2) \in \mathbb{C}^2$, we have

$$\langle \partial \bar{\partial} \rho a, a \rangle = i[(w - \bar{w})g_{z\bar{z}}|a_1|^2 - g_z a_1 \bar{a}_2 + g_{\bar{z}} a_2 \bar{a}_1 + i|a_2|^2].$$

On the manifold $\mathcal{M}$, $|\bar{w} + ig|^2 = g^2 \neq 0$ and so if for $a = (a_1, a_2)$, $\langle \partial \rho, a \rangle = 0$ at a point of $\mathcal{M}$, then

$$a_2 = \left(\frac{i(\bar{w} - w)g_z}{\bar{w} + ig} \right) a_1.$$

It follows that if $\langle \partial \rho, a \rangle = 0$, then

$$\langle \partial \bar{\partial} \rho a, a \rangle = i|a_1|^2 (w - \bar{w}) \left(g_{z\bar{z}} - \frac{2|g_z|^2}{g} \right).$$

The latter, together with the assumptions on g, show that $\mathcal{M} \backslash \Sigma$ is strictly pseudo-convex. Thus $\mathcal{M}$ has one orbit of dimension 2, and all other orbits are of dimension 3. If we make a further assumption on g, say for example, $g(z, \bar{z}) = g(|z|)$, then $\mathcal{M} \backslash \Sigma$ is connected, and hence a single open orbit. When $g(x, y) = x^2 + y^2$, this example appeared in [**BM**]. Our next objective is to analyze the extent to which orbits behave like embedded submanifolds. We begin with:

LEMMA III.1.7. *Let $\mathcal{L}$ be an orbit through p_0 of dimension k, dimension $\mathcal{M} = n$. Then there exists a local chart $(T \times V, \psi)$ on $\mathcal{M}$ about p_0 with T and V neighborhoods of 0 in $\mathbb{R}^k$ and $\mathbb{R}^{n-k}$ respectively, such that*

$$\mathcal{L} \cap \psi(T \times V) = \psi(T \times P_{\mathcal{L}}), \quad \text{where}$$

$$P_{\mathcal{L}} = \{v \in V : \psi(0, v) \in \mathcal{L}\}.$$

PROOF. Let S be a submanifold of $\mathcal{M}$ through p_0 of dimension $n - k$ such that

$$T_{p_0}\mathcal{M} = T_{p_0}S + T_{p_0}\mathcal{L},$$

where we view $T_{p_0}\mathcal{L}$ as a subspace of $T_{p_0}\mathcal{M}$. Let $X_1, \ldots, X_k$ be locally defined vector fields in $\widehat{\Gamma}$ spanning $T_{p_0}\mathcal{L}$ at p_0. After contracting S about p_0 if necessary, we can find a neighborhood T of 0 in $\mathbb{R}^k$ and a neighborhood U of p_0 in $\mathcal{M}$ such that the map

$$F : T \times S \longrightarrow U$$

given by

$$F(t_1, \ldots, t_k; p) = \Phi_{t_1}^{X_1}(\Phi_{t_2}^{X_2}(\ldots \Phi_{t_k}^{X_k}(p)\ldots))$$

is a diffeomorphism. Suppose now that $q \in \mathcal{L} \cap U$. Then $q = F(t, s)$ for a unique $(t, s) \in T \times S$. Hence,

$$F(T \times \{s\}) \subset \mathcal{L} \cap U.$$

Therefore, $\mathcal{L} \cap U = F(T \times P_{\mathcal{L}})$ where $P_{\mathcal{L}} = \mathcal{L} \cap S$. After introducing a chart on S about p, we get the lemma.　　　□

Observe that if an orbit $\mathcal{L}$ is an embedded submanifold, then the sets T and V in Lemma III.1.7 can be chosen so that $P_{\mathcal{L}}$ is a single point. For a general orbit, we will next show that $P_{\mathcal{L}}$ can be chosen to be a countable set. This will follow from:

LEMMA III.1.8. *The topology on an orbit $\mathcal{L}$ is second countable.*

PROOF. For $p \in \mathcal{L}$ we will consider the charts $(T \times V, \psi)$ of Lemma III.1.7. The discussion on the differentiable structure of $\mathcal{L}$ shows that $\psi(T \times V)$ is an open set in $\mathcal{L}$. Since $\mathcal{M}$ is second countable, the subspace topology on $\mathcal{L}$ is second countable. Hence we can get a locally finite open cover for $\mathcal{L}$ of the form

$$\{U_j = \psi(T_j \times V_j)\}_{j=1}^{\infty}.$$

Recall that for each j, $\mathcal{L} \cap \psi(T_j \times V_j) = \psi(T_j \times P_j)$ where $P_j = \{v \in V_j : \psi(0, v) \in \mathcal{L}\}$. If $q \in P_j$, we will call the set $\psi(T \times \{q\})$ a slice of $\mathcal{L}$ in U_j. Fix $p_0 \in U_{j_0} \cap \mathcal{L}$ for some j_0, and hence $p_0 \in \psi(T_{j_0} \times \{p'\}) \subseteq \mathcal{L}$ for some $p' \in V_{j_0}$. For every finite tuple $i = (i_1, \ldots, i_m)$, let A_i be the set of points x in $\mathcal{L}$ such that x can be joined to p_0 by a curve γ consisting of m pieces γ_l where each γ_l lies in U_{i_l}, $l = 1, \ldots, m$. From the definition, it is clear that each A_i is a union of slices in U_{i_m}. The family $\{A_i\}$ where i varies over all finite tuples of positive integers is a countable collection of open subsets of $\mathcal{L}$ which form a basis for the topology of $\mathcal{L}$. Hence we only need to show that each A_i consists of a countable number of slices in U_{i_m}. We will do this by induction on m. When $m = 1$, A_{i_1} contains at most one slice. Suppose the result holds for all tuples $j = (i_1, \ldots, i_{k-1})$ of length $k - 1$. Then A_j is the union of countably many slices in $U_{i_{k-1}}$. Fix a slice Σ in A_j. Since slices are open sets in $\mathcal{L}$, the intersection of Σ with each slice in U_{i_k} is an open set. Moreover, since the slices in U_{i_k} are pairwise disjoint, and Σ is homeomorphic to an open set in $\mathbb{R}^d$, it follows that Σ can intersect only a countable number of slices in U_{i_k}. Thus each A_i is the union of a countable number of slices and therefore $\mathcal{L}$ is second countable. $\qquad\square$

The preceding lemma can be used to show that orbits possess properties not shared by a general immersed submanifold. To see one such property, call an immersed submanifold N of a manifold $\mathcal{M}$ *weakly embedded* if whenever A is a manifold and $f : A \longrightarrow \mathcal{M}$ is smooth with $f(A) \subseteq N$ then $f : A \longrightarrow N$ is smooth. This notion was introduced by Pradines in [**Pr**]. For an example of an immersed submanifold that is not weakly embedded, see remark 6.8 in [**Boo**].

PROPOSITION III.1.9. *An orbit $\mathcal{L}$ in a manifold $\mathcal{M}$ is weakly embedded.*

PROOF. Suppose $f : A \longrightarrow \mathcal{M}$ is C^∞ and $f(A) \subseteq \mathcal{L}$. Let $q \in A$ and $p = f(q)$. Let $\dim \mathcal{L} = k$, $\dim \mathcal{M} = n$, and suppose $(T \times V, \psi)$ is a chart on $\mathcal{M}$ about p as in Lemma III.1.7 with T and V cubes centered about 0 in $\mathbb{R}^k$ and $\mathbb{R}^{n-k}$. Since $f : A \longrightarrow \mathcal{M}$ is C^∞, we can choose a connected neighborhood W of q such that

$$f(W) \subseteq \psi(T \times V).$$

Recall from Lemma III.1.7 and Lemma III.1.8 that

$$\mathcal{L} \cap \psi(T \times V) = \bigcup_{v \in P} \psi(T \times \{v\}),$$

where $P \subseteq V$ is a countable set. The map $\psi^{-1} \circ f : W \longrightarrow T \times V$ is C^∞ and $\psi^{-1} \circ f(W) \subseteq \bigcup_{v \in P} T \times \{v\}$. Since W is connected, there exists a unique $v \in P$ such that $\psi^{-1} \circ f(W) \subseteq T \times \{v\}$. Hence $f : W \longrightarrow \psi(T \times \{v\}) \subseteq \mathcal{L}$ is C^∞. $\qquad\square$

COROLLARY III.1.10. *If $\mathcal{L}$ is an orbit of $\mathcal{M}$, then when topologized with its orbit topology, it has a unique differentiable structure that makes it an immersed submanifold of $\mathcal{M}$.*

Another property of orbits not shared by a general immersed submanifold concerns the propagation of embeddedness. More precisely, we have

PROPOSITION III.1.11. *Let $\mathcal{L}$ be an orbit in $\mathcal{M}$ and suppose for a point p in $\mathcal{L}$, there is a neighborhood W in $\mathcal{M}$ such that $W \cap \mathcal{L}$ is an embedded submanifold of W. Then $\mathcal{L}$ is an embedded submanifold of $\mathcal{M}$.*

PROOF. Let $q \in \mathcal{L}$ and assume $q = \Phi_t^X(p)$ for some $X \in \Gamma$ and $t \in \mathbb{R}$. Set $W' = \Phi_t^X(W)$. Here we may assume W has been contracted enough to lie in the domain of the flow of Φ_t^X. Since $\Phi_t^X : W \longrightarrow W'$ is a diffeomorphism, the submanifold $\Phi_t^X(W \cap \mathcal{L})$ is an embedded submanifold of W'. It is also easy to see that

$$\Phi_t^X(W \cap \mathcal{L}) = W' \cap \mathcal{L}.$$

Hence $\mathcal{L}$ is an embedded submanifold of $\mathcal{M}$. $\qquad\square$

COROLLARY III.1.12. *If an orbit $\mathcal{L}$ is a closed subset of $\mathcal{M}$, then it is an embedded submanifold.*

PROOF. Let $p \in \mathcal{L}$. Choose a chart $(T \times V, \psi)$ about p as in Lemma III.1.7. If such a chart can be selected so that $P_{\mathcal{L}}$ is a finite subset of V, then by Proposition III.1.11, $\mathcal{L}$ is embedded. Otherwise, such a selection is not possible for any point in $\mathcal{L}$. In particular, this means that for any $v \in P_{\mathcal{L}}$, the point $\psi(0, v)$ is an accumulation point of the set $\psi(0 \times P_{\mathcal{L}})$. Hence $v \in P_{\mathcal{L}}$ is an accumulation point of $P_{\mathcal{L}}$. Moreover, since $\mathcal{L}$ is closed, $P_{\mathcal{L}}$ is a closed subset of V. It follows that $P_{\mathcal{L}}$ is a perfect set and hence it is uncountable. This gives rise to the pairwise disjoint, uncountable family of open subsets $\{\psi(T \times \{v\}) : v \in P_{\mathcal{L}}\}$ of $\mathcal{L}$, contradicting the second countability of $\mathcal{L}$. $\quad\square$

III.2 Propagation of support and global unique continuation

This section discusses the relevance of orbits to a variety of global questions of unique continuation in involutive structures. Suppose $\mathcal{V}$ is an involutive structure on $\mathcal{M}$ for which uniqueness for solutions in the (noncharacteristic) Cauchy problem holds, i.e., every solution defined in a neighborhood of a noncharacteristic (with respect to $\mathcal{V}$) hypersurface Σ and whose trace on Σ is zero vanishes in a neighborhood of Σ. The uniqueness results of Chapter II show that an example of such a $\mathcal{V}$ is provided by a locally integrable structure. Our first goal is to present another proof of Corollary II.4.7, which is a result on the propagation of the support of a solution along orbits. Special cases of this theorem were proved by several authors (see the notes). The result stated here is due to Treves ([**T4**]), but the proof is taken from [**BM**].

THEOREM III.2.1. *Assume that V is an involutive structure for which unique-ness in the Cauchy problem holds. If u is a solution, then the support of u is a union of orbits.*

Before we provide the proof, we will recall some definitions and results from a paper of Bony ([**Bo**]).

DEFINITION III.2.2. *Let Ω be an open subset of $\mathbb{R}^n$ and F a closed subset of Ω. A vector v is said to be normal to F at $x_0 \in F$ if there is an open ball $B \subseteq \Omega \backslash F$ centered at x such that $x_0 \in \partial B$ and $v = \lambda(x - x_0)$ for some $\lambda > 0$.*

REMARK III.2.3. By considering cones of varying apertures, it is easy to see that a closed set may have no normals or many normals at a boundary point.

DEFINITION III.2.4. *Suppose Ω is open in $\mathbb{R}^n$ and $F \subseteq \Omega$ is closed. A vector field $X(x)$ is tangent to F if whenever v is normal to x_0 in F, the vector $X(x_0)$ is orthogonal to v.*

In [**Bo**], Bony proved the following:

THEOREM III.2.5. *Suppose Ω is open in $\mathbb{R}^n$ and F a closed subset of Ω. Let $X(x)$ be a Lipschitz vector field in Ω which is tangent to F. If an integral curve of X intersects F at a point, then it is entirely contained in F.*

PROOF OF THEOREM III.2.1. Let π denote the projection map from $T^*\mathcal{M}$ onto $\mathcal{M}$. Suppose u is a solution on $\mathcal{M}$ and F denotes the support of u. Let $\Omega = \mathcal{M} \backslash F$. Define $N(F)$ to be the set of $v \in T^*\mathcal{M} \backslash \{0\}$ over points in F such that there exists f real-valued, smooth, defined near $p = \pi(v)$ and such that $f(p) = 0$, $df(p) = v$ and $f \leq 0$ on F near p. Fix $p \in F$ and suppose $v \in N(F)$ with $\pi(v) = p$. Suppose we show that for any $X = \Re L$ (for some smooth section L of V) defined near p, $\langle v, X \rangle = 0$. Then by Bony's theorem, the integral curve of X through p will lie in F, thus proving the theorem. Let f be chosen as above with $df(p) = v$. Note that near p, the zero set of f is a smooth hypersurface, and $u \equiv 0$ on a side of this hypersurface. Since $p \in F = \operatorname{supp} u$, by the uniqueness in the Cauchy problem, V has to be characteristic to $\{f = 0\}$ at p. Hence, $\langle v, X \rangle = 0$. □

We note that if V is an involutive structure for which uniqueness in the Cauchy problem is not valid, then the support of a solution may not be a union of orbits, as demonstrated by Cohen's celebrated example ([**Co**]).

DEFINITION III.2.6. *A formally integrable structure $(\mathcal{M}, V)$ satisfies the global unique continuation property if every solution that vanishes on an open subset vanishes everywhere on $\mathcal{M}$.*

According to Theorem III.2.1, global unique continuation holds in a locally integrable structure $(\mathcal{M}, \mathcal{V})$ whenever $\mathcal{M}$ is a single orbit for $\mathcal{V}$. However, global unique continuation may hold even when $\mathcal{M}$ is not a single orbit, as shown by the structure on the 2-torus generated by a real vector field each of whose integral curves is dense. The obstruction to the validity of global unique continuation is the presence of proper, closed subsets of $\mathcal{M}$ which are unions of orbits, since by Theorem III.2.1, such sets can potentially be the supports of solutions. We will refer to sets that are unions of orbits as *invariant sets*. In order to check the validity of global unique continuation, one needs to understand when a given proper, closed, invariant set equals the support of a solution. It turns out that in a general locally integrable structure, a proper, closed orbit may not be the support of a solution. This is illustrated by examples below. Some sufficient conditions for the existence of a solution supported on a proper, closed orbit were studied in the work [**BM**]. In particular, the following theorem was proved (see also Theorem III.2.12 below):

THEOREM III.2.7 (Theorem 5.8 in [**BM**].). *Suppose $\mathcal{M}$ is an orientable, connected analytic hypersurface in $\mathbb{C}^n$. If $\mathcal{M}$ is not Levi flat and has a codimension one orbit $\mathcal{L}$, then there is a solution supported on $\mathcal{L}$. Thus, on an analytic, non-Levi flat hypersurface in $\mathbb{C}^n$, the global unique continuation property holds if and only if there is only one orbit.*

EXAMPLE III.2.8. We consider real-analytic vector fields L in the plane that are rotation-invariant. That is, if $\mathcal{V}$ is the bundle generated by L, then

$$dR_\alpha(\mathcal{V}) = \mathcal{V}$$

for every rotation R_α (with angle α) of $\mathbb{R}^2$. In polar coordinates, such an L takes the form (see [**BMe**])

$$L = g(r, \theta)\left(rY(r)\frac{\partial}{\partial r} + iX(r)\frac{\partial}{\partial \theta}\right)$$

where g, X, Y are real-analytic functions, $\dfrac{X}{Y}$ is even in r away from the zeros of Y and we may assume that $X(0) = Y(0) = 1$. The characteristic set $\Sigma = \{\Re(X(r)\overline{Y}(r)) = 0\}$ is a union of circles centered at 0 and $0 \notin \Sigma$. Assume $\Sigma = \{r = 1\}$. If L is of finite type at a point p in Σ, then it is of the same type at every point p in Σ and in this case, $\mathcal{V}$ has only one orbit. Suppose now L is of ∞ type at some and hence every point of Σ. Then (see [**BMe**]) it can be shown that $\mathcal{V}$ is generated by

$$L = \frac{\partial}{\partial \theta} - \sqrt{-1}\,rY(r)\frac{\partial}{\partial r},$$

where $Y(r) = (1 - r^2)^N h(r)$, h is real-analytic, $h(r) \neq 0$, and $h(0) \in \{\pm 1\}$. Without loss of generality, assume $h(0) = 1$. Then, V has three orbits: $\{r < 1\}$, $\{r > 1\}$, and $\Sigma = \{r = 1\}$. We consider next whether Σ can be the support of a distribution solution. When $N \geq 2$, the distribution

$$\langle u, \psi(r, \theta) \rangle = \int_0^{2\pi} \psi(1, \theta) \, d\theta$$

is a solution supported on Σ. Assume $N = 1$. In this case, such a u exists if and only if $h(1)$ is a rational number ([**BMe**]).

Indeed, suppose $Lu = 0$ and u is supported on Σ. Then there exist an integer $k \geq 0$ and $a_j(\theta) \in \mathcal{D}'(S^1)$ $(0 \leq j \leq k)$ such that

$$\langle u, \psi(r, \theta) \rangle = \sum_{m=0}^{k} \int_0^{2\pi} a_m(\theta) \left(\frac{\partial}{\partial r} \right)^m \psi(1, \theta) \, d\theta.$$

Since L is in the tangential direction on Σ, each $a_j(\theta) \in C^\infty(\Sigma)$. Let $\psi_{j,n}(r, \theta) = f_j(r) e^{in\theta}$, where $f_j(r)$ is C^∞ and $f_j^{(l)}(1) = \delta_{jl}$ for $0 \leq j \leq k$. Note that the transpose of L is given by

$${}^t Lw = -\frac{\partial w}{\partial \theta} + irY(r) \frac{\partial w}{\partial r} + i(2Y(r) + rY'(r))w$$

and so

$${}^t L\psi_{k,n} = ie^{in\theta}[rY(r)f_k'(r) + (2Y(r) + rY'(r) - n)f_k].$$

Moreover,

$$\left(\frac{\partial}{\partial r} \right)^m (rY(r)f_k'(r))|_{r=1} = \begin{cases} 0, & m < k \\ kY'(1), & m = k \end{cases}$$

and

$$\left(\frac{\partial}{\partial r} \right)^m [(2Y(r) + rY'(r) - n)f_k(r)]|_{r=1} = \begin{cases} 0, & m < k \\ Y'(1) - n, & m = k. \end{cases}$$

Thus, we get:

$$\begin{aligned} 0 &= \langle Lu, \psi_{k,n} \rangle \\ &= \langle u, {}^t L\psi_{k,n} \rangle \\ &= \left(\int_0^{2\pi} a_k(\theta) e^{in\theta} \, d\theta \right) [(k+1)Y'(1) - n]. \end{aligned} \tag{III.1}$$

Since we may assume that $a_k(\theta)$ does not vanish identically, there is an integer M for which

$$\int_0^{2\pi} a_k(\theta) e^{iM\theta} \, d\theta \neq 0. \tag{III.2}$$

From (III.1) and (III.2) it follows that

$$h(1) = \frac{-M}{2(k+1)} \in \mathbb{Q}, \text{ and}$$

$$a_k(\theta) = ce^{-iM\theta}$$

for some $c \neq 0$. Conversely, suppose $k \geq 0$ and M are integers satisfying

$$2(k+1)h(1) = -M.$$

We will seek a solution u of the form

$$\langle u, \psi(r, \theta) \rangle = \sum_{m=0}^{k} \int_0^{2\pi} b_m(\theta) \left(\frac{\partial}{\partial r} \right)^m \psi(1, \theta) \, d\theta.$$

Set $b_k(\theta) = e^{-iM\theta}$. Each $b_j(\theta)$ can be determined from the equation $\langle Lu, \psi_{j,M} \rangle = 0$. To see this, note that $\langle Lu, \psi_{k-1,M} \rangle = 0$ is equivalent to

$$0 = 2\pi \left(\frac{d}{dr} \right)^k [rYf'_{k-1} + (rY' + 2Y - M)f_{k-1}](1) + \left(\int_0^{2\pi} b_{k-1}(\theta) e^{iM\theta} d\theta \right)$$

$$\times \left(\frac{d}{dr} \right)^{k-1} [rYf'_{k-1} + (rY' + 2Y - M)f_{k-1}](1).$$

The coefficient of $\int_0^{2\pi} b_{k-1}(\theta) e^{iM\theta} d\theta$ in the latter equation is $-2kh(1) - M = 2h(1) \neq 0$, and hence we can get a constant c_{k-1} such that if we set

$$b_{k-1}(\theta) = c_{k-1} e^{-iM\theta}, \text{ then } \langle Lu, \psi_{k-1,M}(r, \theta) \rangle = 0.$$

In general, we can determine $b_l(\theta)$ from $\langle Lu, \psi_{l,M} \rangle = 0$. This leads to $b_l(\theta) = c_l e^{-iM\theta}$ for some constant c_l since

$$\left(\frac{d}{dr} \right)^l [rYf'_l + (2Y + rY' - M)f_l](1) = 2(k-l)h(1) \neq 0.$$

Thus $\langle Lu, \psi_{j,m} \rangle = 0$ for all $m \in \mathbb{Z}$, and all $j = 0, \ldots, k$. Since Lu is a distribution of order k, it follows that $Lu = 0$.

EXAMPLE III.2.9. (See [**BM**].) We denote the coordinates in $\mathbb{R}^3$ by (x, y, s) and we will write $\mathbb{R}^3 = \mathbb{R}_x \times \mathbb{R}_y \times \mathbb{R}_s$. Let $\phi : \mathbb{R} \longrightarrow \mathbb{R}$ be a smooth, 2π-periodic function, $\phi \geq 0$ and ϕ not identically 0. Define

$$L = \frac{\partial}{\partial x} + i \frac{\partial}{\partial y} + \phi(x) \sin(s) \frac{\partial}{\partial s} = X + iY.$$

The coefficients of L are 2π-periodic and so L induces a vector field $\widetilde{L}$ on $\mathbb{T}^3 = S^1 \times S^1 \times S^1$. The involutive structure generated by $\widetilde{L}$ is a Levi flat, locally integrable CR structure. We will show the following:

(1) The orbits of $\widetilde{L}$ through $p_1 = (1, 1, 1)$ and $p_2 = (1, 1, -1)$ are compact but all other orbits are noncompact.
(2) Depending on the value of

$$\int_0^{2\pi} \phi(x) dx,$$

there may not be any solution supported on either of the compact orbits.
(3) Global unique continuation is valid for continuous solutions.

Let $F : \mathbb{R}^3 \longrightarrow \mathbb{T}^3$ be given by

$$F(x, y, s) = (e^{ix}, e^{iy}, e^{is}).$$

Consider the orbit $\mathcal{L}_1$ through the point $p_1 = (1, 1, 1)$. $F(0, 0, 0) = p_1$ and the orbit in $\mathbb{R}^3$ of $\{X, Y\}$ through $(0, 0, 0)$ is $\mathbb{R}_x \times \mathbb{R}_y \times \{0\}$. Therefore, $\mathcal{L}_1 = S^1 \times S^1 \times \{1\}$. Likewise, for the point $p_2 = (1, 1, -1)$, the orbit $\mathcal{L}_2 = S^1 \times S^1 \times \{-1\}$. Consider now a typical point $p = (1, 1, e^{is_0})$ for some $0 < s_0 < \pi$. If $\gamma(t) = (x(t), s(t))$ is the integral curve of X with $\gamma(0) = (0, s_0)$, we will see that the orbit through p is given by

$$\mathcal{L} = \{(e^{it}, e^{iy}, e^{is(t)}) : t, y \in \mathbb{R}\}.$$

Indeed, $x(t) = t$ and $s'(t) = \phi(t) \sin(s(t))$, $s(0) = s_0$. If for some t_0, $s(t_0) = \pi$, then the curves $\gamma(t)$ and $\gamma_1(t) = (t, \pi)$ will both be integral curves of X passing through (t_0, π) at $t = t_0$. This implies that $s(t) \equiv \pi$, contradicting the assumption that $s(0) = s_0 < \pi$. Likewise, $s(t)$ can never equal zero. Thus, $0 \leq s(t) \leq \pi$ and $s'(t) \geq 0$. Suppose

$$\lim_{t \to \infty} s(t) = a < \pi.$$

Then $s_0 \leq s(t) \leq a$ for all $t \geq 0$. Therefore, $s'(t) \geq c\phi(t)$ for some $c > 0$, which in turn leads to

$$\lim_{t \to \infty} s(t) = \infty.$$

Hence,

$$\lim_{t \to \infty} s(t) = \pi$$

and by a similar reasoning,

$$\lim_{t \to -\infty} s(t) = 0.$$

Thus the closure of $\mathcal{L} = \mathcal{L} \cup \mathcal{L}_1 \cup \mathcal{L}_2$.

We consider now the question of existence of a solution supported on a compact orbit, say $\mathcal{L}_1$. Since L is tangent to $\mathcal{L}_1$ and defines a complex structure there, any distribution solution u supported on this orbit has the form

$$u(x, y) = \sum_{l=0}^{N} u_l(x, y) \sigma_l$$

where the u_l are C^∞ on $S^1 \times S^1$ and

$$\langle \sigma_k, g(x, y, s) \rangle = \int_0^{2\pi} \int_0^{2\pi} \partial_s^k g(x, y, 0) \, dx dy.$$

We have

$$\langle L\sigma_k, g(x, y, s) \rangle = \langle \sigma_k, {}^t Lg(x, y, s) \rangle$$

$$= -\int_0^{2\pi} \int_0^{2\pi} \phi(x) \left(\frac{\partial}{\partial s} \right)^k \left[(\sin s) \frac{\partial g}{\partial s} + (\cos s) g \right] (x, y, 0) \, dx dy$$

$$= -\int_0^{2\pi} \int_0^{2\pi} \phi(x) \left(\frac{\partial}{\partial s} \right)^{k+1} [(\sin s) g](x, y, 0) \, dx dy$$

$$= -\sum_{l=0}^{k+1} \int_0^{2\pi} \int_0^{2\pi} \binom{k+1}{l} \left(\frac{\partial}{\partial s} \right)^{k-l} \cos s \left(\frac{\partial}{\partial s} \right)^l g(x, y, 0) \, dx dy.$$

Thus,

$$L\sigma_k = -(k+1)\phi(x)\sigma_k - \phi(x) \sum_{l=0}^{k-2} \binom{k+1}{l} \partial_s^{k-l} \cos s(0) \sigma_l.$$

Let $M_k = \dfrac{\partial}{\partial x} + i \dfrac{\partial}{\partial y} - (k+1)\phi(x)$, for $k = 0, 1, 2, \ldots$ If $v(x, y) \in C^\infty(S^1 \times S^1)$, it follows that

$$L(v\sigma_k) = (M_k v)\sigma_k - v\phi \sum_{l=0}^{k-2} \binom{k+1}{l} \left(\partial_s^{k-l} \cos s \right)(0) \sigma_l.$$

Suppose now

$$u = \sum_{k=0}^{N} u_k(x, y)\sigma_k$$

is a solution. Then

$$Lu = \sum_{l=0}^{N}(M_l u_l)\sigma_l - \phi \sum_{l=0}^{N-2} \sum_{k=l+2}^{N} \binom{k+1}{l} u_k \left(\partial_s^{k-l} \cos s\right)(0)\sigma_l.$$

Let

$$\phi_0 = \frac{1}{2\pi} \int_0^{2\pi} \phi(x)\,dx$$

and define $\phi_1 = \phi - \phi_0$. Since $Lu = 0$ and u_N may be assumed nontrivial, we must have $(N+1)\phi_0 \in \mathbb{Z}$. Thus if ϕ_0 is not a rational number, there are no solutions supported on the orbit $\mathcal{L}_1$. If ϕ_0 is rational, with $\phi_0 = p/q$ where p and q are relatively prime, then $N = q - 1$ is the lowest possible transversal order of a nontrivial solution supported on $\mathcal{L}_1$. This follows from the injectivity of M_l for $l < N$ and the fact that M_N has a nontrivial kernel. Since M_l is also surjective for $l < N$ (as is easily seen using Fourier series), one can correct the 'errors' to obtain a solution u iteratively.

Finally, we remark that there are solutions supported on the closure of any noncompact orbit. This will follow from Theorem III.2.12 as stated below, or can be constructed explicitly as in [**BM**]. Thus, global unique continuation is not valid for distribution solutions. However, it is valid for continuous solutions.

We will now place these two examples in a more general context following [**BM**]. Given a locally integrable structure $(\mathcal{M}, \mathcal{V})$, let Σ be an orbit such that $\dim \Sigma < \dim \mathcal{M} = m + n$, where n is the rank of $\mathcal{V}$. Assume that Σ is an embedded submanifold of $\mathcal{M}$. Fix $p \in \Sigma$ and let $\{Z_1, \ldots, Z_m\}$ be a complete set of first integrals defined in a neighborhood U in $\mathcal{M}$ of p. Let $\{L_1, \ldots, L_n\}$ be smooth, local generators of $\mathcal{V}$ in U such that the brackets $[L_i, L_j] = 0$ for all i, j. Complete this to a basis

$$\{L_1, \ldots, L_n, M_1, \ldots, M_m\}$$

of $\mathbb{C}T\mathcal{M}$ in U such that

(1) $[L_i, M_k] = 0$, and
(2) $M_k Z_i = \delta_{ik}$.

Let $\{\omega_1, \ldots, \omega_n\}$ be smooth, exact one-forms in U such that

$$\{\omega_1, \ldots, \omega_n, dZ_1, \ldots, dZ_m\}$$

is a dual basis to $\{L_1, \ldots, L_n, M_1, \ldots, M_m\}$.

If $\mathcal{V}_\Sigma$ denotes the restriction of $\mathcal{V}$ to Σ, then $\mathcal{V}_\Sigma$ also has rank n. Hence if $\dim \Sigma = k + n$, then after shrinking U about p, the restrictions of exactly k of $\{Z_1, \ldots, Z_m\}$ have linearly independent differentials along Σ. Without loss of generality, assume that $\{Z_1, \ldots, Z_k\}$ have this latter property. It follows that $\{M_{k+1}, \ldots, M_m\}$ is a basis of the complexified normal bundle of Σ in U.

Fix orientations in U and in $U \cap \Sigma$ so that distributions in U (resp. in $U \cap \Sigma$) may be viewed as acting on forms of top degree. We wish to describe all solutions in U that are supported on $U \cap \Sigma$.

Let $M'' = (M_{k+1}, \ldots, M_m)$. If u is any distribution in U that is supported on $U \cap \Sigma$, it is well known that there is an integer N and distributions u_α on $U \cap \Sigma$ for $|\alpha| \leq N$, such that for any $\phi \in C_c^\infty(U)$,

$$\langle u, \phi \omega \wedge dZ \rangle = \sum_{|\alpha| \leq N} \langle u_\alpha, (M'')^\alpha \phi \omega \wedge dZ' \rangle$$

where $\omega = \omega_1 \wedge \cdots \wedge \omega_n$, $dZ = dZ_1 \wedge \ldots dZ_m$, $dZ' = dZ_1 \wedge \cdots \wedge dZ_k$ and $u_\alpha \neq 0$ for some α, $|\alpha| = N$. Here and in what follows, by abusing notations, we are denoting by $\omega \wedge dZ'$ the pullback to Σ.

Observe now that if $h \in C^1(U)$, then

$$dh = \sum_{i=1}^m M_i h dZ_i + \sum_{j=1}^n L_j h \omega_j$$

as can be seen by applying both sides of the equation to the basis

$$\{L_1, \ldots, L_n, M_1, \ldots, M_m\}.$$

Hence if $h \in C^\infty(U)$ and $\phi \in C_c^\infty(U)$, then

$$\begin{aligned}
\langle L_j h, \phi \omega \wedge dZ \rangle &= \int_U (L_j h) \phi \omega \wedge dZ \\
&= (-1)^j \int_U d(h \phi \omega_1 \wedge \cdots \wedge \hat{\omega}_j \cdots \wedge \omega_n \wedge dZ) \\
&\quad - \int_U h(L_j \phi) \omega \wedge dZ \\
&= - \int_U h(L_j \phi) \omega \wedge dZ \\
&= - \langle h, L_j \phi \omega \wedge dZ \rangle \qquad \forall j = 1, \ldots, n.
\end{aligned}$$

It follows that for the distribution u supported on $U \cap \Sigma$ as before, if $\phi \in C_c^\infty(U)$, we have:

$$\langle L_j u, \phi \omega \wedge dZ \rangle = - \langle u, L_j \phi \omega \wedge dZ \rangle$$

$$= - \sum_{|\alpha| \leq N} \langle u_\alpha, (M'')^\alpha (L_j \phi) \, \omega \wedge dZ' \rangle \qquad \forall j = 1, \ldots, n.$$

Assume now that u is also a solution. We will next show that each u_α is a solution of the induced structure $\mathcal{V}_\Sigma$. Fix a point $q \in U \cap \Sigma$. The restrictions of Z_l $(l = k+1, \ldots, m)$ to Σ are solutions of $\mathcal{V}_\Sigma$. By the Baouendi–Treves approximation theorem, for each such Z_l, there is a sequence $\{P_i^l\}_{i=1}^\infty$ of holomorphic polynomials such that

$$Z_l = \lim_{i \to \infty} P_i^l (Z_1, \ldots, Z_k)$$

in $C^\infty(V \cap \Sigma)$ for some neighborhood V of q in $\mathcal{M}$. For each $l = k+1, \ldots, m$, define the sequence $\{f_i^l\}_{i=1}^\infty$ by

$$f_i^l = Z_l - P_i^l (Z_1, \ldots, Z_k).$$

Each $f_i^l \in C^\infty(V)$ and for every l,

$$\lim_{i \to \infty} f_i^l = 0$$

in $C^\infty(V \cap \Sigma)$. Let $f_i = (f_i^{k+1}, \ldots, f_i^m)$ for $i = 1, 2, \ldots$ Fix a multi-index β in $\mathbb{N}^{m-k}$ such that $|\beta| = N$. For any $\phi \in C_c^\infty(V)$ and any $j = 1, \ldots, n$,

$$
\begin{aligned}
0 &= \langle L_j u, f_i^\beta \phi \, \omega \wedge dZ \rangle \\
&= -\langle u, L_j (f_i^\beta \phi) \, \omega \wedge dZ \rangle \\
&= - \sum_{|\alpha| \leq N} \langle u_\alpha, (M'')^\alpha (L_j (f_i^\beta \phi)) \, \omega \wedge dZ' \rangle \\
&= - \sum_{|\alpha| \leq N} \langle u_\alpha, L_j (M'')^\alpha (f_i^\beta \phi) \, \omega \wedge dZ' \rangle \\
&= - \sum_{|\alpha| \leq N} \langle L_j u_\alpha, (M'')^\alpha (f_i^\beta \phi) \, \omega \wedge dZ' \rangle \\
&= -\langle L_j u_\beta, \phi \, \omega \wedge dZ' \rangle + E_i,
\end{aligned}
$$

since $E_i \longrightarrow 0$ on $V \cap \Sigma$ as $i \longrightarrow \infty$ and $M_s f_i^l = \delta_{sl}$. Hence $L_j u_\beta = 0$ whenever $|\beta| = N$. Thus

$$
\begin{aligned}
0 &= \langle L_j u, \psi \, \omega \wedge dZ \rangle \\
&= \sum_{|\alpha| \leq N-1} \langle L_j u_\alpha, (M'')^\alpha (\psi) \, \omega \wedge dZ' \rangle
\end{aligned}
$$

for any $\psi \in C_c^\infty(U)$. Plugging $\psi = f_i^\beta \phi$ with $|\beta| = N-1$ and $\phi \in C_c^\infty(V)$ in these latter equations will likewise lead to

$$L_j u_\beta = 0 \quad \text{whenever } |\beta| = N-1.$$

Continuing this way, we conclude: $L_j u_\alpha = 0 \quad \forall j, \quad \forall \alpha$.

Conversely, it is easy to see that if u has the form

$$\langle u, \phi\, \omega \wedge dZ \rangle = \sum_{|\alpha| \leq N} \langle u_\alpha, (M'')^\alpha \phi\, \omega \wedge dZ' \rangle$$

where each u_α is a solution of $\mathcal{V}_\Sigma$ and some u_α is nontrivial, then u is a solution in U supported on $U \cap \Sigma$. In particular, the distributions σ_α ($\alpha \in \mathbb{N}^{m-k}$) defined by

$$\langle \sigma_\alpha, \phi\, \omega \wedge dZ \rangle = \int_\Sigma (M'')^\alpha \phi\, \omega \wedge dZ'$$

are solutions in U supported on $U \cap \Sigma$. Observe that

$$\sigma_\alpha = (M'')^\alpha \sigma_0.$$

Heuristically speaking then, we may say that each solution u in U supported on $U \cap \Sigma$ can be expressed as

$$u = \sum_{|\alpha| \leq N} u_\alpha \sigma_\alpha$$

where the u_α are solutions of $\mathcal{V}_\Sigma$ in $U \cap \Sigma$.

The distribution σ_0 was introduced by Treves ([**T5**]). The existence of local solutions such as σ_0 supported on a nonopen orbit had previously been established by Baouendi and Rothschild in their proof of the necessity of Tumanov's minimality condition for the holomorphic extension of CR functions into wedges (see Section III.3). We have proved:

THEOREM III.2.10. *Let* $p \in \Sigma$, U, $Z_1, \ldots, Z_m$, $\omega_1, \ldots, \omega_n$ *and* $M'' = (M_{k+1}, \ldots, M_m)$ *be chosen as above. Then,* u *is a solution in* U *supported on* $U \cap \Sigma$ *if and only if* u *can be expressed as*

$$\langle u, \phi\, \omega \wedge dZ \rangle = \sum_{|\alpha| \leq N} \langle u_\alpha, (M'')^\alpha \phi\, \omega \wedge dZ' \rangle,$$

where the u_α *are solutions of* $\mathcal{V}_\Sigma$ *and* u_α *is nontrivial for some* $|\alpha| = N$.

Suppose now u is a distribution supported on Σ. In a chart U about $p \in \Sigma$, write as before

$$\langle u, \phi\, \omega \wedge dZ \rangle = \sum_{|\alpha| \leq N} \langle u_\alpha, (M'')^\alpha \phi\, \omega \wedge dZ' \rangle.$$

Let $N = N(p)$ be the minimum integer for which such a representation is possible. We will call $N(p)$ the *transversal order* of u at p. When u is also a solution, we have:

THEOREM III.2.11. *If u is a solution supported on Σ, the transversal order $N(p)$, $p \in \Sigma$ is constant.*

PROOF. Let $p \in \Sigma$. Choose a chart U as before such that $U \cap \Sigma$ is connected and

$$\langle u, \phi \omega \wedge dZ \rangle = \sum_{|\alpha| \leq N} \langle u_\alpha, (M'')^\alpha \phi \omega \wedge dZ' \rangle$$

where $N = N(p)$. Let $\gamma : [0, 1] \longrightarrow \Sigma$ be an integral curve of X for some smooth section of $\Re V$ such that $\gamma(0) = p$. We consider $N(\gamma(t))$ for those t for which $\gamma(t) \in U$. In any neighborhood of such a $\gamma(t)$, u has the representation above. Moreover, if each u_α for $|\alpha| = N$ vanishes in a neighborhood of such $\gamma(t)$, then since the u_α are solutions for V_Σ, by Theorem III.2.1 the u_α will vanish identically in a neighborhood of p in Σ (for $|\alpha| = N$), leading to the contradiction that $N(p) < N$. Thus whenever $\gamma(t) \in U$, then $N(\gamma(t)) = N(p)$. This argument shows that the set $\{t \in [0, 1] : N(\gamma(t)) = N(p)\}$ is both closed and open, and hence $N(\gamma(1)) = N(p)$. Since any two points of Σ can be joined by a finite number of such γ's, the theorem follows. $\square$

We will continue to assume that the orbit Σ is an embedded orbit. Let $\{U_\alpha\}_\alpha$ be a covering of Σ by open sets in $\mathcal{M}$ such that in each U_α we have a basis $\{L_1^\alpha, \ldots, L_n^\alpha\}$ of V, a basis $\{L_1^\alpha, \ldots, L_n^\alpha, M_1^\alpha, \ldots, M_m^\alpha\}$ of $\mathbb{C}TU_\alpha$, a dual basis

$$\{\omega_1^\alpha, \ldots, \omega_n^\alpha, dZ_1^\alpha, \ldots, dZ_m^\alpha\}$$

where the ω_i^α are exact and $\{Z_1^\alpha, \ldots, Z_m^\alpha\}$ is a complete set of first integrals. We will assume that the restrictions of $\{Z_1^\alpha, \ldots, Z_k^\alpha\}$ to $U_\alpha \cap \Sigma$ form a complete set of first integrals for V_Σ. If u is a solution supported on Σ of transversal order zero, then we know that it is given by distributions u_α in $U_\alpha \cap \Sigma$ in the sense that for any $\phi \in C_c^\infty(U_\alpha)$,

$$\langle u, \phi \, dZ^\alpha \wedge \omega^\alpha \rangle = \langle u_\alpha, \phi \, d(Z^\alpha)' \wedge \omega^\alpha \rangle$$

where in the right-hand side we mean the pullback of the form on Σ. Let $V_\alpha = U_\alpha \cap \Sigma$ and whenever $V_\alpha \cap V_\beta \neq \emptyset$, let $g_{\alpha\beta} \in C^\infty(V_\alpha \cap V_\beta)$ satisfy

$$i^*(dZ_1^\alpha \wedge \cdots \wedge dZ_k^\alpha \wedge \omega^\alpha) = g_{\alpha\beta} \, i^*(dZ_1^\beta \wedge \cdots \wedge dZ_k^\beta \wedge \omega^\beta),$$

where for a form θ in $\mathcal{M}$, $i^*\theta$ denotes the pullback to Σ. Note that the $g_{\alpha\beta}$ are nonvanishing and on $V_\alpha \cap V_\beta$, $g_{\alpha\beta}u_\alpha = u_\beta$. Therefore, $0 = L_j u_\beta = (L_j g_{\alpha\beta})u_\alpha$. If $L_j g_{\alpha\beta}$ is not zero on an open set, then u_α will be zero there. But then u will vanish on this open set and hence on Σ, contradicting the nontriviality of u. Hence the $g_{\alpha\beta}$ are solutions on $V_\alpha \cap V_\beta$. Thus Σ is covered by $\{V_\alpha\}$ and whenever $V_\alpha \cap V_\beta \neq \emptyset$, we have a nonvanishing, smooth solution

$$g_{\alpha\beta} : V_\alpha \cap V_\beta \longrightarrow \mathbb{C}.$$

It follows that we can construct a line bundle $\pi : E \longrightarrow \Sigma$ having the $g_{\alpha\beta}$ as transition functions. In particular, if $(\Sigma, \mathcal{V}_\Sigma)$ is a complex structure, the bundle E becomes a holomorphic line bundle and solutions of $\mathcal{V}$ supported on Σ of transversal order zero correspond to nontrivial holomorphic sections of this bundle. In the situation where Σ is a Stein manifold, it is well known that a holomorphic bundle always has a nontrivial holomorphic section. In other words, we have:

THEOREM III.2.12. *Suppose Σ is an embedded orbit of $\mathcal{V}$ and $(\Sigma, \mathcal{V}_\Sigma)$ is a complex structure. If Σ is a Stein manifold, there are solutions supported on Σ of transversal order 0.*

III.3 The strong uniqueness property for locally integrable solutions

In this section we will consider locally integrable structures $\mathcal{V}$ on an open domain Ω in $\mathbb{R}^N$. The solutions we study will be assumed to be elements of the space L^1_{loc} of locally integrable functions with respect to Lebesgue measure.

DEFINITION III.3.1. *The structure $(\Omega, \mathcal{V})$ satisfies the strong uniqueness property if every solution $u \in L^1_{\text{loc}}(\Omega)$ that is zero on a set of positive measure vanishes identically.*

EXAMPLE III.3.2. Let $\mathcal{V}$ be the structure generated by the Cauchy–Riemann vector fields $\dfrac{\partial}{\partial \bar{z}_j}$ $(1 \leq j \leq n)$ on a domain Ω in $\mathbb{C}^n$. Then $(\Omega, \mathcal{V})$ satisfies the strong uniqueness property.

EXAMPLE III.3.3. Let $\mathcal{V}$ be the structure generated by a real-analytic vector field L on a domain Ω in the plane. Assume that there is only one orbit. Then $(\Omega, \mathcal{V})$ satisfies the strong uniqueness property. Indeed, suppose $u \in L^1_{\text{loc}}(\Omega)$, $Lu = 0$, and u vanishes on a set E of positive measure. Since there is only one orbit, it follows that there is an open set Ω' where L is elliptic and a subset $E' \subseteq \Omega'$ of positive measure where u vanishes. By Corollary I.13.4, the ellipticity of L implies that locally, coordinates can be found in which L

becomes a nonvanishing multiple of the Cauchy–Riemann operator. Hence u vanishes on Ω'. By Theorem III.2.1, u has to vanish on Ω.

EXAMPLE III.3.4. Let $\mathcal{V}$ be the structure generated by $\dfrac{\partial}{\partial x_j}$ $(1 \le j \le n)$ on a domain $\Omega \subseteq \mathbb{R}^N$. It is easy to see that if $n < N$, $(\Omega, \mathcal{V})$ will not satisfy the strong uniqueness property.

It turns out that orbits play a role in the validity of the strong uniqueness property. Before stating the main results, we need to introduce refinements of the concept of an orbit.

DEFINITION III.3.5. *The bundle $\mathcal{V}$ is called minimal at $p \in \Omega$ if, given an open set $p \in U \subseteq \Omega$, there exists a smaller open set $p \in U' \subseteq U$ such that every point in U' can be reached from p by a finite number of integral curves of sections of $\Re\mathcal{V}$ and each integral curve lies in U.*

EXAMPLE III.3.6. If $\mathcal{V}$ is real-analytic and has an open orbit $\mathcal{O}$, then $\mathcal{V}$ is minimal at every $p \in \mathcal{O}$. In this case, we can take $U' = U$.

EXAMPLE III.3.7. Let $\mathcal{V}$ be the structure generated by the vector field

$$ L = \frac{\partial}{\partial x_1} + ig(x_1)\frac{\partial}{\partial x_2} $$

where $g \in C^\infty(\mathbb{R})$, $g > 0$ on $(1, 2)$ and $g \equiv 0$ outside $(1, 2)$. Observe that there is only one orbit in the plane. However, $\mathcal{V}$ is minimal at a point $p = (x_1, x_2)$ if and only if $x_1 \in [1, 2]$.

If $\mathcal{M}$ is a real hypersurface in $\mathbb{C}^n$ with the standard CR structure which is a single orbit, it always has minimal points. This follows from the fact that if there are no minimal points in $\mathcal{M}$, then $\Re\mathcal{V}$ will be closed under Lie brackets leading to a Frobenius foliation of $\mathcal{M}$ by orbits each of dimension $2n - 2$. Each of these orbits is a complex hypersurface. Indeed, the CR bundle $\mathcal{V}$ induces on each orbit $\mathcal{O}$ a locally integrable structure that is CR and elliptic. By Theorem I.10.1, near each $p \in \mathcal{O}$, we can find coordinates $\{x_1, \ldots, x_m, y_1, \ldots, y_m\}$ $(m = n - 1)$ such that the induced structure on $\mathcal{O}$ is generated by $\frac{\partial}{\partial \bar{z}_j}$, $j = 1, \ldots, m$. In particular, any solution on $\mathcal{O}$ is a holomorphic function of the first integrals $(Z_1, \ldots, Z_m)$, $Z_j = x_j + iy_j$. Going back to the complex coordinates $(z_1, \ldots, z_n)$ of $\mathbb{C}^n$, it follows that the restriction to $\mathcal{O}$ of one of these coordinates is holomorphic function of the remaining coordinates. In other words, $\mathcal{O}$ is a complex hypersurface—contradicting the fact that $\mathcal{M}$ is a single orbit. However, there are CR manifolds in $\mathbb{C}^n$ consisting of a single orbit with no minimal points. Examples of such are provided by the following, which appeared in [**Jo1**]:

EXAMPLE III.3.8. Let $\mathcal{M} \subseteq \mathbb{C}^3$ be given by

$$\mathcal{M} = \{(x_1 + iy_1, x_2 + iy_2, x_3 + iy_3) : x_1 = h_1(x_3), x_2 = h_2(x_3)\},$$

where $h_1 \equiv 0$ for $x_3 \geq -\frac{1}{2}$ and h_1 is strictly convex for $x_3 < -\frac{1}{2}$, $h_2 \equiv 0$ for $x_3 \leq \frac{1}{2}$ and h_2 is strictly convex for $x_3 > \frac{1}{2}$. $\mathcal{M}$ is a CR submanifold of codimension 2. It consists of a single orbit but has no minimal points.

The concept of minimality appeared in Tumanov's theorem on the holomorphic extension of CR functions into wedges. Minimality is a necessary and sufficient geometric condition for the holomorphic extension of all CR functions into wedges. In [**Tu1**] Tumanov proved:

THEOREM III.3.9. *Let M be a generic CR submanifold of $\mathbb{C}^N$ and $p \in M$. If M is minimal at p, then for every neighborhood U of p in M there exists a wedge W with edge M centered at p such that every continuous CR function in U extends holomorphically to the wedge W.*

Conversely, if $\mathcal{M}$ is not minimal at p, Baouendi and Rothschild ([**BR**]) proved that there exists a continuous CR function defined in a neighborhood of p in $\mathcal{M}$ which does not extend holomorphically to any wedge of edge $\mathcal{M}$ centered at p.

Tumanov's original definition of minimality was stated differently. He called a CR submanifold of $\mathbb{C}^N$ minimal at p if it contains no proper (i.e., of smaller dimension) CR submanifold of the same CR dimension through p. For the equivalence of the two definitions, we refer the reader to Marson's paper ([**Ma**]).

DEFINITION III.3.10. *Given an involutive structure $\mathcal{V}$ on an open subset Ω of $\mathbb{R}^N$, we say that an orbit $\mathcal{O}$ is a.e. minimal if $\mathcal{V}$ is minimal at p for almost every $p \in \mathcal{O}$ in the sense of Lebesgue measure in $\mathbb{R}^N$.*

Note that if an orbit $\mathcal{O}$ is a.e. minimal, then it is an open orbit.

EXAMPLE III.3.11. If $\mathcal{V}$ is real-analytic and $\mathcal{O}$ is an open orbit, then $\mathcal{O}$ is a.e. minimal since $\mathcal{V}$ is minimal at every $p \in \mathcal{O}$.

Here is a simple example of an a.e. minimal orbit which is not minimal everywhere:

EXAMPLE III.3.12. Let $\mathcal{M} = \mathbb{R}^2$ and $\mathcal{V}$ be the structure generated by

$$L = \frac{\partial}{\partial x} + ib(x, y) \frac{\partial}{\partial y}$$

where $b(x, y)$ is smooth, real-valued, and $b = 0$ only on $(-1, 1) \times \{0\}$. Then $\mathcal{M}$ is minimal exactly at the points in $\mathcal{M} \setminus ((-1, 1) \times \{0\})$.

We can now state the main result on strong uniqueness:

THEOREM III.3.13. *Let $\mathcal{V}$ be a locally integrable structure defined on a connected open set Ω in $\mathbb{R}^N$. Assume that $\Omega = \mathcal{O} \cup F$ where $\mathcal{O}$ is an open a.e. minimal orbit of $\mathcal{V}$ and F is a set of measure zero. Then any solution $u \in L^1_{\text{loc}}(\Omega)$ that vanishes on a set of positive measure must vanish identically.*

Theorem III.3.13 was proved in [**BH2**]. According to the theorem, if Ω satisfies the hypotheses, then almost every point $p \in \Omega$ can be reached from a fixed point $q \in \mathcal{O}$ by a piecewise smooth curve consisting of integral curves of smooth sections of $\Re\mathcal{V}$. We may say that Ω has an a.e. reachability property with respect to $\mathcal{V}$. Thus Ω satisfies the a.e. reachability property if and only if Ω admits a trivial decomposition, that is, if it can be expressed as the union of an open orbit and a set of measure zero. We note, however, that this a.e. reachability condition is not necessary for the conclusion of the theorem. For example, the structure $\mathcal{V}$ generated on the 2-torus $\mathbb{T}^2$ by a real globally hypoelliptic vector field L has the strong uniqueness property although the torus does not admit a trivial decomposition. However, local a.e. reachability is necessary if the conclusion of Theorem III.3.13 is to hold on any base of connected neighborhoods of a given point. Indeed, we have the following [**BH2**] partial converse to Theorem III.3.13:

THEOREM III.3.14. *Let $\mathcal{V}$ be a sub-bundle of $\mathbb{C}T\Omega$ where $\Omega \subseteq \mathbb{R}^N$ is open. Assume there is a base $\{\Omega_j\}_{j=1}^{\infty}$ of connected neighborhoods of p which do not admit a trivial decomposition. Then there is a base of connected neighborhoods $U_k \subseteq \Omega_k$ of p and nontrivial solutions $u_k \in L^1(U_k)$ for which the sets $\{u_k = 0\}$ all have positive measure.*

We remark that in Theorem III.3.14, $\mathcal{V}$ is not assumed to be locally integrable. It is not even assumed that it is involutive. Thus for analytic involutive structures $\mathcal{V}$ (which are always locally integrable), Theorems III.3.13 and III.3.14 establish the local equivalence between a.e. reachability and the uniqueness property that local solutions are determined on sets of positive measure. We will prove Theorem III.3.13 in the important situation where $\mathcal{V}$ is the tangential Cauchy–Riemann bundle of a CR manifold embedded in $\mathbb{C}^N$

(see Theorem III.3.15 below). In fact, by using Marson's ([**Ma**]) trick of embedding a general locally integrable structure $\mathcal{V}$ into a CR structure, one can deduce Theorem III.3.13 from Theorem III.3.15 (see [**BH2**] for the details). Theorem III.3.15 states that the strong uniqueness property that holomorphic functions have—that of being determined on any domain by their values on any subset of positive measure or, equivalently, that their zero sets have measure zero except in a trivial case—is inherited by their boundary values at the edge of the wedge where they are defined. In the particular classical case of a holomorphic function of one variable defined on a disk, this principle is well known and is attributed to Priwaloff and Riesz. Thus Theorem III.3.15 is, to a certain extent, a higher-dimensional version of the theorem of Priwaloff and Riesz.

THEOREM III.3.15. *Let $\mathcal{M} \subseteq \mathbb{C}^N$ be a generic CR manifold of codimension d $(N = n + d)$. Assume that $W \subseteq \mathbb{C}^N$ is a wedge with edge $\mathcal{M}$. Suppose F is a holomorphic function of tempered growth on W with distribution boundary value $f \in L^1_{\text{loc}}(\mathcal{M})$. If f vanishes on a subset E of positive measure, then $f \equiv 0$ in a neighborhood of any Lebesgue density point of E.*

In the proof of Theorem III.3.15, we will use the following lemma where Σ is a smooth hypersurface in $\mathbb{C}^n$, f is a CR function on Σ, and $f \in L^p(\Sigma)$ for some $1 \leq p \leq \infty$. Suppose also that f extends to a holomorphic function F on a side Σ^+, that is, f is the boundary value of F in the distribution sense. Then we have:

LEMMA III.3.16. *For any $\Sigma' \subset\subset \Sigma$, and a sufficiently small ball B in $\mathbb{C}^n$ containing zero, the restrictions of F to the hypersurfaces $\{z \in B : \text{dist}(z, \Sigma') = t\}$ have uniformly bounded L^p norms. In particular, $F \in L^p(B \cap \Sigma^+)$.*

PROOF. Without loss of generality, we may assume that Σ is part of the boundary of a bounded open set D with smooth boundary such that $D \subseteq \Sigma^+$. Let H be harmonic in D with boundary value f on Σ and 0 off Σ. By the classical h^p theory for harmonic functions, the restrictions H_t of H to the hypersurfaces $S_t = \{z \in D : \text{dist}(z, \partial D) = t\}$ (t small) are all in L^p and $\|H_t\|_{L^p(S_t)} \leq \|f\|_{L^p(\Sigma)}$. Moreover, it is well known that 'dist$(z, \partial D)$' can be replaced by any defining function for ∂D. Since F is holomorphic in Σ^+ and has a boundary value on Σ, there exist $C, k > 0$ such that for any $z \in D$,

$$|F(z)| \leq C \, \text{dist}(z, \Sigma)^{-k}.$$

This may require contracting Σ. It follows that F has a boundary value which is a distribution on ∂D. Let

$$u = F - H.$$

u is harmonic in D, has a distributional boundary value bu on ∂D which is 0 on the piece Σ. We wish to show u is smooth up to Σ. Let $G(x, y)$ be the Green's function for D and $P(x, y)$ its Poisson kernel. We recall that

$$P(x, y) = -N_y G(x, y) \text{ for } x \in D, y \in \partial D$$

where $N_y = $ the unit outer normal to D at y. Fix $x \in D$. The function $y \longmapsto G(x, y)$ is 0 on ∂D and positive on $D \backslash \{x\}$. By Hopf's lemma, $N_y G(x, y) \neq 0$ for all $y \in \partial D$. Hence for ϵ small enough, the open sets

$$D_\epsilon = \{y \in D : G(x, y) > \epsilon\}$$

have smooth boundaries. Observe that if $\widetilde{G}_\epsilon(z, y)$ is the Green's function for D_ϵ, then

$$\widetilde{G}_\epsilon(x, y) = G(x, y) - \epsilon.$$

Hence the Poisson kernel $P_\epsilon(z, y)$ for D_ϵ satisfies

$$P_\epsilon(x, y) = -N_y^\epsilon G(x, y),$$

where N_y^ϵ is the unit outer normal to D_ϵ at y. We thus have

$$u(x) = \int_{\partial D_\epsilon} P_\epsilon(x, y) u(y) \, d\sigma_\epsilon(y)$$

$$= \int_{\partial D} P_\epsilon(x, \Pi_\epsilon^{-1}(y)) u(\Pi_\epsilon^{-1}(y)) J_\epsilon(y) \, d\sigma(y),$$

where $\Pi_\epsilon : \partial D_\epsilon \longrightarrow \partial D$ is the normal projection map and J_ϵ is the Jacobian of Π_ϵ^{-1}. Since $P_\epsilon(x, y) = -N_y^\epsilon G(x, y)$, as $\epsilon \longrightarrow 0^+$,

$$P_\epsilon(x, \Pi_\epsilon^{-1}(y)) J_\epsilon(y) \longrightarrow P(x, y)$$

in $C^\infty(\partial D)$. It follows that for any $x \in D$,

$$u(x) = \langle bu, P(x, .) \rangle.$$

This latter formula, together with the vanishing of bu on Σ, tells us that u is C^∞ up to the boundary piece Σ. Since $F = H + u$ and $H \in h^p(D)$, the assertions of the theorem follow. $\square$

COROLLARY III.3.17. [Nontangential Convergence] *Let f and F be as in the lemma and D be as in the proof of the lemma. For $\alpha > 1$ and $A \in \Sigma$, define*

$$\Gamma_\alpha(A) = \{z \in D : |z - A| < \alpha\delta(z)\},$$

where $\delta(z) = \text{dist}(z, \partial D)$. Then

$$\lim_{\Gamma_\alpha(A) \ni z \to A} F(z) = f(A)$$

for almost all A in Σ.

PROOF. Recall from the proof that $F = H + u$. Since u is smooth up to the piece Σ and bu vanishes on Σ, $\lim_{D \ni z \to A} u(z) = 0$ for all $A \in \Sigma$. The corollary therefore follows from the fact that $H \in h^p(D)$ and that on Σ, $H = f$. $\square$

III.4 Proof of Theorem III.3.15

To prove Theorem III.3.15, we may assume that $0 \in \mathcal{M}$ is a density point of E and that $\mathcal{M}$ near 0 is defined by $\Im w = \phi(x, y, \Re w)$, where $z = x + iy \in \mathbb{C}^n$ and $w \in \mathbb{C}^d$, $N = n + d$. The function ϕ is real-valued, smooth, $\phi(0) = 0$, and $d\phi(0) = 0$. We may also assume that the wedge $\mathcal{W}$ contains a wedge of the form

$$\{(z, w) : w = s + i\phi(x, y, s) + iv, |z| < 2\delta, |s| < 2\delta, |v| < 2\delta, v \in \Gamma\}$$

for some open convex cone $\Gamma \subset \mathbb{R}^d$ and $\delta > 0$. We may suppose that $\|d\phi(x, y, s)\| < \frac{1}{4}$ for $|x|, |y|, |s| < 2\delta$. Without loss of generality, assume that

$$\Gamma = \{v = (v', v_d) : |v'| < 2\delta v_d\}.$$

Let

$$\widetilde{\Gamma} = \{(y, t) \in \mathbb{R}^{n+d} : |(y, t')| < \delta t_d, t = (t', t_d)\}.$$

For $|y_0| < \delta$, the set

$$\mathcal{W}_{y_0} = \{(x + iy_0 + iy, s + i\phi(x, y_0, s) + it) : (y, t) \in \widetilde{\Gamma}, |x|, |y|, |s|, |t| < \delta\}$$

is contained in the wedge $\mathcal{W}$. Indeed, this follows from the definitions of Γ and $\widetilde{\Gamma}$ and the assumption on the norm of $d\phi$. Observe that $\mathcal{W}_{y_0}$ is a wedge in $\mathbb{C}^N$ with a maximally totally real edge

$$\mathcal{M}_{y_0} = \{(x + iy_0, s + i\phi(x, y_0, s)) : |x| < \delta, |s| < \delta\}.$$

Fix y_0, $|y_0| < \delta$ such that

$$(x, s) \longmapsto f(x, y_0, s)$$

is in L^1 and the $(n + d)$-dimensional set $\mathcal{M}_{y_0}$ intersects E in a set of positive measure. Note that F is holomorphic and of tempered growth in the wedge $\mathcal{W}_{y_0}$. Hence F has a distribution boundary value bF on $\mathcal{M}_{y_0}$. We will eventually show that bF agrees with f on $\mathcal{M}_{y_0}$ for almost all y_0. Assuming this for now, it is clear that Theorem III.3.15 would follow if we show that $F \equiv 0$ on W_{y_0}. This kind of reduction to a maximally totally real manifold also appears in the proof of theorem 7.2.6 in [**BER**].

We are thus led to consider a maximally totally real submanifold Σ of $\mathbb{C}^m$ given in a neighborhood U of $0 \in \mathbb{C}^m$ by

$$t = \phi(s), \quad s \in U,$$

where $w = s + it$ are standard complex coordinates in $\mathbb{C}^m$, ϕ is a smooth $\mathbb{R}^m$-valued function defined near $0 \in \mathbb{R}^m$, and $\phi(0) = d\phi(0) = 0$. We recall that a $\mathbb{C}^m$-valued analytic disk is a map $A : \overline{\Delta} \to \mathbb{C}^m$ of class $C^{1+\alpha}$ from the closed unit disk of the complex plane which is holomorphic on Δ (here $0 < \alpha < 1$ is fixed once from now on). An analytic disk A is said to be partially attached to Σ at p if (i) $A(e^{i\theta}) \in \Sigma$ for $|\theta| \le \frac{\pi}{2}$ and (ii) $A(1) = p$. The Banach space of $\mathbb{C}^m$-valued analytic disks will be denoted by $\mathcal{A}^m$. We recall theorem 7.4.12 of [**BER**] on the existence of analytic disks partially attached to Σ:

THEOREM III.4.1 ([BER].). *There exist a neighborhood $U \times V$ of the origin $(0,0) \in \mathbb{R}^m \times \mathbb{R}^m$ and a smooth map $U \times V \ni (s,v) \mapsto A_{s,v} \in \mathcal{A}^m$ satisfying the following properties for all $(s,v) \in U \times V$:*

(i) $A_{s,v}(1) = s + i\phi(s);$

(ii) $A_{s,v}(e^{i\theta}) \in \Sigma$ *for* $|\theta| \le \pi/2;$

(iii) $\frac{d}{d\theta}(A_{s,v})(e^{i\theta})|_{\theta=0} = v + i\phi'(s) \cdot v.$

(iv) $\frac{d}{dr}(A_{s,v})(r)|_{r=1} = iv - \phi'(s) \cdot v.$

Notice that we have included (iv) here since it follows from (iii) and the Cauchy–Riemann equations satisfied by $\zeta \mapsto A_{s,v}(\zeta)$ at $\zeta = 1$. The meaning of (i) and (ii) is that $A_{s,v}$ is partially attached to Σ at $p = (s, \phi(s))$ and (iii) implies that we can choose a neighborhood $\tilde{U} \subset U$ of the origin and a small $\epsilon > 0$ such that for every $p = s_0 + i\phi(s_0)$, $s_0 \in \tilde{U}$, the map

$$(0, \epsilon) \times S^{m-1}(0, \epsilon) \quad \ni \quad (\theta, \omega) \longmapsto A_{s_0, \omega}(e^{i\theta}) \in \Sigma$$

yields a $C^{1+\alpha}$ local system of polar coordinates centered at p on Σ, where $S^{m-1}(0, \epsilon)$ denotes the sphere of radius ϵ centered at $0 \in \mathbb{R}^m$. In particular, given $v_0 \in \mathbb{R}^m$, $|v_0| = \epsilon$, and $p_1 = s_1 + i\phi(s_1)$, $s_1 \in \tilde{U}$, we may find $s_0 \in U$ and $\theta_0 \in (0, \epsilon)$ such that

$$p_1 = A_{s_0, v_0}(e^{i\theta_0}).$$

Assume that p_1 is a density point of a measurable set $E \subseteq \Sigma$ (in particular E has positive measure) and let $U_0 \ni s_0$ and $V_0 \ni v_0$ be open sets of diameter $< \epsilon$. Consider the set

$$\widetilde{E}^\epsilon = \left\{ (s, v) \in U_0 \times (S^{m-1}(0, \epsilon) \cap V_0) : \int_0^\epsilon \chi_E(A_{s,v}(e^{i\theta})) \, d\theta > 0 \right\}$$

where χ_E denotes the characteristic function of E. We observe that we may assume without loss of generality that $\widetilde{E}^\epsilon$ has positive $(2m-1)$-dimensional measure. Indeed, the function

$$\theta \longmapsto \int_{\substack{|s-s_0|<\epsilon \\ |v-v_0|<2\epsilon}} \chi_E(A_{s,v}(e^{i\theta})) \, ds \, dv$$

is continuous and assumes a positive value at $\theta = 0$ because $A_{s,v}(1) = s + i\phi(s)$. Hence,

$$\int_{\substack{|s-s_0|<\epsilon \\ |v-v_0|<2\epsilon}} \left(\int_0^\epsilon \chi_E(A_{s,v}(e^{i\theta})) \, d\theta \right) ds \, dv > 0$$

and writing v in polar coordinates we see that for some $0 < \epsilon' < 2\epsilon$ our claim is true for $\widetilde{E}^{\epsilon'}$. We fix such an $\epsilon' > 0$ and, dropping any reference to the dependence on ϵ', simply write $\widetilde{E}^{\epsilon'} = \widetilde{E}$.

Consider now the map

$$U_0 \times (S^{m-1}(0, \epsilon) \cap V_0) \times (1 - \epsilon, 1) \ni (s, v, r) \longmapsto A_{s,v}(r) \in \mathbb{C}^m. \quad \text{(III.3)}$$

Taking account of (iv) we note that this map has rank $2m$ for small $\epsilon > 0$ and maps $\{s\} \times (S^{m-1}(0, \epsilon) \cap V_0) \times (1 - \epsilon, 1)$ onto $B_p \backslash \{p\}$, where B_p is a $C^{1+\alpha}$-differentiable m-ball that intersects Σ orthogonally at $p = s + i\phi(s)$. Indeed, the respective tangent spaces at p are

$$T_p\Sigma = \{s + i\phi(s) + v + i\phi'(s) \cdot v, \quad v \in \mathbb{R}^m\} \quad \text{and}$$
$$T_pB_p = \{s + i\phi(s) + iv - \phi'(s) \cdot v, \quad v \in \mathbb{R}^m\}.$$

Since the map (III.3) is a local diffeomorphism, it takes $\widetilde{E}$ onto a set of positive measure $\widehat{E}$ which is contained in the union of the disks $\bigcup\{A_{s,v} : (s, v) \in \widetilde{E}\}$. We could say that these disks are strongly attached to E in the sense that for any $(s, v) \in \widetilde{E}$ the set of boundary points $\{A_{s,v}(e^{i\theta}) : 0 < \theta < \epsilon\}$ intersects E at a non-negligible set of values of θ. Consider now a holomorphic function F of slow growth defined in a wedge $W = \Sigma \times \Gamma$ with edge Σ possessing a weak trace $f \in L^p(\Sigma)$ and assume that f vanishes on E. Assume furthermore that $v_0 \in \Gamma$. We will now sketch how we try to prove that F must vanish. First one proves that if $\epsilon > 0$ is small enough and $(s, v) \in \widetilde{E}$ the portion $A^\epsilon_{s,v}$ of the disk $A_{s,v}$ described by the inequalities $-\epsilon < \theta < \epsilon$ and $1 - \epsilon < r < 1$ is contained in the wedge $\mathcal{W}$. Then the composition $F(A_{s,v}(re^{i\theta}))$ is defined for $-\epsilon < \theta < \epsilon$, $1 - \epsilon < r < 1$, is holomorphic and has a weak boundary value which, for a.e. $(s, v) \in \widetilde{E}$, is given by—and the proof of this fact is our second step— $f(A_{s,v}(e^{i\theta}))$. The third step is to prove that for a.e. (s, v), the restriction of f to the curve $(\epsilon/2, \epsilon) \ni \theta \mapsto A_{s,v}(e^{i\theta})$ is in L^p. Hence, by Corollary III.3.17 and the classical theorem of Priwaloff, the holomorphic function of one complex variable $F(A_{s,v}(re^{i\theta}))$ vanishes identically for $-\epsilon < \theta < \epsilon$, $1 - \epsilon < r < 1$, in particular for $\theta = 0$. But we know that letting (s, v, r) vary on $\widetilde{E} \times (1 - \epsilon, 1)$ and keeping $\theta = 0$, the union of $\{A_{s,v}(re^{i\theta})\}$ covers $\widehat{E}$. Thus, F vanishes a.e. on $\widehat{E}$ and so must vanish identically. The proof of the second step involves a discussion about the trace which will be developed next.

We begin our considerations by looking at the simplest case of a holomorphic function of one complex variable $F(x + iy)$ defined for $|x| < 1, 0 < y < 2$ which satisfies the inequality

$$|F(x + iy)| \leq C |\log y|, \quad |x| < 1, \quad 0 < y < 2. \tag{III.4}$$

We assume (III.4) for simplicity but the argument below can be iterated to handle the case $|F(x + iy)| \leq C|y|^{-N}$. The standard manner of defining the weak trace f of F as an element of $\mathcal{D}'$ is through the formula

$$\langle f, \psi \rangle = \lim_{\varepsilon \searrow 0} \int F(x + i\varepsilon) \psi(x) \, dx, \quad \psi \in C_c^\infty(-1, 1). \tag{III.5}$$

In formula (III.5) we see that for each fixed x the argument of F describes a straight vertical segment $\varepsilon \mapsto x + i\varepsilon$ that flows toward x as $\varepsilon \to 0$. We wish to see what happens if we change each vertical segment to a curve $\varepsilon \mapsto x + \alpha(x, \varepsilon) + i\varepsilon$. We will assume that $(-1, 1) \times [0, 1) \ni (x, \varepsilon) \mapsto \alpha$ is of class C^2 (we would need class C^{N+2} if we were assuming $|F(x + iy)| \leq C|y|^{-N}$ instead of (III.4)) and that $\alpha(x, 0) = 0$, $|x| < 1$. The latter assumption simply means that the curve $\varepsilon \mapsto x + \alpha(x, \varepsilon) + i\varepsilon$ flows toward x as $\varepsilon \to 0$. Thus,

$$\left(\frac{\partial}{\partial x} \right)^j \alpha(x, 0) = 0, \quad j = 0, 1, 2. \tag{III.6}$$

We now define

$$\langle \widetilde{f}, \psi \rangle = \lim_{\varepsilon \searrow 0} \int F(x + \alpha(x, \varepsilon) + i\varepsilon) \psi(x) \, dx, \quad \psi \in C_c^\infty(-1, 1)$$

and wish to prove that $f = \widetilde{f}$. To that end we write

$$F(x + \alpha(x, \varepsilon) + i\varepsilon) = F(x + \alpha(x, \varepsilon) + i) - i \int_\varepsilon^1 F'(x + \alpha(x, \varepsilon) + it) \, dt.$$

It follows from (III.6) that if x belongs to a compact part of $(-1, 1)$ and ε is small, $|\alpha_x(x, \varepsilon)| < 1/2$. We will assume for simplicity that $|\alpha_x(x, \varepsilon)| < 1/2$ holds everywhere. Then

$$\int F(x+\alpha(x,\,\varepsilon)+i\varepsilon)\,\psi(x)\,dx = \int F(x+\alpha(x,\,\varepsilon)+i)\psi(x)\,dx$$

$$+\,i\int\int_{\varepsilon}^{1} F(x+\alpha(x,\,\varepsilon)+it)\,dt\,\frac{\partial}{\partial x}\left(\frac{\psi(x)}{1+\alpha_x(x,\,\varepsilon)}\right)dx.$$

Letting $\varepsilon \to 0$ and taking account of (III.6), we obtain $\langle f, \psi \rangle = \langle \widetilde{f}, \psi \rangle$ as we wished.

From now on we return to the general situation of a maximally totally real submanifold Σ of $\mathbb{C}^m$ and a holomorphic function F defined on a wedge $W = \Sigma \times \Gamma$ and possessing a trace $f \in L^p(\Sigma)$. We will now take advantage of two facts:

(1) *The formula*

$$\langle f, \psi \rangle = \lim_{\varepsilon \to 0}\int F(s+\alpha(s,\,\varepsilon)+i(\phi(s)+\varepsilon v_0+\beta(s,\,\varepsilon)))\,\psi(s)\,ds$$

is independent of the family of curves

$$\sigma(s,\,\varepsilon) = (s+\alpha(s,\,\varepsilon),\,\varepsilon v_0+\beta(s,\,\varepsilon))$$

as long as all curves $\varepsilon \mapsto \sigma(\varepsilon, s)$ *are contained in* W, *they have the right number of bounded derivatives, and* $\alpha(s,0) = \beta(s,0) = 0$, $s \in U$ *(the assumptions imply that* $v_o \in \Gamma$*).*

(2) *The analytic disks described in theorem 7.4.12 of* **[BER]** *can be taken of class* $C^{k+\alpha}$ *rather than* $C^{1+\alpha}$ *where* k *is a large positive integer.*

The first fact follows from proposition 7.2.22 in **[BER]**. The second fact is true because theorems 6.5.4 and 7.4.12 in **[BER]** are valid with the same proofs if the analytic disks are taken to be in $C^{k,\alpha}$ for a fixed positive integer k. In the proof of theorem 7.4.12, the function γh has to be modified so that one gets a C^k extension.

Set $s' = (s_1, \ldots, s_{m-1})$. We will assume without loss of generality that

(i) For any $\epsilon > 0$ the set

$$\{s' : \quad |s'| < \epsilon \text{ and } (s', 0, v_0) \in \widetilde{E}\} \tag{III.7}$$

has positive measure.

(ii) $v_0 = (0, \ldots, 0, a)$ for some small $a > 0$.

For $|s'| < \epsilon$, $|\theta| < \epsilon$ consider the map

$$(s',\,\theta) \longmapsto A_{(s',0),v_0}(e^{i\theta})$$

which for small ϵ has an injective differential. We consider a family of curves $\sigma(p, \varepsilon)$ defined by

$$p = A_{(s',0),v_0}(e^{i\theta}), \qquad \sigma(p, \varepsilon) = A_{(s',0),v_0}[(1-\varepsilon)e^{i\theta}].$$

Observe that $\sigma(p,0) = p$ and that we are implicitly using (s', θ) as local coordinates. For small ε the curves $\varepsilon \mapsto \sigma(p, \varepsilon)$ are contained in $\mathcal{W}$ and it follows from our assumptions that for any test function ψ with small support around $s = 0$,

$$
\begin{aligned}
\langle f, \psi \rangle &= \lim_{\varepsilon \to 0} \int F(p + \Re\sigma(p, \varepsilon) + i\Im\sigma(p, \varepsilon))\, \psi(p)\, ds \\
&= \lim_{\varepsilon \to 0} \int F(A_{(s',0),v_0}[(1-\varepsilon)e^{i\theta}])\, \psi(s', \theta)\, J(s', \theta)\, ds' d\theta.
\end{aligned}
$$

Assuming that $f \in L^p(\Sigma)$ and using Fubini's theorem in the coordinates (s', θ), we see that for a.e. $|s'| < \epsilon$, the function $\theta \mapsto A_{(s',0),v_0}[e^{i\theta}]$ is in L^p. Fixing such an s' is equivalent to fixing an analytic disk with the property that the restriction of f to a portion of its boundary that is contained in Σ is in L^p. We now take test functions such that ψJ has separated variables, i.e., $\psi(s', \theta)J(s', \theta) = \psi_1(s')\psi_2(\theta)$. Since F has tempered growth, so does the compose $F \circ A_{(s',0),v_0}$ and it follows that

$$\langle \tilde{f}_{s'}, \psi_2 \rangle = \lim_{\varepsilon \to 0} \int F(A_{(s',0),v_0}[(1-\varepsilon)e^{i\theta}])\, \psi_2(\theta)\, d\theta$$

defines a distribution in θ that depends continuously on s' as a parameter (use the usual method to define the trace, integrating by parts with respect to θ). We further have

$$\int \langle \tilde{f}_{s'}, \psi_2 \rangle \psi_1(s')\, ds' = \langle f, \psi \rangle.$$

We may now reason as in Lemma II.3.2 to conclude that for a.e. s', $|s'| < \epsilon$, $\tilde{f}_{s'} \in L^p(-\epsilon, \epsilon)$ and $\tilde{f}_{s'}(\theta) = f(s', \theta)$. If s' is in the set (III.7) and $\tilde{f}_{s'}(\theta) = f(s', \theta)$ holds, then $\zeta \mapsto F(A_{(s',0),v_0}(\zeta))$ has an L^p boundary value that vanishes on a set of positive measure which implies that $\zeta \mapsto F(A_{(s',0),v_0}(\zeta))$ vanishes identically. We conclude that for a.e. s' on the set (III.7), $F(A_{(s',0),v_0}(\zeta)) = 0$, or equivalently, that the set

$$E(0, v_0) = \{s' : \quad |s'| < \epsilon \text{ such that } F \circ A_{(s',0),v_0}(\zeta) \equiv 0\}$$

has positive measure. A similar conclusion could have been reached for the set

$$E(s_m, v) = \{s' : \quad |s'| < \epsilon, \quad \text{such that } F \circ A_{(s',s_m),v}(\zeta) \equiv 0\},$$

where s_m is a small number and $|v - v_0|$ is small. Thus, the set $\{(s, v)\}$ such that $F \circ A_{s,v}(\zeta) \equiv 0$ has positive measure and so does the union of the corresponding partially attached disks.

III.5 Uniqueness for approximate solutions

In this and the following sections we will present uniqueness results for the approximate solutions of two structures: locally integrable structures in the plane defined by vector fields which are of a fixed finite type on their characteristic set and real-analytic structures with $m = 1$. The theorems were proved by Cordaro ([**Cor2**]).

Suppose $\mathcal{V}$ is a locally integrable structure defined on a manifold Ω, dim $(\Omega) = N = m + n$, and the fiber dimension of $\mathcal{V}$ over $\mathbb{C}$ equals n. By going to the quotient, the exterior derivative defines a differential operator

$$C^\infty(\Omega) \stackrel{d_0'}{\to} C^\infty(\Omega, \mathbb{C}T^*\Omega/T')$$

where $T' = \mathcal{V}^\perp$. Equip the manifold $C^\infty(\Omega, \mathbb{C}T^*\Omega/T')$ with a hermitian metric. Observe that a solution for the structure $\mathcal{V}$ is a function or distribution u that satisfies $d_0'u = 0$. If $u \in L^1_{\text{loc}}(\Omega)$, we will say that u is an approximate solution for the structure $\mathcal{V}$ if the coefficients of $d_0'u$ are locally in L^1 and given any $p \in \Omega$, there is a number $M > 0$ such that near p,

$$|d_0'u| \le M|u| \quad \text{a.e. in } U.$$

One way in which approximate solutions may arise is as follows: suppose $F: \Omega \times \mathbb{C} \to C^\infty(\Omega, \mathbb{C}T^*\Omega/T')$ satisfies $|F(p, z) - F(p, z')| \le M|z - z'|$ and u and v are two C^1 solutions of the semilinear equation

$$d_0'w(p) = F(p, w(p)).$$

Then the function $u - v$ is an approximate solution for the structure $\mathcal{V}$. Recall next from Corollary I.10.2 that near a point in Ω, coordinates $(x_1, \ldots, x_m, t_1, \ldots, t_n)$ for Ω and local generators $L_1, \ldots, L_n$ for $\mathcal{V}$ can be chosen so that $dt_j(L_k) = \delta_{jk}$, $j, k = 1, \ldots, n$. With such a choice of coordinates and generators, we can identify the bundle $C^\infty(\Omega, \mathbb{C}T^*\Omega/T')$ with the one spanned by the forms $dt_1, \ldots, dt_n$ and the operator d_0' can be realized as

$$Lu = \sum_{j=1}^n L_j u\, dt_j.$$

Before we discuss the uniqueness results, we will present a description of smooth, planar vector fields which have a uniform finite type on their characteristic set.

PROPOSITION III.5.1. *Let L be a C^∞ nonvanishing vector field defined near the origin in $\mathbb{R}^2$ and let Σ denote its characteristic set. If L is of uniform finite type k on Σ, then Σ is contained in a one-dimensional manifold. Moreover, if Σ is a one-dimensional manifold, then L is never tangent to Σ.*

PROOF. We may choose coordinates (x, y) near 0 so that L is a nonvanishing multiple of

$$L' = \frac{\partial}{\partial y} + ib(x, y)\frac{\partial}{\partial x},$$

with b real-valued and C^∞ near 0. Without loss of generality, let $L = L'$. Then

$$\Sigma = \{p : b(p) = 0\}.$$

The uniform type condition implies that

$$\frac{\partial^j b}{\partial y^j} \equiv 0$$

on Σ for $j < k - 1$ and

$$\frac{\partial^k b}{\partial y^k} \neq 0$$

on Σ. Hence if

$$f(x, y) = \frac{\partial^{k-1} b}{\partial y^{k-1}}(x, y),$$

then Σ is contained in the manifold $\{f(x, y) = 0\}$ which has a parametrization $\{(x, y(x))\}$ for some smooth $y(x)$. $\qquad\square$

PROPOSITION III.5.2. *Suppose L and Σ are as in Proposition III.5.1 and that Σ is a one-dimensional manifold. Assume L is locally integrable in a neighborhood of 0. Then we can find coordinates (s, t) about 0 in which*

$$Z(s, t) = s + i\phi(s, t)$$

is a first integral of L where $\phi(s, t)$ is real-valued and

$$\phi(s, t) = \alpha(s) + t^k \beta(s, t)$$

for some nonvanishing β near 0.

PROOF. We first flatten Σ near the origin so that in coordinates (x, y), $\Sigma = \{(x, 0)\}$. By Proposition III.5.1, L is not tangent to Σ and so if $Z(x, y)$ is a first integral near the origin, then $Z_x(0, 0) \neq 0$. Assume $Z(0, 0) = 0$. Let $s = \Re Z(x, y)$ and $t = y$. Then in (s, t) coordinates,

$$Z = s + i\phi(s, t)$$

and we may take

$$L = \frac{\partial}{\partial t} - \left(\frac{i\phi_t}{1 + i\phi_s}\right)\frac{\partial}{\partial s}.$$

The finite type assumption then implies that

$$\phi(s, t) = \alpha(s) + t^k \beta(s, t), \quad \beta(0) \neq 0,$$

for some smooth β.   □

PROPOSITION III.5.3. *Suppose L and Σ are as in Proposition III.5.2. If the uniform type k is even, then there are coordinates in which $Z(x, y) = x + iy^k$ is a first integral.*

PROOF. By Proposition III.5.2, we have a first integral

$$Z(s, t) = s + i\phi(s, t), \quad \text{where}$$

$$\phi(s, t) = \alpha(s) + t^k \beta(s, t), \quad \beta(0) \neq 0.$$

We may assume $\beta > 0$ near the origin. For ϵ small, let $\Omega = Z(D_\epsilon(0))$ where $D_\epsilon(0)$ denotes the disk centered at 0 of radius ϵ. Let Ω' be a smooth subdomain of Ω such that $0 \in \partial\Omega'$ and the boundary part of Ω' near 0 is $\{(s, \alpha(s))\}$. By the Riemann mapping theorem, there exists a holomorphic function H which is a diffeomorphism up to $\partial\Omega'$ such that

$$H(\Omega') \subset \{(x, y) : y > 0\},$$

and $H(s + i\alpha(s)) \in \mathbb{R}$. Let $W(s, t) = H \circ Z(s, t)$. Then $LW = 0$ and $dW \neq 0$ in a neighborhood of the origin. From the form of ϕ and the fact that $\Im H \circ Z(s, 0) = 0$, we have

$$\Im H \circ Z(s, t) = t^k \tilde{\beta}(s, t),$$

where $\tilde{\beta} > 0$ near the origin. Let $x = \Re H \circ Z(s, t)$ and $y = t\tilde{\beta}(s, t)^{\frac{1}{k}}$. It can easily be checked that these are coordinates near 0 and in these coordinates,

$$W(x, y) = x + iy^k$$

is a first integral.   □

DEFINITION III.5.4. *A locally integrable structure $(\mathcal{M}, \mathcal{V})$ is called hypocomplex if every solution u is locally of the form $H \circ Z$ where H is holomorphic and $Z = (Z_1, \ldots, Z_m)$ is a complete set of first integrals.*

PROPOSITION III.5.5. *Suppose L and Σ are as in Proposition III.5.2. If k is odd, then there are coordinates (x, y) in which $Z(x, y) = x + iy^k$ is a first integral of L if and only if for any first integral W of L, there is a biholomorphism near 0 mapping $W(\Sigma)$ into the real axis.*

PROOF. Since k is odd, if we take the first integral $Z(s, t) = s + i\phi(s, t)$ with $\phi(s, t) = \alpha(s) + t^k \beta(s, t)$, $\beta(0) \neq 0$ as in Proposition III.5.2, we see that L is hypocomplex. Therefore, to prove the necessity, we only need to do it for this first integral. Suppose then (x, y) are coordinates in which $x + iy^k$ is a solution. Let $F = U + iV$ denote this diffeomorphism and we may assume $F(0) = 0$. Then F maps the characteristic set of L to that of

$$\frac{\partial}{\partial y} - iky^{k-1}\frac{\partial}{\partial x},$$

and so $V(s, 0) = 0$ for s near 0. Moreover, by the hypocomplexity of L, there is a holomorphic function H defined near the origin such that

$$U(s, t) + iV(s, t)^k = H(s + i\phi(s, t)).$$

Since $U + iV^k$ and Z are homeomorphisms, H is a biholomorphism (near 0). We also have

$$H(s + i\alpha(s)) = U(s, 0) \in \mathbb{R},$$

showing that $H(Z(\Sigma)) \subseteq \mathbb{R}$. Note also that from the equations

$$\Im H(s, \alpha(s)) = 0 \quad \text{and} \quad d\Im H(0) \neq 0,$$

we conclude that $\alpha(s)$ is real-analytic. In the latter statement, we have assumed as we may that $\alpha'(0) = 0$ and used the consequent fact that $H'(0)$ is real. Conversely, suppose H is a biholomorphism near 0 such that $H \circ Z(\Sigma) \subseteq \mathbb{R}$ where we take $Z(s, t)$ as before. Thus $H(s + i\alpha(s)) \in \mathbb{R}$. Define $F(s, t) = W^{-1} \circ H \circ Z(s, t)$, where $W(x, y) = x + iy^k$. F is a homeomorphism and away from $t = 0$, it is a diffeomorphism. Since $\Re F(s, t) = \Re H \circ Z(s, t)$, $\Re F(s, t)$ is smooth. Next note that since $\Im H \circ Z(s, t)$ vanishes to order k at $t = 0$ and $(\Im F)^k = \Im H \circ Z$, there is a nonvanishing smooth function $g(s, t)$ near the origin such that

$$\Im F(s, t) = g(s, t)t.$$

The latter, together with the fact that $H'(0) \in \mathbb{R}$ (we assume $\alpha'(0) = 0$), implies that F is a diffeomorphism near the origin. Clearly, using $(\Re F, \Im F)$ as new coordinates, we get $x + iy^k$ as a first integral for L. $\qquad\square$

REMARK III.5.6. In Proposition III.5.5, when $Z(s, t) = s + i\phi(s, t)$ with $\phi(s, t) = \alpha(s) + it^k \beta(s, t)$, the proof shows that the two equivalent conditions are equivalent to the real-analyticity of $\alpha(s)$.

Thus, we have:

COROLLARY III.5.7. *Suppose L and Σ are as in Proposition III.5.2 and L is real-analytic. Then there are real-analytic coordinates (x, y) in which the function $x + iy^k$ is a first integral of L. In other words, when L is real-analytic and Σ is a manifold of dimension 1, up to a real-analytic local diffeomorphism (and up to a nonvanishing multiple), there is only one real-analytic vector field of uniform type k.*

The preceding corollary was also proved in [**Me1**]. Proposition III.5.5 can be generalized as follows:

PROPOSITION III.5.8. *Suppose L_1 and L_2 are two vector fields of the same uniform odd type on their respective characteristic sets Σ_1 and Σ_2. Then there exists a local diffeomorphism mapping the structure generated by L_1 to the one generated by L_2 if and only if for any first integrals Z_1 and Z_2 of L_1 and L_2 respectively, there exists a local biholomorphism mapping $Z_1(\Sigma_1)$ onto $Z_2(\Sigma_2)$.*

The proof is similar to that of Proposition III.5.5.

In Proposition III.5.2 and the subsequent discussion, we assumed that Σ is a one-dimensional manifold. However, in general, as the following examples show, Σ may not be one-dimensional.

EXAMPLE III.5.9. Let $\mathcal{V}_1$ be the structure in the plane defined by

$$Z_1 = x + i(x^2 y + y^3).$$

Then the characteristic set $\Sigma_1 = \{(0, 0)\}$ and the type there is 3.

EXAMPLE III.5.10. Let $\mathcal{V}_2$ be the structure defined by

$$Z_2 = x + i(x^4 y + y^3).$$

Again the characteristic set $\Sigma_2 = \{(0, 0)\}$ and the type is 3.

We remark that in any neighborhood of the origin, the structures $\mathcal{V}_1$ and $\mathcal{V}_2$ are not equivalent. More generally, we have:

PROPOSITION III.5.11. *Suppose L is elliptic except at the origin and is of finite type there. Then the type is odd. In particular, L is hypocomplex.*

PROOF. Write $L = \dfrac{\partial}{\partial y} + ib(x, y)\dfrac{\partial}{\partial x}$, with b real-valued. Then $b(p) \equiv 0$ if and only if $p = 0$. Hence b cannot change sign in any neighborhood of the origin.

It follows that the type at the origin is odd. Since b does not change sign, L is locally solvable (Theorem IV.1.6) and hence locally integrable.   □

We are now ready to state and prove the key lemma from [**Cor2**] concerning approximate solutions:

LEMMA III.5.12. *Let L be a locally integrable, planar vector field of uniform finite type on its characteristic set Σ which we assume is a one-dimensional manifold. Assume that u is a nontrivial approximate solution on a side of Σ and that u is continuous up to the boundary piece Σ. Then the set*

$$\{p \in \Sigma : u(p) = 0\}$$

has zero measure with respect to arclength on Σ.

In view of the preceding propositions, Lemma III.5.12 will be a consequence of:

LEMMA III.5.13. *Let L be locally integrable near the origin with a first integral $Z(x, t) = x + i\Phi(x, t)$. Suppose that*

$$\Phi_t(x, t) = a(x, t)t^k,$$

where k is a positive integer and a is never zero for $|x|, |t| \leq \delta$, $(\delta > 0)$. Let u be a nontrivial function satisfying on $|x| < \delta$, $0 < t < \delta$,

$$|Lu(x, t)| \leq M|u(x, t)|$$

and continuous up to $t = 0$. Then the set

$$\{x : |x| < \delta, u(x, 0) = 0\}$$

has zero Lebesgue measure.

PROOF. We may assume that $a(x, t) > 0$ for every (x, t). The map $(x, t) \mapsto (x, \Phi(x, t))$ is a diffeomorphism from the region $|x| < \delta$, $0 < t < \delta$ onto the open set in the plane:

$$\Omega = \{z = x + iy : |x| < \delta, \Phi(x, 0) < y < \Phi(x, \delta)\}.$$

Denote by $z \to (x, \Psi(x, y))$ the inverse of this diffeomorphism and set

$$v(x, y) = u(x, \Psi(x, y)), \quad x + iy \in \Omega. \tag{III.8}$$

By the chain rule, we have

$$\left| \left(\frac{\partial v}{\partial x} + i \frac{\partial v}{\partial y} \right)(x, y) \right| \leq K\Phi_t(x, \Psi(x, y))^{-1}|v(x, y)|. \tag{III.9}$$

Now we have for $t \geq 0$

$$\Phi(x, t) - \Phi(x, 0) = t^{k+1} A(x, t)$$

where $A > 0$ for $|x| \leq \delta$, $0 \leq t \leq \delta$. Hence

$$y - \Phi(x, 0) = \Psi(x, y)^{k+1} A(x, \Psi(x, y)), \quad x + iy \in \Omega.$$

Since

$$\Phi_t(x, t) \geq \epsilon t^k, \quad \epsilon > 0,$$

we get

$$\Phi_t(x, \Psi(x, y)) \geq \epsilon \Psi(x, y)^k = \epsilon \left(\frac{y - \Phi(x, 0)}{A(x, \Psi(x, y))} \right)^{k/(k+1)}.$$

Consequently, (III.9) implies for $x + iy \in \Omega$:

$$\left| \left(\frac{\partial v}{\partial x} + i \frac{\partial v}{\partial y} \right)(x, y) \right| \leq K'[y - \Phi(x, 0)]^{-k/(k+1)} |v(x, y)|. \tag{III.9'}$$

Observe that since $(x, t) \mapsto (x, \Phi(x, t))$ is also a homeomorphism from $|x| < \delta$, $0 \leq t < \delta$ onto

$$\Omega' = \{z = x + iy : |x| < \delta, \Phi(x, 0) \leq y < \Phi(x, \delta)\},$$

the function v is in fact continuous on Ω'.

Fix $0 < \delta' < \delta$ arbitrary. It suffices to show that the Lebesgue measure of the set

$$\{x : |x| < \delta', v(x, \Phi(x, 0)) = 0\}$$

is zero. Consider now a simply connected open subset U of Ω that is bounded by a smooth Jordan curve γ for which there is a decomposition $\gamma = \gamma_1 \cup \gamma_2$ with

$$\gamma_1 = \{x + i\Phi(x, 0) : |x| \leq \delta'\}, \quad \gamma_2 \subset \Omega'.$$

By the Riemann mapping theorem there is a biholomorphism $\zeta = G(z)$ from U onto the unit disk $|\zeta| < 1$. Since G is necessarily a smooth diffeomorphism from $\overline{U}$ onto $|\zeta| \leq 1$, $v'(\zeta) = v(G^{-1}(\zeta))$ will be continuous on $|\zeta| \leq 1$ and will satisfy (III.9'):

$$\left| \frac{\partial v'}{\partial \bar{\zeta}}(\zeta) \right| \leq K(1 - |\zeta|)^{-\frac{k}{k+1}} |v'(\zeta)|, \quad |\zeta| < 1.$$

The lemma now follows from Lemma III.5.14. $\qquad \square$

LEMMA III.5.14. *Let D be the unit disk in the complex z-plane and let* $v \in C(\overline{D})$ *be not identically zero and satisfy*

$$\left| \frac{\partial v}{\partial \overline{z}}(z) \right| \leq K(1 - |z|)^{-\alpha} |v(z)|, \quad z \in D, \qquad (\text{III.10})$$

for some $0 \leq \alpha < 1$. *Then the set*

$$\{\omega \in T : v(\omega) = 0\}$$

has zero Lebesgue measure (here T denotes the boundary of D).

PROOF. The main step in the proof is to show the following property:

There is a solution $S \in \bigcap_{\sigma<1} C^{\sigma}(D)$ *of the equation*

$$\frac{\partial S}{\partial \overline{z}} = \frac{1}{v} \frac{\partial v}{\partial \overline{z}} \quad \text{in D satisfying} \qquad (*)$$

$$\sup_{r<1} \int_0^{2\pi} |S(re^{i\theta})| d\theta < \infty.$$

Let us show right away that $(*)$ implies the conclusion of Lemma III.5.14. Write $v = \exp\{S\}h$ with $h \in \mathcal{O}(D)$. There is $p \in \mathbb{Z}_+$ such that v/z^p is continuous in $\overline{D}$ and does not vanish at the origin. Moreover, (III.10) is satisfied when v/z^p is substituted for v. Summing up, this argument shows that there is no loss of generality in assuming from the outset that $v(0) \neq 0$. Applying Jensen's inequality to the holomorphic function h gives, if $r < 1$,

$$\log|v(0)| \leq \Re S(0) - \frac{1}{2\pi}\int_0^{2\pi} \Re S(re^{i\theta})d\theta + \frac{1}{2\pi}\int_0^{2\pi} \log|v(re^{i\theta})|d\theta$$

and consequently $(*)$ implies

$$\log|v(0)| \leq C + \frac{1}{2\pi}\int_0^{2\pi} \log|v(re^{i\theta})|d\theta \qquad (\text{III.11})$$

where $C > 0$ is independent of r. A standard application of Fatou's lemma in (III.11) shows that $\log^- |v(e^{i\theta})| \in L^1(T)$, whence the sought conclusion.

We now proceed to the proof of $(*)$. To simplify the notation, we set $F = v_{\overline{z}}/v$. We observe that there is $p > 1$ such that $F \in L^p(D)$ (indeed it suffices to take $1 < p < 1/\alpha$). We set

$$S(z) = \frac{1}{\pi}\int\int_D \frac{F(z')}{z - z'} dx' dy'.$$

Then

$$\frac{\partial S}{\partial \overline{z}} = F;$$

moreover, since $p > 1$, it also follows (*cf.* [V], theorem 1.35) that

$$\frac{\partial S}{\partial z} = \Pi(F),$$

where Π denotes the singular integral operator

$$\Pi(g)(z) = -\frac{1}{\pi} \int \int_D \frac{g(z')}{(z-z')^2} \mathrm{d}x' \mathrm{d}y'.$$

Since Π is a bounded linear operator in $L^p(D)$ if $1 < p < \infty$ (*cf.* [V], page 64) we obtain $S \in L_1^p(D)$.

Since $F \in L^\infty(\{|z| < R\})$ for $R < 1$, any solution of the equation $\partial u/\partial \bar{z} = F$ belongs to $\bigcap_{\sigma < 1} C^\sigma(D)$. Hence (*) will follow if we can establish the following property:

$$\sup_{1/2 \le r < 1} \int_0^{2\pi} |S(re^{i\theta})| \mathrm{d}\theta < \infty. \tag{III.12}$$

We observe that $\partial/\partial r = e^{i\theta}\partial/\partial z + e^{-i\theta}\partial/\partial\bar{z}$, $\partial/\partial\theta = ir(e^{i\theta}\partial/\partial z - e^{-i\theta}\partial/\partial\bar{z})$ from which we derive that $(r, \theta) \mapsto S(re^{i\theta})$ belongs to the Sobolev space $L_1^1(]1/4, 1[\times]0, 2\pi[)$. Thus $r \mapsto S(re^{i\theta})$ is absolutely continuous for almost all θ. By first integrating on $[1/2, r]$ and afterwards on $[0, 2\pi]$ we conclude that

$$\int_0^{2\pi} |S(re^{i\theta})| \mathrm{d}\theta \le \int_0^{2\pi} \left|S\left(\frac{1}{2}e^{i\theta}\right)\right| \mathrm{d}\theta + \int_0^{2\pi} \int_{1/2}^1 \left|\frac{\partial S}{\partial r}(r'e^{i\theta})\right| \mathrm{d}r' \mathrm{d}\theta,$$

for every $r \in [1/2, 1]$, from which (III.12) follows. This completes the proof of Lemma III.5.14. $\qquad\qquad\square$

III.6 Real-analytic structures in the plane

We will continue using the notation of the previous section and assume in addition that Φ is real-analytic. If $\Phi_t(0, 0) \neq 0$ then L is elliptic near the origin, and the results we will discuss are well known in this case. We next discuss the case when $\Phi_t(0, 0) = 0$ but Φ_t is not identically zero. We factor out $\Phi_t(x, t) = x^l \Psi(x, t)$, where Ψ is real-analytic and $\Psi(0, \cdot)$ does not vanish identically. Applying the Weierstrass preparation theorem to Ψ allows us to describe the zero set Σ_0 of the function Φ_t as the zero set of $(x, t) \mapsto x^l p(x, t)$, where p is a distinguished polynomial in the t-variable with no multiple factors. Hence we can state:

There is a disjoint decomposition

$$\Sigma_0 = F_0 \cup V_1^+ \cup \cdots \cup V_\alpha^+ \cup V_1^- \cup \cdots \cup V_\beta^- \tag{**}$$

in a small neighborhood of the origin $|x| < \epsilon$, $|t| < \epsilon$, where F_0 is either $\{(0,0)\}$ or is equal to the segment $\{0\} \times (-\epsilon, \epsilon)$ *(according to either $l = 0$ or $l > 0$), and each V_j^+ (resp. V_k^-) is defined by an analytic graph $\{(x, \gamma_j(x)) : 0 < x < \delta\}$ (resp. $\{(x, \sigma_k(x)) : -\delta < x < 0\}$), where $\gamma_1 < \gamma_2 < \cdots < \gamma_\alpha$ (resp. $\sigma_1 < \sigma_2 < \cdots < \sigma_\beta$) and*

$$\lim_{x \to 0^-} \sigma_k(x) = \lim_{x \to 0^+} \gamma_j(x) = 0, \quad \forall j, k.$$

As a consequence we observe that in a neighborhood of each point $(x_0, t_0) \in \Sigma_0 \backslash F_0$ we can write $\Phi_t(x, t) = (t - g(x))^k a(x, t)$, where $k \geq 1$, a and g are real-analytic and a never vanishes.

In what follows, for any set S and a number k, $\mathcal{H}^k(S)$ will denote the k-dimensional Hausdorff measure of S.

We can now prove:

PROPOSITION III.6.1. *Suppose that $\Phi_t(0, 0) = 0$, $\Phi_t \not\equiv 0$. Let u be a nontrivial C^1 function defined for $|x| < \epsilon$, $|t| < \epsilon$ and satisfying:*

$$|Lu(x, t)| \leq M|u(x, t)|$$

and denote its zero set by S. Then:

(1) *If u does not vanish identically on $\{0 < x < \epsilon, |t| < \epsilon\}$, then $S \cap \{x > 0\}$ has a trivial one-dimensional Hausdorff measure (likewise for $x < 0$).*
(2) *If $F_0 = \{0\} \times (-\epsilon, \epsilon)$, then $S \cap F_0 \neq \emptyset \Rightarrow F_0 \subset S$.*
(3) *If $F_0 = \{(0,0)\}$ and if u does not vanish identically then S has a trivial one-dimensional Hausdorff measure.*

PROOF. Assume first that $F_0 = \{0\} \times (-\epsilon, \epsilon)$. Then $L = \partial u / \partial t$ over F_0 (since $Z_t(0, t) = i\Phi_t(0, t) = 0$), which gives

$$\left| \frac{\partial u}{\partial t}(0, t) \right| \leq M|u(0, t)|.$$

By Gronwall's inequality, it follows that if $u(0, t_0) = 0$ for some t_0, then $u(0, t) = 0$ for all t.

Now we consider the general case. Fix a point $(x_0, t_0) \in \Sigma_0 \backslash F_0$ and write $\Phi_t(x, t) = (t - g(x))^k a(x, t)$ in a neighborhood of (x_0, t_0) as before. After the change of variables $x' = x$, $t' = t - g(x)$, the analysis near (x_0, t_0) reduces to the situation treated in Lemma III.5.13. In particular, we obtain that u cannot vanish identically in any component of the set $W^+ = \{(x, t) : 0 < x < \epsilon, |t| < \epsilon, \Phi_t(x, t) \neq 0\}$ and also that the one-dimensional Hausdorff measure of $S \cap (\Sigma_0 \backslash F_0)$ is trivial. Since the vector field L defines a complex structure on W^+, it follows that the one-dimensional Hausdorff measure of $S \cap W^+$ is also trivial. The proof of Proposition III.6.1 follows from these arguments. $\square$

COROLLARY III.6.2. *Suppose u is a C^1-approximate solution defined for $|x| < \epsilon$, $|t| < \epsilon$ and vanishing for $t = 0$. Then u vanishes identically.*

PROOF. Consider the new C^1-approximate solution $\tilde{u}$ defined as u for $t > 0$ and zero for $t \leq 0$. If Φ_t does not vanish identically, it follows from Proposition III.6.1 and the discussion that precedes it that $\tilde{u}$ vanishes identically. If however $\Phi_t \equiv 0$, then $L = \frac{\partial}{\partial t}$ and we reach the same conclusion by applying Gronwall's inequality. $\square$

III.6.1 Real-analytic structures with $m = 1$

As a consequence of Corollary III.6.2 we obtain:

THEOREM III.6.3. *Uniqueness in the Cauchy problem for C^1-approximate solutions holds for real-analytic locally integrable structures with $m = 1$.*

PROOF. Since this is a local statement we can work in local coordinates $(x, t) = (x, t_1, \ldots, t_n)$ centered at the origin for which there is a real-analytic, real-valued function $\Phi(x, t)$ satisfying

$$\Phi(0, 0) = \Phi_x(0, 0) = 0 \qquad \text{(III.13)}$$

such that, if

$$Z(x, t) = x + i\Phi(x, t),$$

then the bundle $\mathcal{V}$ is spanned by the linearly independent, pairwise commuting vector fields

$$L_j = \frac{\partial}{\partial t_j} - \frac{Z_{t_j}}{Z_x} \frac{\partial}{\partial x}, \qquad j = 1, \ldots, n. \qquad \text{(III.14)}$$

Let u be a C^1-approximate solution defined for $|x| < \delta$, $|t| < \delta$:

$$|Lu| \leq M|u|.$$

The conclusion will follow after we show that if u vanishes for $t = 0$ then u vanishes identically.

Fix t_0, $0 < |t_0| < \delta$ and define

$$Z_0(x, s) = Z(x, st_0), \qquad |x| < \delta, \quad |s| < 1.$$

Consider also the vector field

$$L_0 = \frac{\partial}{\partial s} - \frac{Z_{0s}}{Z_{0x}} \frac{\partial}{\partial x}$$

as well as the C^1 function

$$u_0(x, s) = u(x, st_0).$$

We have

$$L_0 u_0(x, s) = \sum_{j=1}^{n} (L_j u)(x, st_0) t_{0j}$$

and thus

$$|L_0 u_0(x, s)| \leq M' |u_0(x, s)|,$$

showing that u_0 is a C^1-approximate solution for the structure defined by L_0 in $|x| < \delta$, $|s| < 1$. Moreover, u_0 vanishes for $s = 0$. Therefore, by Corollary III.6.2 and a standard propagation argument, u_0 vanishes identically for $|x| < \delta$, $|s| < 1$. Hence $u(x, t_0) = 0$ for all $|x| < \delta$. □

Let $\mathcal{V}$ be a real-analytic locally integrable structure over a connected, real-analytic manifold Ω of dimension N. When $m = 1$ (Ω has then dimension $n + 1$) the orbits of the structure $\mathcal{V}$ have either dimension $n + 1$ (open subsets of Ω) or dimension n.

Introduce the projection over Ω of the characteristic set of $\mathcal{V}$:

$$\Sigma = \{p \in \Omega : T'_p \cap \overline{T'_p} \neq 0\}.$$

It is easy to see that Σ is an analytic subset of Ω. Since Ω is connected we either have $\dim \Sigma \leq n$ or $\Sigma = \Omega$.

Assume first that $\Sigma = \Omega$: in this case $\mathcal{V}$ defines a real structure on Ω in the sense that $\mathcal{V} = \mathbb{C} \otimes \mathcal{V}_0$, where $\mathcal{V}_0$ is an involutive vector sub-bundle of $T\Omega$ of rank n. The leaves of the foliation defined by $\mathcal{V}_0$ are precisely the n-dimensional (Nagano) leaves.

Next suppose that the dimension of the analytic set Σ is $\leq n$. On $\Omega \backslash \Sigma$ the bundle $\mathcal{V}$ defines an elliptic structure and every n-dimensional leaf is contained in Σ; in particular, it follows that the union of all n-dimensional leaves is a set of $(n + 1)$-dimensional measure zero. We now prove:

THEOREM III.6.4. *Let $\mathcal{V}$ be a real-analytic locally integrable structure over a connected, real-analytic, $(n + 1)$-dimensional manifold Ω with $m = 1$. Let u be a nontrivial C^1-approximate solution on Ω and let S denote its zero set. Then:*

(1) *If $\mathcal{M}$ is an $(n + 1)$-dimensional leaf, then either $\mathcal{M} \cap S = \mathcal{M}$ or $\mathcal{H}^n(\mathcal{M} \cap S) = 0$.*

(2) *If S has nonempty intersection with some n-dimensional leaf $\mathcal{M}$, then $\mathcal{M} \subset S$.*

PROOF. Suppose that $\Sigma = \Omega$. By the preceding discussion any point $p \in \Omega$ is the center of a system of coordinates $(U; x, t_1, \ldots, t_n)$ over which $\mathcal{V}$ is spanned by the vector fields $\partial/\partial t_j$, $j = 1, \ldots, n$. On U we have $|d_t u| \le M|u|$ and consequently if $u(0, 0) = 0$ then $u(0, t) = 0$ for all t thanks to Gronwall's inequality.

This argument also provides a proof of (2): if $\mathcal{M}$ is an n-dimensional leaf then $\mathcal{V}|_{\mathcal{M}} \subset \mathbb{C}T\mathcal{M}$ and it defines a real structure over $\mathcal{M}$ for which $u|_{\mathcal{M}}$ is also a C^1-approximate solution. Again Gronwall's inequality gives $S \cap \mathcal{M} \ne \phi \Rightarrow \mathcal{M} \subset S$.

Next we observe first that (1) is valid when $n = 1$. Indeed, let $\mathcal{M}$ be a two-dimensional leaf on which u is not identically zero and $p \in \mathcal{M}$. Then p is the center of a system of coordinates (x, t) as in Proposition III.6.1 for which there is $Z(x, t) = x + i\Phi(x, t)$, whose differential spans T' and $\Phi_t \ne 0$. Either $\Phi_t(0, 0) \ne 0$ or $\Phi_t(0, 0) = 0$ and $F_0 = \{(0, 0)\}$. In any of these cases we obtain that the one-dimensional Hausdorff measure of the zero set of the restriction of u to a small neighborhood of p is trivial.

Hence it remains to prove property (1) assuming that the full result has been proved for smaller values of n. Since any $(n + 1)$-dimensional leaf is a connected open subset of Ω, we can assume that Ω itself is a leaf.

Decompose Σ into its regular and singular parts, $\Sigma = \Sigma_r \cup \Sigma_s$. The dimension of Σ_s is $\le n - 1$ and then it follows that $\Omega' := \Omega \backslash \Sigma_s$ is open, connected, and that $\mathcal{H}^n(\Sigma_s) = 0$. This observation allows us to assume from the outset that Σ is an embedded, real-analytic hypersurface of Ω. Denote by $\iota : \mathbb{C}T^*\Omega|_\Sigma \longrightarrow \mathbb{C}T^*\Sigma$ the pullback map, let $\mathcal{N} = \iota(T'|_\Sigma)$ and

$$\Sigma^* = \{p \in \Sigma : \dim \mathcal{N}_p = 1\}.$$

Since any component of Σ cannot be a leaf it follows that $\Sigma \backslash \Sigma^*$ is an analytic subset of Σ of dimension $\le n - 1$ and consequently has trivial n-dimensional Hausdorff measure. Any point $p_0 \in \Sigma^*$ is the center of a system of coordinates $(U_0; x, t_1, \ldots, t_n)$ for which all properties described at the beginning of the proof of Theorem III.6.3 hold and that

$$U_0 \cap \Sigma = \{t_n = 0\}.$$

We make the following claim:

(γ) *If v is a C^1-approximate solution that vanishes on a nonempty open subset of Ω, then v vanishes identically.*

PROOF OF (γ). Let p_l be a sequence of points in Ω, $p_l \to p$ such that v vanishes identically in a neighborhood of each p_l. If $p \notin \Sigma$ then v vanishes identically near p since $\mathcal{V}$ is an elliptic structure in $\Omega\backslash\Sigma$. Suppose now that $p \in \Sigma$ and take a coordinate system $(V, y_1, \ldots, y_{n+1})$, $V = \{|y| < r\}$, centered at p such that $\Sigma \cap V = \{y_1 = 0\}$. Since $\mathcal{V}$ is an elliptic structure in $\{y \in V : y_1 \neq 0\}$ and since $p_l \in V$ for some l it follows necessarily that v vanishes identically on one of the sides $y_1 > 0$ or $y_1 < 0$. Suppose that the first case occurs and take $y^* \in \Sigma^* \cap V$. By Theorem III.6.3 it follows that v vanishes identically in a full neighborhood of y^* and consequently in the whole component $y_1 < 0$.

Since u is a C^1-approximate solution on $\Omega\backslash\Sigma$ with respect to an elliptic structure (with $m = 1$, $\nu = n - 1$ according to the notation of Chapter I), which does not vanish identically on any component of $\Omega\backslash\Sigma$, we have $\mathcal{H}^n(S \cap (\Omega\backslash\Sigma)) = 0$. Hence it suffices to show that $\mathcal{H}^n(S \cap \Sigma^*) = 0$ or, for that matter, that

$$\mathcal{H}^n(S \cap (U_0 \cap \Sigma)) = 0 \qquad (\text{III.15})$$

according to the preceding notation.

The differential of $Z|_{t_n=0}$ defines a locally integrable structure on $U_0 \cap \Sigma$ with $m = 1$. Moreover, the restriction of u to $U_0 \cap \Sigma$ is a C^1-approximate solution for this structure, which is furthermore not identically zero on any n-dimensional leaf thanks to Theorem III.6.3 and (γ). If such a structure is not real, then (III.15) holds by the induction hypothesis. Suppose now that this structure is real, which is the same as saying that $\Phi|_{t_n=0}$ depends only on x. Taking

$$U_0 \cap \Sigma = \{(x, t') : |x| < \delta, |t'| < \delta\},$$

Gronwall's inequality gives

$$S \cap (U_0 \cap \Sigma) = \{x : |x| < \delta, u(x, 0, 0) = 0\} \times \{t' : |t'| < \delta\}. \qquad (\text{III.16})$$

Since moreover Φ_t is not identically zero, there is a line segment $\mathfrak{p}$ through the origin in t-space such that Φ restricted to $(-\delta, \delta) \times \mathfrak{p}$ is not a function of x alone. This means that the differential of the restriction of Z defines a locally integrable structure on $(-\delta, \delta) \times \mathfrak{p}$ which satisfies the hypothesis of Proposition III.6.1. The restriction of u to $(-\delta, \delta) \times \mathfrak{p}$ is a C^1-approximate solution and does not vanish on any nonempty open subset of $(-\delta, \delta) \times \mathfrak{p}$, once more thanks to Theorem III.6.3 and (γ). But then we can apply Proposition III.6.1 in order to infer that the Lebesgue measure of $\{x : |x| < \delta, u(x, 0, 0) = 0\}$ is zero, which according to (III.16) gives (III.15).

The proof of Theorem III.6.4 is now complete. $\qquad\Box$

COROLLARY III.6.5. *Let u be a C^1-approximate solution on Ω. Then $d'_0 u / u$ which can be regarded as a section of $\mathbb{C}T^*\Omega/T'$ with L^∞ coefficients, is d'_1-closed.*

PROOF. We can of course assume that u is not identically equal to zero. By [**HaP**] (corollary 2.4) in conjunction with Theorem III.6.4 (1) it follows that $d_1'(d_0'u/u) = 0$ on the union of all $(n+1)$-dimensional leaves. Now let p be a point belonging to an n-dimensional leaf; we have to show that $d_1'(d_0'u/u) = 0$ in a neighborhood of p.

We can find a coordinate system $(U; x, t_1, \ldots, t_n)$ centered at p, with $U = \{(x, t) : |x| < \delta, |t| < \delta\}$ such that T' is spanned, over U, by the differential of the function $Z(x, t) = x + i\Phi(x, t)$, where Φ is real-valued, real-analytic, and satisfies (III.13). We necessarily have $\Phi(0, \cdot) = 0$, since p belongs to an n-dimensional leaf. We must analyze two cases: either (i) $\Phi \equiv 0$ or else, by taking $\delta > 0$ small, (ii) $\Phi(x, \cdot) \neq 0$ for all $x \in (-\delta, \delta)$, $x \neq 0$.

Under case (i) the complex d' over U equals the complex d_t, and our claim can easily be checked. We consider case (ii). Since $\{(x, t) \in U : x > 0\}$ and $\{(x, t) \in U : x < 0\}$ are contained in $(n+1)$-dimensional leaves, taking into account the representation of the operator d_0' given by

$$Lu = \sum_{j=1}^{n} L_j u \, dt_j,$$

it suffices to show that

$$L\chi_\epsilon \wedge Lu/u \to 0 \quad \text{in } L^1(U, \Lambda^2((\mathbb{C}T^*\Omega/T'))), \tag{III.17}$$

where $\chi_\epsilon \in \mathcal{C}^\infty(\mathbb{R})$ depends only on x and satisfies $\chi_\epsilon = 1$ for $|x| > \epsilon$, $\chi_\epsilon = 0$ for $|x| \leq \epsilon/2$, and $|\chi_\epsilon'| \leq C\epsilon^{-1}$. Now

$$L\chi_\epsilon = -i\chi_\epsilon'(x)\frac{d_t\Phi(x, t)}{Z_x(x, t)}$$

and thus, since $d_t\Phi(x, t) = O(|x|)$, the L^∞ norm of $L\chi_\epsilon$ is bounded uniformly in ϵ. From this (III.17) follows immediately, and the proof is complete. $\square$

III.7 Further applications of Sussmann's orbits

In this chapter, the focus has been on the applications of Sussmann's orbits to a variety of questions on unique continuation. However, Sussmann's orbits have also been applied to several problems in involutive structures. In particular, it is now known that many properties of CR functions propagate along orbits. Here we will very briefly mention some of the results that involved orbits.

As mentioned in Section III.3, orbits were used by Tumanov ([**Tu1**]) and Baouendi and Rothschild ([**BR**]) to prove necessary and sufficient conditions for the holomorphic extension of all CR functions into wedges. In [**Tr**],

Trepreau showed that the wedge extendability of continuous CR functions propagates along the orbits of a CR manifold in $\mathbb{C}^n$. Another proof of this result appeared in [**Jo2**]. In the same paper [**Tr**], Trepreau also described the variation of the direction of extendability along orbits by proving that the wave front set of a CR function is a union of orbits in the conormal bundle with respect to a natural CR structure there. These results were generalized by Tumanov in [**Tu2**], where he showed that CR-extendability of a CR function on a generic CR manifold $\mathcal{M}$ in $\mathbb{C}^n$ propagates along orbits. A CR function on $\mathcal{M}$ is said to be *CR-extendable* at $p \in \mathcal{M}$ if it extends to be CR on some manifold with boundary attached to $\mathcal{M}$ near p. Moreover, Tumanov described the variation of the directions of CR extendability in terms of a certain differential geometric partial connection and the corresponding parallel displacement in a quotient bundle of the normal bundle of $\mathcal{M}$. This description is dual to that of Trepreau. Merker ([**Mer1**]) gave a simplified presentation of Tumanov's connection and used it to prove that if $\mathcal{M}$ is a generic CR manifold consisting of a single orbit, then each continuous CR function on $\mathcal{M}$ is wedge-extendable at each point of $\mathcal{M}$. This result was also obtained by [**Jo2**] independently using a different proof. In Joricke's approach, the key idea is the deformation of the manifold $\mathcal{M}$ so as to produce minimal points in such a way that all points outside a truncated cone C (in suitable local coordinates on $\mathcal{M}$) are left fixed. The cone C has an axis in $\Re V$, a vertex p, and the deformed CR manifold is minimal at the central point p.

The concept of Sussmann's orbit has been used to characterize the first-order linear partial differential operators which are locally solvable (see [**T5**]). Orbits were used by Hounie ([**Ho1**] and [**Ho2**]) in his work on globally solvable and globally hypoelliptic complex vector fields on manifolds.

For tube structures, Hounie and Tavares [**HT5**] have given a necessary and sufficient condition for the validity of Hartog's phenomenon for solutions in terms of the behavior of orbits. Orbits have also been relevant in the study of removable singularities, as shown in numerous works including [**HT2**], [**Jo1**], [**Mer2**], [**KR**], [**MP1**], [**MP2**], and [**MP3**]. The paper [**CR1**] of Chirka and Rea uses orbits to study the regularity of CR mappings. For earlier works exhibiting orbits as propagators of support and singularities, see [**DH**], [**HS**], and [**Z**].

Notes

As indicated in the introduction, the concept of orbits and its basic properties were presented in Sussmann's paper [**Su**]. Lemma III.1.8 and some of its consequences appeared in [BM]. The theorems on the strong unique continuation for L^1_{loc} solutions were proved in [BH2]. The propagation of support for solutions and the link to the uniqueness for the noncharacteristic Cauchy problem has been studied by many mathematicians: Strauss and Treves in [ST], and Cardoso and Hounie in [CH] studied the Cauchy problem for a single smooth vector field satisfying the solvability condition $\mathcal{P}$ of Nirenberg and Treves. Hunt, Polking and Strauss ([HPS]) considered the uniqueness problem for a hypersurface in a complex manifold. Hunt ([Hu]) proved uniqueness for the noncharacteristic Cauchy problem for locally realizable CR manifolds under some hypotheses on the Levi form. Treves proved his theorem on propagation of support along orbits by using the uniqueness theorem for the noncharacteristic Cauchy problem in locally integrable structures—a consequence of the Baouendi–Treves approximation formula.

The description of the zero set of approximate solutions in real-analytic structures where $m = 1$ and for certain planar vector fields follows Cordaro's paper ([Cor2]).

For additional references to the concept of orbits and their applications, we mention the books by Baouendi, Ebenfelt and Rothschild ([BER]), Treves ([T5]), and the manuscript [MP3] by Merker and Porten.

IV

Local solvability of vector fields

In this chapter we study in detail an important class of locally integrable vector fields: those which are locally solvable. The most basic question one can ask concerning the solvability of a vector field L is whether, given a smooth right-hand side f, there exists a solution, at least locally and not subjected to any additional condition, of the equation $Lu = f$. For real vector fields very satisfactory theorems stating local existence of solutions under very mild hypotheses of regularity have been known since long ago, and it came as a surprise when Hans Lewy published in 1956 his now famous example of a nonlocally solvable vector field. Indeed, if $f \in C^\infty(\mathbb{R}^3)$ is conveniently chosen, the equation

$$(\partial_x + i\partial_y - (x + iy)\partial_z)u = f, \qquad (x, y, z) \in \mathbb{R}^3$$

does not have distribution solutions in any open subset of $\mathbb{R}^3$ ([L2]). In the first part of this chapter we focus on vector fields in two variables; in this case, a priori estimates are known to hold under weaker assumptions on the regularity of the coefficients than in the general case. In Section IV.1 we motivate condition $(\mathcal{P})$ with simple examples and prove a priori estimates in L^p and in a mixed norm that involves the Hardy space $h^1(\mathbb{R})$. While the first kind of estimate gives, by duality, local solvability in L^p, $1 < p < \infty$, the latter kind gives local solvability in $L^\infty[\mathbb{R}; \mathrm{bmo}(\mathbb{R})]$ which serves as a substitute for local solvability in L^∞, a property that is not implied by $(\mathcal{P})$, as is shown by the example described at the end of Section IV.1.1. On the other hand, in some applications—this is indeed the case for the similarity principle described in the Epilogue—solvability in the larger space of mixed norm $L^\infty[\mathbb{R}; \mathrm{bmo}(\mathbb{R})]$ suffices. Some technical properties of the space $h^1(\mathbb{R})$ that are useful for the proof of a priori estimates will only be presented later in Appendix A. In Section IV.2 we still consider vector fields in two variables and study the existence of smooth solutions when the right-hand side is smooth. The sufficiency of condition $(\mathcal{P})$ for local solvability in any number of variables is discussed in Section IV.3, while Section IV.4 is devoted to its necessity.

IV.1 Planar vector fields

We shall consider vector fields defined in an open subset $\Omega \subset \mathbb{R}^2$

$$Lu = A(x, t)\frac{\partial u}{\partial t} + B(x, t)\frac{\partial u}{\partial x} \tag{IV.1}$$

with complex coefficients $A, B \in C^{\infty}(\Omega)$ such that

$$|A(x, t)| + |B(x, t)| > 0, \quad (x, t) \in \Omega. \tag{IV.2}$$

Since our point of view is local, most of the time the behavior of L outside a neighborhood of the point under study is irrelevant. This means that we can modify the coefficients of L off that neighborhood in order to assume that they are defined throughout $\mathbb{R}^2$ and we shall often do so. The sort of properties of L we shall be interested in will not change by multiplication of L by a nonvanishing factor. Since (IV.2) implies that either A or B does not vanish in a neighborhood of a given point (assume as well that it is A), we may multiply L by A^{-1} and obtain the new vector field $\tilde{L} = A^{-1}L$ which has the form

$$\tilde{L}u = \frac{\partial u}{\partial t} + \tilde{B}(x, t)\frac{\partial u}{\partial x}. \tag{IV.3}$$

Write $\tilde{B}(x, t) = \tilde{a}(x, t) + i\tilde{b}(x, t)$ with $\tilde{a}$ and $\tilde{b}$ real, and assume that they are defined for $|x| < \rho$, $|t| < \rho$.

LEMMA IV.1.1. *In appropriate new local coordinates $\xi = x$, $s = s(x, t)$ defined in a neighborhood of the origin, the vector field $\tilde{L}$ assumes the form*

$$\tilde{L}u = \frac{\partial u}{\partial s} + ib(\xi, s)\frac{\partial u}{\partial \xi}, \tag{IV.4}$$

with $b(\xi, s)$ real-valued.

PROOF. Consider the ODE

$$\begin{cases} \dfrac{dx}{ds} = \tilde{a}(x, t), & x(0) = \xi, \\[2mm] \dfrac{dt}{ds} = 1, & t(0) = 0, \end{cases}$$

with solution $(x(\xi, s), t(\xi, s))$ given by

$$\begin{cases} x(\xi, s) = \xi + \int_0^s \tilde{a}(x(\xi, \sigma), \sigma) \, d\sigma, \\ t(\xi, s) = s. \end{cases}$$

Observe that $x(\xi, 0) = \xi$ so $(\partial x/\partial \xi)(0, 0) = 1$; also $(\partial t/\partial \xi)(0, 0) = 0$ and $(\partial t/\partial s)(0, 0) = 1$ so the Jacobian determinant $\det[\partial(x, t)/\partial(\xi, s)]$ assumes the value 1 at $x = s = 0$, granting that $(\xi, s) \longleftrightarrow (x, t)$ is, at least locally, a smooth change of variables. The chain rule gives

$$\frac{\partial}{\partial s} = \frac{\partial}{\partial t} + \tilde{a}(x, t) \frac{\partial}{\partial x}, \qquad \frac{\partial}{\partial \xi} = \frac{\partial x}{\partial \xi} \frac{\partial}{\partial x},$$

so in the new coordinates we have $\tilde{L} = \partial_s + i(\tilde{b}/(\partial x/\partial \xi))\partial_\xi = \partial_s + ib\partial_\xi$. $\quad\square$

The reductions just described show that in the study of local problems for a planar vector field L with smooth coefficients we may always assume that L is of the form

$$L = \frac{\partial}{\partial t} + ib(x, t) \frac{\partial}{\partial x} \tag{IV.5}$$

with $b(x, t)$ real and defined for all $(x, t) \in \mathbb{R}^2$.

DEFINITION IV.1.2. *Let L be a vector field defined in an open set $\Omega \subset \mathbb{R}^2$, $p \in \Omega$. We say that L is locally solvable at p if there exists a neighborhood $U = U(p)$ such that for all $f \in C^\infty(\Omega)$ there exists $u \in \mathcal{D}'(\Omega)$ such that $Lu - f$ vanishes identically on U. If L is locally solvable at every point $p \in \Omega$ we say that L is locally solvable in Ω.*

REMARK IV.1.3. Observe that Definition IV.1.2 means that given p there exists a fixed open subset $U \ni p$ such that for every $f \in C^\infty(\Omega)$ there exists $u \in \mathcal{D}'(\Omega)$ such that the equation $Lu = f$ holds on U. A moment's reflection shows that we would get an equivalent definition by requiring instead that for every $f \in C_c^\infty(U)$ there exists $u \in \mathcal{D}'(U)$ such that $Lu = f$ in U. It is less evident that we also get an equivalent definition if we require that for every $f \in C^\infty(\Omega)$ there exists $u \in \mathcal{D}'(\Omega)$ such that $Lu - f$ vanishes on a neighborhood $U(p, f)$ of p that may depend on both f and p. However, a category argument shows that if this happens we may always take U independent of f for fixed p and the apparently weaker requirement is in fact equivalent to that given in Definition IV.1.2 (*cf.* Theorem VII.6.1).

In order to acquire some insight on local solvability let us consider the simpler case in which the coefficient $b(x, t)$ of the vector field (IV.5) is actually independent of x, i.e.,

$$L = \frac{\partial}{\partial t} + ib(t)\frac{\partial}{\partial x},$$

and we wish to study the local solvability of L in a neighborhood of the origin. In other words, we wish to find a distribution u such that $Lu = f$ where $f \in C_c^\infty(\mathbb{R}^2)$ is given. We shall perform a partial Fourier transform in the variable x and denote by $\widehat{u}$ and $\widehat{f}$ the transforms of u and f respectively, so the transformed equation becomes

$$\frac{d\widehat{u}}{dt} - b(t)\xi\widehat{u} = \widehat{f}, \quad \text{where } \widehat{f}(\xi, t) = \int_{\mathbb{R}} e^{-ix\xi} f(x, t)\, dx.$$

Using a standard formula for the linear ODE with parameter ξ, we find a solution $\widehat{u}$

$$\widehat{u}(\xi, t) = \int_{T(\xi)}^{t} e^{(B(t) - B(s))\xi}\widehat{f}(\xi, s)\, ds, \quad \text{where } B(t) = \int_0^t b(\tau)\, d\tau.$$

Changing the endpoint of integration $T(\xi)$ amounts to adding a solution of the homogeneous equation for each value of the parameter ξ. Thus, we see that it is very easy to find (many) solutions of the transformed equation, but in order to get a solution of the original equation we need that $\widehat{u}(\xi, t)$ be tempered in ξ, at least for t in a certain range $|t| < T$, so that we can define u as the inverse partial Fourier transform of $\widehat{u}$. The difficulty comes from the risk of growth at infinity arising from the factor $e^{(B(t) - B(s))\xi}$; notice that since $\xi \mapsto \widehat{f}(\xi, s)$ is in $\mathcal{S}(\mathbb{R})$ uniformly in s its rapid decay can overpower a factor of polynomial growth but to control factors with exponential growth by the decay of $\widehat{f}$ is not possible. A sensible attitude to avoid exponential growth is then to search for conditions that allow—after a convenient choice of $T(\xi)$—that $(B(t) - B(s))\xi \leq 0$ whenever $|t| < T$ and s is in the interval with endpoints $T(\xi)$ and t. Of course, the sign of $(B(t) - B(s))\xi$ does not change if ξ is multiplied by a positive number so we need only define two values for $T(\xi)$: $T(\xi) = T^+$ for $\xi > 0$ and $T(\xi) = T^-$ for $\xi < 0$. Let us concentrate first on the case $\xi > 0$. *We need to find T^+ such that for all $|t| < T$ and s in the interval with endpoints $\{T^+, t\}$ the following inequality holds:*

$$B(t) - B(s) = \int_s^t b(\tau)\, d\tau \leq 0.$$

We immediately see that if $b(\tau) \leq 0$ it will be enough to set $T^+ \doteq -T$ to obtain what we wish! Similarly, if $b(\tau) \geq 0$ the choice $T^+ \doteq T$ does the job, because to require that s be in the interval with endpoints $\{T, t\}$ simply means that $t < s < T$. So, if $b(0) \neq 0$ we may take T small enough so that $b(\tau)$ does not vanish in $(-T, T)$ and then define $T^+ = \pm T$ according to the sign of $b(0)$. Let us assume now that $b(0) = 0$. If $b(\tau)$ does not change sign in $(-T, T)$ for some $T > 0$ we already know how to proceed. What if $b(\tau)$ changes sign in $(-T, T)$? Well, suppose there is a point $t_0 \in (-T, T)$ such that $b(\tau) \geq 0$ for $\tau \in (-T, t_0]$ and $b(\tau) \leq 0$ for $\tau \in [t_0, T]$. In this case, we take $T^+ = t_0$ and notice that $\int_s^t b(\tau) \, d\tau \leq 0$ both for $t_0 < s < t$ and for $t < s < t_0$. It is easy to convince oneself that those are all the cases for which a good choice of T^+ is possible. Indeed, if $b(\tau_0) < 0$ and $b(\tau_1) > 0$ for some $-T < \tau_0 < \tau_1 < T$ no choice of T^+ will work. We would be forced to take $T^+ > \tau_1$ to guarantee that $\int_s^t b(\tau) \, d\tau \leq 0$ for $t < s$, s, t close to τ_1, but this would imply that $\int_s^t b(\tau) \, d\tau > 0$ for $t < s$, s, t close to τ_0. In other words, *we must prevent that $b(t)$ changes sign from minus to plus as t increases.* The analysis of the case $\xi < 0$ and the choice of T^- will tell us that we must as well prevent that $b(t)$ changes sign from plus to minus as t increases and both conclusions imply together that $b(t)$ cannot change sign at all.

REMARK IV.1.4. If we were studying the local solvability of the differential/pseudo-differential operator

$$L = \frac{\partial}{\partial t} - b(t)|D_x|,$$

where $|D_x|$ is the operator defined by $\widehat{|D_x|u}(\xi, t) = |\xi|\widehat{u}(\xi, t)$, this would lead us to consider the ODE

$$\frac{d\widehat{u}}{dt} - b(t)|\xi|\widehat{u} = \widehat{f}$$

and to require that $(B(t) - B(s))|\xi| \leq 0$. This time the sign of ξ does not matter and we are only forced to prevent sign changes of $b(\tau)$ from minus to plus.

Let us return to the problem of finding a solution to the equation $Lu = f$ when the coefficient $b(t)$ does not depend on x and we further assume that $t \mapsto b(t)$ does not change sign for $|t| < T$. Assuming that $b(t) \geq 0$, a solution is given by $u(x, t) = u^+(x, t) + u^-(x, t)$ where

$$u^+(x, t) = \frac{1}{2\pi} \int_0^\infty \int_T^t e^{ix\xi + (B(t) - B(s))\xi} \widehat{f}(\xi, s) \, ds \, d\xi, \tag{IV.6}$$

$$u^-(x, t) = \frac{1}{2\pi} \int_{-\infty}^0 \int_{-T}^t e^{ix\xi + (B(t) - B(s))\xi} \widehat{f}(\xi, s) \, ds \, d\xi, \quad |t| < T. \tag{IV.7}$$

The exponential in the integrals that define u^+ and u^- is bounded by 1 because the exponent always has nonpositive real part. The integrand is bounded by $|\widehat{f}(\xi, s)|$, which is rapidly decreasing in ξ as $|\xi| \to \infty$, in particular, $u(x, t)$

is continuous and bounded. Differentiating under the integral sign we always obtain integrable integrands, showing that our solution $u \in C^\infty(\mathbb{R} \times (-T, T))$.

DEFINITION IV.1.5. *We say that the operator L given by* (IV.5) *satisfies condition* $(\mathcal{P})$ *at* $p = (x_0, t_0)$ *if there is a neighborhood* $(x_0 - \delta, x_0 + \delta) \times (t_0 - \delta, t_0 + \delta)$ *of* p *such that for every* $x \in (x_0 - \delta, x_0 + \delta)$ *the function* $(t_0 - \delta, t_0 + \delta) \ni t \mapsto b(x, t)$ *does not change sign. If L satisfies condition* $(\mathcal{P})$ *at every point of an open set* Ω *we say that L satisfies condition* $(\mathcal{P})$ *in* Ω.

The importance of this definition comes from the following:

THEOREM IV.1.6. *The operator L given by* (IV.5) *is locally solvable at* p *if and only if it satisfies condition* $(\mathcal{P})$ *at* p.

We will not prove Theorem IV.1.6 here. The 'if' part of the theorem will follow from Corollary IV.1.10 presented later in this section while the 'only if' part will be discussed in Section IV.4 under the assumption that L is locally integrable.

REMARK IV.1.7. In the case of a coefficient independent of x, if condition $(\mathcal{P})$ is satisfied in a rectangle Ω, it follows that either $b(t) \geq 0$ in Ω or $b(t) \leq 0$ in Ω, but this is not the general situation. For instance, if $b(x, t) = x$ we see that L satisfies condition $(\mathcal{P})$ in $\mathbb{R}^2$ but b is positive for $x > 0$ and negative for $x < 0$.

REMARK IV.1.8. If L satisfies condition $(\mathcal{P})$ in a rectangle Ω centered at p and $\chi(x, t) \in C_c^\infty(\Omega)$ is identically 1 in a neighborhood of p, replacing $b(x, t)$ by $\chi(x, t)b(x, t)$ gives an operator $\tilde{L}$ that satisfies condition $(\mathcal{P})$ everywhere and coincides with L in a neighborhood of p. Furthermore, it is apparent that L is locally solvable at p if and only if $\tilde{L}$ is locally solvable at p. Thus, when studying the local solvability of an operator that satisfies condition $(\mathcal{P})$ in a neighborhood of p we may assume without loss of generality that $b(x, t)$ is compactly supported and condition $(\mathcal{P})$ is satisfied in $\mathbb{R}^2$.

Returning to the case in which the coefficient $b(t)$ is independent of x, observe that the solution u of $Lu = f$ furnished by (IV.6) and (IV.7) when $b \geq 0$ may be written in operator form as $u = Kf$, $K = K^+ + K^-$. Take a test function $\varphi \in C_c^\infty(\mathbb{R} \times (-T, T))$ and set $f = L\varphi$ and $u = Kf$. We see that $Lu = f = L\varphi$. Moreover, since $\hat{f}(\xi, t) \equiv 0$ for $|t| \geq T$ we see that u is supported in $\mathbb{R} \times [-T, T]$. Thus $w = u - \phi$ satisfies $Lw = 0$ and vanishes for $t \leq -T$. By uniqueness in the Cauchy problem we conclude that $\varphi = KL\varphi$.

Using Parseval's identity it is easy to derive that for fixed $t < s < T$ the $L^2(\mathbb{R})$ norm of

$$x \mapsto (2\pi)^{-1} \int_0^\infty e^{ix\xi + (B(t) - B(s))\xi} \widehat{f}(\xi, s) \, d\xi$$

is bounded by $\|f(\cdot, s)\|_{L^2(\mathbb{R})}$. This implies

$$\|K^+ f(\cdot, t)\|_{L^2(\mathbb{R})} \leq \int_{-T}^T \|f(\cdot, s)\|_{L^2(\mathbb{R})} \, ds, \quad x \in \mathbb{R}, \ |t| < T.$$

Integrating this inequality in t between $-T$ and T we obtain

$$\int_{-T}^T \|K^+ f(\cdot, s)\|_{L^2(\mathbb{R})} \, ds \leq 2T \int_{-T}^T \|f(\cdot, s)\|_{L^2(\mathbb{R})} \, ds.$$

The same inequality holds for K^-, so we obtain the following mixed norm estimate:

$$\|Kf\|_{L^1[(-T,T);L^2(\mathbb{R})]} \leq CT \|f\|_{L^1[(-T,T);L^2(\mathbb{R})]}. \tag{IV.8}$$

Now apply (IV.8) to $f = L\varphi$ and $Kf = \varphi \in C_c^\infty(\mathbb{R} \times (-T, T))$ to get the a priori inequality

$$\|\varphi\|_{L^1[(-T,T);L^2(\mathbb{R})]} \leq CT \|L\varphi\|_{L^1[(-T,T);L^2(\mathbb{R})]}, \quad \varphi \in C_c^\infty(\mathbb{R} \times (-T, T)). \tag{IV.9}$$

Observe that the transpose ${}^t L$ defined by $\langle L\varphi, \psi \rangle = < \varphi, {}^t L\psi >$ for all test functions $\varphi, \psi \in C_c^\infty(\mathbb{R}^2)$ is given by

$${}^t L = -L,$$

so (IV.9) may also be written as

$$\|\varphi\|_{L^1[(-T,T);L^2(\mathbb{R})]} \leq CT \|{}^t L\varphi\|_{L^1[(-T,T);L^2(\mathbb{R})]}, \tag{IV.10}$$

for every $\varphi \in C_c^\infty(\mathbb{R} \times (-T, T))$. It is a remarkable fact that essentially the same formulas that yield an a priori estimate for the simple case in which b is independent of t also give, in spite of technical complications, the same a priori estimate for the case of a general $b(x, t)$. We will prove a priori estimates like (IV.10) for a general vector field (IV.5) that satisfies condition $(\mathcal{P})$. More precisely,

THEOREM IV.1.9. *Let L given by (IV.5) satisfy condition $(\mathcal{P})$ in a neighborhood U of the origin and fix numbers p and q satisfying $1 < p < \infty$, $1 \leq q \leq \infty$. Then, there exist $T_0 > 0$, $a > 0$, and $C > 0$ such that for any $0 < T \leq T_0$ the following a priori estimate holds for every $\varphi \in C_c^\infty((-a, a) \times (-T, T))$:*

$$\|\varphi\|_{L^q[(-T,T);L^p(\mathbb{R})]} \leq CT \|{}^t L\varphi\|_{L^q[(-T,T);L^p(\mathbb{R})]}. \tag{IV.11}$$

Moreover, the constants T_0 and C depend only on p, q, and $\|b_x\|_{L^\infty(U)}$.

Before embarking on the rather long proof of Theorem IV.1.9, let us state a standard consequence that implies the local solvability of L.

COROLLARY IV.1.10. *Let L given by (IV.5) satisfy condition ($\mathcal{P}$) in a neighborhood of the origin, let $1 < p' < \infty$ and $1 \le q' \le \infty$ be given. Then there exist $T_0 > 0$, $C > 0$ such that for any $0 < T \le T_0$ and $f(x, t) \in L^{q'}[\mathbb{R}; L^{p'}(\mathbb{R})]$ there exists $u \in L^{q'}[\mathbb{R}; L^{p'}(\mathbb{R})]$, with norm*

$$\|u\|_{L^{q'}[\mathbb{R}; L^{p'}(\mathbb{R})]} \le CT \|f\|_{L^{q'}[\mathbb{R}; L^{p'}(\mathbb{R})]]}$$

that satisfies the equation

$$Lu = f \quad in \quad \mathbb{R} \times (-T, T). \tag{IV.12}$$

Since $L^{q'}[\mathbb{R}; L^{p'}(\mathbb{R})] \simeq L^{p'}(\mathbb{R}^2)$ when $p' = q'$, L is locally solvable in $L^{p'}$ for any $1 < p' < \infty$.

PROOF. We shall use the notation $\Omega_T = \mathbb{R} \times (-T, T)$. Let p, q be the conjugate exponents, $p = p'/(p'-1)$, $q = q'/(q'-1)$. Take C and T_0 as granted by Theorem IV.1.9 and for some $0 < T < T_0$ consider the linear functional

$${}^t L C_c^\infty(\Omega_T) \ni {}^t L\varphi(x, t) \mapsto \lambda({}^t L\varphi) = \int_{\mathbb{R}^2} f(x, t)\, \varphi(x, t)\, \mathrm{d}x\, \mathrm{d}t.$$

The inequalities

$$|\lambda({}^t L\varphi)| \le \|f\|_{L^{q'}[\mathbb{R}; L^{p'}(\mathbb{R})]} \|\varphi\|_{L^q[\mathbb{R}; L^p(\mathbb{R})]} \le CT \|{}^t L\varphi\|_{L^q[\mathbb{R}; L^p(\mathbb{R})]}$$

show that λ is well-defined and continuous in the $L^q[\mathbb{R}; L^p(\mathbb{R})]$ norm. By the Hahn–Banach theorem this functional can be extended to the whole space $L^q[\mathbb{R}; L^p(\mathbb{R})]$ without increasing its norm that is bounded by CT. By the Riesz representation theorem this extension is represented by integration against a function $u \in L^{q'}[\mathbb{R}; L^{p'}(\mathbb{R})]$ with norm $\|u\|_{L^\infty[\mathbb{R}; L^q(\mathbb{R})]} \le CT$. For $\varphi \in C_c^\infty(\Omega_T)$ we have

$$\int_{\mathbb{R}^2} u(x, t)\, {}^t L\varphi(x, t)\, \mathrm{d}x\, \mathrm{d}t = \lambda({}^t L\varphi) = \int_{\mathbb{R}^2} f(x, t)\, \varphi(x, t)\, \mathrm{d}x\, \mathrm{d}t,$$

which means that $Lu = f$ in Ω_T in the sense of distributions.   □

IV.1.1 A priori estimates in L^p

To prove Theorem IV.1.9 let us start by observing that we may assume without loss of generality that $b(x, t)$ has compact support, L satisfies condition ($\mathcal{P}$) everywhere, and $\|b_x\|_{L^\infty(\mathbb{R})} \le C \|b_x\|_{L^\infty(U)}$. The transpose ${}^t L$ of L is given by ${}^t L = -L - ib_x$ so if an a priori estimate like (IV.11) is proved for L instead of ${}^t L$ it will easily imply the estimate for ${}^t L$ since the contribution of the bounded zero-order term ib_x can be absorbed by taking T_0 small enough. In other words, it is enough to prove (IV.11) with L in the place of $'L$.

When dealing with the case $b(t)$ we already saw the advantage of considering separately the cases $\xi > 0$ and $\xi < 0$ (microlocalization) and this corresponds to writing $1 = H(\xi) + H(-\xi)$, where $H(\xi)$ is the Heaviside function, defined as $H(\xi) = 1$ for $\xi > 0$ and $H(\xi) = 0$ for $\xi < 0$. It will be convenient—although not strictly necessary—to substitute this rough partition of unity by a smooth one, so we consider a test function $\chi \in C_c^\infty(-2, 2)$ such that $\chi(\xi) = 1$ for $|\xi| \leq 1$ and set

$$\psi^+(\xi) = \begin{cases} 1 - \chi(\xi), & \text{if} \quad \xi \geq 0, \\ 0 & \text{if} \quad \xi \leq 0, \end{cases}$$

and

$$\psi^-(\xi) = \begin{cases} 0 & \text{if} \quad \xi \geq 0, \\ 1 - \chi(\xi), & \text{if} \quad \xi \leq 0, \end{cases}$$

so we have $1 = \chi(\xi) + \psi^+(\xi) + \psi^-(\xi)$. Given $\varphi \in \mathcal{S}(\mathbb{R}_x \times \mathbb{R}_t)$, for each fixed t we have a decomposition

$$\begin{aligned} \varphi(\cdot, t) &= P_0\varphi(\cdot, t) + P^+\varphi(\cdot, t) + P^-\varphi(\cdot, t) \\ &= \varphi_0(\cdot, t) + \varphi^+(\cdot, t) + \varphi^-(\cdot, t), \end{aligned} \tag{IV.13}$$

where

$$P_0\varphi(x, t) = \frac{1}{2\pi} \int_{\mathbb{R}} e^{ix\xi} \chi(\xi) \widehat{\varphi}(\xi, t) \, d\xi,$$

$$P^+\varphi(x, t) = \frac{1}{2\pi} \int_{\mathbb{R}} e^{ix\xi} \psi^+(\xi) \widehat{\varphi}(\xi, t) \, d\xi,$$

$$P^-\varphi(x, t) = \frac{1}{2\pi} \int_{\mathbb{R}} e^{ix\xi} \psi^-(\xi) \widehat{\varphi}(\xi, t) \, d\xi.$$

Set $B(x, t) = \int_0^t b(x, \tau) \, d\tau$ and define

$$\begin{aligned} K^+ f(x, t) &= \frac{1}{2\pi} \int_{\mathbb{R}} \int_{T(x)}^t e^{ix\xi + (B(x,t) - B(x,s))\xi} \tilde{\psi}^+(\xi) \widehat{f}(\xi, s) \, ds \, d\xi \\ &= \int_{T(x)}^t \int_{\mathbb{R}} e^{ix\xi + (B(x,t) - B(x,s))|\xi|} \widehat{\tilde{P}^+ f}(\xi, s) \frac{d\xi}{2\pi} \, ds, \end{aligned} \tag{IV.14}$$

where $T(x) = T$ if $\sup_t b(x, t) > 0$ and $T(x) = -T$ if $\inf_t b(x, t) < 0$ (notice that these conditions exclude each other because $t \mapsto b(x, t)$ does not change sign). The function $0 \leq \tilde{\psi}^+(\xi) \leq 1$ is supported in $[0, \infty)$ and chosen so that $\tilde{\psi}^+(\xi) = 1$ for ξ in the support of $\psi^+(\xi)$. This implies that $\tilde{P}^+ P^+ = P^+$. If $\sup_t b(x, t) = \inf_t b(x, t) = 0$ we set $T(x) = T$. In particular, $T(x)$ is constant on the open set $\inf_t b(x, t) < 0$ and also constant in its complement. It follows that if $T(x)$ is not continuous at the point x then $t \mapsto b(x, t)$ vanishes identically. Since the integrand in the definition of K^+ vanishes for $\xi < 1$ we had the right to replace ξ by $|\xi|$ in (IV.14). We now recall that the Fourier transform of the Poisson kernel of the half upper plane $\{(x, y) \in \mathbb{R}^2 : y > 0\}$

$$P_y(x) = \frac{1}{\pi} \frac{y}{y^2 + x^2}$$

is

$$\int_{\mathbb{R}} e^{-ix\xi} P_y(x)\, dx = e^{-y|\xi|}.$$

This is still true for $y = 0$ if we interpret $P_0(x)$ as a limit in the distribution sense: $P_0(x) = \lim_{y \searrow 0} P_y(x) = \delta(x) = $ Dirac's delta. In view of this fact, a common pseudo-differential notation for the convolution $P_y * g$ is $e^{-y|D_x|}g$. Thus, for x, t, s fixed, the inner integral in (IV.14) may be written as the convolution $P_y * \tilde{f}^+$ with $\tilde{f}^+(x, s) = \tilde{P}^+ f(x, s)$ and $y = B(x, s) - B(x, t)$, i.e., as $e^{-(B(x,s)-B(x,t))|D_x|}\tilde{f}^+$. Notice that $B(x, s) - B(x, t) \geq 0$ when s belongs to the interval with endpoints $\{t, T(x)\}$ because of the way $T(x)$ was defined. For any function $g(x)$ in $L^p(\mathbb{R})$, let us write $g^\perp(x) = \sup_{y>0} |P_y * g(x)|$. We thus have

$$|K^+ f(x, t)| \leq \int_{-T}^{T} (\tilde{f}^+)^\perp(x, s)\, ds. \tag{IV.15}$$

It is well known that $g^\perp(x) \leq Mg(x)$, where M denotes the Hardy–Littlewood maximal function

$$Mg(x) = \sup_{r>0} \frac{1}{2r} \int_{x-r}^{x+r} |g(t)|\, dt$$

and that $\|Mg\|_{L^p} \leq C_p \|g\|_{L^p}$, $1 < p < \infty$. This shows that

$$\|K^+ f(\cdot, t)\|_{L^p(\mathbb{R})} \leq C \int_{-T}^{T} \|\tilde{f}^+(\cdot, s)\|_{L^p(\mathbb{R})}\, ds \leq C \int_{-T}^{T} \|f(\cdot, s)\|_{L^p(\mathbb{R})}\, ds$$

where we have used that $\tilde{P}^+$ is bounded in $L^p(\mathbb{R})$ for $1 < p < \infty$ because it is a pseudo-differential operator of order zero. Raising the inequality to the power q and using Hölder's inequality we get

$$\|K^+ f(\cdot, t)\|_{L^p(\mathbb{R})}^q \leq CT^{q/q'} \|f\|_{L^q[(-T,T);L^p(\mathbb{R})]}^q$$

so integrating between $-T$ and T with respect to t and taking the $1/q$th power we obtain

$$\|K^+ f\|_{L^q[(-T,T);L^p(\mathbb{R})]} \leq CT \|f\|_{L^q[(-T,T);L^p(\mathbb{R})]}. \tag{IV.16}$$

Next we have to see the effect of K^+ on $L\varphi$, $\varphi \in C_c^\infty(\Omega_T) = C_c^\infty(\mathbb{R} \times (-T, T))$. Observe that since $\varphi(x, \pm T) \equiv 0$ it follows that $\varphi^+(\xi, \pm T) = P^+\varphi(\xi, \pm T) \equiv 0$, in particular $\varphi^+(\xi, T(x)) = 0$ for any $\xi \in \mathbb{R}$. Let us compute

$$K^+\varphi_t^+(x, t) = \int_{T(x)}^t \int_{\mathbb{R}} e^{ix\xi + (B(x,t) - B(x,s))|\xi|} \frac{d\widehat{\varphi^+}(\xi, s)}{ds} \frac{d\xi}{2\pi} ds.$$

Note that we have used that $\tilde{P}^+\varphi^+ = \varphi^+$. We integrate by parts in s. The boundary term is

$$\frac{1}{2\pi} \int_{\mathbb{R}} e^{ix\xi} \varphi^+(\xi, t) \, d\xi = \varphi^+(x, t)$$

and the integral term is

$$I = \int_{T(x)}^t \int_{\mathbb{R}} e^{ix\xi + (B(x,t) - B(x,s))|\xi|} \, b(x, s) \, |\xi| \widehat{\varphi^+}(\xi, s) \frac{d\xi}{2\pi} ds.$$

Since $|\xi| = \xi$ on the support of φ^+ and $i\xi\widehat{\varphi^+} = \widehat{\varphi_x^+}$ we have

$$K^+\varphi_t^+(x, t) = \varphi^+(x, t) - i \int_{T(x)}^t b(x, s) e^{-(B(x,s) - B(x,t))|D_x|} \varphi_x^+(x, s) \, ds. \tag{IV.17}$$

We may write $b(x, s)e^{-(B(x,s) - B(x,t))|D_x|}\varphi_x^+ = \left[b, e^{-(B(x,s) - B(x,t))|D_x|}\right]\varphi_x^+ + e^{-(B(x,s) - B(x,t))|D_x|}b\varphi_x^+$. Thus, (IV.17) may be rewritten as

$$K^+L\varphi^+(x, t) = \varphi^+(x, t) + R^+\varphi^+(x, t) \tag{IV.18}$$

where

$$R^+\varphi^+(x, t) = \int_{T(x)}^t \left[b, e^{-(B(x,s) - B(x,t))|D_x|}\right] \varphi_x^+(x, s) \, ds. \tag{IV.19}$$

It follows from (IV.16) and (IV.18) that

$$\begin{aligned}
\|\varphi^+\|_{L^q[\mathbb{R}; L^p(\mathbb{R})]} &\leq \|K^+L\varphi^+\|_{L^q[\mathbb{R}; L^p(\mathbb{R})]} + \|R^+\varphi^+\|_{L^q[\mathbb{R}; L^p(\mathbb{R})]} \\
&\leq CT\|L\varphi^+\|_{L^q[\mathbb{R}; L^p(\mathbb{R})]} + \|R^+\varphi^+\|_{L^q[\mathbb{R}; L^p(\mathbb{R})]},
\end{aligned} \tag{IV.20}$$

so (IV.20) will imply (IV.11) for φ^+ with L in the place of ${}^tL$ if the error term $\|R^+\varphi^+\|_{L^q[\mathbb{R}; L^p(\mathbb{R})]}$ can be absorbed. At this point we need

LEMMA IV.1.11. *Let $b(x)$, $x \in \mathbb{R}$, be a Lipschitz function with Lipschitz constant K, $\varphi \in \mathcal{S}(\mathbb{R})$. There is a constant $C > 0$ such that*

$$\left\| \sup_{\varepsilon > 0} \left| \left[b, e^{-\varepsilon|D_x|}\right] \varphi_x \right| \right\|_{L^p(\mathbb{R})} \leq C_p K \|\varphi\|_{L^p(\mathbb{R})}.$$

PROOF. We have

$$[b, e^{-\varepsilon|D_x|}]\varphi_x(x) = \int (b(x) - b(y))P_\varepsilon(x-y)\varphi'(y)\,dy.$$

After an integration by parts we may write

$$[b, e^{-\varepsilon|D_x|}]\varphi_x(x)$$
$$= P_\varepsilon * (b'\varphi)(x) + \int (b(x) - b(y))[P_\varepsilon]'(x-y)\varphi(y)\,dy. \qquad (IV.21)$$

As we already saw, $\sup_{\varepsilon>0} |P_\varepsilon * (b'\varphi)| \leq M(b'\varphi) \leq \|b'\|_{L^\infty} M\varphi$, where M is the Hardy–Littlewood maximal operator. The second term may be majorized by $K|Q_\varepsilon| * |\varphi|(x)$ where

$$Q_\varepsilon(x) = \frac{1}{\varepsilon}Q(x/\varepsilon), \qquad Q(x) = x\frac{dP}{dx}(x) = \frac{-2x^2}{(1+x^2)^2}.$$

Since the function $|Q(x)|$ has an integrable even majorant it follows [S1] that $\sup_{\varepsilon>0} |Q_\varepsilon| * |\varphi| \leq CM(\varphi)$. Therefore, $\left| [b, e^{-\varepsilon|D_x|}]\varphi'(x) \right| \leq CM\varphi(x)$ and the L^p boundedness of the Hardy–Littlewood operator M grants the desired estimate. □

We may now estimate the error term in (IV.20). Since

$$\left| [b, e^{-(B(x,s)-B(x,t))|D_x|}]\varphi_x^+(x,s) \right| \leq \sup_{y>0} \left| [b, e^{-y|D_x|}]\varphi_x^+(x,s) \right|$$

it follows from (IV.19) and Lemma IV.1.11 that

$$\|R^+\varphi^+(\cdot,t)\|_{L^p(\mathbb{R})} \leq C \int_{-T}^{T} \|\varphi^+(\cdot,s)\|_{L^p(\mathbb{R})}\,ds.$$

We already showed that from this inequality follows the estimate

$$\|R^+\varphi^+\|_{L^q[(-T,T);L^p(\mathbb{R})]} \leq CT\|\varphi^+\|_{L^q[(-T,T);L^p(\mathbb{R})]}.$$

Taking account of (IV.20) we obtain

$$\|\varphi^+\|_{L^q[(-T,T);L^p(\mathbb{R})]} \leq CT\|L\varphi^+\|_{L^q[(-T,T);L^p(\mathbb{R})]} + CT\|\varphi^+\|_{L^q[(-T,T);L^p(\mathbb{R})]}.$$

Write $L\varphi^+ = LP^+\varphi = P^+L\varphi + [L, P^+]\varphi = P^+L\varphi + [-bD_x, P^+]\varphi$. Since P^+ is a pseudo-differential of order zero, P^+ is bounded in $L^p(\mathbb{R})$ and so is the commutator $[-bD_x, P^+]$ with norm proportional to $\|b_x\|_{L^\infty}$ (see [S2, page 309], for the continuity in L^2 which implies the L^p continuity, $1 < p < \infty$, by the Calderón–Zygmund theory). Thus,

$$\|\varphi^+\|_{L^q[(-T,T);L^p(\mathbb{R})]} \leq CT\|L\varphi\|_{L^q[(-T,T);L^p(\mathbb{R})]} + CT\|\varphi\|_{L^q[(-T,T);L^p(\mathbb{R})]}.$$

In a similar way, we may prove

$$\|\varphi^-\|_{L^q[(-T,T);L^p(\mathbb{R})]} \le CT\|L\varphi\|_{L^q[(-T,T);L^p(\mathbb{R})]} + CT\|\varphi\|_{L^q[(-T,T);L^p(\mathbb{R})]}.$$

It remains to estimate φ_0, which is easier. We define

$$K_0 f(x,t) = \int_{-T}^{t} \int_{\mathbb{R}} e^{ix\xi + (B(x,t)-B(x,s))\xi} \widehat{P_0 f}(\xi,s) \frac{d\xi}{2\pi} \, ds,$$

and notice that $\widehat{P_0 f}(\xi,s) = \chi(\xi)\widehat{f}(\xi,s)$ is supported in $|\xi| \le 2$ so the exponential remains bounded independently of the sign of the exponent. Reasoning with K_0 as we did with K^+ we derive

$$\|\varphi_0\|_{L^q[(-T,T);L^p(\mathbb{R})]} \le CT\|L\varphi\|_{L^q[(-T,T);L^p(\mathbb{R})]} + CT\|\varphi\|_{L^q[(-T,T);L^p(\mathbb{R})]}.$$

Since $\varphi = \varphi_0 + \varphi^+ + \varphi^-$ we obtain

$$\|\varphi\|_{L^q[(-T,T);L^p(\mathbb{R})]} \le CT\|L\varphi\|_{L^q[(-T,T);L^p(\mathbb{R})]} + CT\|\varphi\|_{L^q[(-T,T);L^p(\mathbb{R})]},$$

which implies, assuming that $CT_0 < 1/2$ and $0 < T < T_0$, that

$$\|\varphi\|_{L^q[(-T,T);L^p(\mathbb{R})]} \le 2CT\|L\varphi\|_{L^q[(-T,T);L^p(\mathbb{R})]}.$$

This proves Theorem IV.1.9.

REMARK IV.1.12. Although the coefficient $b(x,t)$ was assumed to be smooth in the proof of estimate (IV.11), all steps can be carried out assuming only that $b(x,t)$ is continuous and b_x is bounded, so Theorem IV.1.9 and its Corollary IV.1.10 remain valid under these hypotheses.

Consider a finite rectangle $U = (-T,T) \times (-T,T)$. In view of Corollary IV.1.10, for every $f \in L^p(U)$ we may find $u \in L^p(U)$ such that $Lu = f$ in U. Since $L^p(U)$ decreases as p increases from 1 to ∞ the value of p may be considered as a degree of regularity of the functions that belong to $L^p(U)$. If we fix a function $f \in L^\infty(\mathbb{R}^2)$, Corollary IV.1.10 tells us that for any $p < \infty$ we may find a function $u_p \in L^p(U_p)$, $U_p = (-T(p), T(p)) \times (-T(p), T(p))$ solving the equation $Lu_p = f$ in U_p. Unfortunately, $T(p) \to 0$ as $p \to \infty$, so we cannot hope to find a convergent subsequence of the sequence of solutions u_p, $p = 1, 2, 3, \ldots$ The question arises whether we can find a local solution of $Lu = f$ with $u \in L^\infty$. The answer, in general, is no—as the following example shows. Consider the smooth function of one variable

$$B(t) = \begin{cases} \exp(-1/t), & \text{if } t \ge 0, \\ -\exp(1/t), & \text{if } t \le 0, \end{cases}$$

with derivative

$$b(t) = B'(t) = \frac{1}{t^2}\exp(-1/|t|),$$

and define the differential operator on $\mathbb{R}^2$

$$L = \frac{\partial}{\partial t} - ib(t)\frac{\partial}{\partial x}.$$

It is easily verified that L satisfies condition $(\mathcal{P})$ and $L^t = -L$. The function $B(t)$ is strictly increasing for $-\infty < t < \infty$ and has an inverse $\beta(s)$: $(-1, 1) \to (-\infty, \infty)$ given by $\beta(\pm|s|) = \pm 1/|\log|s||$. There is a homeomorphism $\Psi(x, s) = (x, \beta(s)) : \mathbb{R} \times (-1, 1) \to \mathbb{R} \times (-\infty, \infty)$ which is a diffeomorphism for $0 < |s| < 1$. Let $u \in L^\infty(\mathbb{R}^2)$ and $f \in L^\infty(\mathbb{R}^2)$ be such that

$$Lu = f \tag{IV.22}$$

in the sense of distributions and set

$$v(x, s) = u(x, \beta(s)), \qquad g(x, s) = \frac{f(x, \beta(s))}{s\log^2 s}.$$

LEMMA IV.1.13. *Let L, $u(x, t)$, $f(x, t)$, $v(x, s)$, $g(x, s)$ be as above. Then, $v(x, s) \in L^\infty$, $g(x, s) \in L^1_{\text{loc}}$ and*

$$\frac{\partial}{\partial \bar{z}} v = \frac{1}{2} g \qquad \text{for } -1 < s < 1, \tag{IV.23}$$

in the sense of distributions. In particular, if w is any solution of

$$\frac{\partial}{\partial \bar{z}} w = g,$$

in a neighborhood of the origin, w must be essentially bounded in a neighborhood of the origin.

PROOF. If U is an open subset of $\mathbb{R}^2$ and $V = \Psi^{-1}U$, then V is an open subset of $\mathbb{R} \times (-1, 1)$ and its Lebesgue measure is given by

$$m(U) = \int_V \frac{1}{|s|\log^2|s|} \, dx \, ds.$$

It follows that the Borel measure $\mu(X) = m(\Psi(X))$ is absolutely continuous with respect to the Lebesgue measure on $\mathbb{R} \times (-1, 1)$, since $|s|^{-1}\log^{-2}|s|$ is locally integrable in $\mathbb{R} \times (-1, 1)$. Thus, $v(x, s)$ and $g(x, s)$ are measurable, v is bounded, g is locally integrable, and for every $\phi \in C_c^\infty(\mathbb{R} \times (-1, 1))$ we have the identity

$$2\int v(x, s)\frac{\partial \phi(x, s)}{\partial \bar{z}} \, dx \, ds = \int g(x, s)\phi(x, s) \, dx \, ds, \tag{IV.24}$$

as follows from the change of variables $(x, s) = (x, B(t))$ in both integrals. Indeed, $\psi(x, t) = \phi(x, B(t))$ is a test function and (IV.24) becomes $\langle u, L^t\phi \rangle = \langle f, \phi \rangle$, which is precisely (IV.22). Furthermore, if $w/2$ is a local solution of

(IV.24) it follows that $w - 2v$ is holomorphic in a neighborhood of the origin and w must be locally bounded. $\qquad\square$

By the lemma, we will have our example if we show that for an appropriate choice of $f \in L^\infty$, equation (IV.23) has a solution which is not locally bounded in any neighborhood of the origin. We choose f so that $F = f \circ \Psi$ is the characteristic function χ of the sector K described in polar coordinates by $0 \leq r \leq 1/2$, $0 \leq \theta \leq \pi/4$. Hence, $g = \beta' \chi \in L^1_c(\mathbb{R}^2)$ and a solution $w(x, s)$ of (IV.23) is obtained by convolution of $g/2$ with the standard fundamental solution of the Cauchy–Riemann operator. Thus,

$$\Re w(x, s) = \frac{1}{2\pi} \int_K \frac{x - x'}{|x - x'|^2 + |s - s'|^2} \frac{1}{s' \ln^2 s'} \, dx' ds'.$$

We see that for $(x, s) = (0, 0)$, the integral above is given by

$$\frac{1}{2\pi} \int_0^{\pi/4} \int_0^{1/2} \frac{-\cos\theta}{r \sin\theta \log^2(r \sin\theta)} \, dr d\theta = \frac{1}{2\pi} \int_0^{\pi/4} \frac{1}{\sin\theta \log[(\sin\theta)/2]} \, d\theta$$

$$= -\infty$$

and it is easy to conclude that $\Re w$ cannot be essentially bounded in any neighborhood of the origin. Indeed, if (x_n, s_n) is any sequence such that $x_n < 0$ and $(x_n, s_n) \to (0, 0)$, the integrand in $\Re w(x_n, s_n)$ remains negative and by Fatou's lemma $\liminf_{(x_n, s_n) \to (0,0)} \Re w(x_n, s_n) \leq \Re w(0, 0) = -\infty$. Hence, $w(x, s)$ cannot remain essentially bounded in $\{(x, s) : x < 0, \ x^2 + s^2 < \varepsilon^2\}$ for any $\varepsilon > 0$.

Take $q' = \infty$ and write p instead of p' in Corollary IV.1.10. If $f \in L^\infty$ we can obtain local solutions of $Lu = f$ in $L^\infty[(-T, T); L^p(\mathbb{R})]$ for any $1 < p < \infty$ but, as we just saw, we cannot find in general a solution $u \in L^\infty[(-T, T); L^\infty(\mathbb{R})] \simeq L^\infty(\Omega_T)$. Many results in analysis that hold for $1 < p < \infty$ and fail for $p = \infty$ become true if L^∞ is replaced by a space of functions of bounded mean oscillation. In our situation the remedy is to replace the space L^∞ by the space $\mathrm{bmo}(\mathbb{R})$, dual of the semilocal (or localizable) Hardy space $h^1(\mathbb{R})$.

IV.1.2 A priori estimates in h^1

We recall some facts about the real Hardy spaces $H^1(\mathbb{R})$, a particular instance of the spaces introduced by Stein and Weiss in [SW], and its semilocal version $h^1(\mathbb{R})$ introduced by Goldberg [G]. In many situations $H^1(\mathbb{R})$ is an advantageous substitute for $L^1(\mathbb{R})$ ([S2]), as the latter does not behave well in many respects, for instance, concerning the continuity of singular integral operators. Let us choose a function $\Phi \geq 0 \in C_c^\infty([-1/2, 1/2])$, with $\int \Phi dz = 1$. Write $\Phi_\varepsilon(z) = \varepsilon^{-1}\Phi(z/\varepsilon)$, $z \in \mathbb{R}$, and set

$$M_\Phi f(z) = \sup_{0 < \varepsilon < \infty} |(\Phi_\varepsilon * f)(z)|.$$

Then [S2]

$$H^1(\mathbb{R}) = \{f \in L^1(\mathbb{R}) : \quad M_\Phi f \in L^1(\mathbb{R})\}.$$

A space of distributions is called semilocal if it is invariant under multiplication by test functions. The space $H^1(\mathbb{R})$, is not: ψu may not belong to $H^1(\mathbb{R})$ for $\psi \in C_c^\infty(\mathbb{R})$ and $u \in H^1(\mathbb{R})$. A way around this is the definition of the semilocal (or localizable) Hardy space—better suited for the study of PDEs—$h^1(\mathbb{R})$ ([G], [S2]) by means of the truncated maximal function

$$m_\Phi f(z) = \sup_{0 < \varepsilon \leq 1} |(\Phi_\varepsilon * f)(z)|;$$

$$h^1(\mathbb{R}) = \{f \in L^1(\mathbb{R}) : \quad m_\Phi f \in L^1(\mathbb{R})\},$$

which is stable under multiplication by test functions (we will systematically denote by $\mathcal{S}$ the Schwartz space of rapidly decreasing functions and by $\mathcal{S}'$ its dual, i.e., the space of tempered distributions). It turns out that if Φ is substituted in the definition of $h^1(\mathbb{R})$ by any other function $\Phi \in \mathcal{S}(\mathbb{R})$ only subjected to $\int \Phi \neq 0$, this will not change the space $h^1(\mathbb{R})$. Moreover, $h^1(\mathbb{R})$ is a Banach space with the norm

$$\|f\|_{h^1} = \|m_\Phi f\|_{L^1},$$

and $H^1 \subset h^1 \subset L^1$. Of course, this norm depends on the choice of Φ but different Φ's will give equivalent norms, moreover, if $\mathcal{A} \subset \mathcal{S}$ is a bounded subset, there is a constant $C = C(\mathcal{A}) > 0$ such that $\|m_\phi f\|_{L^1} \leq C\|m_\Phi f\|_{L^1}$ for all $f \in \mathcal{S}$ and $\phi \in \mathcal{A}$. In fact, more is true: denoting by $\mathcal{M}f(x) = \sup_{\phi \in \mathcal{A}} m_\phi f(x)$ the grand maximal function associated with $\mathcal{A}$ it follows that $\|\mathcal{M}f\|_{L^1} \leq C\|m_\Phi f\|_{L^1}$.

We now describe the atomic decomposition of $h^1(\mathbb{R})$ ([G], [S2]). An $h^1(\mathbb{R})$ atom is a bounded, compactly supported function $a(z)$ satisfying the following properties: there exists an interval I containing the support of a such that

(1) $|a(z)| \leq |I|^{-1}$, a.e., with $|I|$ denoting the Lebesgue measure of I;
(2) if $|I| < 1$, we further require that $\int a(z) \, dz = 0$.

Any $f \in h^1$ can be written as an infinite linear combination of h^1 atoms, more precisely, there exist scalars λ_j and h^1 atoms a_j such that $\sum_j |\lambda_j| < \infty$ and the series $\sum_j \lambda_j a_j$ converges to f both in h^1 and in $\mathcal{S}'$. Furthermore,

$\|f\|_{h^1} \sim \inf \sum_j |\lambda_j|$, where the infimum is taken over all atomic representations. Another useful fact is that the atoms may be assumed to be smooth functions. A simple consequence of the atomic decomposition is that $h^1(\mathbb{R})$ is stable under multiplication by Lipschitz functions $b(x)$: if a satisfies (1) with $|I| \geq 1$ it follows that $a(x)b(x)/\|b\|_{L^\infty}$ also does. If $|I| < 1$ and the center of I is x_0 we may write $a(x)b(x) = b(x_0)a(x) + (b(x) - b(x_0))a(x) = \beta_1(x) + \beta_2(x)$. Then $\beta_1(x)/\|b\|_{L^\infty}$ satisfies (1) and (2) (with the same I) while $\beta_2(x)/K$ satisfies (1) for the interval I' of center x_0 and length 1, where K is the Lipschitz constant of $a(x)$. It follows that $f \mapsto bf$ is bounded with constant $\leq \|b\|_{L^\infty} + K$ in $h^1(\mathbb{R})$. A refinement of this argument shows that $h^1(\mathbb{R})$ is stable under multiplication by more general continuous functions including Hölder functions, as we now describe. Let ω be a modulus of continuity, meaning that $\omega : [0, \infty) \longrightarrow \mathbb{R}^+$ is continuous, increasing, $\omega(0) = 0$ and $\omega(2t) \leq C\omega(t)$, $0 < t < 1$. Consider the Banach space $C_\omega(\mathbb{R})$ of bounded continuous functions $f : \mathbb{R} \longrightarrow \mathbb{C}$ such that

$$|f|_{C_\omega} \doteq \sup_{x \neq y} \frac{|f(y) - f(x)|}{\omega(|x - y|)} < \infty,$$

equipped with the norm $\|f\|_{C_\omega} = \|f\|_{L^\infty} + |f|_{C_\omega}$. Note that C_ω is only determined by the behavior of $\omega(t)$ for values of t close to 0. We will show in Lemma A.1.1 in the Appendix that if the modulus of continuity $\omega(t)$ satisfies

$$\frac{1}{h} \int_0^h \omega(t) \, dt \leq C \left(1 + \log \frac{1}{h}\right)^{-1}, \qquad 0 < h < 1, \tag{IV.25}$$

then $h^1(\mathbb{R})$ is stable under multiplication by functions $\in C_\omega(\mathbb{R})$. Note that the modulus of continuity $\omega(t) = t^r$, $0 < r < 1$, that defines the Hölder space C^r, satisfies (IV.25).

Consider now a first-order linear differential operator in two variables

$$L = \frac{\partial}{\partial t} + ib(x, t)\frac{\partial}{\partial x} + c(x, t), \qquad x, t \in \mathbb{R}. \tag{IV.26}$$

We assume that for some $0 < r < 1$

(i) $c(x, t) \in C^r(\mathbb{R}^2)$;

(ii) $b(x, t)$ is real and of class C^{1+r}, i.e., for all multi-indexes $|\alpha| \leq 1$, $D^\alpha b$ is bounded and $D^\alpha b \in C^r(\mathbb{R}^2)$;

(iii) for any $x \in \mathbb{R}$ the function $t \mapsto b(x, t)$ does not change sign.

Of course, (iii) means that the operator L given by (IV.26) satisfies condition $(\mathcal{P})$. We now introduce the space $L^1[\mathbb{R}_t; h^1(\mathbb{R}_x)]$ of measurable functions $u(x, t)$ such that, for almost every $t \in \mathbb{R}$, $x \mapsto u(x, t) \in h^1(\mathbb{R})$ and

$$\int_{\mathbb{R}} \|u(\cdot, t)\|_{h^1} \, dt \leq C < \infty.$$

The dual of the space $L^1[\mathbb{R}; h^1(\mathbb{R})]$ is (canonically isomorphic to) the space $L^\infty[\mathbb{R}; \mathrm{bmo}(\mathbb{R})]$ (see page 174).

When proving a priori estimates for norms involving Hardy spaces, the role of the coefficient $c(x, t)$ will be small and its contribution may be absorbed. For that reason, it is convenient to assume initially that $c(x, t) \equiv 0$ and we shall do so for a long time in the computations that follow. We will withdraw the temporary hypothesis only after we have proved our estimates with the additional assumption that $c(x, t) \equiv 0$.

PROPOSITION IV.1.14. *Let the operator L given by (IV.26) with $c(x, t) \equiv 0$ satisfy (ii) and (iii), and let $\alpha > 0$ be given. Then there exist operators $K, R : C_c^\infty((-\alpha, \alpha) \times (-T, T)) \longrightarrow L^1[(-T, T); h^1(\mathbb{R}_x)]$ and constants $C > 0$ and $T_0 > 0$ such that*

$$KLu = u + Ru, \tag{IV.27}$$

$$\|Ku\|_{L^1[(-\alpha,\alpha) \times (-T,T)]} \leq CT \|u\|_{L^1[\mathbb{R}; h^1(\mathbb{R})]}, \tag{IV.28}$$

$$\|Ru\|_{L^1[(-\alpha,\alpha) \times (-T,T)]} \leq CT \|u\|_{L^1[\mathbb{R}; h^1(\mathbb{R})]}, \tag{IV.29}$$

for all $u \in C_c^\infty([-\alpha, \alpha] \times [-T, T])$, $0 < T \leq T_0$.

This is a technical proposition that does not have an immediate duality consequence due to the fact that the norm on the left-hand side of the estimates is a weaker norm than that on the right-hand side and it should be regarded as an intermediate step towards a better estimate to be obtained later. The proof of Proposition IV.1.14 is similar to that of Theorem IV.1.9; in particular, the operators K and R referred to in (IV.27) were implicitly used in its proof, for instance, $K = K^+ + K^- + K_0$, $R = R^+ + R^- + R_0$ with K^+ given by (IV.14), R^+ given by (IV.19) and so on. So the first step will be to prove the analogue of (IV.27) for K^+. This will follow from a slight modification of (IV.15). Let us consider a *restricted* maximal function

$$g^\perp(x) = \sup_{0 < y < 1} |P_y * g(x)|.$$

Notice that the sup is now taken for values of y between 0 and 1 instead of $0 < y < \infty$ as we did in (IV.15), but we keep the same notation $g \mapsto g^\perp$. Assuming without loss of generality that $b(x, t)$ has compact support and taking T small we may assume that $|B(x, t) - B(x, s)| < 1$ in formula (IV.14), so we get

$$|K^+ u(x, t)| \leq \int_{-T}^{T} (\tilde{u}^+)^{\perp}(x, s) \, ds. \qquad \text{(IV.30)}$$

Before we continue with the proof of the estimates, we state and prove some lemmas. The first one deals with the nonlocal space H^1.

LEMMA IV.1.15. *Let* $Q \in C^1(\mathbb{R})$ *be an integrable function such that*

$$|Q'(x)| \leq \frac{C}{1 + |x|^2}, \quad x \in \mathbb{R}$$

for some $C > 0$. *Then, for some* $C > 0$,

$$\int_{\mathbb{R}} M_Q f(x) \, dx = \int_{\mathbb{R}} \sup_{y>0} |Q_y * f(x)| \, dx \leq C \|f\|_{H^1(\mathbb{R})}, \quad f \in H^1(\mathbb{R}).$$

PROOF. By the atomic decomposition we may assume that $f(x) = a(x)$ is an H^1-atom supported in an interval $(x_0 - r, x_0 + r)$. Assume initially that $x_0 = 0$. We have

$$|(Q_y * a)(x)| \leq \|Q_y\|_{L^1} \|a\|_{L^\infty} \leq \frac{C}{r}, \quad x \in \mathbb{R},$$

and we easily derive that

$$\int_{-2r}^{2r} M_Q a(x) \, dx \leq C.$$

Recalling that $\int a(x) \, dx = 0$ we may write

$$(Q_y * a)(x) = \int_{-r}^{r} a(z)(Q_y(x - z) - Q_y(x)) \, dz.$$

By the mean value theorem, we get for $|z| < r$

$$|Q_y(x - z) - Q_y(x)| \leq \frac{1}{y^2} |Q'(\xi/y) z| \leq \frac{Cr}{y^2 + \xi^2} \leq \frac{Cr}{\xi^2}$$

for some $\xi \in (x - r, x + r)$. If $|x| > 2r$, it follows that $|\xi| > |x|/2$. Thus,

$$\sup_{y>0} |(Q_y * a)(x)| \leq \frac{Cr}{x^2}, \quad |x| > 2r,$$

and

$$\int_{|x| \geq 2r} M_Q a(x) \, dx \leq Cr \int_{2r}^{\infty} \frac{dt}{t^2} \leq C'.$$

This shows that $\int M_Q a \, dx \leq C$. In the general case we consider a translated atom $\tilde{a}(x) = a(x + x_0)$ which is centered at the origin and observe that $\|M_Q a\|_{L^1} = \|M_Q \tilde{a}\|_{L^1}$ because $M_Q \tilde{a}(x) = M_Q a(x + x_0)$. $\qquad \square$

We return to the semilocal Hardy space h^1 in the next lemma.

LEMMA IV.1.16. *Let $0 < \alpha < \infty$, let P be the Poisson kernel in $\mathbb{R}^2_+$ and let Q be an integrable function satisfying $|Q'(x)| \leq C/(1+|x|^2)$ as in the previous lemma. There exists $C > 0$ such that*

$$\int_{-\alpha}^{\alpha} \sup_{0<y<1} |P_y * f(x)| \, dx \leq C \|f\|_{h^1(\mathbb{R})}, \quad f \in h^1(\mathbb{R}),$$

$$\int_{-\alpha}^{\alpha} \sup_{0<y<1} |Q_y * f(x)| \, dx \leq C \|f\|_{h^1(\mathbb{R})}, \quad f \in h^1(\mathbb{R}).$$

PROOF. The first inequality follows from the second one, as P satisfies the hypothesis required for Q. To prove the second inequality we need only show that there exists $C > 0$ such that

$$\int_{-\alpha}^{\alpha} \sup_{0<y<1} |Q_y * a| \leq C$$

for all h^1-atoms a. Let a be an h^1-atom supported in the interval $I = (x_0 - r, x_0 + r)$. If $r > 1/2$ we observe that

$$\sup_{0<y<1} |(Q_y * a)(x)| \leq \sup_{0<y<1} \|a\|_{L^\infty} \|Q_y\|_{L^1} \leq |I|^{-1} \|Q\|_{L^1} \leq \|Q\|_{L^1} \leq C,$$

so the integral we must estimate is majorized by $2C\alpha$. If $r \leq 1/2$ the atom a must satisfy the moment condition and it is also an H^1-atom so the required inequality holds even for $\alpha = \infty$ by the proof of Lemma IV.1.15. $\square$

In view of (IV.30) and the first inequality of Lemma IV.1.16 we obtain

$$\|K^+ u\|_{L^1[(-\alpha,\alpha)\times(-T,T)]} \leq CT \|u^+\|_{L^1[\mathbb{R};h^1(\mathbb{R})]}. \tag{IV.31}$$

To obtain a similar inequality for R^+ we use (IV.19) to derive

$$|R^+ u(x,t)| \leq \int_{-T}^{T} \sup_{0<y<1} \left| [b, e^{-y|D_x|}] u_x^+ \right| (x,s) \, ds. \tag{IV.32}$$

We already saw that

$$[b, e^{-y|D_x|}] u^+(x,t)$$

$$= P_y * (b_x u^+)(x,t) + \int \frac{b(x,t) - b(y,t)}{x-y} Q_y(x-y) u^+(y,t) \, dy$$

$$= P_y * (b_x u^+)(x,t) + \int Q_y(x-y) \beta^x(y,t) u^+(y,t) \, dy,$$

with

$$Q_y(x) = \frac{1}{y} Q(x/y), \qquad Q(x) = x \frac{dP}{dx}(x) = \frac{-2x^2}{(1+x^2)^2}$$

and

$$\beta^x(y, t) = \begin{cases} \dfrac{b(x, t) - b(y, t)}{x - y}, & \text{if } y \neq x, \\ b_x(x, s'), & \text{if } y = x. \end{cases}$$

Using once more Lemma IV.1.16 we see that the norm in $L^1[(-\alpha, \alpha) \times (-T, T)]$ of the term $P_y * (b_x u^+)(x, t)$ is dominated by

$$\|b_x u^+\|_{L^1[\mathbb{R}; h^1(\mathbb{R})]} \leq \|u^+\|_{L^1[\mathbb{R}; h^1(\mathbb{R})]}$$

where we have used that multiplication by $b_x \in C^r$ is a bounded operation in $h^1(\mathbb{R})$. Concerning the second term, observe that it may be written as a convolution $Q_y * (\beta^x u^+)(x)$ (note however that the factor β^x depends on the point at which the convolution is evaluated). The main tool to estimate the second term is

LEMMA IV.1.17. *Let* $0 < \alpha < \infty$. *Let* $Q \in C^1(\mathbb{R})$ *satisfy*

$$|Q(x)| + |Q'(x)| \leq \frac{C}{1 + x^2}, \quad x \in \mathbb{R}$$

for some $C > 0$ *and assume that* $\beta \in L^\infty(\mathbb{R}^2)$ *is such that for some* $K > 0$

$$|\beta(x, y) - \beta(x, x_0)| \leq K \frac{|y - x_0|}{|x - x_0|}, \quad \text{if } |x - x_0| \geq 2|y - x_0|.$$

Then there exists $C = C(\beta, Q) > 0$ *such that, for every* $f \in h^1(\mathbb{R})$, *the inequality*

$$\int_{-\alpha}^{\alpha} \sup_{0 < y < 1} |Q_y * (\beta^x f)(x)| \, dx \leq C \|f\|_{h^1(\mathbb{R})} \quad holds,$$

with $\beta^x(y) = \beta(x, y)$.

PROOF. Let a be an h^1-atom, with $s(a) \subset I = (x_0 - r, x_0 + r)$. If $r > 1/2$ we have

$$\int_{-\alpha}^{\alpha} \sup_{0 < y < 1} |Q_y * (\beta^x a)(x)| \, dx = \int_{-\alpha}^{\alpha} \sup_{0 < y < 1} \left| \int Q_y(x - z) \beta^x(z) a(z) \, dz \right| dx$$

$$\leq \int_{-\alpha}^{\alpha} \|\beta\|_{L^\infty} \|a\|_{L^\infty} \|Q_y\|_{L^1} \, dx$$

$$\leq 2\alpha \|\beta\|_{L^\infty} \|Q\|_{L^1}.$$

Let us tackle the case $r \leq 1/2$ assuming initially that $x_0 = 0$. The estimate

$$\int_{-2r}^{2r} \sup_{0 < \varepsilon < 1} |Q_\varepsilon * (\beta^x a)(x)| \, dx \leq Cr \|a\|_{L^\infty} \leq C'$$

is, as usual, easily obtained. Keeping in mind that $\int a(y)\,dy = 0$ and writing

$$Q_\varepsilon(x-y)\beta^x(y) - Q_\varepsilon(x)\beta^x(0) = (Q_\varepsilon(x-y) - Q_\varepsilon(x))\beta^x(0)$$
$$+ Q_\varepsilon(x-y)(\beta^x(y) - \beta^x(0))$$

we get the estimate

$$|Q_\varepsilon * (\beta^x a)(x)| \leq \int_{-r}^{r} |(Q_\varepsilon(x-y)\beta^x(y) - Q_\varepsilon(x)\beta^x(0)|\,|a(y)|\,dy$$
$$\leq \int \frac{1}{\varepsilon^2} \sup_{|y| \leq r} |Q'((x-y)/\varepsilon)|\,\|\beta\|_{L^\infty}\,|y|\,|a(y)|\,dy$$
$$+ \int \frac{K}{\varepsilon} \sup_{|y| \leq r} |Q((x-y)/\varepsilon)|\,\frac{|y|}{|x|}\,|a(y)|\,dy.$$

Since $|x| > 2r$ and $|y| < r$ imply that $|x-y| \geq |x|/2$, using the decay of Q and Q' we see that

$$\frac{1}{\varepsilon^2} \sup_{|y| \leq r} |Q'((x-y)/\varepsilon)| \leq \frac{C}{\varepsilon^2 + x^2} \leq \frac{C}{x^2},$$
$$\frac{1}{\varepsilon} \sup_{|y| \leq r} |Q((x-y)/\varepsilon)| \leq \frac{C\varepsilon}{\varepsilon^2 + x^2} \leq \frac{C}{x^2}$$

for $|x| > 2r$ so

$$|Q_\varepsilon * (\beta^x a)(x)| \leq C(\beta, Q)\frac{r}{x^2} \int_{-r}^{r} |a(y)|\,dy$$
$$\leq C(\beta, Q)\frac{r}{x^2}.$$

Thus,

$$\int_{|x|>2r} \sup_{0<\varepsilon<1} |Q_\varepsilon * (\beta^x a)(x)|\,dx \leq C(\beta, Q)\,r \int_{2r}^{\infty} \frac{1}{t^2}\,dt \leq C(\beta, Q).$$

In the general case, we reason as before with $\tilde{a}(x) = a(x+x_0)$, which is an atom centered at the origin, and $\tilde{\beta}(x, y) = \beta(x+x_0, y+x_0)$, which satisfies the same inequalities as $\beta(x, y)$, then observe that

$$Q_\varepsilon * (\beta^x a)(x) = Q_\varepsilon * (\tilde{\beta}^{x-x_0}\tilde{a})(x - x_0)$$

so

$$\int \sup_{0<\varepsilon<1} |Q_\varepsilon * (\beta^x a)(x)|\,dx = \int \sup_{0<\varepsilon<1} |Q_\varepsilon * (\tilde{\beta}^x\tilde{a})(x)|\,dx \leq C(\beta, Q). \qquad \square$$

REMARK IV.1.18. A function $\beta(x, y)$ satisfying the hypothesis of Lemma IV.1.17 can be obtained by setting

$$\beta^x(y) = \begin{cases} \dfrac{b(x) - b(y)}{x - y}, & \text{if } y \neq x, \\ b'(x), & \text{if } y = x, \end{cases}$$

if $b(x)$ and $b'(x)$ are bounded, as is easily seen.

Returning to the estimate of the second term $Q_y * (\beta^x u^+)(x)$ in the expression of $[b, e^{-y|D_x|}]u^+$ we point out that Lemma IV.1.17 can indeed be applied for any fixed t to $\beta(x, y) = \beta^x(y, t)$, so using Lemma IV.1.17 and (IV.32) we get

$$\|R^+ u\|_{L^1[(-\alpha,\alpha)\times(-T,T)]} \leq CT \|u^+\|_{L^1[\mathbb{R}; h^1(\mathbb{R})]}. \tag{IV.33}$$

Using (IV.31), (IV.33), their analogues for K^-, K_0, R^-, R_0 and the fact that $P^\pm$ and P_0 are pseudo-differential operators of order zero acting on the variable x, so the norm of u^+, u^- and u_0 in $h^1(\mathbb{R})$ are bounded by that of u, we may prove estimates (IV.28) and (IV.29) concluding the proof of Proposition IV.1.14.

Consider now a test function $\varphi \in C_c^\infty([-\alpha, \alpha] \times [-T, T])$. It follows easily from (IV.27) that

$$\|\varphi\|_{L^1[(-\alpha,\alpha)\times(-T,T)]} \leq \|KL\varphi\|_{L^1[(-\alpha,\alpha)\times(-T,T)]} + \|R\varphi\|_{L^1[(-\alpha,\alpha)\times(-T,T)]}$$

which, in view of (IV.28) and (IV.29), implies

$$\|\varphi\|_{L^1[(-\alpha,\alpha)\times(-T,T)]} \leq CT \left(\|L\varphi\|_{L^1[\mathbb{R}; h^1(\mathbb{R})]} + \|\varphi\|_{L^1[\mathbb{R}; h^1(\mathbb{R})]} \right). \tag{IV.34}$$

Notice that we cannot absorb the term $\|\varphi\|_{L^1[\mathbb{R}; h^1(\mathbb{R})]}$ by taking T small because it involves a stronger norm than that of the left-hand side. Thus, we wish to obtain a similar but sharper estimate in which the norm $\|\varphi\|_{L^1[\mathbb{R}; h^1(\mathbb{R})]}$ also appears as well on the left-hand side. To achieve this we make use of the mollified Hilbert transform $\widetilde{H}$ defined by $\widetilde{Hf} = (1 - \chi)\widehat{Hf}$, where H denotes the usual Hilbert transform, $\chi \in C_c^\infty(-2, 2)$, $\phi = 1$, for $|\xi| \leq 1$. Here the usefulness of $\widetilde{H}$, which is a pseudo-differential operator of order zero, derives mainly from the fact that it can be used to define an equivalent norm on $h^p(\mathbb{R})$ without appealing to maximal functions, as granted by the following estimates (*cf.* [G]):

$$C_1 \|\widetilde{H}f\|_{h^1} \leq \|f\|_{h^1} \leq C_2(\|f\|_{L^1} + \|\widetilde{H}f\|_{L^1}), \quad f \in h^1(\mathbb{R}).$$

Another ingredient is the following lemma.

LEMMA IV.1.19. *Let $r(D)$ be a pseudo-differential of order zero with symbol $r(x, \xi) = r(\xi)$ independent of x. Assume that for some $C > 0$ the following inequality holds:*

$$\|f\|_{h^1} \leq C(\|f\|_{L^1} + \|r(D)f\|_{L^1}), \qquad f \in h^1.$$

Let K be the kernel of $r(D)$ and for each $\varepsilon > 0$ write

$$r(D)f(x) = <\chi(\varepsilon(x - \cdot))K, f> + <(1 - \chi(\varepsilon(x - \cdot)))K, f>$$
$$= r_1^\varepsilon(D)f(x) + r_2^\varepsilon(D)f(x),$$

where $\chi \in C_c^\infty(-2, 2)$ with $\chi(y) = 1$ for $|y| \leq 1$. Then there exists ε_0 such that for all $0 < \varepsilon \leq \varepsilon_0$ there exist constants $C_1 = C_1(\varepsilon)$, $C_2 = C_2(\varepsilon) > 0$ such that

$$\|f\|_{h^1} \leq C_1(\|f\|_{L^1} + \|r_1^\varepsilon(D)f\|_{L^1}) \leq C_2\|f\|_{h^1}. \tag{IV.35}$$

PROOF. For each $\varepsilon > 0$, $r_1^\varepsilon(D)$ is a pseudo-differential operator of order zero, thus bounded in h^1, so

$$\|f\|_{L^1} + \|r_1^\varepsilon(D)f\|_{L^1} \leq \|f\|_{h^1} + \|r_1^\varepsilon(D)f\|_{h^1} \leq C_2(\varepsilon)\|f\|_{h^1}.$$

On the other hand, $\|r_2^\varepsilon(D)f\|_{L^1} \leq \|K_2^\varepsilon\|_{L^1}\|f\|_{L^1}$ and $\|K_2^\varepsilon\|_{L^1} \to 0$ as $\varepsilon \to 0$. Therefore, there exists $\varepsilon_0 > 0$ such that $\|K_2^\varepsilon\|_{L^1} \leq 1/2C$ for $0 < \varepsilon \leq \varepsilon_0$. Thus

$$\|f\|_{h^1} \leq C(\|f\|_{L^1} + \|r(D)f\|_{L^1})$$
$$\leq C\left(\|f\|_{L^1(\mathbb{R})} + \|r_1^\varepsilon(D)f\|_{L^1(\mathbb{R})} + \frac{1}{2C}\|f\|_{L^1}\right)$$
$$\leq C(\|f\|_{L^1} + \|r_1^\varepsilon(D)f\|_{L^1}) + \frac{1}{2}\|f\|_{h^1},$$

which implies

$$\|f\|_{h^1} \leq 2C(\|f\|_{L^1} + \|r_1^\varepsilon(D)f\|_{L^1}). \qquad \square$$

REMARK IV.1.20. Notice that $r_1^\varepsilon(D)$ is given by convolution with a distribution supported in the interval $(-2/\varepsilon, 2/\varepsilon)$, in particular if $u \in \mathcal{E}'([-r, r])$— i.e., if u is distribution supported in the interval $[-r, r]$—$r_1^\varepsilon(D)u$ is supported in the interval $[-r - 2\varepsilon^{-1}, r + 2\varepsilon^{-1}]$.

We are now able to prove a stronger estimate. We will show that there exist constants C and $T_0 > 0$ such that for any $0 < T \leq T_0$ and $\varphi \in C_c^\infty((-a, a) \times (-T, T))$,

$$\|\varphi\|_{L^1((-T,T),h^1(\mathbb{R}_x))} \leq CT\|L\varphi\|_{L^1((-T,T),h^1(\mathbb{R}_x))}. \tag{IV.36}$$

Given $\phi \in C_c^\infty((-a, a) \times (-T, T))$ set

$$\widetilde{H}\varphi(\cdot, t)(\xi) = (1 - \chi)(\xi)\widehat{H\varphi}(\cdot, t)(\xi)$$

where H is the Hilbert transform and $\chi \in C_c^\infty(-2, 2)$, $\chi(\xi) = 1$ for $|\xi| \leq 1$. The symbol of $\widetilde{H}$ is equal to $h(\xi) = \psi^+(\xi) - \psi^-(\xi)$, where ψ^+ and ψ^- are the symbols of the operators P^+ and P^- already used. We see that $\widetilde{H}$ is a pseudo-differential operator satisfying the hypotheses of Lemma IV.1.19 and we may write it as a sum $\widetilde{H} = \widetilde{H}_1^\varepsilon + \widetilde{H}_2^\varepsilon$ where $\widetilde{H}_1^\varepsilon : \mathcal{E}'((-a, a)) \to \mathcal{E}'((-a', a'))$ satisfies (IV.35), i.e.,

$$\|\varphi(\cdot, t)\|_{h^1(\mathbb{R}_x)} \leq C(\|\varphi(\cdot, t)\|_{L^1(-a, a)} + \|\widetilde{H}_1^\varepsilon \varphi(\cdot, t)\|_{L^1(-a', a')}), \qquad \text{(IV.37)}$$

for some $C > 0$. Since $\widetilde{H}_1^\varepsilon \varphi(x, t) \in C_c^\infty((-a', a') \times (-T, T))$, applying (IV.34) (with a' in the place of a) to $\widetilde{H}_1^\varepsilon \varphi$ we get

$$\|\widetilde{H}_1^\varepsilon \varphi\|_{L^1((-T,T) \times (-a', a'))}$$
$$\leq C\,T\left(\|L\widetilde{H}_1^\varepsilon \varphi\|_{L^1((-T,T), h^1(\mathbb{R}_x))} + \|\widetilde{H}_1^\varepsilon \varphi\|_{L^1((-T,T), h^1(\mathbb{R}_x))}\right). \qquad \text{(IV.38)}$$

Since $L\widetilde{H}_1^\varepsilon = \widetilde{H}_1^\varepsilon L + [L, \widetilde{H}_1^\varepsilon]$ and, invoking Proposition A.2.2 in the Appendix A, we may claim that $\widetilde{H}_1^\varepsilon$ as well as $[L, \widetilde{H}_1^\varepsilon]$ are bounded operators in $h^1(\mathbb{R}_x)$. It follows from (IV.38) that

$$\|\widetilde{H}_1^\varepsilon \varphi\|_{L^1((-T,T) \times (-a', a'))}$$
$$\leq C\,T(\|L\varphi\|_{L^1((-T,T), h^1(\mathbb{R}_x))} + \|\varphi\|_{L^1((-T,T), h^1(\mathbb{R}_x))}). \qquad \text{(IV.39)}$$

Integrating (IV.37) with respect to t and using (IV.39) we see that

$$\|\varphi\|_{L^1((-T,T), h^1(\mathbb{R}_x))} \leq C(\|\varphi\|_{L^1((-T,T) \times (-a, a))} + \|\widetilde{H}_1^\varepsilon \varphi\|_{L^1((-T,T) \times (-a', a'))})$$
$$\leq CT(\|\varphi\|_{L^1((-T,T), h^1(\mathbb{R}_x))} + \|L\varphi\|_{L^1((-T,T), h^1(\mathbb{R}_x))}).$$

It is now enough to choose T_0 such that $CT \leq 1/2$ if $T \leq T_0$ to get

$$\|\varphi\|_{L^1((-T,T); h^1(\mathbb{R}_x))} \leq 2C\,T\|L\varphi\|_{L^1((-T,T), h^1(\mathbb{R}_x))}$$

as desired. We may now state

THEOREM IV.1.21. *Let the operator L given by (IV.26) satisfy (i), (ii) and (iii) and let $a > 0$. Then there exist constants $C > 0$ and $T_0 > 0$ such that*

$$\|u\|_{L^1((-T,T); h^1(\mathbb{R}_x))} \leq CT\|Lu\|_{L^1((-T,T); h^1(\mathbb{R}_x))}, \qquad \text{(IV.40)}$$

for all $u \in C_c^\infty([-a, a] \times [-T, T])$, $0 < T \leq T_0$.

PROOF. We have already proved (IV.36) assuming that $c(x, t) \equiv 0$ which is the same as (IV.40). In the general case we write $L = L_0 + c$ and since (IV.36) holds for L_0 we obtain

$$\|u\|_{L^1((-T,T);h^1(\mathbb{R}_x))} \leq CT \left(\|Lu\|_{L^1((-T,T);h^1(\mathbb{R}_x))} + \|cu\|_{L^1((-T,T);h^1(\mathbb{R}_x))} \right)$$
$$\leq CT \left(\|Lu\|_{L^1((-T,T);h^1(\mathbb{R}_x))} + C_1 \|u\|_{L^1((-T,T);h^1(\mathbb{R}_x))} \right)$$

as multiplication by a C^r function is a bounded operator in the space $L^1((-T, T); h^1(\mathbb{R}))$. Taking T small so that $CC_1T < 1/2$, we obtain (IV.40).
□

The a priori inequality (IV.40) has a standard duality consequence which we now describe. The dual of $h^1(\mathbb{R})$, denoted by bmo($\mathbb{R}$), may be identified ([**G**]) with the space of locally integrable functions $f(x)$ such that $\sup_{|I|<1} |I|^{-1} \int_I |f - f_I| < \infty$ and $\sup_{|I| \geq 1} |I|^{-1} \int_I |f| < \infty$, where we have denoted by I an arbitrary interval and by f_I the mean of f on I. In particular, bmo($\mathbb{R}$) is contained in BMO ($\mathbb{R}$), the space of bounded mean oscillation functions. Then, (IV.40) implies local solvability in $L^\infty([-T, T], \text{bmo}(\mathbb{R}_x))$ for the formal transpose L^t. Now, L and $-L^t$ have the same principal part, so L and $-L^t$ satisfy simultaneously the hypotheses of Theorem IV.1.21. Summing up,

THEOREM IV.1.22. *Let the operator*

$$L = \frac{\partial}{\partial t} + ib(x, t)\frac{\partial}{\partial x} + c(x, t)$$

satisfy (i), (ii) and (iii). There is a neighborhood $U = (-a, a) \times (-T, T)$ of the origin such that for every function $f \in X = L^\infty(\mathbb{R}_t, \text{bmo}(\mathbb{R}_x))$ there exists a function $u \in X$ which solves $Lu = f$ in U, with norm

$$\|u\|_{L^\infty(\mathbb{R}_t, \text{bmo}(\mathbb{R}_x))} \leq CT \|f\|_{L^\infty(\mathbb{R}_t, \text{bmo}(\mathbb{R}_x))}.$$

In particular, the size of u can be taken arbitrary small by letting $T \to 0$.

We conclude this section by proving consequences of Theorems IV.1.21 and IV.1.22 that can be stated in a more invariant form that does not depend on a special coordinate system. In Theorems IV.1.21 and IV.1.22, the operator L has a special form which is instrumental in obtaining a priori estimates with minimal assumptions on the regularity of the coefficients but, at least heuristically, after a suitable change of variables any first-order operator of principal type has this form as we saw in Lemma IV.1.1. On the other hand, for operators with rough coefficients this change of variables imposes a loss of regularity on the coefficients of the transformed operator. One should also observe the loss of derivatives caused in the process of deriving estimates in terms of the original variables from estimates obtained in the new variables by the behavior of local Hardy norms under composition with diffeomorphisms. For this reason we now deal with operators having C^{2+r} coefficients in the principal part. Since we are dealing with mixed norms, the roles of t and x cannot be interchanged and we must consider changes of variables that preserve the privileged role of t. Consider a general first-order operator defined in an open subset $\Omega \subset \mathbb{R}^2$ that contains the origin

$$Lu = A(x, t)\frac{\partial u}{\partial t} + B(x, t)\frac{\partial u}{\partial x} + C(x, t)\, u$$

with complex coefficients $A, B \in C^{2+r}(\Omega)$, $0 < r < 1$, $C \in C_\omega(\Omega)$. Assume that the lines $t = \text{const.}$ are noncharacteristic, which amounts to saying that $|A(x, t)| > 0$, $(x, t) \in \Omega$. Since the properties we are studying do not change if L is multiplied by a nonvanishing function of class C^{2+r}, we may assume without loss of generality that $A \equiv 1$, i.e.,

$$Lu = \frac{\partial u}{\partial t} + B(x, t)\frac{\partial u}{\partial x} + C(x, t)\, u.$$

Write $B(x, t) = \tilde{a}(x, t) + i\tilde{b}(x, t)$ with $\tilde{a}$ and $\tilde{b}$ real. In convenient new local coordinates $\xi = \xi(x, t)$, $s = t$, the expression of L is

$$\tilde{L} = \partial_s + i(\tilde{b}/(\partial x/\partial \xi))\partial_\xi + C(x(\xi, s), s) = \partial_s + ib\partial_\xi + c,$$

where b is real of class C^{1+r} and $c \in C_\omega$. If L satisfies the Nirenberg–Treves condition $(\mathcal{P})$ so does $\tilde{L}$, due to the invariance of this property that will be discussed in the next section (the coefficients are supposed to be smooth for simplicity in that section but the arguments adapt to the present situation). Multiplying the coefficients b and c by a cut-off function $\chi \geq 0 \in C_c^\infty(\mathbb{R}^2)$ that is identically equal to 1 in the neighborhood of the origin we now have an operator L' with smooth coefficients and globally defined in $\mathbb{R}^2$ that satisfies the hypotheses of Theorem IV.1.21 and agrees with $\tilde{L}$ in a neighborhood of the origin. Thus, the a priori estimate (IV.40) holds for L' in the variables (ξ, s). Let $u'(\xi, s) \in C_c^\infty(\mathbb{R}^2)$ be supported in a sufficiently small neighborhood of the origin and set $u(x, t) = u'(\xi(x, t), t)$, where $(x, t) \mapsto (\xi, s)$ is the inverse of $(\xi, s) \mapsto (x, t)$, thus of class C^{2+r}. Invoking the invariance of $h^1(\mathbb{R})$ under diffeomorphisms of class C^2 discussed in Proposition IV.3.1 we conclude that if u' is supported in a convenient neighborhood of the origin we have

$$C_1 \int_{\mathbb{R}} \|u(\cdot, t)\|_{h^1(\mathbb{R}_x)}\, dt \leq \int_{\mathbb{R}} \|u'(\cdot, s)\|_{h^1(\mathbb{R}_\xi)}\, ds \leq C_2 \int_{\mathbb{R}} \|u(\cdot, t)\|_{h^1(\mathbb{R}_x)}\, dt$$

and this shows that the a priori estimate (IV.40) for L' implies an analogous estimate for L, using the fact that $Lu(x, t) = L'u'(\xi(x, t), t)$. Summing up,

THEOREM IV.1.23. *Let L given by*

$$Lu = A(x, t)\frac{\partial u}{\partial t} + B(x, t)\frac{\partial u}{\partial x} + C(x, t)\, u$$

be defined in a neighborhood of the origin, with complex coefficients $A, B \in C^{2+r}(\Omega)$, $0 < r < 1$, $C \in C_\omega(\Omega)$. *Assume that the level curves* $t = $ constant *are noncharacteristic for L and that L satisfies the Nirenberg–Treves condition* $(\mathcal{P})$. *Then there exist constants* $a > 0$, $C > 0$ *and* $T_0 > 0$ *such that*

$$\|u\|_{L^1(\mathbb{R}_t; h^1(\mathbb{R}_x))} \le CT\|Lu\|_{L^1(\mathbb{R}_t; h^1(\mathbb{R}_x))},$$

for all $u \in C_c^\infty([-a, a] \times [-T, T])$, $0 < T \le T_0$. *Hence, for every function* $f \in X = L^\infty(\mathbb{R}_t, \mathrm{bmo}(\mathbb{R}_x))$ *there exists a function* $u \in X$ *which solves* $Lu = f$ *in a neighborhood U of the origin, with norm*

$$\|u\|_X \le CT\|f\|_X.$$

IV.2 Solvability in C^∞

In the last section we introduced the local solvability condition $(\mathcal{P})$ in Definition IV.1.5 assuming that the vector field L was in the special form

$$L = \frac{\partial}{\partial t} + ib(x, t)\frac{\partial}{\partial x} \tag{IV.41}$$

with $b(x, t)$ real, smooth, and defined for all $(x, t) \in \mathbb{R}^2$. However, to require that $t \mapsto b(x, t)$ does not change sign is not *per se* a coordinate-free definition because we are demanding that a particular coefficient (namely, $b(x, t)$) does not take opposite signs on sets of a special kind (namely, $\{x\} \times \mathbb{R}$). It order to find more invariant ways to formulate condition $(\mathcal{P})$ it is convenient to find larger sets on which $b(x, t)$ keeps its sign unchanged. Assume that L given by (IV.41) satisfies $(\mathcal{P})$. Then the sets

$$A^+ = \{x \in \mathbb{R}: \ \sup_t b(x, t) > 0\} \text{ and } A^- = \{x \in \mathbb{R}: \ \inf_t b(x, t) < 0\}$$

are open and disjoint, and the complement of its union $F = \mathbb{R} \backslash A^+ \cup A^-$ is a closed set with the property that $b(x, t) = 0$ on $F \times \mathbb{R}$. Write A^+ and A^- in terms of their connected components

$$A^+ = \bigcup_j (a_j^+, b_j^+), \qquad A^- = \bigcup_j (a_j^-, b_j^-).$$

If $x \in (a_j^+, b_j^+)$ there exists $t \in \mathbb{R}$ such that $b(x, t) > 0$ so we see that $b(x, t) \ge 0$ on $(a_j^+, b_j^+) \times \mathbb{R}$ and similarly $b(x, t) \le 0$ on $(a_j^-, b_j^-) \times \mathbb{R}$. There is an easy way to describe invariantly the open sets $\Omega_j^+ = (a_j^+, b_j^+) \times \mathbb{R}$ and $\Omega_j^- = (a_j^-, b_j^-) \times \mathbb{R}$: they are the orbits of dimension two of the pair of vector fields $\{X \doteq \Re L, Y \doteq \Im L\}$. Indeed, $\Omega_j^\pm$ is a union of vertical lines, so invariant under the flow of X, and it is also invariant under the flow of Y because Y vanishes on its boundary, so if $p \in \Omega_j^\pm$ the $\mathcal{O}$ orbit $\mathcal{O}(p)$ of $\{X, Y\}$ through p is contained in $\Omega_j^\pm$. Now, $\mathcal{O}(p)$ is an orbit of maximal dimension, thus open and connected, and being invariant under the flow of X it is of the form $(a, b) \times \mathbb{R}$ with $a_j^\pm \leq a < b \leq b_j^\pm$. Since $\{a\} \times \mathbb{R}$ is contained in the boundary of $\mathcal{O}(p)$, $b(x, t)$ must vanish identically on $\{a\} \times \mathbb{R}$ so $a \notin (a_j^\pm, b_j^\pm)$ and similarly $b \notin (a_j^\pm, b_j^\pm)$, which proves that $\Omega_j^\pm = \mathcal{O}(p)$. On the other hand, the sets $\{x\} \times \mathbb{R}$, $x \in F$, are precisely the orbits of dimension one of $\{X, Y\}$. Since $a_j^+, b_j^+, a_j^-, b_j^- \in F$ we see that a two-dimensional orbit is bounded by two one-dimensional orbits in case its orthogonal projection onto the x-axis is a finite interval, by one one-dimensional orbit if its projection has exactly one finite endpoint and, of course, the boundary is empty if the projection is the whole real line. To give a coordinate-free formulation of the fact that $b(x, t)$ does not change sign on two-dimensional orbits we look at $X \wedge Y \in C^\infty(\mathbb{R}^2; \bigwedge^2(T(\mathbb{R}^2)))$. Since $\bigwedge^2(T(\mathbb{R}^2)))$ has a global nonvanishing section $e_1 \wedge e_2$, $X \wedge Y$ is a real multiple of $e_1 \wedge e_2$ and this gives a meaning to the requirement that $\Re L \wedge \Im L$ does not change sign on any two-dimensional orbits of $\{\Re L, \Im L\}$. Note that when L has the form (IV.41) we have seen that this happens if and only if L satisfies $(\mathcal{P})$.

Consider now a vector field defined in an open subset $\Omega \subset \mathbb{R}^2$

$$Lu = A(x, t)\frac{\partial u}{\partial t} + B(x, t)\frac{\partial u}{\partial x} \qquad (IV.42)$$

with complex coefficients $A, B \in C^\infty(\Omega)$ such that

$$|A(x, y)| + |B(x, y)| > 0, \quad (x, t) \in \Omega.$$

DEFINITION IV.2.1. *We say that the operator L given by (IV.42) satisfies condition $(\mathcal{P})$ in Ω if $\Re L \wedge \Im L$ does not change sign on any two-dimensional orbit of L, i.e., on any two-dimensional orbit of the pair of real vector fields $\{\Re L, \Im L\}$.*

The previous discussion shows that the coordinate-free Definition IV.2.1 reduces to Definition IV.1.5 when L is in the form (IV.41).

Let $\varphi(x, t) \in C^\infty(\mathbb{R}^2)$, set

$$Z(x, t) = x + i\varphi(x, t), \qquad (IV.43)$$

and consider the vector field

$$L = \frac{\partial}{\partial t} - \frac{i\varphi_t(x, t)}{1 + i\varphi_x(x, t)} \frac{\partial}{\partial x} = \frac{\partial}{\partial t} - \frac{Z_t}{Z_x} \frac{\partial}{\partial x}. \qquad (\text{IV.44})$$

Thus, $Z(x, t)$ is a global first integral of L, i.e., $LZ = 0$ and $dZ \neq 0$ everywhere.

LEMMA IV.2.2. *Let $Z(x, t)$ and L be given by (IV.43) and (IV.44) respectively. Then, L satisfies $(\mathcal{P})$ in $\mathbb{R}^2$ if and only if $\mathbb{R} \ni t \mapsto \varphi(x, t)$ is monotone for every $x \in \mathbb{R}$.*

PROOF. We have

$$X = \frac{\partial}{\partial t} + \frac{\varphi_t \varphi_x}{1 + \varphi_x^2} \frac{\partial}{\partial x}, \qquad Y = -\frac{\varphi_t}{1 + \varphi_x^2} \frac{\partial}{\partial x},$$

so

$$X \wedge Y = \frac{\varphi_t(x, y)}{1 + \varphi_x^2} \frac{\partial}{\partial x} \wedge \frac{\partial}{\partial t}.$$

Note that X and Y are linearly dependent at a point if and only if φ_t vanishes at that point. Thus, the one-dimensional orbits of L are vertical lines $x = \text{constant}$ on which φ_t vanishes identically. Since the two-dimensional orbits of L are bounded by 0, 1 or 2 one-dimensional orbits we see that each two-dimensional orbit Ω_j, $j = 1, 2, \ldots$, is of the form $(a_j, b_j) \times \mathbb{R}$. If L satisfies $(\mathcal{P})$ then φ_t does not assume opposite signs on Ω_j, say, $\varphi_t \geq 0$ on Ω_j so $t \mapsto \varphi(x, t)$ is monotone increasing for all $a_j < x < b_j$. If $x \notin (a_j, b_j)$ for any j it follows that the point of coordinates $(x, 0)$ belongs to a one-dimensional orbit, so $\varphi_t(x, t) = 0$, $-\infty < t < \infty$, and $t \mapsto \varphi(x, t)$ is constant. This shows that $t \mapsto \varphi(x, t)$ is monotone for every $x \in \mathbb{R}$. Conversely, assume that $t \mapsto \varphi(x, t)$ is monotone for every $x \in \mathbb{R}$ and let $(a_j, b_j) \times \mathbb{R}$ be a two-dimensional orbit. Given $x_0 \in (a_j, b_j)$ we have that $t \mapsto \varphi_t(x_0, t)$ has a consistent sign, say $\varphi_t(x_0, t) \geq 0$. We must show that $\varphi_t(x, t) \geq 0$ for all $a_j < x < b_j$. Indeed, if $\varphi_t(x_1, t) < 0$ for some $x_1 \in (a_j, b_j)$ and $t \in \mathbb{R}$, it is easy to see that there exist an intermediate point x_2 between x_0 and x_1 such that $\varphi_t(x_2, t) = 0$ for all $t \in \mathbb{R}$. Then $\{x_2\} \times \mathbb{R}$ is a one-dimensional orbit and must be disjoint of the two-dimensional orbit $(a_j, b_j) \times \mathbb{R}$, a contradiction to the fact that $x_2 \in (a_j, b_j)$. $\qquad \square$

From now on, we assume that L given by (IV.44) satisfies condition $(\mathcal{P})$ and we wish to find a local solution $Lu = f$ with $u \in C^\infty$ when $f \in C^\infty$. We start from estimate (IV.11) in Theorem IV.1.9, with L in the place of ${}^t L$, $q = p = 2$. There exists $a, T, C > 0$ such that, for every $u \in C_c^\infty((-a, a) \times (-T, T))$,

$$\|u(x, t)\|_{L^2(\mathbb{R}^2)} \leq C \|Lu(x, t)\|_{L^2(\mathbb{R}^2)}. \qquad (\text{IV.45})$$

Modifying $\varphi(x, t)$ outside a neighborhood of the origin as in the proof of Theorem IV.1.9, we may assume that φ_t and φ_x are compactly supported and that $a = \infty$. The a priori estimate (IV.45) may be extended using Friedrichs' lemma to any $u \in L^2_c((-T, T) \times \mathbb{R})$ such that $Lu \in L^2_c((-T, T) \times \mathbb{R})$.

We wish to extend (IV.45) in two ways: first, we want to know that the inequality is still valid when $u(x, t)$ is not regular enough to be in $L^2(\mathbb{R}^2)$ although $Lu(x, t)$ is known be in $L^2(\mathbb{R}^2)$; second, we wish to consider estimates for Sobolev norms. We write

$$M = Z_x^{-1} \partial_x, \qquad D = -L^2 - \lambda M^2,$$

where $\lambda > 0$ is a large parameter. Then L and M commute, which implies that L and D also do so. A consequence of this fact that can be expressed in terms of their respective symbols $\ell(x, t, \xi, \tau) = i(\tau + \Lambda(x, t)\xi)$, $\Lambda = -Z_t/Z_x$, $d(x, t, \xi, \tau) = -(\ell^2 + \lambda m^2)(x, t, \xi, \tau)$, $m(x, t, \xi, \tau) = iZ_x^{-1}(x, t)\xi$, is expressed by the identity

$$\{\ell, d\}(x, t, \xi, \tau) = 0, \qquad (x, t, \xi, \tau) \in \mathbb{R}^4$$

where $\{\ell, d\}$ denotes the Poisson bracket performed in all variables. Note that

$$d(x, t, \xi, \tau) = \tau^2 - 2\frac{Z_t}{Z_x}\xi\tau + \frac{Z_t^2 + \lambda}{Z_x^2}\xi^2$$

so for λ large $|\Im d| \leq C \Re d$ and also $d(x, t, \xi, \tau) = 0$ implies $\tau = \xi = 0$, i.e., D is a uniformly elliptic second-order operator with smooth bounded coefficients.

Consider a pseudo-differential operator $P(x, t, D_x, D_t)$ of order s and type $(\rho, \delta) = (1, 0)$ with symbol $p(x, t, \xi, \tau)$, that is,

$$Pu(x, t) = \frac{1}{(2\pi)^2} \int_{-\infty}^{\infty} e^{i(x\xi + t\tau)} p(x, t, \xi, \tau) \widehat{u}(\xi, \tau) \, d\xi d\tau.$$

The first term in the expansion of the symbol of the commutator $[L, P]$ is given by $-i\{\ell, p\}(x, t, \xi)$ by a well-known formula from the calculus of pseudo-differential operators. Thus, $[L, P]$ is a pseudo-differential operator with the same order s. However, if $p(x, t, \xi) = F(d(x, t, \xi))$ with F holomorphic on the range of d, it follows that

$$\{\ell, p\}(x, t, \xi, \tau) = \{\ell, F \circ d\}(x, t, \xi, \tau) = (F' \circ d)\{\ell, d\}(x, t, \xi, \tau) = 0.$$

We see that in this case $[L, P]$ has order $s - 1$, i.e., it commutes with L to a higher degree than in the general situation, a fact we will explore. We already saw that the range of $d(x, t, \xi, \tau)$ is contained in a closed cone of the complex plane of the form $|\Im z| \leq C \Re z$ and it follows that for any real $\epsilon > 0$ the range of $1 + \epsilon d(x, t, \xi, \tau)$ has positive real part. Consider the pseudo-differential operator $P^\epsilon(x, t, D_x, D_t)$ with symbol

$$p^\epsilon(x, t, \xi, \tau) = \frac{\chi(t)}{(1 + \epsilon d(x, t, \xi, \tau))^{1/2}},$$

where $\chi(t) \in C_c^\infty(-T, T)$ and $\chi(t) = 1$ for $|t| \le (3/4)T$. We point out that $P^\epsilon(x, t, D_x, D_t)$ has order -1 for $\epsilon > 0$ although $\{p^\epsilon\}$ is not a bounded subset of $S_{1,0}^{-1}$. On the other hand, $\{p^\epsilon\}$, $0 < \epsilon < 1$, remains in a bounded subset of $S_{1,0}^0$ which implies that the norm of P^ϵ in $\mathcal{L}(L^2(\mathbb{R}^2))$ is bounded by a constant independent of $0 < \epsilon < 1$, $t \in \mathbb{R}$. By the observations made before, the commutator $[L, P^\epsilon]$ has order -2 for fixed $\epsilon > 0$ on the open set $\mathbb{R} \times (-3T/4, 3T/4)$ and order -1 *uniformly in* $\epsilon > 0$, which implies that $\{[L, P^\epsilon]\}$ is a bounded subset of $\mathcal{L}(L^2(\mathbb{R}^2), H^{-1}(\mathbb{R}^2))$, where H^{-1} denotes the Sobolev space of order -1. Furthermore, $P^\epsilon \to I$ weakly as $\epsilon \to 0$.

Consider now a distribution $u(x, t) \in H_c^{-1}(\mathbb{R}^2)$ supported in $\mathbb{R} \times (-T/2, T/2)$ and assume that

- $Lu \in L^2(\mathbb{R}^2)$.

We will show that $u \in L^2(\mathbb{R}^2)$. Indeed, set $u_\epsilon = P^\epsilon u$. Then, $u_\epsilon \in L^2(\mathbb{R}^2)$ and $Lu_\epsilon = P^\epsilon Lu + [L, P^\epsilon]u \in L^2(\mathbb{R}^2)$. Note that the last inclusion is uniform in ϵ and that $[L, P^\epsilon]u \to 0$ in L^2. Applying (IV.45) to u_ϵ we obtain

$$\|u_\epsilon\|_{L^2(\mathbb{R}^2)} \le C \|Lu_\epsilon\|_{L^2(\mathbb{R}^2)} \le C_1.$$

Since $u_\epsilon \to u$ weakly as $\epsilon \to 0$ we conclude that $\|u\|_{L^2(\mathbb{R}^2)} \le C_1$ and

$$\|u\|_{L^2(\mathbb{R}^2)} \le C \|Lu\|_{L^2(\mathbb{R}^2)}$$

for all $u \in H_c^{-1}(\mathbb{R} \times (-T/2, T/2))$ such that $Lu \in L^2(\mathbb{R}^2)$. Similarly, if $u \in H_c^{s-1}(\mathbb{R} \times (-T/2, T/2))$, $s \in \mathbb{R}$, is such that $Lu \in H_c^s(\mathbb{R} \times (-T/2, T/2))$ we conclude that $u \in H^s(\mathbb{R}^2)$ and

$$\|u\|_{H^s(\mathbb{R}^2)} \le C_s(\|Lu\|_{H^s(\mathbb{R}^2)} + \|u\|_{H^{s-1}(\mathbb{R}^2)}). \tag{IV.46}$$

To prove (IV.46) we apply (IV.45) to $u_\epsilon = B^\epsilon u$ where B^ϵ is the pseudo-differential operator with symbol

$$b^\epsilon(x, t, \xi, \tau) = \frac{\chi(t)(1 + d(x, t, \xi, \tau))^{s/2}}{(1 + \epsilon d(x, t, \xi, \tau))^{1/2}}$$

and reason as before. Note that $b^\epsilon \to b = \chi(1 + d)^{s/2}$ in the symbol space $S_{1,0}^s$ and that $\|u\|_s \sim \|Bu\|_{L^2}$ if B is the pseudo-differential with symbol

b and $u \in H^s_c(\mathbb{R} \times (-T/2, T/2))$. Furthermore, $[L, B]$ has order $s - 1$ on $\mathbb{R} \times (-T/2, t/2)$. Letting $\epsilon \to 0$ we obtain

$$\|Bu\|_{L^2(\mathbb{R}^2)} \leq C(\|BLu\|_{L^2(\mathbb{R}^2)} + \|[L, B]u\|_{L^2(\mathbb{R}^2)}),$$

which gives (46). A consequence of (IV.46) is that

$$u \in \mathcal{E}'(\mathbb{R} \times (-T/2, T/2)) \quad \text{and} \quad Lu \in H^s(\mathbb{R}^2)$$

$$\text{imply that} \quad u \in H^s(\mathbb{R}^2).$$

Indeed, if $u \in \mathcal{E}'(\mathbb{R} \times (-T/2, T/2))$ there exists some $\sigma < s$ such that $s - \sigma = k$ is an integer and $u \in H^\sigma_c(\mathbb{R} \times (-T/2, T/2))$. Then $Lu \in H^s(\mathbb{R}^2) \subset H^{s-k}(\mathbb{R}^2)$ and (IV.46) implies that $u \in H^{s-k+1}(\mathbb{R}^2)$. Repeating this process k times we conclude that $u \in H^s(\mathbb{R}^2)$ as wanted. Observe that this implies that $u \in \mathcal{E}'(\mathbb{R} \times (-T/2, T/2))$ must be smooth if $Lu \in C^\infty$.

Another consequence is that if $u \in \mathcal{E}'(\mathbb{R} \times (-T/2, T/2))$ satisfies $Lu = 0$ it must vanish identically (a fact that also follows from uniqueness in the Cauchy problem). Indeed, $Lu = 0$ implies that $u \in C^\infty_c(\mathbb{R} \times (-T/2, T/2))$ and (IV.45) shows that $u = 0$.

Let K denote a closed ball of radius $r < T/2$ centered at the origin of $\mathbb{R}^2$ and let us prove that for any $s \in \mathbb{R}$

$$\|u\|_{H^s(\mathbb{R}^2)} \leq C(s)\|Lu\|_{H^s(\mathbb{R}^2)}, \quad u \in C^\infty_c(K). \tag{IV.47}$$

Fix $s \in \mathbb{R}$ and assume by contradiction that for every $j = 1, 2, \ldots$, there exists $u_j \in C^\infty_c(K)$ such that $\|u_j\|_{H^s(\mathbb{R}^2)} = 1$ and $\|Lu_j\|_{H^s(\mathbb{R}^2)} \leq 1/j$. Passing through a subsequence we may assume that $u_j \to u$ in $H^{s-1}(\mathbb{R}^2)$ with $Lu = 0$ and this implies that $u = 0$. On the other hand, (IV.46) gives

$$1 \leq \frac{C_s}{j} + C_s \|u_j\|_{H^{s-1}(\mathbb{R}^2)}$$

which, letting $j \to \infty$, contradicts that $u = 0$. Using Friedrichs' lemma we may extend (IV.47) to

$$\|u\|_{H^s(\mathbb{R}^2)} \leq C(s)\|Lu\|_{H^s(\mathbb{R}^2)}, \quad \text{if } u \text{ and } Lu \in H^s_c(K). \tag{IV.48}$$

Let us now prove that for every $f \in C^\infty(\mathbb{R}^2)$ there is $u \in C^\infty(\mathbb{R}^2)$ such that $Lu = f$ in K. Denote by $C^\infty(K)$ the quotient of $C^\infty(\mathbb{R}^2)$ by the subspace of those functions which vanish on K to infinite order. This is a Fréchet space and its dual may be identified with $\mathcal{E}'(K)$, the distributions in $\mathcal{E}'(\mathbb{R}^2)$ supported in K.

In order to identify the dual of $C^\infty(K)$ with $\mathcal{E}'(K)$ it is convenient to introduce the pairing

$$\langle u(x, t), v(x, t) \rangle = \int u(x, t) v(x, t) \, dZ(x, t) \wedge dt$$

$$= \int u(x, t) v(x, t) Z_x(x, t) \, dx \, dt,$$

for which L and $-L$ are formal transposes of each other, i.e., $\langle Lu, v \rangle = -\langle u, Lv \rangle$, $u, v \in C_c^\infty(\mathbb{R}^2)$. This pairing can be extended to $u \in C^\infty(\mathbb{R}^2)$ and $v \in \mathcal{E}'(\mathbb{R}^2)$ and if $v \in \mathcal{E}'(K)$ the value of $\langle Lu, v \rangle$ only depends on the residue class $[u]$ of $u \in C^\infty(\mathbb{R}^2)$ in $C^\infty(K)$ and $[u] \mapsto \langle u, v \rangle$ is clearly continuous. Conversely, given a continuous linear functional λ on $C^\infty(K)$, the continuous linear functional $C^\infty(\mathbb{R}^2) \ni u \mapsto \lambda([u])$ is represented by a compactly supported distribution $v \in \mathcal{E}'(\mathbb{R}^2)$ such that $\lambda([u]) = \langle u, v \rangle$, $u \in C^\infty(\mathbb{R}^2)$. Since $\langle u, v \rangle$ must vanish when u vanishes to infinite order on K we see that v is supported in K. Furthermore, it is clear that $v = 0$ if $\lambda = 0$.

Consider the continuous linear map $T: C^\infty(K) \longrightarrow C^\infty(K)$ defined by $T[u] = [Lu]$, where $[u]$ denotes the residue class of $u \in C^\infty(\mathbb{R}^2)$ in $C^\infty(K)$. Then the range of T is dense; in fact, if μ is a continuous linear functional on $C^\infty(K)$ such that $\langle \mu, T[u] \rangle = 0$, $[u] \in C^\infty(K)$, regarded as an element of $\mathcal{E}'(K)$, μ satisfies the equation $L\mu = 0$ which implies that $\mu = 0$. Thus, to show that T is onto we need only show that the range of T is closed and by the Banach closed range theorem for Fréchet spaces this will follow if we prove that the range of the dual operator T' is closed for the weak* topology. However, $C^\infty(K)$ is reflexive, a consequence of the reflexivity of $C^\infty(\mathbb{R}^2)$, and in this case it is enough to prove that the range of T' is closed for the strong topology (see, e.g., [**T1**], chapter 37). Let the sequence $\mu_j = T'\nu_j = -L\nu_j$, $\nu_j \in \mathcal{E}'(K)$, converge to $\mu \in \mathcal{E}'(K)$. There exist s such that $\{\mu_j\} \subset H^s(\mathbb{R}^2)$ and $\|\mu_j\|_{H^s} \leq C$, $j = 1, 2, \ldots$ This implies that $\nu_j \in H^s(\mathbb{R}^2)$ and by (IV.48)

$$\|\nu_j\|_{H^s} \leq C(s) \|\mu_j\|_{H^s} \leq C'.$$

Passing through a subsequence we may assume that ν_j is convergent in $H^{s-1}(\mathbb{R}^2)$ to some $\nu \in H_c^{s-1}(K)$, showing that $T'\nu = -L\nu = \mu$ so μ is in the range of T'. Thus, the range of T' is closed and so is the range of T, which must be equal to $C^\infty(K)$. In other words, for every $f \in C^\infty(\mathbb{R})$ there is $u \in C^\infty(\mathbb{R})$ such that $Lu - f = 0$ on K. Finally, if $c(x, t)$ is a smooth function we see that we may smoothly solve $Lu + cu = f$ in K. If v, w are smooth, $L(e^{-v}w) = e^{-v}(Lw - wLv)$. If we choose $v \in C^\infty$ such that $Lv = c$ on K and then take $w \in C^\infty$ such that $Lw = e^v f$ on K, we see that $u = e^{-v}w$ satisfies $Lu + cu = f$ on K.

Most of the results we have proved so far in this section are summed up in the following:

THEOREM IV.2.3. *Assume that L is a smooth vector field defined in an open subset Ω of the plane and let $c(x, t) \in C^\infty(\Omega)$. If L satisfies $(\mathcal{P})$ in Ω and it is locally integrable then every point $p \in \Omega$ has a neighborhood U such that the equation*

$$Lu + cu = f, \quad f \in C_c^\infty(U)$$

may be solved with $u \in C^\infty(U)$. Conversely, if L is locally solvable in C^∞ then L is locally integrable.

PROOF. Only the converse part has not been proved already, and we prove it now. Assume that

$$Lu = A(x, t)\frac{\partial u}{\partial t} + B(x, t)\frac{\partial u}{\partial x}$$

with complex coefficients $A, B \in C^\infty(\Omega)$ such that

$$|A(x, t)| + |B(x, t)| > 0, \quad (x, t) \in \Omega$$

is locally solvable in C^∞. Given a point $p \in \Omega$, that we may as well assume to be the origin, we wish to prove the existence of a smooth function Z, defined in a neighborhood of the origin, such that $LZ = 0$ and $dZ \neq 0$. Set

$$d(x, t) = \frac{\partial A(x, t)}{\partial t} + \frac{\partial B(x, t)}{\partial x}$$

and find $u \in C^\infty(\Omega)$ such that $Lu = d$ in a rectangle U centered at the origin. Then the 1-form

$$\omega = B(x, t)e^{-u(x,t)}\, dt - A(x, t)e^{-u(x,t)}\, dx$$

is closed, since

$$\frac{\partial(Be^{-u})}{\partial x} + \frac{\partial(Ae^{-u})}{\partial t} = e^{-u}d - e^{-u}Lu = 0 \quad \text{in } U.$$

Furthermore, ω does not vanish. Since U is simply connected, there exists $Z \in C^\infty(U)$ such that $dZ = \omega$. So $dZ \neq 0$ in U and also $LZ = \langle L, \omega \rangle = e^{-u}\langle A\partial_t + B\partial_x, Bdt - Adx \rangle = 0$. $\square$

REMARK IV.2.4. The assumption in Theorem IV.2.3 that L is locally integrable simplified the construction of smooth solutions but a much more general result is known. In fact, condition $(\mathcal{P})$ alone, formulated in the appropriate way, implies smooth local solvability for operators of principal type of arbitrary order ([**H5**]).

IV.3 Vector fields in several variables

We consider vector fields defined in an open subset $\Omega \subset \mathbb{R}^{n+1}$, $n \geq 1$, that contains the origin,

$$Lu = A(x, t)\frac{\partial u}{\partial t} + \sum_{j=1}^{n} B_j(x, t)\frac{\partial u}{\partial x_j} \qquad (IV.49)$$

with complex coefficients $A, B_1, \ldots, B_n \in C^\infty(\Omega)$ such that

$$|A(x, t)| + \sum_{j=1}^{n} |B_j(x, t)| > 0, \quad (x, t) \in \Omega. \qquad (IV.50)$$

As in the case $n = 1$ discussed in Section IV.1, we may assume locally that $A = 1$ and then apply a several-variables analogue of Lemma IV.1.1, namely

LEMMA IV.3.1. *In appropriate new local coordinates* $x = (x_1, \ldots, x_n)$, t, *defined in a neighborhood of the origin, the vector field L assumes the form*

$$Lu = \frac{\partial u}{\partial t} + i\sum_{j=1}^{n} b_j(x, t)\frac{\partial u}{\partial x_j}, \qquad (IV.51)$$

with $b_j(x, s)$ *real-valued.*

As before, it is useful to write $L = X + iY$ with $X = \Re L$ and $Y = \Im L$ and to refer to the orbits of the pair of real vector fields $\{X, Y\}$ as the orbits of L. Note that since X and Y do not vanish simultaneously then L cannot have any orbits of dimension zero. Let Σ be an orbit of L of dimension two and assume that Σ is orientable. There exists a global nonvanishing section $\rho \in C^\infty(\Sigma; \bigwedge^2(T(\Sigma)))$. Both X and Y are tangent to Σ so they may be considered as sections of the tangent bundle $T(\Sigma) \longrightarrow \Sigma$ that produce a section $X \wedge Y$ of the bundle $\bigwedge^2 T(\Sigma) \longrightarrow \Sigma$. Then $X \wedge Y = b\rho$, where b is a smooth real function defined on Σ. If the real function b does not assume opposite signs on Σ we say that $X \wedge Y$ does not change sign on Σ. Note that if ρ_1 is another nonvanishing section of $\bigwedge^2 T(\Sigma) \longrightarrow \Sigma$ then $\rho_1 = \lambda\rho$ with a smooth real $\lambda \neq 0$ and since Σ is connected either $\lambda > 0$ or $\lambda < 0$. This shows that the notion '$X \wedge Y$ does not change sign on Σ' is independent of the generator ρ.

DEFINITION IV.3.2. *We say that the operator L given by (IV.49) satisfies condition* $(\mathcal{P})$ *in* Ω *if and only if*

(1) *the orbits of L in* Ω *have dimension at most two;*
(2) *the orbits of L of dimension two are orientable and* $\Re L \wedge \Im L$ *does not change sign on any two-dimensional orbit of L.*

It is clear that the above definition is coordinate-free. We will now see that it is invariant under multiplication by a nonvanishing factor.

PROPOSITION IV.3.3. *Let L given by* (IV.49) *satisfy condition* ($\mathcal{P}$) *in* Ω *and let* $h \in C^\infty(\Omega)$ *be a complex nonvanishing function. Then* $L' = hL$ *satisfies* ($\mathcal{P}$) *in* Ω.

PROOF. Write $h = \alpha + i\beta$ with $\alpha, \beta \in C^\infty(\Omega)$ real. Then, $L' = X' + iY'$ with $X' = \alpha X - \beta Y$ and $Y' = \alpha Y + \beta X$. The orbits of L and L' are identical because both L and L' generate the same bundle, so L' has no orbits of dimension higher than two. Let Σ be an orbit of L' of dimension two. Since Σ is also an orbit of L, $X \wedge Y$ does not change sign on Σ and it follows that $X' \wedge Y' = (\alpha^2 + \beta^2)X \wedge Y$ does not change sign on Σ either. $\square$

If L is written in special coordinates in which it has the form (IV.51), condition ($\mathcal{P}$) may be expressed in a more concrete way that extends Definition IV.1.5.

PROPOSITION IV.3.4. *Let L be given by* (IV.51) *in* $\Omega = \{|x| < r\} \times (-T, T)$. *Then L satisfies* ($\mathcal{P}$) *in* Ω *if and only if the following holds:*
for every $x = (x_1, \ldots, x_n) \in \{|x| < r\}$ *and* $\xi = (\xi_1, \ldots, \xi_n) \in \mathbb{R}^n$,

$$\text{the function } (-T, T) \ni t \mapsto \sum_{j=1}^n b_j(x, t)\xi_j \text{ does not change sign.} \quad \text{(IV.52)}$$

PROOF. We begin by showing that if L is given by (IV.51) in Ω the orbits of L of dimensions one and two have a simple description. Since $X = \partial_t$ the orbits of X in Ω are the vertical segments $\{x_0\} \times (-T, T)$. Thus, if (x_0, t_0) belongs to an orbit Σ it follows that $\{x_0\} \times (-T, T) \subset \Sigma$ and this implies that every orbit of L of any dimension may be written as a union of vertical segments. If Σ is a one-dimensional orbit, X and Y are linearly dependent at every point of Σ so $Y = \sum_j b_j \partial_{x_j}$ must vanish identically on Σ, leading to the conclusion that $\Sigma = \{x_0\} \times (-T, T)$ for some $x_0 \in \{|x| < r\}$ such that $b_j(x_0, t) = 0$ for all $1 \le j \le n$, $|t| < T$. Conversely, if $b_j(x_0, t) = 0$ for all $1 \le j \le n$, $|t| < T$ then $\{x_0\} \times (-T, T)$ is a one-dimensional orbit.

We may write $Y = (b_1, \ldots, b_n)$ and denote by $Y \cdot \xi$ the inner product in $\mathbb{R}^n$ of Y and $\xi = (\xi_1, \ldots, \xi_n)$. With this notation (IV.52) states that $t \mapsto Y(x, t) \cdot \xi$ does not change sign.

If Σ is an orbit of dimension ≥ 2 that contains the point (x_0, t_0) there must be a point $(x_0, t_1) \in \Sigma$ such that $Y(x_0, t_1) \ne 0$ for otherwise $(x_0, t_0) \times (-T, T)$ would be a one-dimensional orbit intersecting Σ, which is not possible. Consider the maximal integral curve γ in $\{|x| < r\}$ through the point x_0

of the vector field $Y(x, t_1)$, $x \in \{|x| < r\}$. Then $\gamma \times (-T, T)$ is a closed subset of Σ which is also a two-dimensional manifold. Thus, if the dimension of Σ is two we conclude by connectedness that $\Sigma = \gamma \times (-T, T)$, in particular every two-dimensional orbit of L is orientable. Observe that $Y(\cdot, t_1)$ does not vanish in γ (otherwise γ would reduce to a single point) and set $\mathbf{v}(x) = Y(x, t_1)$. Then $\rho = \mathbf{v} \wedge \partial_t \in \bigwedge^2(\Sigma)$ never vanishes.

Assume now that L satisfies $(\mathcal{P})$ and we wish to prove (IV.52) for some x_0 and ξ fixed. If (x_0, t_0) belongs to a one-dimensional orbit for some $t_0 \in (-T, T)$, then $Y(x_0, t) = 0$ for $|t| < T$ and obviously $t \mapsto Y(x_0, t) \cdot \xi$ cannot change sign. Hence we may assume that $Y(x_0, t_0) \neq 0$ for some $t_0 \in (-T, T)$, so $(x_0, t_0) \in \Sigma$ where Σ is an orbit of L of dimension two on which $X \wedge Y$ does not change sign. Let γ be the integral curve of $\mathbf{v}(x) = Y(x, t_0)$ in $\{|x| < r\}$ through the point x_0. Then $\Sigma = \gamma \times (-T, T)$ and $\rho = \mathbf{v} \wedge \partial_t$ generates $\bigwedge^2(\Sigma)$ at every point of Σ. Let $(x_0, t) \in \Sigma$. Since Y is a horizontal vector tangent to $\gamma \times (-T, T)$ we see that $Y(x_0, t) = \lambda(x_0, t)\mathbf{v}(x_0)$. Furthermore, $X \wedge Y(x_0, t) = \partial_t \wedge \lambda(x_0, t)\mathbf{v}(x_0) = \lambda(x_0, t)\rho(x_0, t)$, so either $\lambda(x_0, t) \geq 0$ on $(-T, T)$ or $\lambda(x_0, t) \leq 0$ on $(-T, T)$. This proves that the vector-valued map $(-T, T) \ni t \mapsto Y(x_0, t)$ does not change direction and $t \mapsto Y(x_0, t) \cdot \xi$ does not change sign for any $\xi \in \mathbb{R}^n$ and $|x_0| < r$.

Conversely, let us prove that (IV.52) implies condition $(\mathcal{P})$. Fix a point $(x_0, t_0) \in \{|x| < r\} \times (-T, T)$ and assume that it belongs to an orbit Σ of dimension ≥ 2. If $Y(x_0, t) = 0$ for all $|t| < T$ then the dimension of Σ would be one, so changing t_0 we may as well assume that $Y(x_0, t_0) \neq 0$. Let γ be the integral curve through x_0 of the vector field $\mathbf{v}(x) = Y(x, t_0)$ in $\{|x| < r\}$. Then, for every $x \in \gamma$, $Y(x, t) = \lambda(t, x)\mathbf{v}(x)$ with $\lambda \geq 0$. Indeed, if for some $x \in \gamma$ and $t_1 \in (-T, T)$ the vectors $Y(x, t_1)$ and $\mathbf{v}(x)$ were not parallel or were parallel but pointing in opposite directions, they would lie on different half-spaces determined by a hyperplane $\{\eta : \eta \cdot \xi = 0\}$, i.e., $Y(x, t_0) \cdot \xi$ and $Y(x, t_1) \cdot \xi$ would have opposite signs, contradicting (IV.52). In particular, this shows that both X and Y are tangent to $\gamma \times (-T, T)$, which makes $\gamma \times (-T, T)$ invariant under the flow of X and Y. This shows that $\Sigma \subset \gamma \times (-T, T)$ and since the orbit has dimension ≥ 2 and the latter set is connected we conclude that $\Sigma = \gamma \times (-T, T)$, which shows that there are no orbits of dimension > 2. Also, $X \wedge Y(x, t) = \lambda(x, t)\partial_t \wedge \mathbf{v}(x)$, $(x, t) \in \Sigma$, so $X \wedge Y$ does not change sign on Σ. $\square$

We are now able to extend Theorem IV.1.9 to any number of variables.

THEOREM IV.3.5. *Let L given by* (IV.49) *satisfy* (IV.50) *and condition* $(\mathcal{P})$ *in a neighborhood of the origin and fix $1 < p < \infty$. Then, there exist a* *neighborhood U of the origin and a constant C > 0 such that the following a priori estimate holds for every $\varphi \in C_c^\infty(U)$:*

$$\|\varphi\|_{L^p(\mathbb{R}^{n+1})} \leq C \operatorname{diam}(\operatorname{supp} \varphi) \|L\varphi\|_{L^p(\mathbb{R}^{n+1})}. \qquad (\text{IV.53})$$

Moreover, the constant C depends only on p and the L^∞ norms of the derivatives of order at most two of the coefficients of L. Furthermore, a similar inequality holds with ${}^tL$ in the place of L.

PROOF. The proof of this theorem requires six steps. Since Theorem IV.3.5 follows from Theorem IV.1.9 when $n = 1$, we will assume in the proof that $n \geq 2$.

The first step. Renaming coordinates if necessary we may assume that $A(0, 0) \neq 0$. Then, dividing by A in a neighborhood of the origin and applying Lemma IV.3.1 we put L in the form (IV.51). The new vector field thus obtained still satisfies condition $(\mathcal{P})$ by its invariance under multiplication by nonvanishing factors and change of coordinates. If φ is a test function supported in a small neighborhood of the origin and Φ is the diffeomorphism induced by the change of variables, the L^p norm of φ and the L^p norm of $\varphi \circ \Phi$ are comparable because the Jacobian determinant $\det(\Phi')$ satisfies $c_1 \leq |\det(\Phi')| \leq c_2$ in a neighborhood of the origin for some positive constants c_1, c_2. Note that the derivatives of order k of the coefficients b_j, $j = 1, \ldots, n$, may be estimated in terms of bounds for the derivatives of order up to $k + 1$ of the original coefficients $A, B_1, \ldots, B_n$, as one extra derivative is consumed by the change of coordinates. Furthermore, by multiplying the coefficients b_j, $j = 1, \ldots, n$, by a non-negative cut-off function equal to 1 on a neighborhood of the origin, we may assume that $b_1, \ldots, b_n \in C_c^\infty(\mathbb{R}^{n+1})$. Hence, it is enough to prove the theorem when L is given by (IV.51) and its coefficients are compactly supported, provided that we prove that the constant C in (IV.53) depends only on p and the L^∞ norms of the derivatives of order at most *one* of the coefficients of L.

The second step. We assume that L is given by (IV.51) and its coefficients are compactly supported, then denote by $\vec{b}(x, t)$ the vector field in $\mathbb{R}^n$ given by $\sum_{j=1}^n b_j(x, t)\partial/\partial x_j$. In view of Proposition IV.3.4 and its proof, the fact that L verifies $(\mathcal{P})$ implies that there exists a unit vector field $\vec{v}(x)$ defined on $\mathbb{R}^n$ such that

$$\vec{b}(x, t) = |\vec{b}(x, t)|\vec{v}(x), \qquad x \in \mathbb{R}^n, t \in \mathbb{R}.$$

Note that $\vec{v}(x_0)$ may be defined arbitrarily if $\vec{b}(x_0, t) = 0$ for all t. Set

$$\mathbb{N} = \left\{ x \in \mathbb{R}^n : \quad \vec{b}(x, t) = 0, \quad |t| < 1 \right\} \qquad (\text{IV.54})$$

and

$$\rho(x) = \sup_{|t|<1} |\vec{b}(x,t)|, \qquad x \in \mathbb{R}^n,$$

so that $\mathbb{N}$ is precisely the set where $\rho(x)$ vanishes. From now on we use the notations $\Omega = \mathbb{R}^n \times (-1,1)$ and $\Omega_T = \mathbb{R}^n \times (-T,T)$, $0 < T < 1$.

LEMMA IV.3.6. *Let χ be the characteristic function of $\mathbb{N}$. Then $L\chi = 0$ in the sense of distributions.*

PROOF. Let $\varphi \in C_c^\infty(\Omega)$. Then

$$\langle \chi, {}^tL\varphi \rangle = -\int_{\mathbb{N}\times(-1,1)} \varphi_t + i\sum_{j=1}^n b_j \partial_{x_j}\varphi + i\varphi \sum_{j=1}^n \partial_{x_j} b_j \, dx dt$$

$$= -\int_{\mathbb{N}} \int_{-1}^1 \varphi_t(x,t) \, dt \, dx = 0$$

where we have used that $\sum_{j=1}^n \partial_{x_j} b_j$ vanishes a.e. on $\mathbb{N} \times (-1,1)$. Indeed, if $(\partial b_j/\partial x_j)(x_0, t_0) \neq 0$ for some $1 \leq j \leq n$ and $(x_0, t_0) \in \mathbb{N} \times (-1,1)$, by the implicit function theorem there is an $\epsilon > 0$ such that the set $\{x: b_j(x, t_0) = 0\} \cap \{|x - x_0| < \epsilon\}$ is a hypersurface. Thus, $\rho(x) > 0$ a.e. in $\{|x - x_0| < \epsilon\}$. This shows that $\{\rho = 0\} \cap \{\sum_j \partial_{x_j} b_j \neq 0\}$ has measure zero. $\qquad\square$

In view of Lemma IV.3.6, $[L, \chi] = 0$ so to obtain (IV.53) it is enough to prove separately the inequalities

$$\|\chi\varphi\|_{L^p(\mathbb{R}^{n+1})]} \leq CT\|L\chi\varphi\|_{L^p(\mathbb{R}^{n+1})]}, \qquad \varphi \in C_c^\infty(\Omega_T), \tag{IV.55}$$

$$\|(1-\chi)\varphi\|_{L^p(\mathbb{R}^{n+1})} \leq CT\|L(1-\chi)\varphi\|_{L^p(\mathbb{R}^{n+1})}, \qquad \varphi \in C_c^\infty(\Omega_T). \tag{IV.56}$$

The third step. We prove inequality (IV.55). The proof of (IV.55) is easy because $L\chi\varphi = \chi L\varphi = \chi\varphi_t$, so

$$\chi(x)\varphi(x,t) = \int_{-T}^t L(\chi\varphi)(x,s) \, ds.$$

Hence,

$$\|\chi(\cdot)\varphi(\cdot,t)\|_{L^p(\mathbb{R}^n)} \leq \int_{-T}^t \|L(\chi\varphi)(\cdot,s)\|_{L^p(\mathbb{R}^n)} \, ds$$

$$\leq (2T)^{1/p'} \|L(1-\chi)\varphi\|_{L^p(\mathbb{R}^{n+1})},$$

with $p'^{-1} + p^{-1} = 1$. Raising both sides to the power p and integrating with respect to t between $-T$ and T we obtain (IV.55) with $C = 2$.

The fourth step. We introduce a partition of unity that reduces the proof of inequality (IV.56) to the proof of local estimates for test functions. Note that the function $(1 - \chi)\varphi$ is not even continuous which, of course, is a source of trouble. The main idea to overcome this difficulty is to write $(1 - \chi)$ as a series of convenient test functions supported in $\Omega \backslash \mathbb{N}$.

We start by proving some lemmas.

LEMMA IV.3.7. *Let $\rho(x)$ and $\mathbb{N}$ be as defined above.*

(1) *The function $\rho(x)$ is Lipschitz and*

$$\|\nabla \rho\|_{L^\infty} \leq \|\nabla_x \vec{b}\|_{L^\infty}. \tag{IV.57}$$

(2) *Outside $\mathbb{N}$ the vector $\vec{v}(x)$ is locally Lipschitz and satisfies*

$$|\nabla \vec{v}(x)| \leq \frac{2\|\nabla_x \vec{b}\|_{L^\infty}}{\rho(x)} \quad \text{for } x \notin \mathbb{N}. \tag{IV.58}$$

PROOF. Let $x, y \in \mathbb{R}^n$ and let $t \in [-1, 1]$ such that $\rho(x) = |\vec{b}(x, t)|$. Then

$$\rho(x) = |\vec{b}(x, t)| \leq |\vec{b}(y, t)| + |\vec{b}(y, t) - \vec{b}(x, t)|$$

$$\leq \rho(y) + \|\nabla_x \vec{b}\|_{L^\infty} |x - y|.$$

This shows that $\rho(x) - \rho(y) \leq \|\nabla_x \vec{b}\|_{L^\infty} |x - y|$ and interchanging x and y we are led to $|\rho(x) - \rho(y)| \leq \|\nabla_x \vec{b}\|_{L^\infty} |x - y|$ for all $x, y \in \mathbb{R}^n$. This implies (IV.57).

Next, given $x_0 \notin \mathbb{N}$ select $|t| \leq 1$ such that $\rho(x_0) = |\vec{b}(x_0, t)| > 0$. Then $|\vec{b}(x, t)|$ is positive and differentiable in a neighborhood of x_0, so

$$|\nabla \vec{v}(x_0)| = \left| \nabla_x \frac{\vec{b}}{|\vec{b}|} (x_0, t) \right| \leq \left| \frac{\nabla_x \vec{b}}{|\vec{b}|} + \vec{b} \otimes \frac{\nabla_x |\vec{b}|}{|\vec{b}|^2} \right| \leq \frac{2\|\nabla \vec{b}\|_{L^\infty}}{\rho(x)}$$

where we have used that $|\nabla_x |\vec{b}|| \leq |\nabla_x \vec{b}|$. This proves (IV.58). $\square$

In the sequel, *cube* will mean a closed cube in $\mathbb{R}^n$, with sides parallel to the axes. Two such cubes will be said to be *disjoint* if their interiors are disjoint. If Q is a cube with side length ℓ and $\lambda > 0$ is a positive number, λQ will denote the cube with the same center as Q and side length equal to $\lambda \ell$.

LEMMA IV.3.8. *Let $f : \mathbb{R}^n \longrightarrow \mathbb{R}^+$ be a Lipschitz continuous function with Lipschitz constant $0 < \mu \leq 1$, i.e., $|f(x) - f(y)| \leq \mu |x - y|$, $x, y \in \mathbb{R}^n$. Assume that $F = f^{-1}\{0\}$ is not empty and set $\mathcal{O} = \{x \in \mathbb{R}^n : f(x) > 0\}$. There exists a collection of cubes $\mathcal{F} = \{Q_1, Q_2, \ldots\}$ such that*

(1) $\bigcup_j Q_j = \mathcal{O} = \mathbb{R}^n \backslash F$;

(2) *the $Q_j \in \mathcal{F}$ are mutually disjoint;*

(3) $\operatorname{diam}(Q_j) \leq \inf_{Q_j} f(x) \leq \sup_{Q_j} f(x) \leq 5 \operatorname{diam}(Q_j)$.

PROOF. Let $\mathcal{Q}_0$ denote the family of cubes with side length one and vertices with integral coordinates. For every integer k we define

$$\mathcal{Q}_k = \{2^{-k}Q: \quad Q \in \mathcal{Q}_0\}$$

so the cubes in $\mathcal{Q}_k$ form a mesh of cubes of side length 2^{-k} and diameter $\sqrt{n}2^{-k}$. Each cube $\in \mathcal{Q}_k$ gives rise to 2^n cubes $\in \mathcal{Q}_{k+1}$ by bisecting the sides. Set for any integer k

$$\mathcal{O}_k = \{x \in \mathbb{R}^n: \quad 2\sqrt{n}2^{-k} < f(x) \leq 4\sqrt{n}2^{-k}\}.$$

Note that $\mathcal{O}_k \subset \mathcal{O}$ and $\mathcal{O} = \bigcup_k \mathcal{O}_k$.

We now define

$$\mathcal{F}_0 = \bigcup_k \{Q \in \mathcal{Q}_k: \quad Q \cap \mathcal{O}_k \neq \emptyset\}.$$

Let $Q \in \mathcal{F}_0 \cap \mathcal{Q}_k$. There exists $x \in Q$ such that $2\sqrt{n}2^{-k} < f(x) \leq 4\sqrt{n}2^{-k}$. Given $y \in Q$ we have

$$f(x) - \mu|y - x| \leq f(y) \leq f(x) + \mu|y - x|,$$

so using that $\mu \leq 1$ and $|y - x| \leq \sqrt{n}2^{-k} = \operatorname{diam}(Q)$ we get

$$\operatorname{diam}(Q) \leq \inf_Q f(x) \leq \sup_Q f(x) \leq 5\operatorname{diam}(Q).$$

Since $f(y) > 0$ on Q it follows that $Q \subset \mathcal{O}$. Also, given $y \in \mathcal{O}$ there exists a unique k such that $y \in \mathcal{O}_k$ and y also belongs to some $Q \in \mathcal{Q}_k$ because $\bigcup\{Q \in \mathcal{Q}_k\} = \mathbb{R}^n$, so $y \in Q$ and $Q \in \mathcal{F}_0$, which shows that $\bigcup\{Q \in \mathcal{F}_0\} = \mathcal{O}$. Thus, the cubes of $\mathcal{F}_0$ satisfy (1) and (3) although they may not be disjoint. To obtain the required collection $\mathcal{F}$ we must discard from $\mathcal{F}_0$ the superfluous cubes, which is easy because if two distinct cubes in $\mathcal{F}_0$ are not disjoint one contains the other. Namely, if $Q_1, Q_2 \in \mathcal{F}_0$ are not disjoint, then $Q_1 \in \mathcal{Q}_{k_1}$ and $Q_2 \in \mathcal{Q}_{k_2}$ with $k_1 \neq k_2$, so if, say, $k_1 > k_2$ it turns out that $Q_1 \subset Q_2$. Hence, if $Q \in \mathcal{F}_0$ is contained in some other cube $Q' \in \mathcal{F}_0$ we discard Q and apply the same procedure to Q', discarding it if it is contained in a bigger cube of $\mathcal{F}_0$ and keeping it in the opposite case. For a fixed cube Q, this process stops after a finite number of steps, otherwise the cubes $Q \subset Q' \subset Q'' \subset \cdots$ would fill $\mathbb{R}^n$, contradicting that $F \neq \emptyset$. Thus, each cube $Q \in \mathcal{F}_0$ is contained in a maximal cube of $\mathcal{F}_0$ and the collection $\mathcal{F}$ of those cubes of $\mathcal{F}_0$ which are maximal satisfies (1), (2), and (3). $\square$

We now need a more detailed discussion of the family $\mathcal{F}$ defined in the previous lemma. Although two distinct cubes Q_1 and $Q_2 \in \mathcal{F}$ are always disjoint in the sense that they have disjoint interior their intersection may be nonempty, as they could share a vertex, an edge, or some k-dimensional face, $k < n$. In this case we say that Q_1 and Q_2 *touch*.

PROPOSITION IV.3.9. *If two cubes* $Q_1, Q_2 \in \mathcal{F}$ *touch, then*

$$\frac{1}{4}\operatorname{diam}(Q_2) \leq \operatorname{diam}(Q_1) \leq 4\operatorname{diam}(Q_2).$$

PROOF. Let Q_1 and $Q_2 \in \mathcal{F}$ have a common point x in their boundaries and assume without loss of generality that $\operatorname{diam}(Q_1) \geq \operatorname{diam}(Q_2)$, so their respective sides ℓ_1 and ℓ_2 are related by $\ell_2 = 2^{-k}\ell_1$ for some integer $k \geq 0$. If $z \in Q_2$ we have

$$f(z) \leq f(x) + \mu\sqrt{n}\ell_2 \leq \sqrt{n}\ell_1(5 + 2^{-k}) \leq 6\sqrt{n}\ell_1,$$

where we have used that Q_1 satisfies (3) of Lemma IV.3.8 to estimate $f(x)$. Now, (3) applied to Q_2 gives $\operatorname{diam}(Q_2) \leq \sup_{z \in Q_2} f(z) \leq 6\operatorname{diam}(Q_1)$. Since the quotient $\operatorname{diam}(Q_2)/\operatorname{diam}(Q_1)$ is a power of 2, the latter estimate implies that $\operatorname{diam}(Q_2)/\operatorname{diam}(Q_1) \leq 4$. □

PROPOSITION IV.3.10. *If* $Q \in \mathcal{F}$, *less than* 12^n *cubes of* $\mathcal{F}$ *touch* Q.

PROOF. Let $Q \in \mathcal{F}$ have side $\ell = 2^{-k}$. There are exactly $3^n - 1$ cubes in $\mathcal{Q}_k$ that touch Q and each one of them contains at most 4^{n-1} cubes that belong to $\mathcal{Q}_{k+2}$ and touch Q. Since by Proposition IV.3.9 the cubes of $\mathcal{F}$ that touch Q may only have the side lengths ℓ, $\ell/2$, or $\ell/4$ it is easily seen that the total number of cubes of $\mathcal{F}$ that touch Q is $\leq (3^n - 1)4^{n-1} < 12^n$. □

The family $\mathcal{F}$ that disjointly fills up $\mathcal{O}$ with closed cubes gives rise to a cover by open cubes that has the bounded intersection property. We fix $0 < \varepsilon < 1/4$ and for any $Q \in \mathcal{F}$ denote by Q^* the cube with the same center as Q but with side dilated by the factor $1 + \varepsilon$. Let Q_1 and $Q_2 \in \mathcal{F}$ do not touch. We claim that Q_1^* and Q_2 cannot intersect. Indeed, the union of Q_1 with all the cubes of $\mathcal{F}$ that touch Q_1 (among which Q_2 is not) contains, by Proposition IV.3.9, the cube $(5/4)Q_1$ whose interior contains Q_1^*. This shows that $Q_1^* \cap Q_2 = \emptyset$. Consider now a point $x \in \mathcal{O}$ and select $Q \in \mathcal{F}$ such that $x \in Q$. If $x \in Q_j^*$ for some $Q_j \in \mathcal{F}$ then $Q \cap Q_j^* \neq \emptyset$, which implies that Q and Q_j touch. Then Proposition IV.3.10 shows that x belongs to at most 12^n cubes Q_j^*. If $z \in Q^*$ then $f(z) \geq \inf_Q f - \mu\varepsilon\operatorname{diam}(Q) \geq (3/4)\operatorname{diam}(Q) \geq (3/5)\operatorname{diam}(Q^*)$. Similarly, $f(z) \leq 5\operatorname{diam}(Q) + \mu\varepsilon\operatorname{diam}(Q) \leq 5\operatorname{diam}(Q^*)$. Thus, for every $Q \in \mathcal{F}$ we have

$$\frac{1}{2}\operatorname{diam}(Q^*) \leq \inf_{Q^*} f(x) \leq \sup_{Q^*} f(x) \leq 5\operatorname{diam}(Q^*). \tag{IV.59}$$

This estimate implies that $Q^* \subset \mathcal{O}$ and since the interior $\mathrm{Int}\,(Q^*) \supset Q$ we see that $\{\mathrm{Int}\,(Q^*)\}$ is an open cover of $\mathcal{O}$ with the bounded intersection property.

LEMMA IV.3.11. *Let* $\mathbb{N} \subset \mathbb{R}^n$ *be the closed set defined in* (IV.54) *and let* $0 < \mu \le 1$, $1 < p < \infty$. *There exists a covering of* $\mathbb{R}^n \backslash \mathbb{N}$ *by open cubes with sides parallel to the coordinate axes* $\{\mathrm{Int}\,(Q_j^*)\}$, $j = 1, 2, \ldots$, *such that the intersection of* 12^n *cubes of the family is always empty and for any* $j = 1, 2, \ldots$, *we have the estimate:*

$$\frac{1}{2}\mathrm{diam}\,(Q_j^*) \le \mu \inf_{Q_j^*} \rho(x) \le \mu \sup_{Q_j^*} \rho(x) \le 5\,\mathrm{diam}\,(Q_j^*). \qquad (\mathrm{IV.60})$$

Furthermore, there are functions $\phi_j \in C_c^\infty(\mathbb{R}^n \backslash \mathbb{N})$ *such that* $\{\phi_j^p\}$ *is a partition of unity in* $\mathbb{R}^n \backslash \mathbb{N}$ *subordinated to the covering* $\{\mathrm{Int}\,(Q_j^*)\}$ *and for a certain constant* $C > 0$,

$$\|\nabla \phi_j\|_{L^\infty} \le \frac{C}{\mathrm{diam}\,(Q_j^*)}, \qquad j = 1, 2, \ldots \qquad (\mathrm{IV.61})$$

PROOF. From now on we assume without loss of generality that $\|\nabla_x \vec{b}\|_{L^\infty} \le 1$. We apply Lemma IV.3.8 with $f(x) = \mu\rho(x)$ so $F = \mathbb{N}$. The hypotheses are satisfied because the Lipschitz constant of $\rho(x)$ is 1 by (IV.57) and the complement of $\mathbb{N}$ is bounded so $\mathbb{N} \ne \emptyset$. Thus we obtain the collection $\mathcal{F}$ of disjoint cubes $\{Q_j\}$ which, dilated by the factor $1 + \varepsilon$, yields the associated collection $\{Q_j^*\}$ of cubes whose interiors cover $\mathbb{R}^n \backslash \mathbb{N}$, have the bounded intersection property, and satisfy (IV.59). This proves (IV.60). Fix a function $0 \le \psi \in C_c^\infty(\mathbb{R}^n)$ supported in $|x| < (1 + \varepsilon)/2$ such that $\psi^p(x)$ is smooth and $\psi(x) = 1$ if $|x| \le 1/2$ (such a function is easily constructed). If $Q_j \in \mathcal{F}$, denote by x_j its center and by ℓ_j its side length. Then $\psi_j(x) = \psi((x - x_j)/\ell_j) \in C_c^\infty(\mathrm{Int}\,(Q_j^*))$ and $\psi_j(x) = 1$ on Q_j. We have

$$\|\nabla \psi_j\|_{L^\infty} \le \frac{\|\nabla \psi\|_{L^\infty}}{\ell_j} \le \frac{C}{\mathrm{diam}\,(Q_j^*)}. \qquad (\mathrm{IV.62})$$

Note that $\Psi = \sum_j \psi_j^p$ is smooth and ≥ 1 in $\mathbb{R}^n \backslash \mathbb{N}$. Let us estimate $\nabla \Psi(x)$ on the support of ψ_j. If $x \in Q_j^*$ and $\psi_k(x) \ne 0$ for some $k \in \mathbb{Z}^+$ it follows that $Q_j^* \cap Q_k^* \ne \emptyset$. We know that Q_j^* is contained in the union of Q_j with those cubes of $\mathcal{F}$ which touch it and the same can be said about Q_k. This implies that there are cubes $Q_{j'}$ and $Q_{k'}$ in $\mathcal{F}$ such that

(1) $Q_{j'}$ touches Q_j;
(2) $Q_{k'}$ touches Q_k;
(3) $Q_{j'} \cap Q_{k'} \ne \emptyset$ so they either coincide or touch.

Applying Proposition IV.3.9 three times we obtain that $\mathrm{diam}\,(Q_k) \geq 4^{-3}$ $\mathrm{diam}\,(Q_j)$ and Proposition IV.3.10 tells us that there are less than $N = 12^{3n}$ integers k such that $Q_j^* \cap Q_k^* \neq \emptyset$. This shows that at most N terms $\psi_k^p(x)$ of the infinite sum that defines $\Psi(x)$ are not zero if $x \in \mathrm{supp}\,(\psi_j)$. Thus, using the analogue for ψ_k^p of (IV.62) we obtain

$$\sup_{Q_j^*} |\nabla \Psi(x)| \leq \sum_k \sup_{Q_j^*} |\nabla \psi_k^p(x)| \leq \sum_k \frac{C}{\mathrm{diam}\,(Q_k^*)} \leq \frac{4^3 NC}{\mathrm{diam}\,(Q_j^*)}. \qquad \text{(IV.63)}$$

Since

$$|\nabla \Psi^{-1/p}(x)| \leq \frac{1}{p} \|\Psi^{-1-1/p}\|_{L^\infty} |\nabla \Psi(x)| \leq |\nabla \Psi(x)|,$$

because $\Psi \geq 1$, (IV.63) implies

$$\sup_{Q_j^*} |\nabla \Psi^{-1/p}(x)| \leq \frac{C}{\mathrm{diam}\,(Q_j^*)}. \qquad \text{(IV.64)}$$

Set

$$\phi_j(x) = \frac{\psi_j(x)}{\Psi^{1/p}(x)}.$$

Then, $\{\phi_j^p\}$ is a partition of unity in $\mathbb{R}^n \setminus \mathbb{N}$ with the required properties. Indeed, to prove (IV.61) we use the Leibniz rule and invoke (IV.62) and (IV.64). $\qquad \square$

The fifth step. We prove estimate (IV.56) when $\varphi(x, t)$ is supported in $Q_j^* \times (-T, T)$, $Q_j \in \mathcal{F}$. Assume that φ is supported in $Q_k^* \times (-T, T)$ for a certain cube $\in \mathcal{F}$; the value of $T < 1$ will be chosen momentarily. Since we are assuming that $\|\nabla_x \vec{b}\|_{L^\infty} \leq 1$, (IV.58) yields

$$|\nabla \vec{v}(x)| \leq \frac{2}{\rho(x)} \qquad \text{for } x \notin \mathbb{N}.$$

This shows, in view of (IV.60), that $|\nabla \vec{v}(x)| \leq 4\mu/\mathrm{diam}\,(Q_j^*)$ on Q_j^*. Furthermore, $\mathbb{R}^n \setminus \mathbb{N}$ is bounded so $\mathrm{diam}\,(Q_j^*) \leq C$, $j \in \mathbb{Z}$. Hence, $\vec{v}(x)$ is approximately constant on Q_j^* if μ is small; this allows us to rectify its flow as follows. Since $\vec{v}$ is a unit vector, we may assume without loss of generality that at the center x_j of Q_j^* we have $v_1(x_j) \geq 1/\sqrt{n}$. Then, $|v_1(x_j) - v_1(x)| \leq 4\mu < 1/(2\sqrt{n})$ for μ fixed once for all, small but independent of j, and we may assume that $v_1(x) \geq 1/(2\sqrt{n})$ on Q_k^*. Solving the differential equations

$$\frac{\mathrm{d}x_j}{\mathrm{d}y_1} = \frac{v_j(x)}{v_1(x)}, \qquad x_j(0) = y_j, \qquad j = 2, \dots, n \qquad \text{(IV.65)}$$

we obtain a change of variables on a neighborhood of Q_k^* given by $x_1 = y_1$, $x_j = x_j(y_1; y_2, \dots, y_n)$, $1 < j \leq n$, where the right-hand side denotes the solution of (IV.65). In the new coordinates $\vec{v}(x(y)) = v_1(x(y))\partial/\partial y_1$ and L assumes the form

$$\frac{\partial}{\partial t} - i b_1(x(y), t)\frac{\partial}{\partial y_1}$$

with $b_1 > 0$, since $(b_1(x(y), t), 0, \ldots, 0) = \vec{b}(x(y), t)$ implies $b_1(x(y), t) = |\vec{b}(x(y), t)|$. Set $B(y, t) = b_1(x(y), t)$. Then, by the chain rule,

$$\|\nabla_y B\|_{L^\infty} \leq C\|\nabla_x b_1\|_{L^\infty} \leq C'$$

because the Lipschitz constant of the change of variables $y \mapsto x(y)$ is bounded by a constant independent of j, as follows from the fact that the right-hand side of the ODE (IV.65) is bounded by $C\mu$. Now we apply Theorem IV.1.9 with $p = q$ to the vector field

$$L_1 = \frac{\partial}{\partial t} - i B(y, t)\frac{\partial}{\partial y_1} \tag{IV.66}$$

that we regard as a vector field in two variables depending on a parameter $y' = (y_2, \ldots, y_n)$. For some constants C and T_0 whose size only depends on $\|\nabla_x b_1\|_{L^\infty}$ we get for any $0 < T \leq T_0$

$$\|\varphi(\cdot, \cdot, y')\|_{L^p}^p \leq CT\|L_1\varphi(\cdot, \cdot, y')\|_{L^p}^p, \quad \varphi \in C_c^\infty(Q_j^\dagger \times (-T, T)),$$

where the L^p norms are taken in the variables (y_1, t) and the map $y \mapsto x(y)$ takes $Q_j^\dagger$ onto Q_j^*. Integrating this estimate with respect to y' we get

$$\|\varphi\|_{L^p}^p \leq CT\|L_1\varphi\|_{L^p}^p, \quad \varphi \in C_c^\infty(Q_j^\dagger \times (-T, T)).$$

Observing that the absolute value of the Jacobian determinant of $y \mapsto x(y)$ is close to 1 uniformly in $j \in \mathbb{Z}^+$, the latter estimate implies in the original variables (x, t)

$$\|\varphi\|_{L^p}^p \leq CT\|L\varphi\|_{L^p}^p, \quad \varphi \in C_c^\infty(Q_j^* \times (-T, T)), \tag{IV.67}$$

which may be regarded as estimate (IV.56) for $\varphi \in C_c^\infty(Q_j^* \times (-T, T))$.

The sixth step. We prove (IV.56) in general. Let $\varphi \in \Omega_T$ and set $\varphi_j = \phi_j\varphi$ where $\{\phi_j\}$ is the collection of functions described by Lemma IV.3.11. We have

$$(1 - \chi(x))|\varphi(x, t)|^p = \sum_j |\phi_j(1 - \chi(x))\varphi(x, t)|^p.$$

Integrating this identity and taking account of (IV.67),

$$\|(1 - \chi)\varphi\|_{L^p}^p = \sum_j \|\phi_j(1 - \chi)\varphi\|_{L^p}^p \leq CT\sum_j \|L(\phi_j\varphi)\|_{L^p}^p$$

$$\leq CT\|(1-\chi)L\varphi\|_{L^p}^p + CT\sum_j \|(L\phi_j)(1-\chi)\varphi\|_{L^p}^p$$

where we have used the Leibniz rule and the fact that $\sum_j \phi_j^p = 1$. The second term on the right-hand side is dominated by $CT\|(1-\chi)\varphi\|_{L^p}^p$. Indeed,

$$|L\phi_j(x)| = |\vec{b}(x,t)\cdot\nabla_x\phi_j(x)| \leq \sup_{Q_j^*} |\vec{b}|\,|\nabla_x\phi_j| \leq \frac{C}{\mu} \leq C_1$$

in view of the definition of ρ, (IV.60) and (IV.61). Hence, $|L\phi_j(x)|^p \leq C$ and since $|L\phi_j(x)|^p = 0$ except for at most 12^n values of j we also have $\sum_j |L\phi_j(x)|^p \leq C$. Thus,

$$\|(1-\chi)\varphi\|_{L^p}^p \leq CT\|(1-\chi)L\varphi\|_{L^p}^p + CT\|(1-\chi)\varphi\|_{L^p}^p$$

and the last term can be absorbed as soon as $CT < 1/2$. This proves (IV.56). We have already seen in steps 1 and 2 that (IV.53) follows in general once (IV.55) and (IV.56) are proved for L of the form (IV.51), so the proof of Theorem IV.3.5 is now complete for L and we may also replace L by $-L+c(x,t)$ in (IV.53) if $c(x,t)$ is any bounded function provided we shrink the neighborhood U of the origin, in particular, we may replace L by the transpose operator ${}^tL = -L - i\mathrm{div}_x\,\vec{b}$. $\qquad\square$

As usual, we obtain by duality

COROLLARY IV.3.12. *Let L given by (IV.49) satisfy (IV.50) and condition $(\mathcal{P})$ in a neighborhood of the origin and fix $1 < p < \infty$. Then, there exist R_0 and $C > 0$ such that for every $0 < R < R_0$ and $f \in L^p(\mathbb{R}^{n+1})$ there exists $u \in L^p(\mathbb{R}^{n+1})$ with norm*

$$\|u\|_{L^p(\mathbb{R}^{n+1})} \leq C\,R\|f\|_{L^p(\mathbb{R}^{n+1})}$$

that satisfies the equation

$$Lu = f \quad \text{for } |x|^2 + t^2 < R^2. \tag{IV.68}$$

Moreover, the constants C and R_0 depend only on p and the L^∞ norms of the derivatives of order at most two of the coefficients of L.

Let us assume now that we are dealing with a locally integrable vector field L in an open set of $\mathbb{R}^{n+1}$ that contains the origin. After an appropriate local change of coordinates (x,t) we may assume that there are functions $Z_j(x,t)$, $j = 1,\ldots,n$ defined on a neighborhood of the origin of the form

$$Z_j(x,t) = x_j + i\varphi_j(x,t), \quad j = 1,\ldots,n,$$

with $\varphi_j(x, t)$ smooth and real satisfying

$$\varphi_j(0, 0) = \nabla_x \varphi_j(0, 0) = 0, \quad j = 1, \ldots, n,$$

such that

$$LZ_j = 0, \quad j = 1, \ldots, n.$$

We denote by Z the function $Z = (Z_1, \ldots, Z_n)$ with values in $\mathbb{C}^n$ and similarly write $\varphi = (\varphi_1, \ldots, \varphi_n)$, so $Z(x, t) = x + i\varphi(x, t)$. The $n \times n$ matrix

$$\varphi_x = \begin{pmatrix} \partial\varphi_1/\partial x_1 & \cdots & \partial\varphi_1/\partial x_n \\ \vdots & \ddots & \vdots \\ \partial\varphi_n/\partial x_1 & \cdots & \partial\varphi_n/\partial x_n \end{pmatrix}$$

vanishes at the origin and after modification of L outside a neighborhood of the origin we may assume that the functions $\varphi_j(x, t)$ are defined throughout $\mathbb{R}^{n+1}$, have bounded derivatives of all orders, and satisfy

$$\|\varphi_x(x, t)\| \leq \frac{1}{2}, \qquad (x, t) \in \mathbb{R}^{n+1}.$$

This implies that the matrix $Z_x = I + i\varphi_x$ is everywhere invertible and we write $Z_x^{-1}(x, t) = (\mu_{jk}(x, t))$. Then the vector fields

$$M_j = \sum_{k=1}^{n} \mu_{jk}(x, t) \frac{\partial}{\partial x_k}, \quad j = 1, \ldots, n \tag{IV.69}$$

commute pairwise and the vector field

$$L_1 = \frac{\partial}{\partial t} - \sum_{k=1}^{n} \lambda_k(x, t) \frac{\partial}{\partial x_k}$$

commutes with $M_1, \ldots, M_n$ and is proportional to L if

$$\lambda_k(x, t) = -i \sum_{j=1}^{n} \mu_{kj}(x, t) \frac{\partial\varphi_j}{\partial t}(x, t).$$

Furthermore, $M_1, \ldots, M_n, L$ are linearly independent at every point and generate $T\mathbb{R}^{n+1}$. Multiplying L by a nonvanishing factor we may assume that $L = L_1$.

We now extend Theorem IV.2.3 to several variables.

THEOREM IV.3.13. *Assume that L is a smooth vector field defined in an open subset $\Omega \subset \mathbb{R}^{n+1}$ and let $c(x, t) \in C^\infty(\Omega)$. If L satisfies $(\mathcal{P})$ in Ω and is locally integrable then every point $p \in \Omega$ has a neighborhood U such that the equation*

$$Lu + cu = f, \quad f \in C_c^\infty(U)$$

*may be solved with $u \in C^\infty(U)$. Conversely, if L is locally solvable in C^∞
then L is locally integrable.*

PROOF. The construction of smooth solutions is a straightforward extension
of the two-dimensional case. We write

$$D = -L^2 - \lambda(M_1^2 + \cdots + M_n^2),$$

where $M_1, \ldots, M_n$ are given by (IV.69) and $\lambda > 0$ is a large parameter. Since
$L = L_1$ and M_j commute, $j = 1, \ldots, n$, it follows that L and D commute. If
$\ell(x, t, \xi, \tau)$ denotes the symbol of L, $m_j(x, t, \xi)$ denotes the symbol of M_j
and $d(x, t, \xi, \tau) = -(\ell^2 + \lambda(m_1^2 + \cdots + m_n^2))(x, t, \xi, \tau)$ is the principal symbol
of D, we have

$$\{\ell, d\}(x, t, \xi, \tau) = 0, \qquad (x, t, \xi, \tau) \in \mathbb{R}^{2(n+1)}.$$

For large $\lambda > 0$, D is a uniformly elliptic second-order differential operator.
Consider, for fixed $s \in \mathbb{R}$, the pseudo-differential operator

$$B^\epsilon u(x, t) = \frac{1}{(2\pi)^{n+1}} \int_{\mathbb{R}^{N+1}} e^{i(x \cdot \xi + t\tau)} p(x, t, \xi, \tau) \hat{u}(\xi, \tau) \, d\xi d\tau$$

with symbol

$$b^\epsilon(x, t, \xi, \tau) = \frac{\chi(t)(1 + d(x, t, \xi, \tau))^{s/2}}{(1 + \epsilon d(x, t, \xi, \tau))^{1/2}}$$

where $\chi(t) \in C_c^\infty(-T, T)$ and $\chi(t) = 1$ for $|t| \leq (3/4)T$. Here we choose T so
that the estimate

$$\|u(x, t)\|_{L^2(\mathbb{R}^{n+1})} \leq C\|Lu(x, t)\|_{L^{n+1}(\mathbb{R}^2)} \qquad \text{(IV.70)}$$

holds for every $u \in C_c^\infty(\mathbb{R}^n \times (-T, T))$ for some $C > 0$, as guaranteed by
the proof of Theorem IV.3.5. The estimate can be extended to any $u \in L_c^2(\mathbb{R}^n$
$(-T, T) \times (-T, T))$ such that $Lu \in L_c^2(\mathbb{R}^n(-T, T) \times (-T, T)$ by Friedrich's
lemma. It follows that $b^\epsilon \to b = \chi(1 + d)^{s/2}$ in the symbol space $S_{1,0}^s$ and that
$\|u\|_s \sim \|Bu\|_{L^2}$ if B is the pseudo-differential with symbol b and $u \in H_c^s(\mathbb{R}^n \times$
$(-T/2, T/2))$. Furthermore, $[L, B]$ has order $s - 1$ on $\mathbb{R} \times (-T/2, t/2)$. If
$u \in H_c^{s-1}(\mathbb{R}^n \times (-T/2, T/2))$ is such that $Lu \in H_c^s(\mathbb{R}^n \times (-T/2, T/2))$ we
may apply (IV.70) to $B^\epsilon u$. Letting $\epsilon \to 0$ we obtain

$$\|Bu\|_{L^2(\mathbb{R}^{n+1})} \leq C(\|BLu\|_{L^2(\mathbb{R}^{n+1})} + \|[L, B]u\|_{L^2(\mathbb{R}^{n+1})})$$

which implies that $u \in H^s(\mathbb{R}^{n+1})$ and

$$\|u\|_{H^s(\mathbb{R}^{n+1})} \leq C_s(\|Lu\|_{H^s(\mathbb{R}^{n+1})} + \|u\|_{H^{s-1}(\mathbb{R}^{n+1})}). \qquad \text{(IV.71)}$$

Once (IV.71) is known, general arguments lead to an a priori estimate

$$\|u\|_{H^s(\mathbb{R}^{n+1})} \leq C_s \|Lu\|_{H^s(\mathbb{R}^{n+1})},$$ (IV.72)

if $u \in H_c^{s-1}(\mathbb{R}^n \times (-T/2, T/2))$ is such that $Lu \in H_c^s(\mathbb{R} \times (-T/2, T/2))$ and to the existence of local smooth solutions, as described in the proof of Theorem IV.2.3. We leave details to the reader.

While the method to obtain smooth solutions starting from the existence of L^2 solutions is essentially the same independently of the number of variables, the proof that smooth local solvability implies local integrability is rather different if $n = 1$ or $n > 2$. In the proof of Theorem IV.2.3 it was shown that, for $n = 1$, solving $Lu = f$ for a specific f obtained from the coefficients of L was enough to produce locally a smooth Z such that $LZ = 0$ and $dZ \neq 0$. Nothing like this is available if $n > 1$ and we must proceed indirectly. Assume that L given by (IV.51) is locally solvable in C^∞ and we wish to find n first integrals with linearly independent differentials defined in a neighborhood of a given point p that we may as well assume to be the origin. The first step is to find a complete set of *approximate* first integrals, namely, n smooth functions $Z_j^\#$, $j = 1, \dots, n$, such that $LZ_j^\# = f_j$ vanishes to infinite order at the origin—i.e., $f_j(x) = O(|x|^k)$, $k = 1, 2, \dots$—and $dZ_1^\#(0), \dots, dZ_n^\#(0)$ are linearly independent. To find $Z_j^\#$ we solve first the noncharacteristic Cauchy problem

$$\begin{cases} LU_j &= 0, \\ U_j(x, 0) &= x_j, \end{cases}$$

in the sense of formal power series. The coefficients of the formal series U_j corresponding to monomials that do not contain t are determined by the initial condition $U_j(x, 0)$, i.e., they are all zero with the exception of the coefficient of x_j which is 1. The coefficients of monomials of the form $t^\ell x^\alpha$ are determined from $LU_j = 0$ inductively on ℓ. Once the formal series U_j has been found we take as $Z_j^\#$ any smooth function that has U_j as its Taylor series at the origin (the existence of such a function is usually called Borel's lemma). By their very definition $Z_1^\#, \dots, Z_n^\#$ are approximate first integrals. To obtain exact first integrals by correction of $Z_1^\#, \dots, Z_n^\#$ we must solve the equations $Lu_j = f_j$, $j = 1, \dots, n$, in a neighborhood of the origin and then define $Z_j = Z_j^\# - u_j$. Clearly, $LZ_j = 0$, so the problem is now to verify that $dZ_1(0), \dots, dZ_n(0)$ are linearly independent. This will be guaranteed if we can make sure that $|du_j(0)|$ is small. Let K be a ball centered at the origin such that $LC^\infty(K) = C^\infty(K)$ and let $\mathcal{H}$ denote the subspace of $C^\infty(K)$ of the (equivalence classes of) functions h such that $Lh = 0$. Then L defines a continuous linear map from $C^\infty(K)/\mathcal{H}$ onto $C^\infty(K)$ which, by the open mapping theorem for Fréchet spaces, has a continuous inverse. This means, in particular, that given $\epsilon > 0$ there exists $\delta > 0$ and $m \in \mathbb{Z}^+$ such that for every $f \in C^\infty(K)$ such that $\|D^\beta f\|_{L^\infty(K)} < \delta$ for all $|\beta| \leq k$ there exist $u \in C^\infty(K)$ such that $Lu = f$ and $\|du\|_{L^\infty(K)} < \epsilon$. Let $\chi(x, t) \in C_c^\infty(\mathbb{R}^{n+1})$ be equal to 1 for $|x|^2 + t^2 < 1$ and set $f_{j,\rho}(x, t) = f_j(x, t)\chi(\rho x, \rho t)$. Since f_j vanishes to infinite order at the origin we see that, choosing ρ big enough, $\|D^\beta f_{j,\rho}\|_{L^\infty} < \delta$ for all $|\beta| \leq k$. Choose now u_j such that $Lu_j = f_{j,\rho}$ and $\|du_j\|_{L^\infty(K)} < \epsilon$. Since $f_{j,\rho} = f_j$ for $|x|^2 + t^2 < 1/\rho$ we see that the functions $Z_j = Z_j^\# - u_j$, $j = 1, \ldots, n$ form a complete set of first integrals in a neighborhood of the origin if ϵ is taken small enough. $\qquad \square$

IV.4 Necessary conditions for local solvability

In this section we discuss the necessity of condition $(\mathcal{P})$ for the local solvability of a locally integrable vector field. Assume that L defined in $\Omega \subset \mathbb{R}^{n+1}$ by (IV.49) is locally solvable in the sense of Definition IV.1.2. We will show that L must satisfy condition $(\mathcal{P})$ in Ω. In doing so, due to the local nature of the problem, we may assume that L is given by (IV.51) and that $\Omega = B \times (-T, T)$ where $B \subset \mathbb{R}^n$ is a ball centered at the origin. We may also assume that there is a vector-valued function $Z(x, t) = (Z_1(x, t), \ldots, Z_n(x, t))$ defined in a neighborhood U of $\overline{\Omega}$ such that $LZ_j = 0$, $j = 1, \ldots, n$ and $\|I - Z_x\| < 1/2$ in Ω, where I denotes the identity matrix. In particular, the form $dZ_1 \wedge \cdots \wedge dZ_n$ does not vanish in Ω and the pairing

$$C_c^\infty(\Omega) \times C_c^\infty(\Omega) \ni (f, v) \mapsto \int fv \det(Z_x)\, dxdt$$

is nondegenerate. The formula

$$\int f\, Lv\, \det(Z_x)\, dxdt = -\int Lf\, v\, \det(Z_x)\, dxdt, \quad v, f \in C_c^\infty(\Omega)$$

means that L and $-L$ are each other's formal transpose with respect to this pairing. The formula is also valid by continuity if $v \in \mathcal{D}'(\Omega)$ provided that we replace the integration by the standard duality between distributions and test function, i.e.,

$$\langle Lv, f \det(Z_x) \rangle = -\langle v, Lf \det(Z_x) \rangle, \quad f \in C_c^\infty(\Omega), \ v \in \mathcal{D}'(\Omega). \quad \text{(IV.73)}$$

One of the basic tools in the study of necessary conditions for local solvability is Hörmander's lemma ([**H6**]), of which we give the following version.

LEMMA IV.4.1. *Let L be as described above and suppose that for every $f \in C_c^\infty(\Omega)$ there exists $u \in \mathcal{D}'(\Omega)$ such that $Lu = f$. Then, for any compact set $K \subset \Omega$ there exist constants $C > 0$, $M \in \mathbb{Z}^+$ such that*

$$\left| \int fv \det(Z_x) \, dx dt \right| \le C \sum_{|\alpha| \le M} \|D_{x,t}^\alpha f\|_{L^\infty} \sum_{|\beta| \le M} \|D_{x,t}^\beta Lv\|_{L^\infty} \qquad (IV.74)$$

for all $f, v \in C_c^\infty(K)$.

PROOF. Let $K \subset\subset \Omega$ with nonempty interior be given and consider the bilinear form (IV.73) restricted to pairs $(f, v) \in C_c^\infty(K) \times C_c^\infty(K)$. Endow the first factor with the topology defined by the seminorms $\|D_{x,t}^\alpha f\|_{L^\infty}$—so it becomes a Fréchet space—and the second factor with the countable family of seminorms $\|D_{x,t}^\beta Lv\|_{L^\infty}$. Our solvability hypothesis implies that the latter topology is Hausdorff, indeed, if $v \in C_c^\infty(K)$ is such that $Lv = 0$ we may choose for any $f \in C_c^\infty(K)$ a distribution $u \in \mathcal{D}'(\Omega)$ such that $Lu = f$, so we have

$$\langle f, v \det(Z_x) \rangle = \langle Lu, v \det(Z_x) \rangle = -\langle u, Lv \det(Z_x) \rangle = 0$$

for any $f \in C_c^\infty(K)$, which implies that $v = 0$. For fixed v, the bilinear form clearly depends continuously on f. The solvability hypothesis implies that the dependence on v is also continuous for f fixed. Indeed, we may assume that $f = Lu$ for some $u \in \mathcal{D}'(\Omega)$. Hence

$$\int fv \det(Z_x) \, dx dt = \langle Lu, f \det(Z_x) \rangle = -\langle \det(Z_x) u, Lf \rangle$$

in view of (IV.73), which shows the continuity with respect to f for fixed v. A bilinear form defined on the product of a Fréchet space and a metrizable space which is separately continuous is continuous in both variables. This proves (IV.74). $\qquad \square$

The last lemma shows that in order to prove that L is not solvable it is enough to violate the a priori inequality (IV.74). We now describe a method to violate (IV.74) provided we find a solution h of the homogenous equation $Lh = 0$ with certain geometric property. Let $g \in C^0(\Omega)$ be a real function and $K \subset\subset \Omega$ be compact. We say that g *assumes a local minimum over K* if there exists $a \in \mathbb{R}$ and V open, $K \subset V \subset \Omega$ such that

(1) $g \equiv a$ on K;
(2) $g > a$ on $V \backslash K$.

Note that we may always replace the open set V with one of its open subsets with compact closure that contains K. In this case, still denoting the new set by V we have

$$\inf_{\partial V} g = a_1 > a.$$

Then, taking $a < b < a_1$ we see that the set $W = \{g < b\} \cap V$ has compact closure contained in V and $g \geq b > a$ on $\overline{V} \backslash W$.

The proof of the next lemma shows how (IV.74) may be violated.

LEMMA IV.4.2. *Assume that there exists $h \in C^\infty(\Omega)$ such that*

(i) $Lh = 0$;
(ii) $\Re h$ assumes a local minimum over some $K_1 \subset\subset \Omega$.

Then there exists $f \in C_c^\infty(\Omega)$ such that $Lu \neq f$ for all $u \in \mathcal{D}'(\Omega)$.

PROOF. By Lemma IV.4.1 it will be enough to show that for a convenient choice of $K \subset\subset \Omega$, (IV.74) cannot hold for all $f, v \in C_c^\infty(K)$ whatever the choice of $M \in \mathbb{Z}^+$ and $C > 0$. By hypothesis $\Re h$ assumes a local minimum over $K_1 \subset\subset \Omega$ for some homogeneous solution h. Subtracting a constant we may assume that $\Re h = 0$ on K_1 and $\Re h \geq \varepsilon > 0$ on $\overline{V} \backslash W$ for some open sets $V \supset W \supset K_1$ such that $K \doteq \overline{V} \subset\subset \Omega$. Select $\zeta \in C_c^\infty(K)$, $0 \leq \zeta \leq 1$, such that $\zeta = 1$ on W and set, for a large parameter $\rho > 0$,

$$v_\rho(x, t) = \zeta(x, t) e^{-\rho h(x,t)}.$$

Since $e^{-\rho h(x,t)}$ is a homogeneous solution, $Lv_\rho = e^{-\rho h} L\zeta$. Furthermore, $L\zeta$ is supported in $K \backslash W$ so it follows that

$$\sum_{|\beta| \leq M} \|D_{x,t}^\beta Lv_\rho\|_{L^\infty} \leq C\rho^M e^{-\varepsilon\rho}. \tag{IV.75}$$

Next, choose $\psi \in C_c^\infty(V)$, $0 \leq \psi \leq 1$, such that $\psi = 1$ on K_1 and $\Re h(x, t) < \varepsilon/2$ on the support of ψ. Define

$$f_\rho(x, t) = \frac{\psi(x, t)}{\det(Z_x(x, t))} e^{\rho h(x,t)}.$$

Then

$$\sum_{|\alpha| \leq M} \|D_{x,t}^\alpha f_\rho\|_{L^\infty} \leq C\rho^M e^{\varepsilon\rho/2}. \tag{IV.76}$$

On the other hand, since ζ and ψ are positive in a neighborhood of K_1,

$$\int f_\rho v_\rho \det(Z_x) \, dxdt = \int \zeta(x, t)\psi(x, t) \, dxdt = c > 0$$

which together with (IV.75) and (IV.76) shows that (IV.74) cannot hold for the pair $(f_\rho, v_\rho) \in C_c^\infty(K) \times C_c^\infty(K)$ if ρ is large enough. $\square$

Our next task is to produce solutions of the homogeneous equation $Lh = 0$ whose real part assumes a local minimum over a compact set assuming that condition $(\mathcal{P})$ does not hold. We will first discuss this in the case $n = 1$, which is technically simpler and the geometric ideas involved are easier to spot. Suppose $n = 1$, $L = \partial_t - (Z_t/Z_x)\partial_x$, $Z = x + i\varphi(x, t)$, $(x, t) \in \mathbb{R}^2$. We know by Lemma IV.2.2 that if $(\mathcal{P})$ does not hold then $t \mapsto \varphi(x_0, t)$ is not monotone for some x_0, or equivalently that $t \mapsto \varphi_t(x_0, t)$ takes opposite signs and, in particular, vanishes for some t_0. The simplest situation occurs when $\varphi_t(x_0, t_0) = 0$ and $\varphi_{tt}(x_0, t_0) \neq 0$. If $\varphi_{tt}(x_0, t_0) \doteq A > 0$, $\varphi_{xt}(x_0, t_0) \doteq B$ and $\varphi_{xx}(x_0, t_0) \doteq C$ set, for $\lambda > 0$ to be chosen later,

$$w(x, t) = \frac{x - x_0 + i(\varphi(x, t) - \varphi(x_0, t_0))}{1 + i\varphi_x(x_0, t_0)}$$

$$h(x, t) = w^2(x, t) - i\lambda w(x, t).$$

Note that $w(x_0, t_0) = 0$, $w_t(x_0, t_0) = 0$, $w_x(x_0, t_0) = 1$—which implies that $\Im w_x(x_0, t_0) = 0$—and it is also clear that $Lh = 0$. Let us write $u(x, t) \doteq \Re h(x, t)$, so

$$u(x, t) = (\Re w(x, t))^2 - (\Im w(x, t))^2 + \lambda \Im w(x, t)$$

and it follows that $u(x_0, t_0) = u_t(x_0, t_0) = u_x(x_0, t_0) = 0$. Then,

$$u_{xx}(x_0, t_0) = 2\,(\Re w_x(x_0, t_0))^2 + c\lambda\, C = 2 + c\lambda\, C,$$
$$u_{tt}(x_0, t_0) = \lambda\,\Im w_{tt}(x_0, t_0) = c\lambda\,\varphi_{tt}(x_0, t_0) = c\lambda\, A > 0,$$
$$u_{xt}(x_0, t_0) = c\lambda\,\varphi_{xt}(x_0, t_0) = c\lambda\, B,$$

where $c = (1 + \varphi_x^2(x_0, t_0))^{-1} > 0$, which shows that the Hessian of u at (x_0, t_0) is positive definite if $\lambda > 0$ is small enough. Then $\Re h$ has a strict local minimum at (x_0, t_0), i.e., the hypotheses of Lemma IV.4.2 are satisfied if we choose $K_1 = \{(x_0, t_0)\}$. If $\varphi_{tt}(x_0, t_0) = A < 0$ we reason similarly, taking $\lambda < 0$ and small.

The previous discussion shows that when looking for a homogeneous solution h whose real part assumes a local minimum over a compact set we may work under the assumption that

$$\varphi_t(x, t) = 0 \implies \varphi_{tt}(x, t) = 0. \tag{IV.77}$$

Assume that condition $(\mathcal{P})$ does not hold in any square centered at $(0, 0)$. Then given $\epsilon > 0$ we may find points (x_*, t_1), (x_*, t_2) in the cube Q centered at the origin with side length ϵ such that, say, $t_1 < t_2$, $\varphi_t(x_*, t_1) < 0$, and $\varphi_t(x_*, t_2) > 0$. We consider homogeneous solutions of the form

$$h(x, t; x_0) = (Z(x, t) - Z(x_0, 0))^2 - i\lambda\,\frac{Z(x, t) - Z(x_0, 0)}{Z_x(x_0, 0)}$$

and the difficulty is to show under assumption (IV.77) that for an appropriate choice of $|\lambda| \le 1$ and $|x_0| \le 1$ our function h assumes a local minimum over a compact set. Writing h in terms of its real and imaginary parts,

$$h(x, t; x_0) = u^{x_0}(x, t) + iv^{x_0}(x, t),$$

we obtain

$$u^{x_0}(x, t) = (x - x_0)^2 - [\varphi(x, t) - \varphi(x_0, 0)]^2$$
$$+ \lambda c[\varphi(x, t) - \varphi(x_0, 0) - \varphi_x(x_0, 0)(x - x_0)] \qquad \text{(IV.78)}$$

where $c = (1 + \varphi_x^2(x_0, 0))^{-1} > 0$. A straightforward computation shows that $\partial_x u^{x_0}(x_0, 0) = u_x^{x_0}(x_0, 0) = 0$. Since $u_{xx}^{x_0}(0, 0) = 2 + \lambda \varphi_{xx}(0, 0)$ we may assume, taking λ small but fixed and shrinking Q, that $u_{xx}^{x_0} > 0$ on $\overline{Q}$. Then the connected component γ_{x_0} that contains the point $(x_0, 0)$ of the level set

$$\{(x, t) : \quad u_x^{x_0}(x, t) = 0\}$$

is a smooth curve that intersects transversally the x-axis at $(x_0, 0)$. Hence, the curves γ_{x_0} foliate a neighborhood of the origin and shrinking $\epsilon > 0$ if necessary we may assume $\bigcup_{|x_0| \le \epsilon} \gamma_{x_0} \supset Q$. From now on we will assume that $|x_0| \le \epsilon$. Note that the vector field

$$\ell = \frac{\partial}{\partial t} - \frac{u_{xt}^{x_0}}{u_{xx}^{x_0}} \frac{\partial}{\partial x} \qquad \text{(IV.79)}$$

is tangent to the curve γ^{x_0} along γ^{x_0} so this curve may be realized as the graph of a function $x = x^{x_0}(t)$, $|t| < \epsilon_0$. Let us take a closer look at the behavior of u^{x_0} on the curve γ_{x_0}. For any $(x', t') \in \gamma_{x_0}$ we have that $u_x^{x_0}(x', t') = 0$ and $u_{xx}^{x_0}(x', t') > 0$ so $x \mapsto u^{x_0}(x, t')$ attains a strict minimum precisely at $x = x'$ (geometrically, the graph of $x \mapsto u^{x_0}(x, t')$ looks like a parabola pointing upwards with vertex at x'). Hence, there is a tubular neighborhood V of γ_{x_0} such that

$$\min_V u^{x_0}(x, t) = \min_{\gamma_{x_0}} u^{x_0}(x, t).$$

Thus, if we can find points $(x_1', t_1'), (x_0', t_0'), (x_2', t_2')$ in γ_{x_0} such that $t_1' < t_0' < t_2'$ and

$$u^{x_0}(x_1', t_1') > u^{x_0}(x_0', t_0'),$$
$$u^{x_0}(x_2', t_2') > u^{x_0}(x_0', t_0')$$

it follows that there is a compact set $K \subset \gamma^{x_0}$ such that $u^{x_0}(x, t)$ assumes a local minimum over K. To study the variation of u^{x_0} along γ_{x_0} we consider the parameterization $\gamma_{x_0}(s) = (x^{x_0}(s), s)$ and differentiate

$$u^{x_0}(x^{x_0}(s), s)$$

with respect to s. Since $u_x^{x_0}(x^{x_0}(s), s) \equiv 0$, we obtain

$$\frac{d}{ds} u^{x_0} \circ \gamma_{x_0}(s) = u_t^{x_0} \circ \gamma_{x_0}(s) = [\varphi_t (c\lambda - 2(\varphi - \varphi(x_0, 0))^2)] \circ \gamma_{x_0}(s).$$

Shrinking $\epsilon < \epsilon_0$ we may assume that $2|\varphi(x, t) - \varphi(x_0, 0)|^2 < c|\lambda|/2$. Thus, u^{x_0} is *monotone along* γ_{x_0} *if and only if* φ_t *does not change sign on* γ_{x_0}. Hence, if for some curve γ_{x_0} we find points (x_1', t_1'), (x_2', t_2') in γ_{x_0} such that $t_1' < t_2'$, $\varphi_t(x_1', t_1') < 0$, $\varphi_t(x_2', t_2') > 0$, then for $\lambda > 0$ and small the curve γ_{x_0} will contain a compact subset K over which u^{x_0} assumes a local minimum; if, instead, $\varphi_t(x_1', t_1') > 0$ and $\varphi_t(x_2', t_2') < 0$ we take $\lambda < 0$ in the definition of h to achieve the desired homogeneous solution. To see that such γ_{x_0} exists, consider the quadrilateral $Q^\flat$ having as horizontal sides the segments $t = \pm\epsilon$ and as 'vertical' sides the curves γ_{x_0} with $x_0 = \pm\epsilon$. Then $Q^\flat$ is the union of the curves γ_{x_0}, $-\epsilon < x_0 < \epsilon$. Assume by contradiction that φ_t does not change sign along any of these curves. We may decompose $Q^\flat$ into three disjoint sets: the union $Q_+^\flat$ of the curves γ_{x_0} that contain at least one point on which $\varphi_t > 0$, the union $Q_-^\flat$ of the curves γ_{x_0} that contain at least one point on which $\varphi_t < 0$, and the union $Q_0^\flat$ of the curves γ_{x_0} on which φ_t vanishes identically. Observe that $Q_+^\flat$ and $Q_-^\flat$ are open sets and neither $Q_+^\flat$ nor $Q_-^\flat$ can be empty, for this would imply that φ_t does not change sign on some square containing the origin and condition $(\mathcal{P})$ would be satisfied in that square, contradicting our assumptions. Since $Q_+^\flat$ and $Q^\flat \setminus Q_+^\flat$ are invariant sets (i.e., they are a union of the curves γ_{x_0} that intersect them) so is the boundary of $Q_+^\flat$. Let p be a boundary point of $Q_+^\flat$ and let γ_{x_0} be the curve passing through p. We claim that γ_{x_0} is a vertical segment. Indeed, $\gamma_{x_0} \subset Q_0^\flat$ since it cannot meet $Q_+^\flat \cup Q_-^\flat$. So φ_t vanishes identically on γ_{x_0} and also does φ_{tt} because of (IV.77). Let $q \in \gamma_{x_0}$. If $\varphi_{xt}^{x_0}(q) \neq 0$ the set $S = \{\varphi_x = 0\}$ is a smooth curve in a neighborhood of q and since $\varphi_{tt} = 0$ on S we conclude that the intersection of S with a neighborhood of q must be a vertical segment, in particular, the tangent to γ_{x_0} at q is vertical. Assume now that $\varphi_{xt}^{x_0}(q) = 0$. Differentiating twice (IV.78), first with respect to x, then with respect to t and evaluating the result at q we get $u_{xt}^{x_0}(q) = 0$ because $\varphi_t(q) = \varphi_{xt}(q) = 0$. Then the vector field ℓ given by (IV.79) reduces to ∂_t at q. Thus the velocity vector of γ_{x_0} is always vertical and γ_{x_0} is itself the vertical segment $\{x_0\} \times (-\epsilon, \epsilon)$.

Let us return to the points (x_*, t_1), (x_*, t_2) in the cube Q centered at the origin with side length ϵ such that $t_1 < t_2$, $\varphi_t(x_*, t_1) < 0$ and $\varphi_t(x_*, t_2) > 0$. Then trivially $(x_*, t_1) \in Q_-^\flat$ and $(x_*, t_2) \in Q_+^\flat$ so there exists a point $(x_*, t_0) \in \partial Q_+^\flat$ such that $t_1 < t_0 < t_2$. But, as we have seen, this implies that $\gamma_{x_*} = \{x_*\} \times (-\epsilon, \epsilon)$ and $\varphi_t(x_*, t) = 0$ for $|t| < \epsilon$, which is a contradiction. Thus, for some $|x_0| < \epsilon$, φ_t assumes opposite signs on γ_{x_0}, u^{x_0} is not monotone on γ_{x_0}, and $h(x, t; x_0)$ is a homogeneous solution whose real part assumes a local minimum over a compact set.

Essentially the same approach works in a higher number of variables although the proofs are technically more involved. The following elementary lemma about real quadratic forms in $\mathbb{R}^2$ will be useful:

LEMMA IV.4.3. *Assume that the real quadratic form*

$$q_1(x, y) = Ax^2 + 2Bxy + Cy^2, \qquad (x, y) \in \mathbb{R}^2, A, B, C \in \mathbb{R}$$

has positive trace $A + C > 0$ and set

$$q_2(x, y) = \Re\left[\left(\frac{C-A}{2} + iB\right)(x+iy)^2\right] = \frac{C-A}{2}(x^2 - y^2) - 2Bxy.$$

Then

$$q_1(x, y) + q_2(x, y) = \frac{A+C}{2}(x^2 + y^2)$$

is diagonal and positive definite.

PROOF. The assertion is self-evident. $\qquad\square$

We consider a vector field L given by (IV.51) defined on

$$\Omega = B \times (-T, T) \subset \mathbb{R}^n \times \mathbb{R}, \qquad B = \{x \in \mathbb{R}^n : \ |x| < \delta\}$$

and assume that there exist n first integrals $Z_1, \ldots, Z_n$, $LZ_j = 0$, $j = 1, \ldots, n$, with $dZ_1, \ldots, dZ_n$ linearly independent in Ω. We write

$$Z = (Z_1, \ldots, Z_n)$$

and further assume that $\det(Z_x) \neq 0$ in Ω, $Z(0, 0) = 0$ and $Z_x(0, 0) = I$. We also use the notation

$$\vec{b}(x, t) = (b_1(x, t), \ldots, b_n(x, t)).$$

LEMMA IV.4.4. *Assume that there exists $(x_0, t_0) \in \Omega$ and $\xi \in \mathbb{R}^n$ such that*

(i) $\vec{b}(x_0, t_0) \cdot \xi = 0$;
(ii) $\vec{b}_t(x_0, t_0) \cdot \xi \neq 0$.

Then there exists $f \in C_c^\infty(\Omega)$ such that $Lu \neq f$ for all $u \in \mathcal{D}'(\Omega)$.

PROOF. By Lemma IV.4.2 we need only show that there exists a solution h of $Lh = 0$ such that $\Re h$ assumes a local minimum at $p = (x_0, t_0)$. Set $Z' = (Z'_1, \ldots, Z'_n) = Z_x^{-1}(x_0, t_0)[Z(x, t) - Z(x_0, t_0)]$. Then $LZ'_j = 0$, $j = 1, \ldots, n$, $Z'(p) = 0$, $Z'_x(p) = I$. Then, the change of coordinates $x' = x - x_0$, $t' = t - t_0$, shows that there is no loss of generality in assuming from the start that $(x_0, t_0) = (0, 0)$. Write $\Phi_j(x, t) = Z_j(x, t) - x_j$, so $\Phi_j(0, 0) = \partial_x \Phi_j(0, 0) = 0$, $j = 1, \ldots, n$. Set

$$W(x, t) = Z(x, t) \cdot \xi = \sum_{j=1}^{n} \xi_j Z_j(x, t).$$

Then $LW = 0$ and in view of (i) we get

$$0 = LW(0, 0) = \sum_{j=1}^{n} \xi_j \left(\frac{\partial \Phi_j}{\partial t}(0, 0) + i b_j(0, 0) \right) = i \Phi_t(0, 0) \cdot \xi$$

where $\Phi = (\Phi_1, \ldots, \Phi_n)$. Hence, $\Phi(0, 0) = \Phi_t(0, 0) = \Phi_x(0, 0) = 0$. We distinguish two cases.

Case 1. $\vec{b}(0, 0) = 0$. Differentiating with respect to t the equation $LW = 0$ we obtain $\Phi_{tt}(0, 0) \cdot \xi + i \vec{b}_t(0, 0) \cdot \xi = 0$ and using (ii) we derive

$$\Im \Phi_{tt}(0, 0) \cdot \xi \neq 0.$$

Set

$$h(x, t) = Z_1^2(x, t) + \cdots + Z_n^2(x, t) - i\lambda W(x, t),$$
$$u(x, t) \doteq \Re h(x, t) = |x + \Re \Phi|^2 - |\Im \Phi(x, t)|^2 + \lambda \Im \Phi(x, t) \cdot \xi.$$

Thus, $u(0, 0) = 0$, $u_t(0, 0) = 0$, $\nabla_x u(0, 0) = 0$ and if we choose λ with the same sign as $\gamma = \Im \Phi_{tt}(0, 0) \cdot \xi$ it follows that the Taylor series of u at the origin is

$$u(x, y) = x_1^2 + \cdots + x_n^2 + |\lambda \gamma| t^2 + \lambda \sum_{j=1}^{n} c_j x_j t + \cdots$$

where the dots indicate terms of order > 2. Thus, the Hessian of u at the origin with respect to (x, t) is positive definite and u has a strict local minimum at the origin for $|\lambda|$ small.

Case 2. $\vec{b}_j(0, 0) \neq 0$ for some $1 \leq j \leq n$. After a linear change in the x-variables we may assume that

$$\begin{cases} b_1(0, 0) = 1, \\ b_j(0, 0) = 0, \quad j = 2, \ldots, n, \\ \xi = (0, \xi_2, \ldots, \xi_n). \end{cases}$$

Since (ii) implies that $\xi \neq 0$ this case can only occur if $n \geq 2$. Set

$$W(x, t) = iZ(x, t) \cdot \xi = i \sum_{j=2}^{n} \xi_j Z_j(x, t).$$

Proceeding as in Case 1 we obtain $\Re W_t(0,0) = \Re W_{x_j}(0,0) = 0$, for $j = 1, \ldots, n$. Differentiating the equation $LW = 0$ with respect to t we obtain

$$\partial_{tt}^2 W(0,0) + i\partial_{tx_1}^2 W(0,0) = \vec{b}_t(0,0) \cdot \xi \neq 0$$

while differentiation with respect to x_1 gives

$$\partial_{tx_1}^2 W(0,0) + i\partial_{x_1x_1}^2 W(0,0) = 0.$$

Using both equations to eliminate the term $\partial_{tx_1}^2 W(0,0)$ and replacing ξ by $-\xi$ if necessary we obtain

$$\partial_{tt}^2 \Re W(0,0) + \partial_{x_1x_1}^2 \Re W(0,0) \doteq 2\gamma > 0.$$

Applying Lemma IV.4.3 to the quadratic form

$$q_1(x_1, t) = \partial_{tt}^2 \Re W(0,0)t^2 + 2\partial_{tx_1}^2 \Re W(0,0)tx_1 + \partial_{x_1x_1}^2 \Re W(0,0)x_1^2$$

we find a complex number ζ such that $q_1(x_1, t) + \Re[\zeta(x_1 + it)^2]$ is positive definite. Since $\partial_{x_1} Z_1(0,0) = 1$ and it follows from $LZ_1(0) = 0$ that $\partial_t Z_1(0,0) = i$ the Taylor expansion in the variables (x_1, t) of ζZ_1^2 is

$$\zeta Z_1^2(x_1, 0, \ldots, 0, t) = \Re[\zeta(x_1 + it)^2] + \cdots$$

Thus,

$$\Re(W + \zeta Z_1^2)(x_1, 0, \ldots, 0, t) = \gamma(t^2 + x_1^2) + \cdots$$

If we now set

$$h(x, t) = Z_1^2(x, t) + \cdots + Z_{n-1}^2(x, t) + \lambda\left(W(x, t) + \zeta Z_1^2\right),$$
$$u(x, t) \doteq \Re h(x, t)$$

we may check as in case (i) that $Lh = 0$ and that for $\lambda > 0$ small $u = \Re h$ has a positive definite Hessian at the origin. $\qquad\square$

REMARK IV.4.5. Lemma IV.4.4 has the following geometric interpretation. Writing $L = X + iY$ with X and Y real we have that $X = \partial_t$, $Y = \vec{b}$, and $[X, Y] = \vec{b}_t$. Then conditions (i) and (ii) at $p = (x_0, t_0)$ mean that $[X, Y](p)$, $X(p)$, and $Y(p)$ are not linearly dependent. Indeed, if $AX(p) + BY(p) + C[X, Y](p) = 0$, the obvious fact $\vec{b} \cdot X = \vec{b}_t \cdot X = 0$ implies that $A = 0$ so $[X, Y](p)$ and $Y(p)$ would be collinear, contradicting (i) and (ii). This implies that the orbit Σ of the pair of vectors $\{X, Y\}$ that passes through p cannot have dimension ≤ 2. In fact, the three vectors $[X, Y](p)$, $X(p)$, and $Y(p)$ belong to $T_p(\Sigma)$ so $\dim \Sigma \leq 2$ would force a linear relationship between them. Hence, (i) and (ii) of Lemma IV.4.4 imply that $\dim \Sigma \geq 3$, which violates (1) of condition $(\mathcal{P})$ in Definition IV.3.2.

In order to find a solution h of $Lh = 0$ with the property that its real part assumes a local minimum over a compact set we need only worry about those cases not covered by Lemma IV.4.4, i.e., we may always assume that

$$\varphi_t(x, t) \cdot \xi = 0 \implies \varphi_{tt}(x, t) \cdot \xi = 0, \quad (x, t) \in \Omega, \quad \xi \in \mathbb{R}^n. \qquad \text{(IV.80)}$$

Let us assume that L does not satisfy condition $(\mathcal{P})$ in any cube centered on the origin and let us try to produce the required homogeneous solution h. As in the case of two variables we will look for solutions $h = u + iv$ such that the Hessian matrix u_{xx} is everywhere positive definite and the critical points of $x \mapsto u(x, t)$ are located on a certain curve γ so that when looking for a local minimum of u we only need to direct our attention to the restriction $u|_\gamma$. Then, assuming by contradiction that u is monotone on γ and that this happens for all the functions u of this type, we must conclude that L is forced to satisfy $(\mathcal{P})$ in some neighborhood of the origin. The first step is then to show the abundance of solutions of this type, which is taken care of by the next lemma that describes a family of solutions depending on two parameters, $x_0 \in B$ and $\eta \in \mathbb{R}^n$. The general form of these solutions is based on the function h introduced in case (i) of Lemma IV.4.4.

LEMMA IV.4.6. *If T and δ are small enough there exists a smooth function $h \in C^\infty(\Omega \times B \times \mathbb{R}^n)$,*

$$h(x, t; x_0, \eta) = u(x, t; x_0, \eta) + iv(x, t; x_0, \eta)$$

with u and v real such that

(i) $Lh = 0$ *in Ω for all $(x_0, \eta) \in B \times \mathbb{R}^n$;*

(ii) $u_x(x_0, 0; x_0, \eta) = 0$ *and* $v_x(x_0, 0; x_0, \eta) = \eta$;

(iii) $u_{xx}(x, t; x_0, \eta)$ *is positive definite at all points $(x, t) \in \Omega$ for all $(x_0, \eta) \in B \times \mathbb{R}^n$.*

PROOF. Set

$$h(x, t; x_0, \eta) = \lambda \, (1 + |\eta|^2)^{1/2} \sum_{j=1}^{k} (Z(x, t) - Z(x_0, 0))^2$$
$$+ i\eta \cdot Z_x^{-1}(x_0, 0)[Z(x, t) - Z(x_0, 0)].$$

Since h is a polynomial in $Z_1, \ldots, Z_n$ it is apparent that (i) holds. Differentiating h with respect to x and evaluating the result at $(x, t) = (x_0, 0)$ we get $h_x(x_0, 0, x_0, \eta) = i\eta$ which shows (ii). Finally, write $(1 + |\eta|^2)^{-1/2} u = F = \lambda f + g$. Then f is independent of η and $f_{xx}(0, 0, 0, \eta) = 2\lambda I, I = $ identity matrix, so f_{xx} has n eigenvalues $\geq \lambda > 0$ on $\Omega \times B \times \mathbb{R}^n$ if T and δ are chosen small.

Since g_{xx} is uniformly bounded in $\Omega \times B \times \mathbb{R}^n$, taking λ large we obtain that F_{xx} is positive definite in $\Omega \times B \times \mathbb{R}^n$, which implies the positivity of u_{xx}. $\qquad \square$

We regard the function h defined in Lemma IV.4.6 primarily as a function in the variables (x, t) that depends on the parameters (x_0, η), whose geometric meaning is furnished by (ii). To the function h we associate the real vector field V defined for $(x, t, x_0, \eta, \xi) \in \Omega \times B \times \mathbb{R}^n \times \mathbb{R}^n$ by

$$V = \frac{\partial}{\partial t} - A(x, t, x_0, \eta) \cdot \frac{\partial}{\partial x} + B(x, t, x_0, \eta) \cdot \frac{\partial}{\partial \xi}$$

where $A = (A_1, \ldots, A_n)$, $B = (B_1, \ldots, B_n)$ are defined by

$$A = u_{xx}^{-1} u_{xt},$$
$$B = v_{tx} - v_{xx}A.$$

Note that the jth component of Vu_x is $(Vu_x)_j = u_{tx_j} - (u_{xx}A)_j = u_{tx_j} - u_{tx_j} = 0$, $j = 1, \ldots, n$ so u_x is constant along the integral curves of V. A similar computation shows that $(V(\xi - v_x))_j = 0$, $j = 1, \ldots, n$ so $\xi - v_x$ is also constant along the integral curves of V. It follows that V is tangent to the submanifold of $\Omega \times B \times \mathbb{R}^n \times \mathbb{R}^n$ of dimension $2n + 1$

$$\Sigma = \{(x, t, x_0, \eta, \xi): \quad u_x(x, t; x_0, \eta) = 0, \quad \xi = v_x(x, t; x_0, \eta)\}.$$

Since $(x_0, 0, x_0, \eta, \eta) \in \Sigma$ by (ii) of Lemma IV.4.3 the partial derivative of

$$(x, t, x_0, \eta, \xi) \mapsto (u_x(x, t, x_0, \eta), \xi - v_x(x, t, x_0, \eta))$$

with respect to (x_0, η) at $(0, 0, 0, 0, 0, 0)$ is the identity. Thus, Σ may be parameterized by (x, t, ξ) for $|x| < \delta_1$ $|t| < T_1$, $|\xi| < \delta_1$ as the graph of a smooth map

$$(x, t, \xi) \mapsto (x_0(x, t, \xi), \eta(x, t, \xi))$$

with values in $\{|x_0| < \delta_2\} \times \{|\eta| < \delta_2\}$. We may assume, if δ and T are further shrunken, that the image of $|x| < \delta_1$ $|t| < T_1$, $|\xi| < \delta_1$ by the map

$$(x, t, \xi) \mapsto (x_0(x, t, \xi), t, \eta(x, t, \xi))$$

covers $\Omega \times B \times \{|\xi| < \delta\}$. Thus, the vector field

$$V_* = \frac{\partial}{\partial t} - \alpha(x, t, \xi) \cdot \frac{\partial}{\partial x} + \beta(x, t, \xi) \cdot \frac{\partial}{\partial \xi} \qquad \text{(IV.81)}$$

where

$$\alpha(x, t, \xi) = A(x, t, x_0(x, t, \xi), \eta(x, t, \xi)),$$
$$\beta(x, t, \xi) = B(x, t, x_0(x, t, \xi), \eta(x, t, \xi))$$

agrees with V on Σ—in particular, V_* is tangent to Σ—and its coefficients do not depend on x_0 and η. Fix $x_0 \in B$ and $|\eta| < \delta$ and consider the function of $u(x, t, x_0, \eta)$ as a function of (x, t). By (iii) of Lemma IV.4.6 the roots of the equations $u_x(x, t; x_0, \eta) = 0$, $\xi - v_x(x, t; x_0, \eta) = 0$ determine a smooth curve $\tilde{\gamma}_{x_0\eta}$ in (x, t, ξ)-space contained in Σ that passes through the point $(x_0, 0, \eta)$. The curves $\tilde{\gamma}_{x_0\eta}$ may be parameterized as $\tilde{\gamma}_{x_0\eta}(s) = (x(s; x_0, \eta), s, \xi(s; x_0, \eta))$ and they foliate Σ as x_0, η vary. The vector field V_* is tangent to $\tilde{\gamma}_{x_0\eta}$ at every point of $\tilde{\gamma}_{x_0\eta}$ so we may parameterize $\tilde{\gamma}_{x_0\eta}$ so that its velocity vector is V_*. The projection of $\tilde{\gamma}_{x_0\eta}$ on (x, t)-space gives a curve $\gamma_{x_0\eta}$ passing through $(x_0, 0)$ on which u_x vanishes and u_{xx} is positive definite. Hence, there is a tubular neighborhood V of $\gamma_{x_0\eta}$ such that

$$\min_V u(x, t; x_0, \eta) = \min_{\gamma_{x_0\eta}} u(x, t; x_0, \eta).$$

Thus, if the restriction of u to $\gamma_{x_0\eta}$ assumes a local minimum over a compact subarc K of $\gamma_{x_0\eta}$ we will also have that u itself assumes a local minimum over K. In order for the restriction of u to $\gamma_{x_0\eta}$ to assume a local minimum over a compact subarc K we must find points $t_1 < t_2$ such that

$$\frac{d}{ds}[u(x(s; x_0, \eta), s)](t_1) < 0 \quad \text{and} \quad \frac{d}{ds}[u(x(s; x_0, \eta), s)](t_2) > 0.$$

Now, writing $x(s; x_0, \eta) = x(s)$ and $\xi(s; x_0, \eta) = \xi(s)$ to simplify the notation,

$$\frac{d}{ds}u(x(s), s) = u_x(x(s), s; x_0, \eta)\frac{d}{ds}x(s) + u_t(x(s), s; x_0, \eta)$$
$$= u_t(x(s), s; x_0, \eta)$$
$$= \vec{b}(x(s), s) \cdot v_x(x(s), s; x_0, \eta)$$
$$= \vec{b}(x(s), s) \cdot \xi(s).$$

Note that the identity $u_x = \vec{b} \cdot v_x$ is just the real part of the equation $Lh = 0$. This reduced the problem of finding a homogeneous solution $h = u + iv$ whose real part assumes a local minimum over a compact set for an appropriate choice of (x_0, η) to the problem of finding a curve $\tilde{\gamma}_{x_0\eta}$ such that the function $q(x, t, \xi) = \vec{b}(x, t) \cdot \xi$ changes from negative to positive along $\tilde{\gamma}_{x_0\eta}$. Thus, from the fact that $(\mathcal{P})$ is not satisfied in any neighborhood of the origin—which amounts to saying that any cube centered at the origin contains an integral curve of $X = \partial_t$ along which $q(x, t, \xi)$ changes sign—we must derive that there exists an integral curve of V_* along which $q(x, t, \xi)$ changes sign. The tool to compare the changes of sign of a function along the integral curves of two different vector fields is provided by

LEMMA IV.4.7. *Let $U \subset \mathbb{R}^N$ be an open set, X and V_* Lipschitz vector fields in U and $q \in C^1(U)$ a real function such that*

(1) $q(x) = 0$ implies $Xq(x) \leq 0$;
(2) $q(x) = 0$ and $dq(x) = 0$ imply that $X(x) = V_*(x)$.

Assume that the integral curves γ of V_* have the following property:

(•) *if $q(x) < 0$ for some $x \in \gamma$ then $q(y) \leq 0$ for all points $y \in \gamma$ that lie ahead of x in the order determined by the flow.*

Then, the integral curves of X also satisfy property (•).

We postpone the proof of Lemma IV.4.7 and continue our reasoning. We apply the lemma with U given by $|x| < \delta_1$, $|t| < T_1$, $|\xi| < \delta_1$, $|x_0| < \delta$, $|\eta| < \delta$, $N = 4n + 1$, $X = \partial_t$, V_* given by (IV.81) and $q(x, t, \xi) = \vec{b}(x, t) \cdot \xi$. Let us check that hypotheses (1) and (2) in the lemma are satisfied. From (IV.80) we get (1). Assume now that $q(x, t, \xi) = dq(x, t, \xi) = 0$ at some point (x, t, ξ). Since q is independent of (x_0, η) we may say q and dq vanish at $p = (x, t, x_0, \eta, \xi) \in \Sigma$, $x_0 = x_0(x, t, \xi)$, $\eta = \eta(x, t, \xi)$ and since $V_* = V$ on Σ and X and V_* do not depend on (x_0, η) we need only prove that $V(p) = X(p)$. From $q(x, t, \xi) = dq(x, t, \xi) = 0$ we derive that $\vec{b}(x, t) = 0$, $\vec{b}_t(x, t) \cdot \xi = 0$, $\vec{b}_{x_j}(x, t) \cdot \xi = 0$, $j = 1, \dots, n$. The real part of $Lh = 0$ is $u_t = \vec{b} \cdot \nabla v$ which, differentiated with respect to x_j, gives $u_{tx_j}(x, t, \xi) = 0$, so the coefficient A_j of $\partial/\partial x_j$ in V satisfies $A_j(x, t, x_0, \eta) = 0$. Similarly, differentiating $v_t + \vec{b} \cdot \nabla u = 0$ we get that $B = v_{xt} - v_{xx}A = v_{xt} = -\vec{b}_x \cdot u_x - u_{xx}(\vec{b}) = 0$ at (x, t, ξ) so $V_*(p) = V(p) = \partial_t = X(p)$ which proves (2). Since L does not satisfy $(\mathcal{P})$ there is an integral curve of X contained in U on which q changes sign from minus to plus. Then, by Lemma IV.4.7, V_* cannot possess property (•) showing the existence of a curve $\tilde{\gamma}_{x_0 \eta}$ along which $\vec{b} \cdot \xi$ changes sign from minus to plus as required to show that $u(x, t; x_0, \eta)$ assumes a local minimum over a compact set of Ω, which, by Lemma IV.4.2 implies that L is not solvable in Ω. Summing up,

THEOREM IV.4.8. *Assume that L, given by (IV.49), is locally solvable in Ω. Then every point $p \in \Omega$ has a neighborhood U such that L satisfies condition $(\mathcal{P})$ in U.*

To complete the proof of the theorem we must prove Lemma IV.4.7.

We start by recalling that if $f : (a, b) \to \mathbb{R}$ is a continuous function we define

$$D^+ f(x) = \limsup_{\varepsilon \searrow 0} \frac{f(x + \varepsilon) - f(x)}{\varepsilon}$$

which may vary in the range $[-\infty, \infty]$. The mean value inequality states that if $f \in C^0[a, b]$ there exists $c \in (a, b)$ such that $f(b) - f(a) \leq D^+ f(c)(a - b)$. If $f(a) = f(b)$ it is enough to choose $c \in (a, b)$ so that $f(c) = \inf f(x)$ and the general case is reduced to this one by subtracting the affine function $f(a) + (x - a)(f(b) - f(a))/(b - a)$. It follows that if $D^+ f(x) \leq 0$, $x \in (a, b)$, then $f(x)$ is monotone nonincreasing.

Let V be a Lipschitz vector field in $U \subset \mathbb{R}^N$, that is, $|V(x) - V(y)| \leq K|x - y|$, $x, y \in U$. We denote by $\Phi_t(x)$, the forward flow of V stemming from x, i.e., the solution $\Phi_t(x)$ defined in a maximal interval $0 \leq t < T(x)$ of the ODE

$$\begin{cases} \frac{d}{dt}\Phi_t(x) = V(\Phi_t(x)), \\ \quad\;\; \Phi_t(0) = x. \end{cases}$$

Let $F \subset U$ be a closed set. We say that F is positively V-invariant, or just V-invariant for brevity, if

$$x \in F \implies \Phi_t(x) \in F \quad \text{for all} \quad t \in [0, T(x)).$$

The characterization of V-invariant sets given below is due to Brézis ([**Br**]).

The following properties are equivalent:

(i) F is positively V-invariant;

(ii) $\forall x \in F, \quad \lim_{\varepsilon \searrow 0} \dfrac{\text{dist}\,(x + \varepsilon V(x), F)}{\varepsilon} = 0.$

Indeed, assume (i). Then

$$\frac{\text{dist}\,(x + \varepsilon V(x), F)}{\varepsilon} \leq \frac{|x + \varepsilon V(x) - \Phi_\varepsilon(x)|}{\varepsilon}$$

$$= \left| V(x) - \frac{\Phi_\varepsilon(x) - x}{\varepsilon} \right|$$

and the right-hand side converges to 0 as $\varepsilon \searrow 0$.

Conversely, assume that (ii) holds. To prove (i) it is enough to show that the Lipschitz continuous function $f : [0, T(x)) \to [0, \infty)$ defined by

$$f(t) = \text{dist}\,(\Phi_t(x), F)$$

vanishes identically. This will follow if we prove that $e^{-At} f(t)$ is nonincreasing for some $A > 0$, since $f(0) = 0$. Thus, it is enough to show that

$$D_t^+ (e^{-At} f)(t) \leq e^{-At} (D_t^+ f(t) - A f(t)) \leq 0,$$

which in turn is implied by $D_t^+ f(t) \leq A f(t)$. Fix $t \in (0, T(x))$ and choose $z_t \in F$ such that $f(t) = |\Phi_t(x) - z_t|$. For small $\varepsilon > 0$ we have

$$f(t + \varepsilon) = \text{dist}\,(\Phi_{t+\varepsilon}(x), F)$$

$$\leq |\Phi_{t+\varepsilon}(x) - \Phi_\varepsilon(z_t)| + |\Phi_\varepsilon(z_t) - z_t - \varepsilon V(z_t)|$$
$$+ \operatorname{dist}(z_t + \varepsilon V(z_t), F).$$

Now $|\Phi_{t+\varepsilon}(x) - \Phi_\varepsilon(z_t)| = |\Phi_\varepsilon(\Phi_t(x)) - \Phi_\varepsilon(z_t)|$, so by Gronwall's inequality,

$$|\Phi_\varepsilon(\Phi_t(x)) - \Phi_\varepsilon(z_t)| \leq e^{K\varepsilon}|\Phi_t(x) - z_t| = e^{K\varepsilon} f(t),$$

for $\varepsilon > 0$ small, where K is the Lipschitz constant of V. Thus,

$$\frac{f(t+\varepsilon) - f(t)}{\varepsilon} \leq \frac{(e^{K\varepsilon} - 1) f(t)}{\varepsilon}$$
$$+ \left| \frac{\Phi_\varepsilon(z_t) - z_t}{\varepsilon} - V(z_t) \right| + \frac{\operatorname{dist}(z_t + \varepsilon V(z_t), F)}{\varepsilon}$$

and letting $\varepsilon \searrow 0$ we get $D_t^+ f(t) \leq K f(t)$, since the right-hand side's middle term obviously $\to 0$ and the last one also does because we are assuming that (ii) holds. This shows that $e^{-Kt} f(t)$ is nonincreasing and proves (i).

We now prove Lemma IV.4.7.

PROOF. Let U^- be the V_*-flow out of the set $\{x \in U : q(x) < 0\}$, i.e., a point $x \in U^-$ if $x = \Phi_t(y)$ for some $y \in U$ with $q(y) < 0$ and $0 \leq t < T(y)$, where Φ_t is the flow of V_*. Hence, U^- is an open set and $\{q(x) < 0\} \subset U^- \subset \{q(x) \leq 0\}$ because of ($\bullet$). By its very definition, U^- is positively V_*-invariant and so is its closure $F = \overline{U^-}$. Indeed, if $x \in F$ there exist a sequence $(x_j) \subset U^-$ such that $x_j \to x$. If $0 < t < T(x)$, then $0 < t < T(x_j)$ for large j because $s \mapsto T(x)$ is lower semicontinuous. Then $\Phi_t(x_j) \in U^-$ by the V_*-invariance of U^- and $\Phi_t(x) = \lim_j \Phi_t(x_j) \in \overline{U^-}$.

To prove the lemma we will show that F is X-invariant, which clearly implies that X has property ($\bullet$) because $F \subset \{q(x) \leq 0\}$. We must show that

$$\lim_{\varepsilon \searrow 0} \frac{\operatorname{dist}(x + \varepsilon X(x), F)}{\varepsilon} = 0, \qquad x \in F. \qquad \text{(IV.82)}$$

If $q(x) < 0$ this is trivially true, since $x + \varepsilon X(x) \in F$ for small $\varepsilon > 0$. If $q(x) = dq(x) = 0$, (2) implies that $X(x) = V_*(x)$ so

$$\frac{\operatorname{dist}(x + \varepsilon X(x), F)}{\varepsilon} = \frac{\operatorname{dist}(x + \varepsilon V_*(x), F)}{\varepsilon}$$

and the right-hand side $\to 0$ as $\varepsilon \searrow 0$ because F is V_*-invariant.

If $q(x) = 0$ and $dq(x) \neq 0$, the set $\{q(y) = 0\} \cap W$ is a C^1 manifold where W is a convenient ball centered at x. It is easy to find a smooth unit vector field $N(y)$ that meets $\{q(y) = 0\}$ transversally and points toward $\{q(y) < 0\}$, so $Nq < 0$ on $W \cap \{q(y) = 0\}$. Let $\Psi_t(y, \lambda)$ denote the flow of the vector $X^\lambda \doteq X + \lambda N$, $\lambda \geq 0$. Then, (1) implies that $X^\lambda q(y) < 0$ on $W \cap \{q(y) = 0\}$ for any $\lambda > 0$.

Note that no integral curve of X^λ, $\lambda > 0$, that stems from a point in $W \cap \{q(y) < 0\}$ can cross $W \cap \{q(y) = 0\}$ (this would amount to traveling against the flow at $W \cap \{q(y) = 0\}$) and this implies that $q(\Psi_t(x, \lambda)) < 0$ for $\lambda > 0$, $t > 0$ small, in particular $\Psi_t(x, \lambda) \in U^-$. Hence, $\Psi_t(x, 0) = \lim_{\lambda \searrow 0} \Psi_t(x, \lambda) \in F$ where the limit holds by the continuous dependence on the parameter λ. Thus, the flow $\Psi_t(x, 0)$ of X does not exit F for small values of $t > 0$, which easily implies (IV.82), as in the proof of '(i) $\Longrightarrow$ (ii)' of the characterization of flow-invariant sets. $\square$

Notes

A few years after the publication of Hans Lewy's example [**L1**], Hörmander ([**H6**], [**H7**]) shed new light on the nonsolvability phenomenon explaining it in a novel way. Although his results are set in the framework of general order operators of principal type we will describe its consequences for vector fields. He proved that if a (nonvanishing) vector field L is locally solvable in Ω then the principal symbol of the commutator $[L, \overline{L}]$ between L and its conjugate must vanish at every zero of the principal symbol $\ell(x, \xi)$ of L. A vector field with this property is said to satisfy condition $(\mathcal{H})$. For the Lewy operator condition $(\mathcal{H})$ is violated at every point. If the coefficients of L are real or constant $[L, \overline{L}]$ vanishes identically. This was a most remarkable advance because it explained a phenomenon that had appeared as an isolated example in terms of very general geometric properties of the symbol, an invariantly defined object. However, it turns out that condition $(\mathcal{H})$ does not tell apart the solvable vector fields from the nonsolvable ones among some examples considered by Mizohata ([**M**]), which we now describe. Let k be a positive integer and consider the vector field in $\mathbb{R}^2$ defined by

$$M_k = \frac{\partial}{\partial y} - iy^k \frac{\partial}{\partial x}.$$

If $k = 1$ condition $(\mathcal{H})$ is violated at all points of the x-axis so, in particular, M_1 is not locally solvable at the origin. For $k \geq 2$ condition $(\mathcal{H})$ is satisfied everywhere. On the other hand, it follows from relatively simple arguments that M_k is locally solvable at the origin if and only if k is even ([**Gr**], [**Ga**]). The principal symbol of M_k is $m_1 = -i(\eta - iy^k\xi)$. The crucial difference between k odd and k even is that in the first case the function y^k changes sign and in the second case it doesn't. Nirenberg and Treves ([**NT**]) elaborated these examples and identified a property that turned out to be the right condition for local solvability of vector fields, i.e., condition $(\mathcal{P})$. When L

satisfies $(\mathcal{P})$ the arguments in [**NT**] allow $Lu = f$ to be solved locally with u in the Sobolev space $L^{2,-1}$ for $f \in L^2$. This result was improved by Treves ([**T2**]) to L^2 solvability, i.e., u can be taken in L^2. Concerning the regularity of the coefficients, it was shown in [**Ho1**] that if L is in the canonical form

$$Lu = \frac{\partial u}{\partial t} + i \sum_{j=1}^{n} b_j(x, t) \frac{\partial u}{\partial x_j}, \tag{a}$$

with b_j real-valued and Lipschitz and satisfies $(\mathcal{P})$ then it is locally solvable in L^2. Since there is loss of one derivative in the process of obtaining coordinates in which L has this form one must require, in general, that derivatives up to order one of the coefficients of L be Lipschitz. However, in two variables (i.e., when $n = 1$) it is possible to prove L^2 solvability directly without assuming that L is in the special form (a) ([**HM1**]). Hence, planar vector fields with Lipschitz coefficients that satisfy $(\mathcal{P})$ are locally solvable in L^2. This result is essentially sharp in the sense that there are counterexamples to L^2 solvability and to the existence of L^2 a priori estimates if the coefficients are only restricted to belong to the Hölder class C^α for any $0 < \alpha < 1$ ([**J1**], [**HM1**], [**HM2**]). Whether any vector field with Lipschitz coefficients that satisfies $(\mathcal{P})$ in three or more variables is locally solvable in L^2 is an open problem at the time of this writing.

It is a characteristic feature of locally solvable operators of order one that the L^2 a priori estimates that they satisfy can be extended to L^p estimates for $1 < p < \infty$, a fact that turns out to be false for second-order operators in three or more variables (for results in that direction see [**Li**], [**K**], [**KT1**], [**KT2**], [**Gu**], [**Ch1**]). Solvability in L^p for vector fields was first considered in [**HP**], where the method involved pseudo-differential operators and demanded smooth coefficients. On the other hand, using the method of H. Smith ([**Sm**]), L^p a priori estimates in the range $1 < p < \infty$ can be proved in one stroke under the same regularity hypothesis on the coefficients initially known to guarantee just L^2 estimates ([**HM2**]). This is the point of view used in the presentation of a priori estimates in this book, although for simplicity we have not included the proof that in two variables L^p estimates for vector fields with Lipschitz coefficients are valid without assuming they are in the canonical form (a) ([**HM2**]). The proof of a priori estimates in several variables is reduced, thanks to the geometry of $(\mathcal{P})$ that prevents the existence of orbits of dimension higher than 2, to two-dimensional a priori estimates that are glued by a partition of unity associated with a convenient Whitney decomposition in cubes. The presentation in this chapter owes much to the discussion in [**S1**] about decomposition of open sets in cubes.

While it is true that for any locally solvable vector field L and $1 < p < \infty$ the equation $Lg = f$ can locally be solved in L^p if f is in L^p, this is false, in general, for $p = \infty$ as we saw in the example after Remark IV.1.12 that was taken from [**HT2**]. This difficulty can be dealt with by introducing the space $X = L^\infty(\mathbb{R}_t; \mathrm{bmo}(\mathbb{R}_x))$ of measurable functions $u(x, t)$ such that, for almost every $t \in \mathbb{R}$, $x \mapsto u(x, t) \in \mathrm{bmo}(\mathbb{R})$ and $\|u(t, \cdot)\|_{\mathrm{bmo}} \le C < \infty$ for a.e. $t \in \mathbb{R}$, where $\mathrm{bmo}(\mathbb{R})$ is a space of bounded mean oscillation functions, dual to the semilocal Hardy space $h^1(\mathbb{R})$ of Goldberg. This was first observed in [**BHS**], where it is proved that for a substantial subclass of the class of locally solvable vector fields L, the equation $Lu = f$ can be locally solved with $u \in X$ if $f \in L^\infty$. This result was later improved by showing that for any locally solvable vector field L the equation $Lu = f$ can be locally solved with $u \in X$ for any $f \in X$ ([**daS**], [**HdaS**]) which can be regarded as an ersatz for $p = \infty$ of the L^p local solvability valid for $1 < p < \infty$. The presentation in Section IV.1.2 follows closely [**HdaS**] but replaces lemma 4.5 of that paper—which is true but incorrectly proved—by Lemma IV.1.17 which is sharper.

A priori estimates in L^2 easily give a priori estimates in $L^{2,s}$ for any $s \in \mathbb{R}$ but the absorption of lower-order terms requires shrinking of the neighborhood in which the estimate holds in a way that makes its diameter tend to zero when $|s| \to \infty$. Therefore, the technique of a priori estimates gives solutions of arbitrary high but finite regularity for smooth right-hand sides. Using a different approach, Hörmander ([**H9**]) proved solvability for differential operators of arbitrary order that satisfy $(\mathcal{P})$ by studying the propagation of singularities of the equation $Pu = 0$ mod (C^∞), showing the existence of semiglobal solutions, i.e., solutions defined on a full compact set under the geometric assumption that bicharacteristics do not get trapped in the given compact set. Furthermore, the solutions can be taken smooth if f is smooth. In Sections IV.2 and IV.3 of this chapter, the construction of smooth solutions is simplified by the assumption that the vector fields are locally integrable. Since vector fields that satisfy $(\mathcal{P})$ are indeed locally integrable, the local integrability hypothesis is superfluous, however this fact depends on the difficult and long theorems on smooth solvability by Hörmander ([**H9**], [**H5**]). Thus, it would be interesting to have a shorter *ad hoc* proof of the local existence of smooth solutions for vector fields that satisfy $(\mathcal{P})$ without invoking local integrability.

Concerning the necessity of $(\mathcal{P})$, Nirenberg and Treves had shown in their seminal paper [**NT**] that local solvability implies $(\mathcal{P})$ for vector fields with real-analytic coefficients and conjectured the same implication should hold for smooth coefficients. This state of affairs remained unchanged for 15 years until Moyer ([**Mo**]) removed in 1978 the analyticity hypothesis for operators in two variables in a never published manuscript. His ideas, however, were applied by Hörmander [**H4**] to extend the result for operators in any number of variables with smooth coefficients. The discussion of the necessity of ($\mathcal{P}$) in Section IV.4 of this chapter is again simplified by the assumption of local integrability and follows the presentation in [**T3**] (see also [**T5**] and [**CorH2**]).

V

The FBI transform and some applications

This chapter begins with a discussion of certain submanifolds in CR and hypoanalytic manifolds. We then introduce the FBI transform which is a nonlinear Fourier transform that characterizes analyticity. We also present a more general version of this transform which characterizes hypoanalyticity. We will discuss several applications of the FBI transform to the study of the regularity of solutions in locally integrable structures.

V.1 Certain submanifolds of hypoanalytic manifolds

This section discusses certain submanifolds of hypoanalytic manifolds. We begin with a discussion of CR manifolds in $\mathbb{C}^N$. CR manifolds are good models for hypoanalytic manifolds. Later in the chapter we will see that a hypoanalytic structure can be locally embedded in a CR structure. This can sometimes be useful in reducing problems about general vector fields in hypoanalytic structures to CR vector fields. We will first recall the concept of a complex linear structure on a real vector space and apply it to the real tangent bundle of real submanifolds in $\mathbb{C}^N$. Let $\mathbb{V}$ be a vector space over $\mathbb{R}$ and suppose $J : \mathbb{V} \longrightarrow \mathbb{V}$ is a linear map such that $J^2 = -\mathrm{Id}$ (where Id = the identity). Clearly J is an isomorphism and $\dim \mathbb{V}$ is even since $(\det J)^2 = \det(-\mathrm{Id}) = (-1)^{\dim \mathbb{V}}$. The map J is called a *complex structure* on $\mathbb{V}$. Indeed, with such a J, $\mathbb{V}$ becomes a complex vector space by defining $(a + ib)v = av + b(Jv)$ for $a, b \in \mathbb{R}, v \in \mathbb{V}$. Conversely, if $\mathbb{V}$ is a complex vector space, it is also a vector space over $\mathbb{R}$ and the map $Jv = iv$ is an $\mathbb{R}$-linear map with $J^2 = -\mathrm{Id}$. If $\{v_1, ..., v_N\}$ is a basis of $\mathbb{V}$ over $\mathbb{C}$, then $\{v_1, ..., v_N, Jv_1, ..., Jv_N\}$ is a basis of $\mathbb{V}$ over $\mathbb{R}$.

EXAMPLE V.1.1. In $\mathbb{C}^N$ let $z_j = x_j + iy_j$, $1 \le j \le N$, denote the coordinates. We will identify $\mathbb{C}^N$ with $\mathbb{R}^{2N}$ by means of the map

$$(z_1, \ldots, z_N) \longmapsto (x_1, y_1, \ldots, x_N, y_N).$$

Multiplication by i in $\mathbb{C}^N$ then induces a map $J : \mathbb{R}^{2N} \longrightarrow \mathbb{R}^{2N}$ given by

$$J(x_1, y_1, \ldots, x_N, y_N) = (-y_1, x_1, \ldots, -y_N, x_N).$$

Note that $J^2 = -\mathrm{Id}$ and so J is a complex structure on $\mathbb{R}^{2N}$, called the standard complex structure on $\mathbb{R}^{2N}$.

EXAMPLE V.1.2. With notation as in the previous example, for $p \in \mathbb{C}^N$, a basis of the real tangent space $T_p\mathbb{C}^N$ is given by $\frac{\partial}{\partial x_1}\big|_p, \frac{\partial}{\partial y_1}\big|_p, \ldots, \frac{\partial}{\partial x_N}\big|_p, \frac{\partial}{\partial y_N}\big|_p$. This basis can be used to identify $T_p\mathbb{C}^N$ with $\mathbb{R}^{2N}$ by choosing the usual basis

$$e_1 = (1, 0, \ldots, 0), \ldots, e_{2N} = (0, \ldots, 0, 1) \quad \text{of} \quad \mathbb{R}^{2N}.$$

This leads to a complex structure $J : T_p\mathbb{C}^N \longrightarrow T_p\mathbb{C}^N$ given by

$$J\left(\frac{\partial}{\partial x_j}\big|_p\right) = \frac{\partial}{\partial y_j}\bigg|_p \quad \text{and} \quad J\left(\frac{\partial}{\partial y_j}\big|_p\right) = -\frac{\partial}{\partial x_j}\bigg|_p, \quad j = 1, \ldots, N.$$

This complex structure is independent of the choice of the holomorphic coordinates $(z_1, \ldots, z_N)$. To see this, suppose $w = F(z)$ is a biholomorphic map defined near 0 with $F(0) = 0$ where we are assuming as we may that $p = 0$. Write $F = U + iV$ and let $w_j = u_j + iv_j$, $j = 1, \ldots, N$. We need to show that $\mathrm{d}F_0 \circ J = J \circ \mathrm{d}F_0$. We have:

$$\mathrm{d}F_0\left(J\left(\frac{\partial}{\partial x_j}\right)\right) = \mathrm{d}F_0\left(\frac{\partial}{\partial y_j}\right) = \sum_l \frac{\partial U_l}{\partial y_j}\frac{\partial}{\partial u_l} + \sum_l \frac{\partial V_l}{\partial y_j}\frac{\partial}{\partial v_l}$$

and

$$J\left(\mathrm{d}F_0\left(\frac{\partial}{\partial x_j}\right)\right) = J\left(\sum_l \frac{\partial U_l}{\partial x_j}\frac{\partial}{\partial u_l} + \sum_l \frac{\partial V_l}{\partial x_j}\frac{\partial}{\partial v_l}\right)$$
$$= \sum_l \frac{\partial U_l}{\partial x_j}\frac{\partial}{\partial v_l} - \sum_l \frac{\partial V_l}{\partial x_j}\frac{\partial}{\partial u_l},$$

where everything is to be evaluated at 0. Thus an application of the Cauchy–Riemann equations to the U_j and V_j shows that $\mathrm{d}F(J(\frac{\partial}{\partial x_j})) = J(\mathrm{d}F(\frac{\partial}{\partial x_j}))$. The equality also holds in the same fashion for $\frac{\partial}{\partial y_j}$. Thus, J is independent of the holomorphic coordinates. This also means that J can be defined on the real tangent space of any complex manifold. Note that J extends to a $\mathbb{C}$-linear map from $\mathbb{C}T_p\mathbb{C}^N$ into itself and the extension still satisfies $J^2 = -\mathrm{Id}$. We will also denote this extension by J. The fact that $J^2 = -\mathrm{Id}$ implies that

$J: \mathbb{C}T_p\mathbb{C}^N \longrightarrow \mathbb{C}T_p\mathbb{C}^N$ has only two eigenvalues: i and $-i$. Define $T_p^{1,0}\mathbb{C}^N$ to be equal to the eigenspace associated with i, and $T_p^{0,1}\mathbb{C}^N$ will be the eigenspace associated with $-i$. We get corresponding vector bundles $T^{1,0}$ and $T^{0,1}$. Observe that $T^{1,0}$ is generated by $\frac{\partial}{\partial z_1}, \ldots, \frac{\partial}{\partial z_N}$. Hence $T^{1,0}$ is the bundle of holomorphic vector fields introduced in Chapter I (see the discussion preceding Theorem I.5.1). Likewise, $T^{0,1}$ is generated by $\frac{\partial}{\partial \bar{z}_1}, \ldots, \frac{\partial}{\partial \bar{z}_N}$.

DEFINITION V.1.3. *Let $\mathcal{M}$ be a real submanifold of $\mathbb{C}^N$. For $p \in \mathcal{M}$, define*

$$\mathcal{V}_p(\mathcal{M}) = \mathbb{C}T_p\mathcal{M} \cap T_p^{0,1}\mathbb{C}^N.$$

DEFINITION V.1.4. *Let $\mathcal{M}$ be a real submanifold of $\mathbb{C}^N$ and $p \in \mathcal{M}$. The complex tangent space of $\mathcal{M}$ at p denoted $T_p^c\mathcal{M}$ is defined by*

$$T_p^c\mathcal{M} = T_p\mathcal{M} \cap J(T_p\mathcal{M})$$

It is easy to see that $T_p^c\mathcal{M} = \{v \in T_p\mathcal{M} : J(v) \in T_p\mathcal{M}\}$. Observe that $J: T_p^c\mathcal{M} \longrightarrow T_p^c\mathcal{M}$ and so $T_p^c\mathcal{M}$ is equipped with a complex vector space structure. It is also evident that $J: \mathbb{C}T_p^c\mathcal{M} \longrightarrow \mathbb{C}T_p^c\mathcal{M}$.

EXAMPLE V.1.5. Let $\mathcal{M}$ be a hypersurface in $\mathbb{C}^N$ through the point 0. Let ρ be a defining function for $\mathcal{M}$ near 0. Since $d\rho(0) \neq 0$ and ρ is real-valued, $\partial\rho(0) \neq 0$. After a complex linear change of coordinates, we may assume that

$$\frac{\partial\rho}{\partial z}(0) = (0, \ldots, 0, 1).$$

That is, we have coordinates $(z, w), z = x + iy \in \mathbb{C}^{N-1}, w = s + it \in \mathbb{C}$, such that $\frac{\partial\rho}{\partial x_j}(0) = \frac{\partial\rho}{\partial y_j}(0) = 0, j = 1, \ldots, N-1, \frac{\partial\rho}{\partial s}(0) = 0$ and $\frac{\partial\rho}{\partial t}(0) \neq 0$. These conditions on the partial derivatives of ρ allow us to apply the implicit function theorem and conclude that near 0 the submanifold $\mathcal{M}$ is given by

$$\mathcal{M} = \{(z, s + i\phi(z, s))\},$$

where ϕ is real-valued, $\phi(0, 0) = 0$, and $d\phi(0, 0) = 0$. Hence $T_0\mathcal{M} =$ span at 0 of

$$\left\{ \frac{\partial}{\partial x_j}, \frac{\partial}{\partial y_j}, \frac{\partial}{\partial s}, \quad j = 1, \ldots, N-1 \right\},$$

while $\mathcal{V}_0(\mathcal{M}) =$ the $\mathbb{C}$-span at 0 of

$$\left\{ \frac{\partial}{\partial \bar{z}_j} : j = 1, \ldots, N-1 \right\}$$

and $T_0^c\mathcal{M} =$ the $\mathbb{R}$-span at 0 of

$$\left\{ \frac{\partial}{\partial x_j}, \frac{\partial}{\partial y_j} : j = 1, \ldots, N-1 \right\}.$$

The spaces $T_p^c \mathcal{M}$ and $\mathcal{V}_p(\mathcal{M})$ are related. To see this, we recall the following result from [**BER**] where $\Re\mathcal{V}_p(\mathcal{M})$ denotes the real parts of elements of $\mathcal{V}_p(\mathcal{M})$:

PROPOSITION V.1.6. *For $p \in \mathcal{M}$,*

(a) $\Re\mathcal{V}_p(\mathcal{M}) = T_p^c \mathcal{M}$;
(b) $\mathbb{C}T_p^c \mathcal{M} = \mathcal{V}_p(\mathcal{M}) \oplus \overline{\mathcal{V}_p(\mathcal{M})}$;
(c) $\mathcal{V}_p(\mathcal{M}) = \{x + iJ(x) : x \in T_p^c \mathcal{M}\}$.

PROOF. Observe first that for any $x \in T_p\mathbb{C}^N$, $x + iJ(x) \in T_p^{0,1}\mathbb{C}^N$. Let $x \in T_p^c \mathcal{M}$. Then x and $J(x) \in T_p\mathcal{M}$ and so $x + iJ(x) \in \mathcal{V}_p(\mathcal{M})$. Thus $x \in \Re\mathcal{V}_p(\mathcal{M})$. Conversely, if $x \in \Re\mathcal{V}_p(\mathcal{M})$, then there is $y \in T_p\mathbb{C}^N$ such that $x + iy \in \mathcal{V}_p(\mathcal{M}) \subseteq \mathbb{C}T_p\mathcal{M}$ implying that $x \in T_p^c\mathcal{M}$ since $y = J(x)$ and $y \in T_p\mathcal{M}$. We have thus proved (a) and (b) follows from (a) trivially. The proof of (c) is also contained in that of (a). □

From Proposition V.1.6 we see that

$$\dim T_p^c \mathcal{M} = 2\dim_{\mathbb{C}} \mathcal{V}_p(\mathcal{M}).$$

DEFINITION V.1.7. *A submanifold $\mathcal{M}$ of $\mathbb{C}^N$ is called* CR *(for Cauchy–Riemann) if $\dim_{\mathbb{C}}\mathcal{V}_p(\mathcal{M})$ is constant as p varies in $\mathcal{M}$. In this case, $\dim_{\mathbb{C}}\mathcal{V}_p(\mathcal{M})$ is called the CR dimension of $\mathcal{M}$.*

DEFINITION V.1.8. *A CR submanifold $\mathcal{M}$ of $\mathbb{C}^N$ is called* totally real *if its CR dimension is* 0.

EXAMPLE V.1.9. The copy of $\mathbb{R}^N$ in $\mathbb{C}^N$ given by

$$\{x + iy \in \mathbb{C}^N : y = 0\}$$

is a totally real submanifold.

EXAMPLE V.1.10. Let k and N be positive integers, $1 \leq k \leq N$. Write the coordinates in $\mathbb{C}^N$ as (z, w), $z = x + iy \in \mathbb{C}^k$ and $w = u + iv \in \mathbb{C}^{N-k}$. Let $\phi : \mathbb{R}^k \mapsto \mathbb{R}^k$ and $\psi : \mathbb{R}^k \mapsto \mathbb{C}^{N-k}$ be smooth functions with $\phi(0) = 0$, $d\phi(0) = 0$, $\psi(0) = 0$, and $d\psi(0) = 0$. Then the submanifold

$$\mathcal{M} = \{(x + i\phi(x), \psi(x)) : x \in \mathbb{R}^k\}$$

is totally real near the point $0 \in \mathcal{M}$. Conversely, if $\mathcal{N}$ is any totally real submanifold of $\mathbb{C}^N$, then near each of its points, there are holomorphic coordinates in which $\mathcal{N}$ takes the form of $\mathcal{M}$ above (see proposition 1.3.8 in [**BER**]). If $\mathcal{N}$ is also real-analytic, holomorphic coordinates can be found so that $\phi \equiv 0$ and $\psi \equiv 0$.

LEMMA V.1.11. *Suppose $\mathcal{M}$ is a submanifold of $\mathbb{C}^N$ of real codimension d. Then*

$$2N - 2d \leq \dim T_p^c \mathcal{M} \leq 2N - d.$$

PROOF. Since $T_p^c \mathcal{M} \subseteq T_p \mathcal{M}$,

$$\dim T_p^c \mathcal{M} \leq \dim T_p \mathcal{M} = 2N - d.$$

On the other hand, $T_p \mathcal{M} + J(T_p \mathcal{M}) \subseteq T_p \mathbb{C}^N$ and so

$$\dim T_p^c \mathcal{M} = 2 \dim T_p \mathcal{M} - \dim(T_p \mathcal{M} + J(T_p \mathcal{M})) \geq 2N - 2d. \qquad \square$$

EXAMPLE V.1.12. A hypersurface $\mathcal{M} \subseteq \mathbb{C}^N$ is a CR submanifold of CR dimension $N - 1$. Indeed, this follows from the lemma since $T_p^c \mathcal{M}$ always has even real dimension, which when $p \in \mathcal{M}$ has to equal $2N - 2$.

EXAMPLE V.1.13. Let $\mathcal{M}$ be a complex submanifold of $\mathbb{C}^N$ of complex dimension n. Then $\mathcal{M}$ is a CR submanifold of CR dimension n. This follows from the J-invariance of $T_p \mathcal{M}$. To see this, let $X \in T_p \mathcal{M}$. If $f_j = u_j + i v_j$ $(1 \leq j \leq N - n)$ are local holomorphic defining functions near $p \in \mathcal{M}$, then by the Cauchy–Riemann equations we have $J(X)u_j = J(X)v_j = 0$ for all j. Hence $J(X) \in T_p \mathcal{M}$.

Suppose $\mathcal{M} \subseteq \mathbb{C}^N$ has codimension d and is locally defined by $\rho_j = 0, j = 1, \ldots, d$. Then a vector $v = \sum_{j=1}^N v_j \frac{\partial}{\partial \bar{z}_j} \in \mathcal{V}_p(\mathcal{M})$ if and only if $\sum_{j=1}^N v_j \frac{\partial \rho_l}{\partial \bar{z}_j} = 0$ for all l. Hence $\dim \mathcal{V}_p \mathcal{M} = N - r$ where $r =$ the dimension of the $\mathbb{C}$-span of $\{\bar{\partial}\rho_1(p), \ldots, \bar{\partial}\rho_d(p)\}$.

EXAMPLE V.1.14. Let $\mathcal{M}$ be the two-dimensional submanifold of $\mathbb{C}^2$ defined by $\rho_1 = x_2 - x_1^2 = 0$ and $\rho_2 = y_2 - y_1^2 = 0$. Then by calculating $\bar{\partial}\rho_1(p)$ and $\bar{\partial}\rho_2(p)$, we easily see that $\dim \mathcal{V}_p \mathcal{M} = 0$ or 1 depending on whether $p \in \mathcal{M} \cap \{x_1 = y_1\}$. Hence $\mathcal{M}$ is not a CR manifold.

If $\mathcal{M} \subseteq \mathbb{C}^N$ has codimension d, since $2 \dim_{\mathbb{C}} \mathcal{V}_p \mathcal{M} = \dim T_p^c \mathcal{M}$, Lemma V.1.11 tells us that the minimum possible value of $\dim_{\mathbb{C}} \mathcal{V}_p \mathcal{M} = N - d$. This minimum value is attained precisely when the forms $\{\bar{\partial}\rho_1(p), \ldots, \bar{\partial}\rho_d(p)\}$ are linearly independent. The CR submanifolds for which $\dim \mathcal{V}_p \mathcal{M}$ has such minimal value are the generic ones introduced in Chapter I. It will be convenient to present here an equivalent definition.

DEFINITION V.1.15. *A CR submanifold $\mathcal{M} \subseteq \mathbb{C}^N$ of codimension d is called generic if for $p \in \mathcal{M}$, $\dim_{\mathbb{C}} \mathcal{V}_p \mathcal{M} = N - d$.*

EXAMPLE V.1.16. A hypersurface of $\mathbb{C}^N$ is a generic CR submanifold.

EXAMPLE V.1.17. A complex submanifold of $\mathbb{C}^N$ that is not an open subset is a nongeneric CR submanifold.

EXAMPLE V.1.18. Let (z, w) denote the coordinates of $\mathbb{C}^{n+d}$ where $z = x + iy \in \mathbb{C}^n$ and $w = s + it \in \mathbb{C}^d$. Let $\phi_1(z, s), \ldots, \phi_d(z, s)$ be smooth, real-valued functions. Then

$$\mathcal{M} = \{(z, w) : t_j - \phi_j(z, s) = 0, 1 \leq j \leq d\}$$

is a generic CR manifold. This is easily checked by noting that

$$\rho_j = \frac{w_j - \overline{w}_j}{2i} - \phi_j(z, s)$$

are defining functions with $\{\bar{\partial}\rho_1(p), \ldots, \bar{\partial}\rho_d(p)\}$ linearly independent at each point.

REMARK V.1.19. Conversely, as we saw in Chapter I, given any generic CR submanifold $\mathcal{M}$, in appropriate holomorphic coordinates, $\mathcal{M}$ takes the form in the example.

Let $\mathcal{M}$ be a CR submanifold of $\mathbb{C}^N$ of codimension d that is not generic and assume $0 \in \mathcal{M}$. We will show that in a certain sense, near the point 0, $\mathcal{M}$ can be viewed as a generic CR submanifold of $\mathbb{C}^L$ for some $L < N$. Define

$$\mathcal{S} = T_0\mathcal{M} + J(T_0\mathcal{M}).$$

Let Y be a subspace of $T_0\mathcal{M}$ such that

$$T_0\mathcal{M} = T_0^c\mathcal{M} \oplus Y.$$

Note that $J(Y) \cap T_0\mathcal{M} = 0$. Let $\{v_1, \ldots, v_n\}$ be a $\mathbb{C}$-basis of the complex space $T_0^c\mathcal{M}$. Then: $\{v_1, \ldots, v_n, J(v_1), \ldots, J(v_n)\}$ is an $\mathbb{R}$-basis of $T_0^c\mathcal{M}$. Complete this to a basis $\{v_1, \ldots, v_n, J(v_1), \ldots, J(v_n), u_1, \ldots, u_r\}$ of $T_0\mathcal{M}$ where $2n + r + d = 2N$. Then since $J(Y) \cap T_0\mathcal{M} = 0$, it follows that

$$\mathcal{B} = \{v_1, \ldots, v_n, J(v_1), \ldots, J(v_n), u_1, \ldots, u_r, J(u_1), \ldots, J(u_r)\}$$

is a basis of $\mathcal{S}$. Extend $\mathcal{B}$ to a basis

$$\mathcal{B}' = \mathcal{B} \cup \{u'_1, \ldots, u'_l, J(u'_1), \ldots, J(u'_l)\}$$

of $T_0\mathbb{C}^N$, $N = n + r + l$. Split the coordinates in $\mathbb{C}^N = \mathbb{C}^{n+r+l}$ as (z, w, p) where $z = x + iy \in \mathbb{C}^n$, $w = s + it \in \mathbb{C}^r$, and $p = s' + it' \in \mathbb{C}^l$. Define the map $A : T_0\mathbb{C}^N \to T_0\mathbb{C}^N$ by $A(v_i) = \frac{\partial}{\partial x_i}$, $A(Jv_i) = \frac{\partial}{\partial y_i}$, $1 \leq i \leq n$; $A(u_k) = \frac{\partial}{\partial s_k}$, $A(Ju_k) = \frac{\partial}{\partial t_k}$, $1 \leq k \leq r$; $A(u'_j) = \frac{\partial}{\partial s'_j}$, $A(Ju'_j) = \frac{\partial}{\partial t'_j}$, $1 \leq j \leq l$. Note that the map A commutes with J and hence after a complex linear map (see Remark V.1.20 below) we are in coordinates $(z, w, p) \in \mathbb{C}^{n+r+l}$ where

$$\mathcal{S} = \text{span of} \quad \left\{ \frac{\partial}{\partial x_j}, \frac{\partial}{\partial y_j}, \frac{\partial}{\partial s_k}, \frac{\partial}{\partial t_k} : 1 \le j \le n, 1 \le k \le r \right\}$$

and $T_0\mathcal{M} = \text{span of } \{\frac{\partial}{\partial x_j}, \frac{\partial}{\partial y_j}, \frac{\partial}{\partial s_k}\}$. It follows that near 0, $\mathcal{M}$ can be expressed as a graph of the form:

$$\mathcal{M} = \{(x + iy, s + if(x, y, s), g(x, y, s))\}$$

where f is valued in $\mathbb{R}^r$ and g is valued in $\mathbb{C}^l$. The components of the functions $s + if(x, y, s)$ and $g(x, y, s)$ are CR functions. Observe that the projection $\pi : \mathbb{C}^{n+r+l} \to \mathbb{C}^{n+r}$, $\pi(z, w, p) = (z, w)$ is a diffeomorphism of $\mathcal{M}$ onto the generic CR submanifold $\pi(\mathcal{M})$ of $\mathbb{C}^{n+r}$.

REMARK V.1.20. Recall the identification of $\mathbb{C}^N$ with $\mathbb{R}^{2N}$ of Example V.1.1 given by $(z_1, \ldots, z_N) \mapsto (x_1, y_1, \ldots, x_N, y_N)$. With this identification, it is easy to see that a real linear map $A : \mathbb{R}^{2N} \to \mathbb{R}^{2N}$ induces a $\mathbb{C}$-linear map on $\mathbb{C}^N$ if and only if A commutes with the operator J.

PROPOSITION V.1.21. *If $\mathcal{M}$ is a totally real submanifold of $\mathbb{C}^N$ of codimension d, then $d \ge N$ and hence dim $\mathcal{M} \le N$. If $\mathcal{M}$ is also generic, then $d = N$. Thus, a totally real submanifold of maximal dimension has dimension $= N$.*

PROOF. Let $p \in \mathcal{M}$ and $\rho_1, \ldots, \rho_d$ be defining functions of $\mathcal{M}$ near p. Since $\mathcal{V}_p(\mathcal{M}) = \{0\}$, we must have:

$$\text{span}_{\mathbb{C}}\{\partial\rho_1, \cdots, \partial\rho_d\} = \text{span}_{\mathbb{C}}\{dz_1, \ldots, dz_N\}$$

at the point p. Hence $d \ge N$. If $\mathcal{M}$ is also generic, then $\partial\rho_1, \ldots, \partial\rho_d$ are linearly independent and so $d = N$. $\qquad\qquad\square$

The map J can be used to characterize CR, generic CR, and totally real submanifolds.

PROPOSITION V.1.22. *Let $\mathcal{M}$ be a submanifold of $\mathbb{C}^N$. Then*

(i) *$\mathcal{M}$ is CR if and only if $\dim(T_p\mathcal{M} \cap J(T_p\mathcal{M}))$ is constant as p varies in $\mathcal{M}$.*

(ii) *$\mathcal{M}$ is totally real if and only if $T_p\mathcal{M} \cap J(T_p\mathcal{M}) = \{0\}$ for all $p \in \mathcal{M}$.*

(iii) *$\mathcal{M}$ is a generic CR submanifold if and only if*

$$T_p\mathcal{M} + J(T_p\mathcal{M}) = T_p\mathbb{C}^N \quad \text{for all } p \in \mathcal{M}.$$

PROOF. (i) follows from the definition of $T_p^c\mathcal{M}$ and Proposition V.1.6. (ii) also follows from Proposition V.1.6. To prove (iii), if $\mathcal{M}$ is generic and $\rho_1, ..., \rho_d$ are local defining functions, then the linear independence of $\partial\rho_1, ..., \partial\rho_d$ is equivalent to:

$$\dim_{\mathbb{C}}\mathcal{V}_p(\mathcal{M}) = N - d.$$

Hence, by Proposition V.1.6, dim $(T_p\mathcal{M} \cap J(T_p\mathcal{M})) = 2(N - d)$. But then dim $(T_p\mathcal{M} + J(T_p\mathcal{M})) = 2N$, implying that $T_p\mathcal{M} + J(T_p\mathcal{M}) = T_p\mathbb{C}^N$ for all $p \in \mathcal{M}$. Conversely, if $T_p\mathcal{M} + J(T_p\mathcal{M}) = T_p\mathbb{C}^N$, then $\dim(T_p\mathcal{M} \cap J(T_p\mathcal{M})) = 2(N - d)$ and so by Proposition V.1.6, $\dim_{\mathbb{C}}\mathcal{V}_p(\mathcal{M}) = N - d$ showing that $\mathcal{M}$ is generic. $\qquad\square$

We will next describe certain submanifolds in hypoanalytic structures that play important roles in the analysis of the solutions of the sections of the associated vector bundle. Let $(\mathcal{M}, \mathcal{V})$ be a hypoanalytic structure. $\mathcal{M}$ is a smooth manifold of dimension N and $\mathcal{V}$ is an involutive sub-bundle of $\mathbb{C}TM$ of fiber dimension n whose orthogonal bundle T' in $\mathbb{C}T^*M$ is locally generated by the differentials of $m = N - n$ smooth functions. Recall from Chapter I that $T^0 = \bigcup_{p\in\mathcal{M}} T_p^0$ denotes the characteristic set of the structure $(\mathcal{M}, \mathcal{V})$.

DEFINITION V.1.23. *A submanifold* $\mathcal{Y}$ *is called* noncharacteristic *if*

$$T_p\mathcal{M} = T_p\mathcal{Y} + \Re(\mathcal{V}_p) \qquad \forall p \in \mathcal{Y}.$$

DEFINITION V.1.24. *A submanifold* $\mathcal{N}$ *is called* strongly noncharacteristic *if*

$$\mathbb{C}T_p\mathcal{M} = \mathbb{C}T_p\mathcal{N} + \mathcal{V}_p \qquad \forall p \in \mathcal{N}.$$

DEFINITION V.1.25. *A submanifold* $\mathcal{X}$ *of* $\mathcal{M}$ *is called* maximally real *if*

$$\mathbb{C}T_p\mathcal{M} = \mathbb{C}T_p\mathcal{X} \oplus \mathcal{V}_p \qquad \forall p \in \mathcal{X}.$$

Clearly, a maximally real submanifold is strongly noncharacteristic. If $\mathcal{N}$ is strongly noncharacteristic, then dim $\mathcal{N} \geq m$, while if $\mathcal{X}$ is maximally real, dim $\mathcal{X} = m$. A strongly noncharacteristic submanifold is a noncharacteristic submanifold. A noncharacteristic hypersurface in $\mathcal{M}$ is strongly noncharacteristic.

EXAMPLE V.1.26. Denote the coordinates in $\mathbb{R}^3$ by (x, y, t) and consider the structure generated by $L = \frac{\partial}{\partial t} + i\frac{\partial}{\partial y}$. The orthogonal of L is generated by dZ_1 and dZ_2 where $Z_1 = x$ and $Z_2 = t + iy$. If $S = \{(x, 0, 0)\}$, then S is a noncharacteristic submanifold that is not strongly noncharacteristic.

A CR submanifold of $\mathbb{C}^N$ is strongly noncharacteristic if and only if it is generic. It is maximally real precisely when it is totally real of maximal dimension.

The proofs of the following propositions are left to the reader.

PROPOSITION V.1.27. *A submanifold $\mathcal{Y}$ of $\mathcal{M}$ is noncharacteristic if and only if the natural map $T^*\mathcal{M}|_{\mathcal{Y}} \longrightarrow T^*\mathcal{Y}$ maps T^0 injectively into $T^*\mathcal{Y}$.*

PROPOSITION V.1.28. *A submanifold $\mathcal{N}$ of $\mathcal{M}$ is strongly noncharacteristic if and only if the natural map $\mathbb{C}T^*\mathcal{M}|_{\mathcal{N}} \longrightarrow \mathbb{C}T^*\mathcal{N}$ maps $T'|_{\mathcal{N}}$ injectively into $\mathbb{C}T^*\mathcal{N}$.*

PROPOSITION V.1.29. *A submanifold $\mathcal{X}$ of $\mathcal{M}$ is maximally real if and only if the natural map $\mathbb{C}T^*\mathcal{M}|_{\mathcal{X}} \longrightarrow \mathbb{C}T^*\mathcal{X}$ induces a bijection of $T'|_{\mathcal{X}}$ onto $\mathbb{C}T^*\mathcal{X}$.*

Distribution solutions have traces on a noncharacteristic submanifold of $\mathcal{M}$ (see proposition 1.4.3 in [**T5**]). In particular, a solution can always be restricted to a maximally real manifold. The local and microlocal regularity of solutions are often studied by analyzing their restrictions to maximally real submanifolds. Instances of this will occur later in this chapter.

V.2 Microlocal analyticity and the FBI transform

The Paley–Wiener Theorem (see Theorem V.3.1 in the next section) characterizes the smoothness of a tempered distribution u in terms of the rapid decay of its Fourier transform $\hat{u}(\xi)$. This characterization is very useful in studying the local and microlocal regularity of solutions of partial differential equations with smooth coefficients. There is also a characterization of analyticity in terms of the Fourier transform ([**H8**]). However, the latter is based on estimates using a sequence of cut-off functions making it more difficult in applications. The FBI transform is a nonlinear Fourier transform which characterizes analyticity (see Theorem V.2.4 below).

DEFINITION V.2.1. *Let $u \in \mathcal{E}'(\mathbb{R}^m)$. Define the FBI transform of u by*

$$F_u(x, \xi) = \int e^{i(x-y)\cdot\xi - |\xi||x-y|^2} u(y)\, dy \qquad (V.1)$$

for $(x, \xi) \in \mathbb{R}^m \times \mathbb{R}^m$, where

$$(x-y)\cdot\xi = \sum_{j=1}^{m}(x_j - y_j)\xi_j.$$

The integral is to be understood in the duality sense.

THEOREM V.2.2. (Inversion with the FBI.) *Let $u \in C_c(\mathbb{R}^m)$. Then*

$$u(x) = \lim_{\epsilon \to 0^+} \frac{1}{(4\pi^3)^{\frac{m}{2}}} \iint F_u(t, \xi) e^{i(x-t)\cdot\xi - \epsilon|\xi|^2} |\xi|^{\frac{m}{2}} \, dt d\xi$$

where the convergence is uniform.

REMARK V.2.3. If $u \in \mathcal{E}'(\mathbb{R}^m)$, the theorem also holds where convergence is understood in the distribution sense.

PROOF. From the Fourier transform of the Gaussian, we have:

$$\int_{\mathbb{R}^m} e^{i(x-y)\cdot\xi - \epsilon|\xi|^2} \, d\xi = \left(\frac{\pi}{\epsilon}\right)^{\frac{m}{2}} e^{\frac{-|x-y|^2}{4\epsilon}}.$$

Hence

$$\frac{1}{(2\pi)^m} \iint e^{i(x-y)\cdot\xi - \epsilon|\xi|^2} u(y) \, d\xi dy = \frac{1}{2^m(\pi\epsilon)^{\frac{m}{2}}} \int e^{\frac{-|x-y|^2}{4\epsilon}} u(y) \, dy$$

$$= \frac{1}{\pi^{\frac{m}{2}}} \int e^{-t^2} u(x - 2\sqrt{\epsilon}t) \, dt$$

$$\to u(x)$$

uniformly on $\mathbb{R}^m$ since $u \in C_c(\mathbb{R}^m)$. Thus

$$u(x) = \lim_{\epsilon \to 0} \frac{1}{(2\pi)^m} \iint e^{i(x-y)\cdot\xi - \epsilon|\xi|^2} u(y) \, dy d\xi$$

$$= \frac{1}{(4\pi^3)^{\frac{m}{2}}} \lim_{\epsilon \to 0} \iiint e^{i(x-y)\cdot\xi - |\xi||t-y|^2 - \epsilon|\xi|^2} |\xi|^{\frac{m}{2}} u(y) \, dt dy d\xi$$

$$= \frac{1}{(4\pi^3)^{\frac{m}{2}}} \lim_{\epsilon \to 0} \iint F_u(t, \xi) e^{i(x-t)\cdot\xi - \epsilon|\xi|^2} |\xi|^{\frac{m}{2}} \, dt d\xi. \qquad \square$$

The following characterization of analyticity by means of an exponential decay of the FBI transform may be viewed as an analogue of the Paley–Wiener Theorem.

THEOREM V.2.4. *Let $u \in \mathcal{E}'(\mathbb{R}^m)$. The following are equivalent:*

(i) *u is real-analytic at $x_0 \in \mathbb{R}^m$.*
(ii) *There exist a neighborhood V of x_0 in $\mathbb{R}^m$ and constants $c_1, c_2 > 0$ such that*

$$|F_u(x, \xi)| \leq c_1 e^{-c_2|\xi|} \quad \text{for } (x, \xi) \in V \times \mathbb{R}^m.$$

PROOF. We will assume that u is continuous and leave the general case for the reader.

(i) $\Rightarrow$ (ii) Suppose u is real-analytic at x_0. Let $0 \le \phi \le 1$, $\phi \in C_0^\infty(\mathbb{R}^m)$, $\phi \equiv 1$ near x_0, and $\mathrm{supp}(\phi) \subseteq \{x : u$ is analytic at $x\}$. The integrand in $F_u(x, \xi)$ has a holomorphic extension in a neighborhood of $y = x_0$ in $\mathbb{C}^m$. We will denote by u the holomorphic extension of u near x_0. In the integration defining $F_u(x, \xi)$, we deform the contour from $\mathbb{R}^m$ to the image of $\mathbb{R}^m$ under the map $y \mapsto \theta(y) = y - is\phi(y)\frac{\xi}{|\xi|}$ where s is chosen small enough so that u is defined on the image $\theta(\mathbb{R}^m)$. We then have

$$F_u(x, \xi) = \int_{\mathbb{R}^m} e^{Q(x,y,\xi)} u(\theta(y)) \det \theta'(y) \, dy \qquad (V.2)$$

where $Q(x, y, \xi) = i(x - \theta(y)) \cdot \xi - |\xi|(x - \theta(y))^2$. Observe that

$$\Re Q(x, y, \xi) = -s|\xi|\phi(y)[1 - s\phi(y)] - |\xi||x - y|^2. \qquad (V.3)$$

Let $\delta > 0$ such that $\phi(y) \equiv 1$ when $|y - x_0| \le \delta$. Choose $s = \frac{\delta}{4}$. With these choices, (V.2) and (V.3), we get:

$$|F_u(x, \xi)| \le c \int_{|y-x_0| \le \delta} e^{\Re Q(x,y,\xi)} \, dy + c \int_{y \in \mathrm{supp}(u), |y-x_0| \ge \delta} e^{\Re Q(x,y,\xi)} \, dy$$
$$= I_1(x, \xi) + I_2(x, \xi).$$

Note then that

$$I_1(x, \xi) \le c \int_{|y-x_0| \le \delta} e^{-s|\xi|\phi(y)[1-s\phi(y)]} \, dy$$
$$\le c e^{\frac{-\delta}{8}|\xi|}$$

for any ξ, and for $|x - x_0| \le \frac{\delta}{2}$. Moreover,

$$I_2(x, \xi) \le c \int_{|y-x_0| \ge \delta} e^{-|\xi||x-y|^2} \, dy$$
$$\le c e^{-(\frac{\delta}{2})^2|\xi|}.$$

Hence, for $|x - x_0| \le \frac{\delta}{2}$ and any $\xi \in \mathbb{R}^m$,

$$|F_u(x, \xi)| \le c_1 e^{-c_2|\xi|} \text{ for some } c_1, c_2 > 0.$$

(ii)$\Rightarrow$(i) Assume without loss of generality that $x_0 = 0$. Suppose then that for some $c_1, c_2 > 0$,

$$|F_u(x, \xi)| \le c_1 e^{-c_2|\xi|}$$

for all $\xi \in \mathbb{R}^m$, and for all x near 0. We will use the inversion given by Theorem V.2.2. Write

$$\iint F_u(t, \xi) e^{i(x-t)\cdot\xi - \epsilon|\xi|^2} |\xi|^{\frac{m}{2}} \, dt d\xi = I_1^\epsilon(x) + I_2^\epsilon(x) + I_3^\epsilon(x) + I_4^\epsilon(x)$$

where for some A_1, A_2, B to be chosen later,

$I_1^\epsilon(x) =$ the integral over $\{(t, \xi) : |t| \le A_1, \xi \in \mathbb{R}^m\}$,

$I_2^\epsilon(x) =$ the integral over $\{(t, \xi) : A_1 \le |t| \le A_2, |\xi| \le B\}$,

$I_3^\epsilon(x) =$ the integral over $\{(t, \xi) : |t| \ge A_2, \xi \in \mathbb{R}^m\}$,

$I_4^\epsilon(x) =$ the integral over $\{(t, \xi) : A_1 \le |t| \le A_2, |\xi| \ge B\}$.

Our goal is to show that there is a neighborhood of the origin in $\mathbb{C}^m$ to which the I_j^ϵ extend as holomorphic functions and for each j, $I_j^\epsilon(z)$ converges uniformly on this neighborhood as $\epsilon \to 0$. Consider first I_1^ϵ. Recall that $x_0 = 0$. Choose $A_1 > 0$ so that

$$|F_u(x, \xi)| \le c_1 e^{-c_2 |\xi|} \quad \text{for} \quad |x| \le A_1.$$

If we complexify x to $z = x + iy$ in the integrand of I_1^ϵ, we see that the integrand is bounded by a constant multiple of

$$|\xi|^{\frac{m}{2}} e^{(-c_2 + |y|)|\xi|}$$

which therefore has an integrable majorant for $|y| \le \frac{c_2}{2}$. Hence, as $\epsilon \to 0$, the entire functions $I_1^\epsilon(z)$ converge uniformly on a neighborhood of 0 to a holomorphic function. The functions I_2^ϵ easily extend as entire functions of z and converge uniformly on compact subsets to an entire function as $\epsilon \to 0$. Next choose A_2 so that

$$\text{supp}(u) \subseteq \left\{ y : |y| \le \frac{A_2}{4} \right\}.$$

Then note that when $|t| \ge A_2$,

$$|F_u(t, \xi)| \le c e^{-|\xi| \left(|t| - \frac{A_2}{4} \right)^2}$$
$$\le c e^{-|\xi| \left(\frac{|t|^2}{4} + \frac{A_2^2}{16} \right)}.$$

Using the latter we see that after integrating in t, the integrand in I_3^ϵ is uniformly bounded by a constant multiple of

$$e^{-\frac{A_2^2}{16} |\xi|}.$$

This allows us to complexify as in I_1^ϵ to conclude that $I_3^\epsilon(z)$ converges uniformly to a holomorphic function in a neighborhood of 0. Write

$$I_4^\epsilon(x) = \iiint_R e^{i(x-y)\cdot\xi - |\xi||t-y|^2 - \epsilon|\xi|^2} |\xi|^{\frac{m}{2}} u(y) \, dy \, dt \, d\xi$$

where

$$R = \{(y, t, \xi) : |\xi| \ge B, A_1 \le |t| \le A_2, y \in \text{supp } u\}.$$

Note that the function $\xi \mapsto |\xi|$ has a holomorphic extension $\langle \zeta \rangle$ in the region $|\Im \xi| < |\Re \xi|$, where

$$\langle \zeta \rangle = \left(\sum_{j=1}^{m} \zeta_j^2 \right)^{\frac{1}{2}}$$

and an appropriate branch of the square root is taken. We change the contour in the ξ integration from $\mathbb{R}^m$ to its image under the map $\zeta(\xi) = \xi + is|\xi|(x-y)$ for s small, $s > 0$. The number s is chosen to be small enough to ensure that for $\xi \neq 0$, $|\Im \zeta(\xi)| < |\Re \zeta(\xi)|$. We then have, modulo entire functions that converge uniformly to an entire function,

$$I_4^\epsilon(x) = \iiint e^{P(x,y,t,\xi,\epsilon)} \langle \zeta(\xi) \rangle^{\frac{m}{2}} u(y) \, dy \, dt \, d\zeta$$

where

$$P(x, y, t, \xi, \epsilon) = i(x-y) \cdot \xi - s|x-y|^2 |\xi| - \langle \zeta(\xi) \rangle |t - y|^2 - \epsilon \zeta(\xi)^2.$$

Note that for s small, $\Re \zeta(\xi)^2 \geq \frac{|\xi|^2}{2}$ and $\Re \langle \zeta(\xi) \rangle \geq \frac{|\xi|}{2}$. Hence the crucial exponential term can be bounded as follows:

$$\left| e^{P(x,y,t,\xi,\epsilon)} \right| \leq e^{-s|x-y|^2 |\xi| - \frac{|t-y|^2}{2} |\xi| - \frac{\epsilon}{2} |\xi|^2}.$$

In particular, when $x = 0$, since $|t| \geq A_1$, there is a constant $c > 0$ so that

$$\left| e^{P(0,y,t,\xi,\epsilon)} \right| \leq e^{-c|\xi|} \quad \text{for all } \xi.$$

This gives us enough freedom to complexify x to z and vary z near 0 to conclude that $I_4^\epsilon(z)$ converges uniformly to a holomorphic function near 0. $\quad \square$

We consider now the boundary values of holomorphic functions defined on wedges with flat edges, that is, edges that are open subsets of $\mathbb{R}^m$. Let $\Gamma \subseteq \mathbb{R}^m \setminus 0$ be an open convex cone with vertex at the origin, $V \subseteq \mathbb{R}^m$ open. For $\delta > 0$, let

$$\Gamma_\delta = \Gamma \cap \{ v : |v| < \delta \}.$$

If Γ' is another cone, we write $\Gamma' \subset\subset \Gamma$ if $\overline{\Gamma'} \cap S^{m-1} \subset \Gamma \cap S^{m-1}$ where S^{m-1} denotes the unit sphere in $\mathbb{R}^m$.

DEFINITION V.2.5. *A holomorphic function $f \in \mathcal{O}(V + i\Gamma_\delta)$ is said to be of tempered growth if there is an integer k and a constant c such that*

$$|f(x+iy)| \leq \frac{c}{|y|^k}.$$

For $f \in \mathcal{O}(V + i\Gamma_\delta)$, $\varphi \in C_0^\infty(V)$, and $v \in \Gamma$, set

$$\langle f_v, \varphi \rangle = \int f(x + iv)\varphi(x)dx.$$

THEOREM V.2.6. *Suppose* $f \in \mathcal{O}(V + i\Gamma_\delta)$ *is of tempered growth and* k *is as in the definition above. Then*

$$bf = \lim_{v \to 0, v \in \Gamma'} f_v$$

exists in $\mathcal{D}'(V)$ *and is of order* $k + 1$.

PROOF. Assume that

$$|f(x + iy)| \leq \frac{c}{|y|^k}.$$

We may assume that $\Gamma = \{y = (y_1, \ldots, y_m) : |y| < C_1 y_m\}$ for some $C_1 > 0$. Fix $y^0 \in \Gamma$. Let $\delta_0 = \frac{|y^0|}{2C_1}$. If $y \in \Gamma_{\delta_0}$, we have

(a) $y_m^0 \geq \frac{|y^0|}{C_1} \geq 2|y| \geq 2y_m$ and

(b) $|y_m^0 - y_m| \geq |y_m^0| - |y_m| \geq \frac{|y^0|}{C_1} - |y| > \frac{|y^0|}{2C_1}$.

Fix $\phi \in C_0^\infty(V)$. For $y \in \Gamma_\delta$, let

$$h(y) = \int f(x + iy)\phi(x)\,dx.$$

Using the growth condition on f and the fact that f is holomorphic, we can integrate by parts and arrive at

$$|D^\alpha h(y)| \leq \frac{CC_\alpha}{|y|^k} \quad \text{for all } \alpha \quad \text{where } C_\alpha = \sup|D^\alpha \phi|.$$

Let $|\beta| = k$. We will estimate $D^\beta h(y)$ on Γ_{δ_0}. Assume first that $k \geq 2$. For $y \in \Gamma_{\delta_0}$,

$$|D^\beta h(y) - D^\beta h(y^0)| = \left| \int_0^1 D(D^\beta h)(ty + (1-t)y^0) \cdot (y - y^0)\,dt \right|$$

$$\leq C \left(\sum_{|\alpha| = k+1} C_\alpha \right) \int_0^1 \frac{1}{|ty + (1-t)y^0|^k}\,dt$$

$$\leq C \left(\sum_{|\alpha| = k+1} C_\alpha \right) \int_0^1 \frac{1}{(ty_m + (1-t)y_m^0)^k}\,dt$$

$$\leq C \left(\sum_{|\alpha| = k+1} C_\alpha \right) \frac{1}{y_m^0 - y_m} \left(\frac{1}{y_m^{k-1}} - \frac{1}{(y_m^0)^{k-1}} \right)$$

$$\leq C \left(\sum_{|\alpha|=k+1} C_\alpha \right) \frac{1}{|y|^{k-1}}.$$

We have used (a) and (b) and the fact that since Γ is convex, $ty+(1-t)y^0 \in \Gamma$. Thus there is $C > 0$ such that for all β, $|\beta| = k$,

$$|D^\beta h(y)| \leq C \left(\sum_{|\alpha|=k+1} C_\alpha \right) \frac{1}{|y|^{k-1}} \quad \text{whenever } y \in \Gamma_{\delta_0}.$$

Continuing this way, we get δ_{k-2}, $C > 0$ such that

$$|D^2 h(y)| \leq C \left(\sum_{|\alpha|=k+1} C_\alpha \right) \frac{1}{|y|} \quad \text{whenever } y \in \Gamma_{\delta_{k-2}}.$$

Note that this inequality also holds when $k = 1$. Fix $y^{k-2} \in \Gamma_{\delta_{k-2}}$. Let $\delta_{k-1} = \frac{|y^{k-2}|}{2C_1}$. Using the preceding inequality, for $y \in \Gamma_{\delta_{k-1}}$, we can easily get:

$$|Dh(y)| \leq C \left(\sum_{|\alpha|=k+1} C_\alpha \right) |\log|y|| \quad \text{for some } C > 0.$$

Let now $y, y' \in \Gamma_{\delta_{k-1}}$. We have:

$$
\begin{aligned}
|h(y) - h(y')| &\leq \left| \int_0^1 Dh(ty+(1-t)y') \cdot (y-y') \, dt \right| \\
&\leq C \left(\sum_{|\alpha|=k+1} C_\alpha \right) \left| \int_0^1 |\log|ty+(1-t)y'| \, dt \right| |y-y'| \\
&\leq C \left(\sum_{|\alpha|=k+1} C_\alpha \right) \left(\int_0^1 \frac{1}{[ty_m+(1-t)y'_m]^{\frac{1}{2}}} \, dt \right) |y-y'| \\
&\leq C \left(\sum_{|\alpha|=k+1} C_\alpha \right) \left(\frac{y_m+y'_m}{\sqrt{y_m}+\sqrt{y'_m}} \right) \\
&\leq C \left(\sum_{|\alpha|=k+1} C_\alpha \right) (|y|+|y'|).
\end{aligned}
$$

Hence $\lim_{\Gamma \ni y \to 0} h(y)$ exists and as $\Gamma \ni y \to 0$,

$$|h(y)| \leq C \sum_{|\alpha| \leq k+1} ||D^\alpha \phi||_{L^\infty}$$

with C independent of ϕ. $\qquad\qquad\qquad\qquad\qquad\qquad\qquad\qquad\qquad\square$

REMARK V.2.7. We note here that when $m = 1$, the theorem above says that if a holomorphic function f defined on a rectangle $Q = (-a, a) \times (0, b)$ satisfies the growth condition $|f(x+iy)| \leq \frac{c}{y^k}$, then the traces $f(.+iy)$ converge in $\mathcal{D}'(-a,a)$ to a distribution of order $k+1$.

EXAMPLE V.2.8. Consider $f(x,y) = \frac{1}{x+iy}$ which is holomorphic and of tempered growth in the upper half-plane $y > 0$. By the theorem, f has a boundary value $bf \in \mathcal{D}'(\mathbb{R})$. It is not hard to show that in fact,

$$bf = \mathrm{pv}\left(\frac{1}{x}\right) - i\pi\delta_0$$

where pv denotes the Cauchy principal value.

Distributions which are boundary values of holomorphic functions of tempered growth arise quite naturally. Indeed, we have:

THEOREM V.2.9. *Any $u \in \mathcal{E}'(\mathbb{R}^m)$ can be expressed as a finite sum $\sum_{j=1}^n bf_j$ where each $f_j \in \mathcal{O}(\mathbb{R}^m + i\Gamma_j')$ for some cones $\Gamma_j' \subseteq \mathbb{R}^m$, and the f_j are of tempered growth.*

PROOF. Let $u \in \mathcal{E}'(\mathbb{R}^m)$. There exist an integer N and a constant $c > 0$ such that the Fourier transform $\hat{u}(\xi)$ satisfies the estimate $|\hat{u}(\xi)| \leq c(1+|\xi|)^N$. Let $\mathcal{C}_j, 1 \leq j \leq k$ be open, acute cones such that

$$\mathbb{R}^m = \bigcup_{j=1}^k \overline{\mathcal{C}_j}$$

and $\overline{\mathcal{C}_j} \cap \overline{\mathcal{C}_l}$ has measure zero when $j \neq l$. Define the cones

$$\Gamma_j = \{v \in \mathbb{R}^m : v \cdot \xi > 0 \quad \forall \xi \in \overline{\mathcal{C}_j}\}.$$

For each $j = 1, \ldots, k$, define

$$f_j(x+iy) = \frac{1}{(2\pi)^m} \int_{\mathcal{C}_j} e^{i(x+iy)\cdot\xi} \hat{u}(\xi) d\xi.$$

Note that f_j is holomorphic on $\mathbb{R}^m + i\Gamma_j$. Let Γ_j' be a cone, $\Gamma_j' \subset\subset \Gamma_j$. Then there exists $c > 0$ such that $y \cdot \xi \geq c|y||\xi| \quad \forall y \in \Gamma_j', \forall \xi \in \mathcal{C}_j$. Hence for $x+iy \in \mathbb{R}^m + i\Gamma_j'$,

$$|f_j(x+iy)| \leq \int e^{-c|y||\xi|} |\hat{u}(\xi)| d\xi$$

$$\leq c \int_{\mathbb{R}^m} e^{-c|y||\xi|} (1+|\xi|)^N d\xi$$

$$\leq \frac{c_j}{|y|^{m+N}}.$$

Thus each f_j is of tempered growth on $\mathbb{R}^m + i\Gamma'_j$ and so by Theorem V.2.6 the f_j have boundary values $bf_j \in \mathcal{D}'(\mathbb{R}^m)$. To prove $u = \sum_{j=1}^k bf_j$, let $\varphi \in C_0^\infty(V)$. Then

$$
\begin{aligned}
\langle bf_j, \varphi \rangle &= \lim_{y \to 0, y \in \Gamma'_j} \int_{\mathbb{R}^m} f_j(x+iy)\varphi(x)dx \\
&= \lim_{y \to 0, y \in \Gamma'_j} \int_{\mathbb{R}^m} \int_{\mathcal{C}_j} e^{i(x+iy)\cdot\xi} \; \varphi(x)\widehat{u}(\xi)\frac{d\xi}{(2\pi)^m}dx \\
&= \lim_{y \to 0, y \in \Gamma'_j} \frac{1}{(2\pi)^m} \int_{\mathcal{C}_j} e^{-y\cdot\xi} \; \widehat{u}(\xi)\widehat{\varphi}(-\xi)d\xi \\
&= \frac{1}{(2\pi)^m} \int_{\mathcal{C}_j} \widehat{u}(\xi)\widehat{\varphi}(-\xi)d\xi.
\end{aligned}
$$

Hence

$$
\langle u, \varphi \rangle = \sum_{j=1}^k \langle bf_j, \varphi \rangle. \qquad \square
$$

EXAMPLE V.2.10. Let $f_1(x, y) = \frac{1}{x+iy}$ for $y > 0$ and $f_2(x, y) = -\frac{1}{x+iy}$ for $y < 0$. Then it is not hard to show that

$$
-2\pi i\delta_0 = bf_1 + bf_2.
$$

Granted this, since $u * \delta_0 = u$ for any $u \in \mathcal{E}'(\mathbb{R})$, we get an explicit decomposition of u as a sum of two distributions each of which is the boundary value of a tempered holomorphic function on a half-plane.

DEFINITION V.2.11. *Let* $u \in \mathcal{D}'(\mathbb{R}^m)$, $x_0 \in \mathbb{R}^m$, $\xi^0 \in \mathbb{R}^m \backslash \{0\}$. *We say that u is microlocally analytic at* (x_0, ξ^0) *if there exist a neighborhood V of x_0, cones* $\Gamma^1, \dots, \Gamma^N$ *in* $\mathbb{R}^m \backslash \{0\}$, *and holomorphic functions* $f_j \in \mathcal{O}(V + i\Gamma_\delta^j)$ *(for some $\delta > 0$) of tempered growth such that* $u = \sum_{j=1}^N bf_j$ *near x_0 and* $\xi^0 \cdot \Gamma^j < 0$ $\forall j$.

REMARK V.2.12. When $m = 1$, if we take $x_0 = 0$ and $\xi^0 = -1$, then u is microlocally analytic at $(0, -1)$ if there is a tempered holomorphic f on some rectangle $(-a, a) \times (0, b)$ such that $u = bf$ on $(-a, a)$.

DEFINITION V.2.13. *The analytic wave front set of a distribution u, denoted* $WF_a(u)$, *is defined by*

$$
WF_a(u) = \{(x, \xi) : u \text{ is not microlocally analytic at } (x, \xi)\}.
$$

Observe that from Definition V.2.13 it can easily be shown that the analytic wave front set is invariant under an analytic diffeomorphism, and hence microlocal analyticity can be defined on any real-analytic manifold. The following theorem provides a very useful criterion for microlocal analyticity in terms of the FBI transform:

THEOREM V.2.14. *Let* $u \in \mathcal{D}'(\mathbb{R}^m)$, $x_0 \in \mathbb{R}^m$, $\xi^0 \in \mathbb{R}^m \backslash \{0\}$. *Then* $(x_0, \xi^0) \notin WF_a(u)$ *if and only if there is a neighborhood* V *of* x_0 *in* $\mathbb{R}^m$, *an open cone* $\Gamma \subset \mathbb{R}^m \backslash 0$, $\xi^0 \in \Gamma$ *and constants* $c_1, c_2 > 0$ *such that*

$$|F_u(x, \xi)| \leq c_1 e^{-c_2|\xi|} \quad \forall (x, \xi) \in V \times \Gamma.$$

The proof uses the inversion formula of Theorem V.2.2 and ideas similar to those in the proof of Theorem V.2.4 (see also Theorem V.3.7). The reader is referred to [**Sj1**] for the proof of this theorem.

COROLLARY V.2.15. *A distribution* u *is analytic near* x_0 *if and only if for every* $\xi^0 \in \mathbb{R}^m \backslash \{0\}$, $(x_0, \xi^0) \notin WF_a(u)$.

COROLLARY V.2.16. (The edge-of-the-wedge theorem.) *Let* $V \subset \mathbb{R}^m$ *be a neighborhood of the point* p, *and* Γ^+, Γ^- *be cones such that* $\Gamma^- = -\Gamma^+$. *Suppose for some* $\delta > 0$, $f^+ \in \mathcal{O}(V + i\Gamma_\delta^+)$, $f^- \in \mathcal{O}(V + i\Gamma_\delta^-)$ *are both of tempered growth and* $bf^+ = bf^-$. *Then there exists a holomorphic function* f *defined in a neighborhood of* p *that extends both* f^+ *and* f^-. *In particular,* bf^+ *is analytic at* p.

EXAMPLE V.2.17. Let

$$u(x) = \begin{cases} |x|^{\frac{3}{2}}, & x \geq 0 \\ i|x|^{\frac{3}{2}}, & x \leq 0. \end{cases}$$

Then $u(x) = bf(x)$ where $f(x, y) = (x + iy)^{\frac{3}{2}}$ for $y > 0$ and we take the principal branch of the fractional power. Since f is holomorphic for $y > 0$, it follows that $(0, -1) \notin WF_a(u)$. On the other hand, since u is not analytic (it is not even C^2), by Corollary V.2.15, $(0, 1) \in WF_a(u)$.

EXAMPLE V.2.18. Let (x, t) denote the variables in $\mathbb{R}^{m+n}$, $x \in \mathbb{R}^m$ and $t \in \mathbb{R}^n$. Let $\phi(t) = (\phi_1(t), \ldots, \phi_m(t))$ be real-analytic functions near the origin and consider the associated tube structure generated by

$$L_k = \frac{\partial}{\partial t_k} - i \sum_{j=1}^{m} \frac{\partial \phi_j}{\partial t_k} \frac{\partial}{\partial x_j}, \quad k = 1, \ldots, n.$$

It was shown in [**BT5**] that this system is analytic hypoelliptic at 0, i.e., every solution u of $L_k u = 0, k = 1, \ldots, n$ is analytic at 0 if and only if, for every $\xi \in \mathbb{R}^m$, the function

$$t \mapsto \phi(t) \cdot \xi$$

does not have a local minimum at 0. This result was proved using the FBI transform. The authors also proved a microlocal version of this result.

When a distribution u is a solution of a partial differential equation with analytic coefficients, the analyticity or microlocal analyticity of the solution can sometimes be established by using the FBI transform. Sections V.4 and V.5 contain results in this direction. The notes at the end of this chapter contain several references to such applications of the FBI transform.

V.3 Microlocal smoothness

In this section we introduce the concept of the C^∞ wave front set which is a refined way of describing the singularities of distributions. It is well known that a distribution u of compact support is C^∞ if and only if its Fourier transform $\hat{u}(\xi)$ decays rapidly as $|\xi| \to \infty$. More precisely, we recall Paley–Wiener's Theorem:

THEOREM V.3.1. (Theorem 7.3.1 in [H2].) *A distribution u with support in the ball $\{x \in \mathbb{R}^m : |x| \le R\}$ is C^∞ if and only if $\hat{u}(\zeta)$ is entire on $\mathbb{C}^m$ and for each positive integer k there is C_k such that*

$$|\hat{u}(\zeta)| \le C_k \frac{e^{R|\Im \zeta|}}{(1+|\zeta|)^k} \quad \forall \zeta \in \mathbb{C}^m.$$

DEFINITION V.3.2. *Let $u \in \mathcal{D}'(\Omega)$, $\Omega \subseteq \mathbb{R}^m$ open, $x_0 \in \Omega$, and $\xi^0 \in \mathbb{R}^m \backslash \{0\}$. We say u is microlocally smooth at (x_0, ξ_0) if there exists $\phi \in C_0^\infty(\Omega)$, $\phi \equiv 1$ near x_0 and a conic neighborhood $\Gamma \subseteq \mathbb{R}^m \backslash \{0\}$ of ξ^0 such that for all $k = 1, 2, \ldots$ and for all $\xi \in \Gamma$,*

$$|\widehat{\phi u}(\xi)| \le \frac{C_k}{(1+|\xi|)^k} \text{ on } \Gamma.$$

DEFINITION V.3.3. *The C^∞ wave front set of a distribution u denoted $WF(u)$ is defined by*

$$WF(u) = \{(x, \xi) : u \text{ is not microlocally smooth at } (x, \xi)\}.$$

It is easy to see that a distribution u is C^∞ if and only if $WF(u) = \emptyset$. When a distribution u is a solution of a linear partial differential equation with smooth coefficients, its wave front set is constrained. We quote here a basic result along this line:

THEOREM V.3.4. (Theorem 8.3.1 in [H2].) *Let $P = \sum_{|\alpha| \le k} a_\alpha(x) D^\alpha$ be a smooth linear partial differential operator on an open set $\Omega \subset \mathbb{R}^m$ and suppose $u \in \mathcal{D}'(\Omega)$. Then*

$$WF(u) \subset \text{char } P \cup WF(Pu)$$

where the characteristic set

$$\text{char } P = \left\{ (x, \xi) \in \Omega \times \mathbb{R}^m \backslash \{0\} : \sum_{|\alpha|=k} a_\alpha(x) \xi^\alpha = 0 \right\}.$$

In particular, if Pu is smooth, then $WF(u) \subset \text{char } P$. If Pu is smooth, and P is elliptic, then u has to be smooth. In Section V.5 we will consider an analogous result for solutions of first-order nonlinear partial differential equations.

DEFINITION V.3.5. *Let* $f \in C^\infty(\Omega)$, $\Omega \subseteq \mathbb{R}^m$ *open, and suppose* $\tilde{\Omega}$ *is a neighborhood of* Ω *in* $\mathbb{C}^m$. *A function* $\tilde{f}(x, y) \in C^\infty(\tilde{\Omega})$ *is called an almost analytic extension of* $f(x)$ *if* $\tilde{f}(x, 0) = f(x) \, \forall x \in \Omega$ *and for each* $j = 1, \ldots, m$,

$$\frac{\partial \tilde{f}}{\partial \bar{z}_j}(x, y) = O(|y|^k) \text{ for } k = 1, 2, \ldots$$

REMARK V.3.6. Lemma V.5.1 in Section V.5 shows that each smooth function of one real variable has an almost analytic extension. Such extensions also exist in higher dimensions (see [**GG**]).

The following theorem characterizes microlocal smoothness in terms of almost analytic extendability in certain wedges.

THEOREM V.3.7. *Let* $u \in \mathcal{D}'(\mathbb{R}^m)$. *Then* $(x_0, \xi^0) \notin WF(u)$ *if and only if there exist a neighborhood* V *of* x_0, *open acute cones* $\Gamma^1, \ldots, \Gamma^N$ *in* $\mathbb{R}^m \backslash \{0\}$, *and almost analytic functions* f_j *on* $V + i\Gamma_\delta^j$ *(for some* $\delta > 0$*) of tempered growth such that* $u = \sum_j^N bf_j$ *near* x_0 *and* $\xi^0 \cdot \Gamma^j < 0$ *for all* j.

PROOF. Suppose $(x_0, \xi^0) \notin WF(u)$. Let $\phi \in C_0^\infty(\mathbb{R}^m)$, $\phi \equiv 1$ near x_0 such that $\widehat{\phi u}(\xi)$ decays rapidly in a conic neighborhood of ξ^0. By the Fourier inversion formula,

$$\phi u = \frac{1}{(2\pi)^m} \int e^{ix \cdot \xi} \widehat{\phi u}(\xi) \, d\xi$$

where the formula is understood in the duality sense, that is, for $\psi \in C_0^\infty(\mathbb{R}^m)$,

$$\langle \phi u, \psi \rangle = \frac{1}{(2\pi)^m} \int \left(\int e^{ix \cdot \xi} \psi(x) \, dx \right) \widehat{\phi u}(\xi) \, d\xi.$$

Let $\mathcal{C}_j, 1 \leq j \leq N$ be open, acute cones such that

$$\mathbb{R}^m = \bigcup_{j=1}^N \overline{\mathcal{C}}_j$$

and $\overline{\mathcal{C}}_j \cap \overline{\mathcal{C}}_k$ has measure zero when $j \neq k$. We may assume that $\xi^0 \in \mathcal{C}_1$ and $\xi^0 \notin \overline{\mathcal{C}}_j$ for $j \geq 2$. This implies that we can get acute, open cones $\Gamma^j, 2 \leq j \leq N$

and a constant $c > 0$ such that

$$\xi^0 \cdot \Gamma^j < 0 \quad \text{and} \quad y \cdot \xi \geq c|y||\xi| \; \forall y \in \Gamma^j, \; \forall \xi \in \mathcal{C}_j.$$

For each $j = 2, \ldots, N$, define

$$f_j(x + iy) = \frac{1}{(2\pi)^m} \int_{\mathcal{C}_j} e^{i(x+iy)\cdot\xi} \widehat{\phi u}(\xi) \, d\xi$$

and set

$$g_1(x) = \frac{1}{(2\pi)^m} \int_{\mathcal{C}_1} e^{ix\cdot\xi} \widehat{\phi u}(\xi) \, d\xi.$$

For $j \geq 2$, f_j is holomorphic on $\mathbb{R}^m + i\Gamma^j$ and as we saw in the proof of Theorem V.2.9, it is of tempered growth and hence has a boundary value $bf_j \in \mathcal{D}'(\mathbb{R}^m)$. Since $(x_0, \xi^0) \notin WF(u)$, we may assume that $\widehat{\phi u}(\xi)$ decays rapidly in the cone $\mathcal{C}_1$. It is then easy to see that g_1 is C^∞ on $\mathbb{R}^m$. By Remark V.3.6, the function g_1 has an almost analytic extension f_1 which is smooth on $\mathbb{C}^m$. It follows that $u = \sum_j^N bf_j$ near x_0 with the f_j's as asserted. For the converse, we may assume that on some neighborhood V of x_0, $u = bf$ where f is almost analytic and of tempered growth on $V + i\Gamma$, Γ is an open cone, and $\xi^0 \cdot \Gamma < 0$. Let $\phi \in C_0^\infty(V)$, $\phi \equiv 1$ near x_0. We have

$$\widehat{\phi u}(\xi) = \langle u, \phi(x)e^{-ix\cdot\xi} \rangle = \lim_{\Gamma \ni y \to 0} \int_{\mathbb{R}^m} f(x+iy)e^{-ix\cdot\xi}\phi(x) \, dx.$$

Let $\Phi(x, y)$ be an almost analytic extension of $\phi(x)$. Fix $y_0 \in \Gamma$ and let

$$D = \{x + ity_0 \in \mathbb{C}^m : x \in V, \, 0 \leq t \leq 1\}.$$

Consider the m-form

$$f(x+iy)e^{-i(x+iy)\cdot\xi}\Phi(x, y) \, dz_1 \wedge \cdots \wedge dz_m$$

where each $z_j = x_j + iy_j$, $1 \leq j \leq m$. By Stokes' theorem,

$$\widehat{\phi u}(\xi) - \int_V f(x+iy_0)e^{-i(x+iy_0)\cdot\xi}\Phi(x, y_0) \, dx = \sum_{j=1}^m \int_D \frac{\partial}{\partial \bar{z}_j}(f\Phi)e^{-i(x+iy)\cdot\xi}$$

$$d\bar{z}_j \wedge dz_1 \wedge \cdots \wedge dz_m.$$

After contracting Γ if necessary, we may assume that for some $c > 0$, $y_0 \cdot \xi \leq -c|\xi|$ for all $\xi \in \Gamma$. This latter inequality, together with the almost analyticity of f and Φ, and the tempered growth of f, imply that on D, for any integer $k \geq 0$, we can find a constant C_k such that

$$\left| \frac{\partial}{\partial \bar{z}_j}(f\Phi)(x+ity_0) \right| \left| e^{-i(x+ity_0)\cdot\xi} \right| \leq C_k'|ty_0|^k |e^{ty_0\cdot\xi}| \leq \frac{C_k}{|\xi|^k}.$$

Observe also that the inequality $y_0 \cdot \xi \leq -c|\xi|$ $(\xi \in \Gamma)$ implies that the integral

$$\int_V f(x + iy_0) e^{-i(x+iy_0)\cdot\xi} \Phi(x, y_0) \, dx$$

decays rapidly in Γ. It follows that $(x_0, \xi^0) \notin WF(u)$. $\square$

COROLLARY V.3.8. *Let* $u \in \mathcal{D}'(\mathbb{R}^m)$. *If* $(x_0, \xi^0) \in WF(u)$, *then* $(x_0, \xi^0) \in WF_a(u)$.

V.4 Microlocal hypoanalyticity and the FBI transform

A *hypoanalytic structure* (or manifold) is an involutive structure $(\mathcal{M}, \mathcal{V})$ with charts (U_α, Z^α) where the U_α form an open covering of $\mathcal{M}$, and the $Z^\alpha = (Z_1^\alpha, \ldots, Z_m^\alpha)$ are a complete set of first integrals on U_α that are determined on the overlaps up to a local biholomorphism of $\mathbb{C}^m$. A basic example is a generic CR submanifold $\mathcal{M}$ of $\mathbb{C}^m$. A function f on a hypoanalytic manifold is said to be *hypoanalytic* if in a neighborhood of each point p it is of the form $f = h(Z_1, \ldots, Z_m)$ for some holomorphic function h defined in a neighborhood of $(Z_1(p), \ldots, Z_m(p))$ in $\mathbb{C}^m$. In the case of generic CR submanifolds of $\mathbb{C}^m$, the hypoanalytic functions are the restrictions to $\mathcal{M}$ of holomorphic functions defined in a neighborhood of $\mathcal{M}$. Hypoanalytic structures will be discussed some more in the epilogue. For more details on hypoanalytic structures, the reader is referred to [BCT] and [T5]. In this section we will briefly discuss the notion of the hypoanalytic wave front set. This notion is a generalization of the concept of microlocal analyticity we discussed in Section V.2 and the reader is referred to the work [BCT] for more details. We begin with the concept of a wedge in $\mathbb{C}^N$ whose edge is a generic CR manifold. Let $\mathcal{M}$ be a generic CR manifold in $\mathbb{C}^N$ of codimension d. Then dim $\mathcal{M} = 2n + d$, $m = n + d = N$ and the bundle $T' = T'\mathcal{M}$ is generated by the differentials of the restrictions to $\mathcal{M}$ of the N complex coordinates in $\mathbb{C}^N$. Fix $p \in \mathcal{M}$ and let $h = (h_1, \ldots, h_d)$ be smooth defining functions of $\mathcal{M}$ in a neighborhood U of p in $\mathbb{C}^N$.

DEFINITION V.4.1. *For* Γ *an open convex cone with vertex at* $0 \in \mathbb{R}^d$, *the set*

$$\mathcal{W}(U, h, \Gamma) = \{z \in U : h(z) \in \Gamma\}$$

is called a wedge with edge $\mathcal{M}$. *The wedge is said to be centered at* p *and to point in the direction of* Γ.

Observe that $\mathcal{W}(U, h, \Gamma)$ is an open set in $\mathbb{C}^N$ and $\mathcal{M} \cap U$ lies in its boundary. When $\mathcal{M}$ is a hypersurface, $\Gamma = (0, \infty)$ or $(-\infty, 0)$ and so a wedge with edge $\mathcal{M}$ in this case is simply a side of $\mathcal{M}$. Although the definition of a wedge involves the defining functions, the following proposition shows some independence from the defining functions.

PROPOSITION V.4.2. (Proposition 7.1.2 in [**BER**].) *Assume that* $h = (h_1, ..., h_d)$ *and* $g = (g_1, ..., g_d)$ *are two defining functions for* $\mathcal{M}$ *near* p. *Then there is a* $d \times d$ *real invertible matrix* B *such that for every* U *and* Γ *as above, the following holds: for any open convex cone* $\Gamma_1 \subseteq \mathbb{R}^d$ *with* $B\Gamma_1 \cap S^{d-1}$ *relatively compact in* $\Gamma \cap S^{d-1}$ (S^{d-1} *denotes the unit sphere in* $\mathbb{R}^d$), *there exists a neighborhood* U_1 *of* p *in* $\mathbb{C}^N$ *such that*

$$\mathcal{W}(U_1, g, \Gamma_1) \subseteq \mathcal{W}(U, h, \Gamma).$$

The reader is referred to [**BER**] for the proof of this proposition. We mention that if $a(z)$ is a $d \times d$ smooth invertible matrix satisfying $g = ah$ near p, then the matrix $B = [a(p)]^{-1}$.

DEFINITION V.4.3. *A holomorphic function* f *defined on a wedge* $\mathcal{W} = \mathcal{W}(U, h, \Gamma)$ *is said to be of* tempered growth *if there exists a constant* $c > 0$ *and an integer* k *such that*

$$|f(z)| \leq \frac{c}{|h(z)|^k} \qquad \forall z \in \mathcal{W}. \tag{V.4}$$

By using a diffeomorphism that flattens $\mathcal{M}$ near p, it is easy to see that the growth condition (V.4) is equivalent to

$$|f(z)| \leq \frac{c'}{\text{dist}(z, \mathcal{M})^k} \qquad \forall z \in \mathcal{W}.$$

Recall from Chapter I that for the generic $\mathcal{M}$ we can find complex coordinates $(z_1, ..., z_n, w_1, ..., w_d)$ vanishing at $p \in \mathcal{M}$, $z = x + iy \in \mathbb{C}^n$, $w = s + it \in \mathbb{C}^d$, and smooth real-valued functions $\phi_1, ..., \phi_d$ defined near $(0, 0)$ in (z, s) space with $\phi_k(0) = 0$, $d\phi_k(0) = 0$, $1 \leq k \leq d$ such that near 0, $\mathcal{M}$ is given by

$$\rho_k(z, w) = \phi_k(z, s) - t_k = 0, \quad 1 \leq k \leq d.$$

That is, near 0, $\mathcal{M} = \{(z, s + i\phi(z, s))\}$. By Proposition V.4.2, there exist $\epsilon > 0$ and a convex open cone $\Gamma' \subseteq \mathbb{R}^d$ such that if

$$\mathcal{W}' = \{(z, s + i\phi(z, s) + iv) : |z| < \epsilon, |s| < \epsilon, |v| < \epsilon, v \in \Gamma'\}$$

then $\mathcal{W}' \subseteq \mathcal{W}(U, h, \Gamma)$. The description of $\mathcal{W}'$ makes it clear what a wedge with edge $\mathcal{M}$ means. Observe also that a holomorphic function $f(z, w)$ on $\mathcal{W}'$ is of tempered growth if and only if it satisfies an estimate of the form

$$|f(z, s + i\phi(z, s) + iv)| \leq \frac{c}{|v|^k}$$

for $v \in \Gamma'$ small and $(z, s) \in \mathbb{C}^n \times \mathbb{R}^d$ near $(0, 0)$. Holomorphic functions of tempered growth in a wedge have distributional boundary values on the edge of the wedge. We have:

THEOREM V.4.4. (Theorem 7.2.6 in [**BER**].) *Let $f(z, w)$ be a holomorphic function of tempered growth in a wedge W' as above. Then there exists a CR distribution $u = bf$ defined in a neighborhood of 0 in $\mathcal{M}$ by*

$$\langle u, \psi \rangle = \lim_{\Gamma \ni v \to 0} \int_{\mathbb{R}^{2n+d}} f(z, s + i\phi(z, s) + iv)\psi(x, y, s)\mathrm{d}x\mathrm{d}y\mathrm{d}s$$

for any smooth function ψ of sufficiently small compact support near the origin in $\mathbb{R}^{2n+d}$.

PROOF. The proof will use arguments similar to those used in the proof of Theorem V.2.6. For $\psi(x, y, s)$ smooth, supported near the origin, set

$$h(v) = \int_{\mathbb{R}^{2n+d}} f(z, s + i\phi(z, s) + iv)\psi(x, y, s)\mathrm{d}x\mathrm{d}y\mathrm{d}s$$

for $v \in \Gamma'$, $|v| < \epsilon$. We will estimate the derivatives of h. We have

$$\frac{\partial h}{\partial v_j}(v) = \int_{\mathbb{R}^{2n+d}} i\frac{\partial f}{\partial w_j}(z, s + i\phi(z, s) + iv)\psi(x, y, s)\mathrm{d}x\mathrm{d}y\mathrm{d}s$$

for each $j = 1, \ldots, d$. Observe that since

$$\frac{\mathrm{d}}{\mathrm{d}s_m} f(z, s + i\phi(z, s) + iv) = \sum_{k=1}^{d} \frac{\partial f}{\partial w_k}(z, s + i\phi(z, s) + iv)\left(\delta_{km} + i\frac{\partial\phi_k}{\partial s_m}\right),$$

and the matrix $I + i\phi_s$ is invertible near the origin, there are smooth functions $a_{jm}(z, s)$ such that for each $k = 1, \ldots, d$,

$$\frac{\partial f}{\partial w_k}(z, s + i\phi(z, s) + iv) = \sum_{m=1}^{d} a_{km}(z, s)\frac{\mathrm{d}}{\mathrm{d}s_m}f(z, s + i\phi(z, s) + iv).$$

It follows that

$$\frac{\partial h}{\partial v_j}(v) = \sum_{m=1}^{d}\int_{\mathbb{R}^{2n+d}} \frac{\mathrm{d}}{\mathrm{d}s_m}f(z, s + i\phi(z, s) + iv)a_{jm}(z, s)\psi(x, y, s)\mathrm{d}x\mathrm{d}y\mathrm{d}s.$$

We can thus integrate by parts and iterate the procedure to conclude that for some constant $C > 0$ and every multi-index α,

$$|D^\alpha h(v)| \le \frac{CC_\alpha}{|v|^k}$$

where $C_\alpha = \sup |D^\alpha\psi|$. It then follows, as in the proof of Theorem V.2.6, that $h(v)$ has a limit as $\Gamma' \ni v \to 0$. Set $\langle u, \psi \rangle = \lim_{v \to 0} h(v)$. Note that u

is CR since it is the distributional limit of the CR functions $\mathcal{M} \ni (z, s) \longmapsto f(z, s + i\phi(z, s) + iv)$. □

The reader is referred to [**BER**] for an invariant formulation of Theorem V.4.4 (corollary 7.2.9 in [**BER**]).

Suppose now X is a hypoanalytic structure of codimension 0. Such an X often arises as a maximally real submanifold in a hypoanalytic structure. The structure bundle of X is all of $\mathbb{C}T^*X$ and since $\mathcal{V}$ is empty, any distribution is a solution. Fix $p \in X$ and let $Z = (Z_1, \ldots, Z_m)$ be a hypoanalytic chart near p. In a neighborhood V of p in X, the map $Z : V \longrightarrow \mathbb{C}^m$ is a diffeomorphism onto $Z(V)$. $Z(V)$ is a generic submanifold of $\mathbb{C}^m$ which is totally real of maximal dimension. In what follows, we will identify V with $Z(V)$.

DEFINITION V.4.5. *A distribution $u \in \mathcal{D}'(X)$ is* microlocally hypoanalytic *at $\sigma \in T_p^*X \backslash \{0\}$ if there exist open convex cones $\mathcal{C}_1, \ldots, \mathcal{C}_k$ in T_pX satisfying $\sigma(v) < 0 \quad \forall v \in \mathcal{C}_j, (1 \le j \le k)$ and wedges $\mathcal{W}_1, \ldots, \mathcal{W}_k$ in $\mathbb{C}^m$ with edge $Z(V)$ centered at p and pointing in the directions of $\Gamma_1, \ldots, \Gamma_k$ respectively such that $J\mathcal{C}_j \subseteq \Gamma_j$ and for each j, there is a holomorphic function of tempered growth u_j on $\mathcal{W}_j$ such that $u = \sum_{j=1}^k bu_j$ in V.*

DEFINITION V.4.6. *The* hypoanalytic wave front *set of u, denoted $WF_{ha}u$ is defined by*

$$WF_{ha}u = \{\sigma \in T^*X \backslash \{0\} : u \text{ is not microlocally hypoanalytic at } \sigma\}.$$

The hypoanalytic wave front set for solutions in structures of positive codimension is defined by restriction to a maximally real submanifold as follows ([**BCT**]). Let $(\mathcal{M}, \mathcal{V})$ be a hypoanalytic structure and u a distribution solution near $p \in \mathcal{M}$. Select a maximally real submanifold $\mathcal{X}$ through p. We recall that the restriction $u|_{\mathcal{X}}$ is well-defined and by Proposition V.1.29 $\mathcal{X}$ inherits a hypoanalytic structure of codimension 0. Hence the hypoanalytic wave front set $WF_{ha}(u|_{\mathcal{X}})$ is defined and lives in $T^*\mathcal{X} \backslash \{0\}$. Since $\mathcal{X}$ is maximally real, by Propositions V.1.27 and V.1.29, the inclusion $i_{\mathcal{X}} : \mathcal{X} \to \mathcal{M}$ induces an injection $i_{\mathcal{X}}^* : T^0|_{\mathcal{X}} \to T^*\mathcal{M}$. We will say a covector $\sigma \in T_p^0 \backslash \{0\}$ is in the hypoanalytic wave front set of u if $i_{\mathcal{X}}^*\sigma \in WF_{ha}(u|_{\mathcal{X}})$. This set will be denoted by $WF_{ha,p}^{\mathcal{M}}(u)$. This definition is independent of the choice of the maximally real submanifold $\mathcal{X}$ through p (see [**BCT**] for the proof) and thus for any such $\mathcal{X}$, we have a bijection $i_{\mathcal{X}}^* : WF_{ha,p}^{\mathcal{M}}(u) \to WF_{ha,p}(u|_{\mathcal{X}})$, where $WF_{ha,p}$ denotes the hypoanalytic wave front set at p.

We will next recall the FBI transform of [**BCT**] which gives a very useful Fourier transform criterion for microlocal hypoanalyticity. X is a hypoanalytic structure of codimension 0 as above. If $p \in X$, by the results in Chapter I (see for example Corollary I.10.2), we may choose local coordinates $x_1, \ldots, x_m$ for X vanishing at p so that locally, X becomes a neighborhood U of 0 in $\mathbb{R}^m$ and we may assume that a hypoanalytic chart has the form

$$Z_j = x_j + i\phi_j(x), \qquad 1 \leq j \leq m,$$

$\phi = (\phi_1, \ldots, \phi_m)$ real-valued. For $\kappa > 0$ and $u \in \mathcal{E}'(U)$, define

$$F^\kappa(u, z, \zeta) = \int_U e^{i\zeta \cdot (z - Z(y)) - \kappa \langle \zeta \rangle [z - Z(y)]^2} u(y) \, dZ(y)$$

where $z \in \mathbb{C}^m$, $[w]^2 = w_1^2 + \cdots + w_m^2$, and for any $\zeta \in \mathbb{C}^m$ with $|\Im \zeta| < |\Re \zeta|$, $\langle \zeta \rangle = (\zeta_1^2 + \cdots + \zeta_m^2)^{\frac{1}{2}}$ (the principal branch of the square root).

DEFINITION V.4.7. $F^\kappa(u, z, \zeta)$ *is called the FBI transform of u (with parameter κ).*

In [**BCT**] the authors characterized microlocal hypoanalyticity in terms of an exponential decay of the FBI transform. In particular, when $\phi(0) = 0$ and $d\phi(0) = 0$, they proved:

THEOREM V.4.8. *There is a universal constant $M > 0$ such that if $\kappa > M$* $\sup\limits_{x \in U, |\alpha| = 2} \partial^\alpha \phi(x)$, *the following holds: for $\sigma \in \mathbb{R}^m \backslash \{0\}$, $u \in \mathcal{E}'(U)$, V a neighborhood of 0 in $\mathbb{C}^m$, $\Gamma \subseteq \mathbb{C}^m \backslash \{0\}$ a complex conic neighborhood of σ, if*

$$|F^\kappa(u, z, \zeta)| \leq c_1 e^{-c_2 |\zeta|}, \qquad \forall z \in V, \quad \forall \zeta \in \Gamma$$

and for some $c_1, c_2 > 0$, then $(0, \sigma) \notin WF_{ha} u$.

Here U is a neighborhood of 0 in $\mathbb{R}^m$.

V.5 Application of the FBI transform to the C^∞ wave front set of solutions of nonlinear PDEs

In this section the FBI transform will be used to prove a result on the C^∞ wave front set of solutions of first-order nonlinear PDEs. Suppose $u = u(x, t)$ is a C^2 solution of a nonlinear pde

$$u_t = f(x, t, u, u_x)$$

where $f(x, t, \zeta_0, \zeta)$ is complex-valued, C^∞ in all the variables, and holomorphic in (ζ_0, ζ). Here x varies in an open set in $\mathbb{R}^m$, t in an interval of $\mathbb{R}$, and (ζ_0, ζ) in an open set in $\mathbb{C}^{m+1}$. We will present Asano's ([**A**]) proof of

Chemin's ([**Che**]) result that the C^∞ wave front set of any C^2 solution is contained in the characteristic set of the linearized operator

$$L^u = \frac{\partial}{\partial t} - \sum_{j=1}^{m} \frac{\partial f}{\partial \zeta_j}(x, t, u, u_x) \frac{\partial}{\partial x_j}.$$

We begin with some lemmas about linear vector fields:

LEMMA V.5.1. *Let*

$$L = \frac{\partial}{\partial t} + \sum_{j=1}^{N} a_j(x, t, \zeta) \frac{\partial}{\partial x_j} + \sum_{k=1}^{M} b_k(x, t, \zeta) \frac{\partial}{\partial \zeta_k}$$

where the coefficients a_j and b_k are C^∞ in the variables $(x, t) \in \Omega \times J \subset \mathbb{R}^N \times \mathbb{R}$ and holomorphic in the variable $\zeta \in \mathcal{N} \subset \mathbb{C}^M$, $\mathcal{N}$ open. Let $f(x, \zeta)$ be a C^∞ function defined on $\Omega \times \mathcal{N}$, holomorphic in ζ. There exists a C^∞ function $u(x, t, \zeta)$ holomorphic in ζ which is an approximate solution of $Lu = 0$ in the sense that

$$Lu(x, t, \zeta) = O(t^k) \qquad for \quad k = 1, 2, \ldots$$

and such that $u(x, 0, \zeta) = f(x, \zeta)$.

PROOF. The conditions that u has to satisfy determine the Taylor coefficients of the formal series

$$u(x, t, \zeta) = \sum_{j=0}^{\infty} u_j(x, \zeta) t^j$$

where $u_j(x, \zeta) = \frac{\partial_t^j u(x, 0, \zeta)}{j!}$. Set $u_0(x, \zeta) = f(x, \zeta)$. For each j, since we want $Lu = O(t^{j+1})$, we must have $\partial_t^{j-1}(Lu)(x, 0, \zeta) = 0$. This then leads to

$$u_j(x, \zeta) = -\frac{1}{j} \sum_{p+q=j-1} \frac{1}{q!} \left[\sum_{k=1}^{N} \frac{\partial u_p}{\partial x_k}(x, \zeta) \frac{\partial^q a_k}{\partial t^q}(x, 0, \zeta) \right.$$
$$\left. + \sum_{k=1}^{M} \frac{\partial u_p}{\partial \zeta_k}(x, \zeta) \frac{\partial^q b_k}{\partial t^q}(x, 0, \zeta) \right]$$

for $j \geq 1$. Note that the functions $u_j(x, \zeta)$ are C^∞ and holomorphic in ζ. Let $\chi \in C_0^\infty(\mathbb{R})$ be such that $\chi \geq 0$, $\chi \equiv 1$ in $[-\frac{1}{2}, \frac{1}{2}]$ and supp $\chi \subset [-1, 1]$. Then there exists a sequence $R_j > 1$, $R_j \nearrow +\infty$ such that the series

$$u(x, t, \zeta) = \sum_{j=1}^{\infty} \chi(R_j t) u_j(x, \zeta) t^j$$

is convergent in C^∞. It follows that u is C^∞ in all the variables and holomorphic in ζ. Moreover, from the way the functions u_j are defined, u is an approximate solution of $Lu = 0$ with the property that $u(x, 0, \zeta) = f(x, \zeta)$. $\quad\square$

In the following lemma, WF denotes the C^∞ wave front set.

LEMMA V.5.2. *Let* $X \subset \mathbb{R}^m$ *be open, U an open neighborhood of $X \times \{0\}$ in* $\mathbb{R}^{m+1}$, $U_+ = U \cap \mathbb{R}^{m+1}_+$. *Let*

$$L = \frac{\partial}{\partial t} + \sum_{j=1}^m a_j(x, t) \frac{\partial}{\partial x_j}$$

be a C^l vector field in U for some positive integer l. Assume $f \in C^1(\overline{U_+})$ satisfies

$$|Lf(x, t)| = O(t^k), \quad k = 1, 2, \ldots$$

uniformly on compact subsets of X. Suppose there exist C^l functions

$$\Psi_1(x, t), \ldots, \Psi_m(x, t)$$

on U such that $Z(x, t) = x + t\Psi(x, t)$ satisfies $Z(x, 0) = x$ and

$$LZ(x, t) = O(t^k), \quad k = 1, 2, \ldots$$

Let $a(x, t) = (a_1(x, t), \ldots, a_m(x, t))$. Assume

$$\partial_t^j a(x, 0) = 0 \quad \forall j < l, \quad \forall x \in X$$

and that

$$\langle \partial_t^l \Im a(x_0, 0), \xi^0 \rangle > 0 \quad \text{for some } x_0 \in X, \quad \xi^0 \in \mathbb{R}^m.$$

Then $(x_0, \xi^0) \notin WF(f(x, 0))$.

REMARK V.5.3. If L is C^∞, then Lemma V.5.1 insures that the Z_j exist and the proof below will show that in this case, we only need to assume that $f \in C([0, T], \mathcal{D}'(\mathbb{R}^m))$.

PROOF. Without loss of generality, we may assume that $x_0 = 0$. For $j = 1, \ldots, m$ let $M_j = \sum_{k=1}^m b_{jk}(x, t) \frac{\partial}{\partial x_k}$ be vector fields satisfying

$$M_j Z_k = \delta_j^k, \quad [M_j, M_k] = 0.$$

Note that for each j,

$$[M_j, L] = \sum_{s=1}^m c_{js} M_s \tag{V.5}$$

where each $c_{js} = O(t^k)$, $k = 1, 2, \ldots$ Indeed, the latter can be seen by expressing $[M_j, L]$ in terms of the basis $\{L, M_1, \ldots, M_m\}$ and applying both sides to the $m + 1$ functions $\{t, Z_1, \ldots, Z_m\}$. For any C^1 function g, observe that the differential

$$dg = \sum_{k=1}^{m} M_k(g) dZ_k + \left(Lg - \sum_{k=1}^{m} M_k(g) LZ_k \right) dt. \qquad (V.6)$$

This is verified by evaluating each side at the basis vector fields

$$\{L, M_1, \ldots, M_m\}.$$

Using (V.6) we get:

$$d(g dZ_1 \wedge \cdots \wedge dZ_m) = \left(Lg - \sum_{k=1}^{m} M_k(g) LZ_k \right) dt \wedge dZ_1 \wedge \cdots \wedge dZ_m. \qquad (V.7)$$

For $\xi \in \mathbb{R}^m$, $s \in \mathbb{R}^m$, let

$$E(s, \xi, x, t) = i\xi \cdot (s - Z(x, t)) - |\xi|[s - Z(x, t)]^2,$$

where for $w \in \mathbb{C}^m$, we write $[w]^2 = \sum_{j=1}^{m} w_j^2$. Let B denote a small ball centered at 0 in $\mathbb{R}^m$ and $\phi \in C_0^\infty(B)$, $\phi \equiv 1$ near the origin. We will apply (V.7) to the function

$$g(s, \xi, x, t) = \phi(x) f(x, t) e^{E(s, \xi, x, t)}$$

where (s, ξ) are parameters. We get:

$$d(g dZ) = \left\{ L(\phi f) + (\phi f) LE - \sum_{k=1}^{m} (M_k(\phi f) + \phi f(M_k E)) LZ_k \right\} e^E dt \wedge dZ, \qquad (V.8)$$

where $dZ = dZ_1 \wedge \cdots \wedge dZ_m$. Next by Stokes' theorem we have, for $t_1 > 0$ small:

$$\int_B g(s, \xi, x, 0) dx = \int_B g(s, \xi, x, t_1) d_x Z(x, t_1) + \int_0^{t_1} \int_B d(g dZ). \qquad (V.9)$$

We will estimate the two integrals on the right in (V.9). Write

$$Z = (Z_1, \ldots, Z_m) = x + t\Psi(x, t), \quad \text{and} \quad \Psi = \Psi_1 + i\Psi_2.$$

Since the Z_j are approximate solutions of L, we have

$$\Psi + t\Psi_t + (I + t\Psi_x) \cdot a = O(t^k), \quad k = 1, 2, \ldots$$

and hence

$$\partial_t^j \Psi(x, 0) = 0, \quad j < l \text{ and } \langle \partial_t^l \Psi_2(x, 0), \xi^0 \rangle < 0 \qquad (V.10)$$

for x in a neighborhood V of $\overline{B}$ (after shrinking B, if necessary). Observe that

$$\Re E(s, \xi, x, t) = t\xi \cdot \Psi_2(x, t) - |\xi|((s - x - t\Psi_1)^2 - t^2\Psi_2(x, t)^2).$$

Because of (V.10), continuity and homogeneity in ξ, we can get $c_1 > 0$ such that

$$\Re E(s, \xi, x, t) \leq -c_1|\xi|t^{l+1}, \quad \text{for } x \in V, \quad 0 \leq t \leq t_1 \tag{V.11}$$

$s \in \mathbb{R}^m$ and ξ in a conic neighborhood Γ of ξ^0. Going back to the integrals in (V.9), we clearly have

$$\left| \int_B g(s, \xi, x, t_1) d_x Z(x, t_1) \right| \leq e^{-c_2|\xi|},$$

for some $c_2 > 0$, for $s \in \mathbb{R}^m$ and $\xi \in \Gamma$. To estimate $\int_0^{t_1} \int_B d(g dZ)$, we use (V.8) and look at each term that appears there. We first consider the term $L(\phi f)e^E$. For any k,

$$|\phi(Lf)e^E| \leq C_k t^{lk} e^{-c_1 t^l|\xi|} \leq \frac{C'_k}{|\xi|^k}.$$

Moreover, for the x-integral

$$\int_B (L\phi) f e^E dZ = \langle f(., t), (L\phi)e^E \rangle$$

after decreasing t_1, we can get $\delta > 0$ such that if $|s| \leq \delta$ and $\xi \in \Gamma$,

$$|\langle f(., t), (L\phi)e^E \rangle| \leq C e^{-c|\xi|}$$

for some constants $c, C > 0$. In the latter, we have used the constancy of ϕ near 0. It follows that the integral

$$\int_B \int_0^{t_1} L(\phi f)e^E dt \wedge dZ$$

decays rapidly in ξ. The term $(\phi f)LEe^E$ is estimated using the fact that for any k, $|LE| \leq c_k t^k|\xi|$ for some constant c_k and that $|e^E| \leq e^{-c_1 t^l|\xi|}$. This shows that

$$\int_B \int_0^{t_1} (\phi f)LEe^E dt \wedge dZ$$

decays rapidly in ξ. The integrals of the terms $\phi f(M_k E)LZ_k e^E$ and $(M_k(\phi f))LZ_k e^E$ are estimated in the same fashion. Thus

$$\int_B \int_0^{t_1} d(g dZ)$$

has a rapid decay in ξ, and going back to (V.9), we have shown:

$$F(s, \xi) = \int_B e^{i\xi \cdot (s-x) - |\xi||[s-x]^2} \phi(x) f(x, 0) dx \qquad (V.12)$$

decays rapidly for $|s| \leq \delta$ in $\mathbb{R}^m$ and ξ in a conic neighborhood Γ of ξ^0. The function $F(s, \xi)$ is the standard FBI transform of the distribution $\phi(x) f(x, 0)$. To conclude the proof, we will exploit the inversion formula for the FBI, namely,

$$\phi(x) f(x, 0) = \lim_{\epsilon \to 0^+} c_m \iint e^{i(x-s) \cdot \xi - \epsilon |\xi|^2} F(s, \xi) |\xi|^{\frac{n}{2}} ds d\xi \qquad (V.13)$$

where c_m is a dimensional constant. Assume now that $\phi(x)$ is supported in the ball centered at the origin with radius M. We will study the inversion integral in (V.13) by writing it as a sum of three pieces: $I_1(\epsilon)$, $I_2(\epsilon)$, and $I_3(\epsilon)$. The first piece consists of integration over the region $\{(\xi, s) : |s| \geq 2M\}$. In the second piece we integrate over $\{(\xi, s) : \delta \leq |s| < 2M\}$, and in the third piece over $\{(\xi, s) : |s| \leq \delta\}$. For the integral $I_1(\epsilon)$, after integrating in s, one gets an exponential decay in ξ independent of ϵ, and hence $\lim_{\epsilon \to 0^+} I_1(\epsilon)$ is in fact a holomorphic function near the origin in $\mathbb{C}^m$. To study the second piece, we write it as

$$I_2(\epsilon) = c_m \int_{\{(y, \xi, s) : \delta \leq |s| < 2M\}} e^{i(x-y) \cdot \xi - |\xi||[s-y]^2 - \epsilon |\xi|^2} \phi(y) f(y, 0) |\xi|^{\frac{m}{2}} dy ds d\xi.$$

We will use the holomorphic function $\langle \zeta \rangle = (\zeta_1^2 + \cdots + \zeta_m^2)^{\frac{1}{2}}$ where we take the principal branch of the square root in the region $|\Im \zeta| < |\Re \zeta|$. Observe that this function is a holomorphic extension of $|\xi|$ away from the origin. In the ξ integration above, we can deform the contour to the image of

$$\zeta(\xi) = \xi + i\beta |\xi| (x - y)$$

where β is chosen sufficiently small. In particular, we choose β so that when x varies near the origin and y stays in the support of ϕ, then $|\Im \zeta(\xi)| < |\Re \zeta(\xi)|$, away from $\xi = 0$. In the integrand of $I_2(\epsilon)$, if $|x| \leq \frac{\delta}{4}$, we get an exponential decay independent of ϵ. It follows that this piece is also holomorphic near the origin in $\mathbb{C}^m$ after setting $\epsilon = 0$. Finally, for the third piece, let $\Gamma_1, \ldots, \Gamma_n$ be convex cones such that with $\Gamma_0 = \Gamma$,

$$\mathbb{R}^m = \bigcup_{j=0}^n \Gamma_j,$$

and for each $j \geq 1$ there exists a vector v_j satisfying $v_j \cdot \overline{\Gamma_j} > 0$ and $v_j \cdot \xi^0 < 0$. We now write

$$I_3(\epsilon) = \sum_{j=0}^n K_j(\epsilon),$$

where K_j equals the integral over Γ_j. The decay in the FBI established in (V.12) shows us that K_0 is a smooth function even after setting $\epsilon = 0$. Each of the remaining functions K_j, after setting $\epsilon = 0$, is a boundary value of a tempered holomorphic function in a wedge whose inner product with ξ^0 is negative. Hence

$$(0, \xi^0) \notin WF_a(K_j(0+)),$$

where WF_a denotes the analytic wave front set. By Corollary V.3.8, the latter implies that

$$(0, \xi^0) \notin WF(K_j(0+)).$$

We have thus proved that

$$(0, \xi^0) \notin WF(f(x, 0)).$$

$\square$

Consider now the vector field

$$L = \frac{\partial}{\partial t} + \sum_{j=1}^{m} a_j(x, t) \frac{\partial}{\partial x_j}$$

where the a_j are C^1 on an open set $\Omega \subset \mathbb{R}_x^m \times \mathbb{R}_t$. To L we associate vector fields

$$L^\theta = \frac{\partial}{\partial s} - e^{-i\theta} L$$

where $s \in \mathbb{R}$ is a new variable and $\theta \in [0, 2\pi)$ is a parameter. Suppose that for each $\theta \in [0, 2\pi)$ there exist C^1 functions

$$\Psi_1^\theta(x, t, s), \ldots, \Psi_m^\theta(x, t, s)$$

defined on $\Omega \times J$ ($J \subset \mathbb{R}$ is an open interval centered at the origin) such that

$$Z_j^\theta(x, t, s) = x_j + s\Psi_j^\theta(x, t, s), \quad j = 1, \ldots, m$$

are approximate solutions of $L^\theta Z_j^\theta = 0$ in the sense that $L^\theta Z_j^\theta$ are s-flat at $s = 0$. Define also $\Psi_{m+1}^\theta(x, t, s) = e^{-i\theta}$ and $Z_{m+1}^\theta(x, t, s) = t + e^{-i\theta}s$ and note that $L^\theta Z_{m+1}^\theta = 0$. If we write $\Psi^\theta = (\Psi_1^\theta, \ldots, \Psi_{m+1}^\theta)$ and $Z^\theta = (Z_1^\theta, \ldots, Z_{m+1}^\theta)$, then

$$Z_s^\theta(0, 0, 0) = \Psi^\theta(0, 0, 0) = -\begin{pmatrix} e^{-i\theta} a(0, 0) \\ e^{-i\theta} \end{pmatrix} = e^{-i\theta}\begin{pmatrix} \Psi(0, 0) \\ -1 \end{pmatrix}$$

and

$$\begin{pmatrix} \xi \\ \tau \end{pmatrix} \cdot \Im\Psi^\theta(0, 0, 0) = \begin{pmatrix} \xi \\ \tau \end{pmatrix} \cdot \begin{pmatrix} \Im\Psi(0, 0) \cos\theta - \Re\Psi(0, 0) \sin\theta \\ \sin\theta \end{pmatrix}$$

$$= \xi \cdot \Im\Psi(0, 0) \cos\theta + (\tau - \xi \cdot \Re\Psi(0, 0)) \sin\theta.$$

So the condition

$$\begin{pmatrix} \xi \\ \tau \end{pmatrix} \cdot \Im \Psi^\theta(0,0,0) \neq 0$$

for some $\theta \in [0, 2\pi)$ is equivalent to saying that $(0, 0, \xi, \tau)$ is not in the characteristic set of L. Suppose now $h(x, t)$ is a C^1 function with the following property: there exist C^1 functions $h^\theta(x, t, s)$ such that $h^\theta(x, t, 0) = h(x, t)$ and $L^\theta h^\theta$ is s-flat at $s = 0$. If $(0, 0, \xi^0, \tau^0)$ is not in the characteristic set of L, we know that there is $\theta \in [0, 2\pi)$ such that

$$\begin{pmatrix} \xi^0 \\ \tau^0 \end{pmatrix} \cdot \Im \Psi^\theta(0,0,0) \neq 0.$$

By replacing θ by $\theta + \pi$ or $\theta - \pi$ if necessary, we may assume that

$$\begin{pmatrix} \xi^0 \\ \tau^0 \end{pmatrix} \cdot \Im \Psi^\theta(0,0,0) < 0$$

and we can apply what we saw in the proof of Lemma V.5.2 to an FBI in (x, t)-space to conclude the following: there exist a conic neighborhood Γ of (ξ^0, τ^0) in $\mathbb{R}^{m+1} \backslash \{0\}$ and a neighborhood $\mathcal{O}$ of the origin in $\mathbb{R}^{m+1}$ such that

$$Fh^\theta(0; x', t', \xi, \tau)$$

$$= \int_{B \times J} e^{i[\xi \cdot (x'-x) + \tau(t'-t)] - |(\xi,\tau)|[\langle x'-x \rangle^2 + (t'-t)^2]} h^\theta(x, t, 0) dx dt$$

$$= Fh(x', t', \xi, \tau)$$

is rapidly decreasing for $(\xi, \tau) \in \Gamma$ and $(x', t') \in \mathcal{O}$. We have thus proved:

LEMMA V.5.4. *For each $\theta \in [0, 2\pi)$ let $L^\theta = \frac{\partial}{\partial s} - e^{-i\theta} L$ and suppose there exist $\Psi_1^\theta, \ldots, \Psi_{m+1}^\theta \in C^1(\Omega \times J)$ such that $Z^\theta = (x, t) + s\Psi^\theta(x, t, s)$ is an approximate solution of $L^\theta Z^\theta = 0$ in the sense that $L^\theta Z^\theta$ is s-flat at $s = 0$. Suppose moreover that there exist $h^\theta \in C^1(\Omega \times J)$ such that $h^\theta(x, t, 0) = h(x, t)$ and $L^\theta h^\theta$ is s-flat at $s = 0$. Then*

$$WF(h)\big|_0 \subset (\mathrm{char} L)\big|_0 .$$

The preceding linear results will next be applied to a nonlinear equation. Let $\Omega \subset \mathbb{R}^{m+1}$ be a neighborhood of the origin and suppose $u \in C^2(\Omega)$ is a solution of

$$u_t = f(x, t, u, u_x) \tag{V.14}$$

where $f(x, t, \zeta_0, \zeta)$ is a C^∞ function in the variables $(x, t) \in \Omega$ and holomorphic in the variables

$$(\zeta_0, \zeta) \in \mathcal{N} \subset \mathbb{C} \times \mathbb{C}^M, \quad (a, \omega) = (u(0,0), u_x(0,0)) \in \mathcal{N}.$$

Consider

$$\mathcal{L} = \frac{\partial}{\partial t} - \sum_{j=1}^{m} \left(\frac{\partial f}{\partial \zeta_j} \right)(x, t, \zeta_0, \zeta) \frac{\partial}{\partial x_j} \tag{V.15}$$

and

$$L^u = \frac{\partial}{\partial t} - \sum_{j=1}^{m} \left(\frac{\partial f}{\partial \zeta_j} \right)(x, t, u, u_x) \frac{\partial}{\partial x_j}.$$

Let $v = (u, u_x)$. It is easy to check that v solves the quasi-linear system

$$L^u v = g(x, t, v) \tag{V.16}$$

where

$$g_0(x, t, \zeta_0, \zeta) = f(x, t, \zeta_0, \zeta) - \sum_{j=1}^{m} \zeta_j \frac{\partial f}{\partial \zeta_j}(x, t, \zeta_0, \zeta)$$

and

$$g_i(x, t, \zeta_0, \zeta) = f_{x_i}(x, t, \zeta_0, \zeta) - \zeta_i \frac{\partial f}{\partial \zeta_0}(x, t, \zeta_0, \zeta), \qquad i = 1, \ldots, m.$$

Consider now the principal part of the holomorphic *Hamiltonian* of (V.16)

$$H = \mathcal{L} + g_0 \frac{\partial}{\partial \zeta_0} + \sum_{j=1}^{m} g_j \frac{\partial}{\partial \zeta_j}.$$

For $\Psi(x, t, \zeta_0, \zeta)$ a C^∞ function in (x, t, ζ, ζ_0) and holomorphic in the variables $(\zeta_0, \zeta) \in \mathcal{N}$, set $\Psi^v(x, t) = \Psi(x, t, v(x, t))$ and let $\mathcal{L}^p$ denote the vector field in Ω obtained by plugging $p(x, t)$ for (ζ, ζ_0) in the coefficients of $\mathcal{L}$. Note that $\mathcal{L}^v = L^u$. Equation (V.16) implies that

$$\mathcal{L}^v \Psi^v = (H\Psi)^v$$

where $\Psi(x, t, \zeta_0, \zeta)$ is any C^∞ function in $(x, t) \in \Omega$ and holomorphic in $(\zeta_0, \zeta) \in \mathcal{N}$. Let $Z_j(x, t, \zeta_0, \zeta), j = 1, \ldots, m$, and $\Xi_k(x, t, \zeta_0, \zeta), k = 0, \ldots, m$ be t-flat solutions of $H\Psi = 0$ such that $Z_j(x, 0, \zeta_0, \zeta) = x_j, j = 1, \ldots, m$, and $\Xi_k(x, 0, \zeta_0, \zeta) = \zeta_k, k = 0, \ldots, m$. Let $\check{Z}(z, t, \zeta_0, \zeta)$ and $\check{\Xi}(z, t, \zeta_0, \zeta)$, $z = x + iy \in \mathbb{R}^m \oplus i\mathbb{R}^m$ be almost analytic extensions of $Z(x, t, \zeta_0, \zeta)$ and $\Xi(x, t, \zeta_0, \zeta)$ respectively, i.e., $\check{Z}(x, t, \zeta_0, \zeta) = Z(x, t, \zeta_0, \zeta)$, $\check{\Xi}(x, t, \zeta_0, \zeta) = \Xi(x, t, \zeta_0, \zeta)$ and for all $k \in \mathbb{N}$ there exists $C_k > 0$ such that for $j = 1, \ldots, m$ we have

$$\left| \frac{\partial}{\partial \bar{z}_j} \check{Z}(z, t, \zeta_0, \zeta) \right| \leq C_k |\Im z|^k$$

and

$$\left| \frac{\partial}{\partial \bar{z}_j} \tilde{\Xi}(z, t, \zeta_0, \zeta) \right| \leq C_k \left| \Im z \right|^k.$$

Since the Jacobian

$$\frac{\partial(\Re \tilde{Z}, \Im \tilde{Z}, \Re \tilde{\Xi}, \Im \tilde{\Xi})}{\partial(\Re z, \Im z, \Re \zeta_0, \Im \zeta_0, \Re \zeta, \Im \zeta)}$$

is nonsingular near $t = 0$, we may solve

$$\begin{cases} \tilde{Z}(z, t, \zeta_0, \zeta) & = \tilde{Z}, \\ \tilde{\Xi}(z, t, \zeta_0, \zeta) & = \tilde{\Xi} \end{cases}$$

with respect to (z, ζ_0, ζ) in a neighborhood of $(0, a, \omega)$ by the implicit function theorem and get

$$\begin{cases} z & = P(\tilde{Z}, t, \tilde{\Xi}), \\ (\zeta_0, \zeta) & = Q(\tilde{Z}, t, \tilde{\Xi}) \end{cases}$$

with $P(0, 0, a, \omega) = 0$ and $Q(0, 0, a, \omega) = (a, \omega)$. We get

$$\begin{cases} \tilde{Z}(P(\tilde{Z}, t, \tilde{\Xi}), t, Q(\tilde{Z}, t, \tilde{\Xi})) & = \tilde{Z}, \\ \tilde{\Xi}(P(\tilde{Z}, t, \tilde{\Xi}), t, Q(\tilde{Z}, t, \tilde{\Xi})) & = \tilde{\Xi} \end{cases}$$

and differentiating with respect to $\bar{\tilde{Z}}$ we obtain

$$\frac{\partial(\tilde{Z}, \tilde{\Xi})}{\partial(z, \zeta_0, \zeta)} (P(\tilde{Z}, t, \tilde{\Xi}), t, Q(\tilde{Z}, t, \tilde{\Xi})) \frac{\partial(P, Q)}{\partial \bar{\tilde{Z}}} (\tilde{Z}, t, \tilde{\Xi})$$

$$+ \frac{\partial(\tilde{Z}, \tilde{\Xi})}{\partial(\bar{z}, \bar{\zeta_0}, \bar{\zeta})} (P(\tilde{Z}, t, \tilde{\Xi}), t, Q(\tilde{Z}, t, \tilde{\Xi})) \frac{\partial(\overline{P}, \overline{Q})}{\partial \bar{\tilde{Z}}} (\tilde{Z}, t, \tilde{\Xi}) = 0.$$

If $A(z, t, \zeta_0, \zeta)$ denotes a generic entry of the matrix

$$\frac{\partial(\tilde{Z}, \tilde{\Xi})}{\partial(\bar{z}, \bar{\zeta_0}, \bar{\zeta})} (z, t, \zeta_0, \zeta),$$

then $|A(z, t, \zeta_0, \zeta)| \leq C_k |\Im z|^k$ for all k. It follows that for each k

$$\left| \frac{\partial Q_0}{\partial \bar{\tilde{Z}}_j} (\tilde{Z}, t, \tilde{\Xi}) \right| \leq C_k' \left| \Im P(\tilde{Z}, t, \tilde{\Xi}) \right|^k \qquad \forall j = 1, \dots, m$$

and Q_0 is holomorphic in (ζ_0, ζ). Now consider

$$\Psi(z, t, \zeta_0, \zeta) = Q_0(\tilde{Z}(z, t, \zeta_0, \zeta), 0, \tilde{\Xi}(z, t, \zeta_0, \zeta))$$

and observe that

$$
\begin{aligned}
\Psi^v(x, 0) &= \Psi(x, 0, u(x, 0), u_x(x, 0)) \\
&= Q_0(\tilde{Z}(x, 0, u(x, 0), u_x(x, 0)), 0, \tilde{\Xi}(x, 0, u(x, 0), u_x(x, 0))) \\
&= u(x, 0).
\end{aligned}
$$

Observe that $H\tilde{Z}(x, t, \zeta_0, \zeta)$ and $H\tilde{\Xi}(x, t, \zeta_0, \zeta)$ are t-flat at $t = 0$. We will next show that

$$
H\Psi = \sum_{j=1}^{m}\left(\frac{\partial Q_0}{\partial \tilde{Z}_j}H\tilde{Z}_j + \frac{\partial Q_0}{\partial \overline{\tilde{Z}}_j}H\overline{\tilde{Z}}_j\right) + \sum_{k=0}^{m}\left(\frac{\partial Q_0}{\partial \tilde{\Xi}_k}H\tilde{\Xi}_k + \frac{\partial Q_0}{\partial \overline{\tilde{\Xi}}_k}H\overline{\tilde{\Xi}}_k\right)
$$

is t-flat. Note that

$$
\begin{aligned}
P(x, 0, \zeta_0, \zeta) &= P(Z(x, 0, \zeta_0, \zeta), 0, \Xi(x, 0, \zeta_0, \zeta)) \\
&= P(\tilde{Z}(x, 0, \zeta_0, \zeta), 0, \tilde{\Xi}(x, 0, \zeta_0, \zeta)) \\
&= x.
\end{aligned}
$$

This implies that for some $C > 0$,

$$
\left|\Im P(\tilde{Z}(x, t, \zeta_0, \zeta), 0, \tilde{\Xi}(x, t, \zeta_0, \zeta))\right| \le C|t|.
$$

Hence $\frac{\partial Q_0}{\partial \overline{\tilde{Z}}_j}(\tilde{Z}(x, t, \zeta_0, \zeta), 0, \tilde{\Xi}(x, t, \zeta_0, \zeta))$ is t-flat at $t = 0$, which in turn implies that for all $k \in \mathbb{N}$, there exists $C_k'' > 0$ such that

$$
|(H\Psi)(x, t, \zeta_0, \zeta)| \le C_k''|t|^k.
$$

Hence $L^u \Psi^v = \mathcal{L}^v \Psi^v = (H\Psi)^v$ is t-flat at $t = 0$, and so we have found $h(x, t) = \Psi^v(x, t)$ such that $L^u h$ is t-flat at $t = 0$ and $h(x, 0) = u(x, 0)$. Now $u(x, t)$ is also a solution of the equation

$$
u_s = e^{-i\theta}(u_t - f(x, t, u, u_x))
$$

which is of the same kind as (V.14), and the associated vector field $\mathcal{L}^\theta$ as in (V.15) is given by

$$
\mathcal{L}^\theta = \frac{\partial}{\partial s} - e^{-i\theta}\mathcal{L}
$$

with $\mathcal{L}$ as before. Note that

$$
\left(\mathcal{L}^\theta\right)^v = \frac{\partial}{\partial s} - e^{-i\theta}\mathcal{L}^v = \frac{\partial}{\partial s} - e^{-i\theta}L^u = \left(L^u\right)^\theta.
$$

It follows that there exists a C^1 function $h^\theta(x, t, s)$ such that

$$
(L^u)^\theta h^\theta = \left(\frac{\partial}{\partial s} - e^{-i\theta}L^u\right)h^\theta
$$

is s-flat at $s = 0$ and $h^{\theta}(x, t, 0) = u(x, t)$. We apply Lemma V.5.4 and conclude that $WF(u)\mid_0 \subset$ char $L^u\mid_0$. By translation we may apply the same argument to all points of Ω and state

THEOREM V.5.5. *Let $u \in C^2(\Omega)$ be a solution of (V.14). Then the C^{∞} wave front set of u is contained in the characteristic set of the linearized operator L^u.*

V.6 Applications to edge-of-the-wedge theory

Consider now a hypoanalytic structure $(\mathcal{M}, \mathcal{V})$, dim $\mathcal{M} = N$, fiber dimension of $\mathcal{V} = n$ and $m = N - n$. If $\mathcal{N}$ is a strongly noncharacteristic submanifold of $\mathcal{M}$, then Proposition V.1.28 shows that $\mathcal{V}$ induces a hypoanalytic structure on $\mathcal{N}$ by taking as the structure bundle in $\mathcal{N}$ the image of T' under the natural map

$$\mathbb{C}T^*\mathcal{M}\mid_{\mathcal{N}} \to \mathbb{C}T^*\mathcal{N}.$$

The associated bundle of vector fields will be denoted by $\mathcal{V}\mathcal{N}$ and we have $\mathcal{V}\mathcal{N} = \mathcal{V} \cap \mathbb{C}T\mathcal{N}$. Note that for any $p \in \mathcal{N}$, $\dim_{\mathbb{C}} \mathcal{V}_p\mathcal{N} = \dim \mathcal{N} - m$. For $p \in \mathcal{N}$ define

$$\mathcal{V}_p^{\mathcal{N}} = \left\{ L \in \mathcal{V}_p : \Re L \in T_p\mathcal{N} \right\}.$$

LEMMA V.6.1. *$\mathcal{V}^{\mathcal{N}}$ is a real sub-bundle of $\mathcal{V}\mid_{\mathcal{N}}$ of rank $n+$ dim $\mathcal{N} - m$. The map $\Im$ which takes the imaginary part induces an isomorphism between $\mathcal{V}^{\mathcal{N}}/\mathcal{V}\mathcal{N}$ and $T\mathcal{M}\mid_{\mathcal{N}}/T\mathcal{N}$.*

PROOF. Let $p \in \mathcal{N}$. The map $\Im : \mathcal{V}_p^{\mathcal{N}} \to T_p\mathcal{M}$ induces a map $\Im : \mathcal{V}_p^{\mathcal{N}} \to T_p\mathcal{M}/T_p\mathcal{N}$. This latter map is surjective. Indeed, given $v \in T_p\mathcal{M}$, since $\mathcal{N}$ is strongly noncharacteristic, we can find $L \in \mathcal{V}_p$ and $w \in \mathbb{C}T_p\mathcal{N}$ such that $iv = L + w$. Taking real and imaginary parts, we see that $L \in \mathcal{V}_p^{\mathcal{N}}$ and $v = \Im L + \Im w$ as desired. Since the kernel of the map $\Im : \mathcal{V}_p^{\mathcal{N}} \to T_p\mathcal{M}/T_p\mathcal{N}$ is $\mathcal{V}_p\mathcal{N}$, we get an isomorphism as asserted in the lemma. Hence, dim $\mathcal{V}_p^{\mathcal{N}} =$ dim $T_p\mathcal{M} -$ dim $T_p\mathcal{N} + \dim_{\mathbb{R}} \mathcal{V}_p\mathcal{N} = n+$ dim $\mathcal{N} - m$ for any $p \in \mathcal{N}$. $\square$

DEFINITION V.6.2. *Let E be a submanifold of $\mathcal{M}$, dim $\mathcal{M} = r + s$, dim $E = r$. We say a subset W is a wedge in $\mathcal{M}$ at $p \in E$ with edge E if the following holds: there exists a diffeomorphism φ of a neighborhood V of 0 in $\mathbb{R}^{r+s}$ onto a neighborhood U of p in $\mathcal{M}$ with $\varphi(0) = p$ and a set $B \times \Gamma \subseteq V$ with B a ball centered at $0 \in \mathbb{R}^r$ and Γ a truncated open convex cone in $\mathbb{R}^s$ with vertex at 0 such that $\varphi(B \times \Gamma) = W$ and $\varphi(B \times \{0\}) = E \cap U$.*

If $E, \mathcal{M}, \mathcal{W}$ and $p \in E$ are as in the previous definition, the direction wedge $\Gamma_p(\mathcal{W}) \subseteq T_p(\mathcal{M})$ is defined as the interior of

$$\{ c'(0) \mid c : [0, 1] \to \mathcal{M} \text{ smooth, } c(0) = p, \quad c(t) \in \mathcal{W} \quad \forall t > 0 \}.$$

If φ is as in Definition V.6.2, $\Gamma_p(\mathcal{W}) = \{ d\varphi(\mathbb{R}^r \times \{ \lambda v \mid v \in \Gamma, \ \lambda > 0 \}) \}$. Note that $\Gamma_p(\mathcal{W})$ is a linear wedge in $T_p \mathcal{M}$ with edge equal to $T_p E$. Set

$$\Gamma(\mathcal{W}) = \bigcup_{p \in E \cap U} \Gamma_p(\mathcal{W}).$$

Suppose now $\mathcal{N}$ is a strongly noncharacteristic submanifold of $\mathcal{M}$ and $\mathcal{W}$ is a wedge in $\mathcal{M}$ with edge $\mathcal{N}$. Let $u \in \mathcal{D}'(\mathcal{W})$ be a solution of $\mathcal{V}$ and let $f \in \mathcal{D}'(\mathcal{N})$. In a neighborhood of $p \in \mathcal{N}$ we may choose coordinates (x, y) vanishing at p such that $y = 0$ defines $\mathcal{N}$ locally and $\mathcal{W}$ has the form $B \times \Gamma$ with B a ball centered at 0 in x-space and Γ a truncated cone in y-space with vertex at 0. Since u is a solution and $\mathcal{N}$ is noncharacteristic, by proposition 1.4.3 in [**T5**], $u(x, y)$ is a smooth function of $y \in \Gamma$ valued in $\mathcal{D}'(B)$. We say f is the boundary value of u and write $bu = f$ if $\Gamma \ni y \mapsto u(., y)$ extends continuously to $\Gamma \cup \{0\}$ with $u(., 0) = f$, and that this is true for any $p \in \mathcal{N}$. In this case, since $\mathcal{VN} = \mathcal{V} \cap \mathbb{C}T\mathcal{N}$, it is readily seen that f is a solution of $\mathcal{VN}$, i.e., of the induced structure on $\mathcal{N}$. If the codimension of $\mathcal{N}$ is 1, then a wedge $\mathcal{W}$ with edge $\mathcal{N}$ is simply a side of $\mathcal{N}$ and distribution solutions in $\mathcal{W}$ in this case with boundary values in $\mathcal{N}$ were studied in [**T5**]. We continue to assume that $\mathcal{W}$ is a wedge in $\mathcal{M}$ with edge $\mathcal{N}$ which is strongly noncharacteristic. For $p \in \mathcal{N}$, define

$$\Gamma_p^{\mathcal{V}}(\mathcal{W}) = \left\{ L \in \mathcal{V}_p^{\mathcal{N}} : \Im L \in \Gamma_p(\mathcal{W}) \right\},$$

and

$$\Gamma_p^T(\mathcal{W}) = \left\{ \Re L : L \in \Gamma_p^{\mathcal{V}}(\mathcal{W}) \right\}.$$

$\Gamma_p^T(\mathcal{W})$ is an open cone in $\Re \mathcal{V}_p \mathcal{M} \cap T_p \mathcal{N}$. To see this, fix $p \in \mathcal{N}$ and let $\{ L_1, \dots, L_l \}$ be an $\mathbb{R}$-basis for $\mathcal{V}_p \mathcal{N}$ and complete this to an $\mathbb{R}$-basis

$$\{ L_1, \dots, L_l, V_1, \dots, V_k \}$$

of $\mathcal{V}_p^{\mathcal{N}}$. Observe that $\Re \mathcal{V}_p \mathcal{M} \cap T_p \mathcal{N}$ is spanned by

$$\Re L_1, \dots, \Re L_l, \Re V_1, \dots, \Re V_k.$$

Note also that $\Gamma_p(\mathcal{W})$ is a linear wedge in $T_p \mathcal{M}$ and hence is translation invariant by elements of $T_p \mathcal{N}$. Therefore

$$\Gamma_p^T(\mathcal{W}) = \left\{ \sum_1^l a_i \Re L_i + \sum_1^k b_j \Re V_j : a_i \in \mathbb{R}, b_j \in \mathbb{R}, \sum_1^k b_j \Im V_j \in \Gamma_p(\mathcal{W}) \right\}.$$

This description shows that $\Gamma_p^T(\mathcal{W})$ is an open cone in $\mathfrak{R}\mathcal{V}_p\mathcal{M} \cap T_p\mathcal{N}$.

LEMMA V.6.3. *Let* $(\mathcal{M}, \mathcal{V})$ *be a CR structure,* $p \in \mathcal{M}$ *and* $v \in T_p\mathcal{M}$. *Then there is a maximally real submanifold* $\mathcal{X} \subseteq \mathcal{M}$ *with* $p \in \mathcal{X}$ *and* $v \in T_p\mathcal{X}$.

PROOF. Recall from Chapter I that there are local coordinates

$$(x_1, \ldots, x_n, y_1, \ldots, y_n, s_1, \ldots, s_d)$$

vanishing at p and smooth, real-valued $\phi_1, \ldots, \phi_d$ defined near the origin such that the differentials of

$$z_j = x_j + iy_j, \qquad j = 1, \ldots, n;$$

$$w_k = s_k + i\phi_k(x, y, s), \qquad k = 1, \ldots, d$$

span T' in a neighborhood of the origin, $\phi(0) = 0$ and $d\phi(0) = 0$. Let

$$v = \sum_{k=1}^n a_k \frac{\partial}{\partial x_k} + \sum_{k=1}^n b_k \frac{\partial}{\partial y_k} + \sum_{k=1}^d c_k \frac{\partial}{\partial s_k}$$

be a real tangent vector at the origin, $v \neq 0$. If $a_j = 0 = b_j$ for all j, we can take $\mathcal{X} = \{(x, y, s) : y = 0\}$. Otherwise, assume without loss of generality that $a_1 + ib_1 \neq 0$. Consider the subspace S of the tangent space at the origin generated by the $n + d$ linearly independent vectors $v, \frac{\partial}{\partial s_1}, \ldots, \frac{\partial}{\partial s_d}, \frac{\partial}{\partial x_2}, \ldots, \frac{\partial}{\partial x_n}$. Let $\mathcal{X}$ be a submanifold of dimension $m = n + d$ through the origin so that $T_0\mathcal{X} = S$ (can take $\mathcal{X}$ to be a linear space). We claim that $\mathcal{X}$ is maximally real near the origin. To see this, suppose a one-form $\theta = \sum_{j=1}^n A_j dz_j(0) + \sum_{k=1}^d B_k ds_k$ is orthogonal to $T_0\mathcal{X}$. Then

$$\left\langle \theta, \frac{\partial}{\partial s_j} \right\rangle = 0 \qquad \forall j$$

and so $B_j = 0$ $\forall j$. Moreover, since $\left\langle \theta, \frac{\partial}{\partial x_l} \right\rangle = 0$ $\forall l \geq 2$, we get $A_j = 0$ for $j \geq 2$. Finally, note that $0 = \langle \theta, v \rangle = A_1(a_1 + ib_1)$ and so since $a_1 + ib_1 \neq 0$, $A_1 = 0$ showing that $\theta = 0$. Hence $\mathcal{X}$ is maximally real near 0. $\quad\square$

We observe that Lemma V.6.3 is not valid for a general hypoanalytic structure $(\mathcal{M}, \mathcal{V})$ which has a section L in $\mathcal{V}$ such that at a point $p \in \mathcal{V}$, L_p is a real vector field.

Recall next Marson's technique of locally embedding a hypoanalytic structure into a generic CR manifold ([**Ma**]). Suppose $(\mathcal{M}, \mathcal{V})$ is a hypoanalytic structure with the integers m and n having their usual meaning. Let $d = \dim T_p^0$ for some $p \in \mathcal{M}$. Choose a coordinate system $(x_1, \ldots, x_m, y_1, \ldots, y_n)$

vanishing at p and smooth, real-valued functions $\phi_1, \ldots, \phi_d$ defined in a neighborhood U of the origin and satisfying

$$\phi_k(0) = 0, \qquad d\phi_k(0) = 0 \qquad \forall k = 1, \ldots, d$$

such that T' over U is spanned by the differentials of

$$z_j = x_j + iy_j, \qquad j = 1, \ldots, \nu;$$

$$z_{\nu+k} = x_{\nu+k} + i\phi_k(x, y, s), \qquad k = 1, \ldots, d.$$

Let $U' = U \times \mathbb{R}^{n-\nu}$ and suppose $(x_{m+1}, \ldots, x_{m+n-\nu})$ are the coordinates for $\mathbb{R}^{n-\nu}$. Define

$$z_{m+k} = x_{m+k} + iy_{\nu+k}, \qquad \text{for } k = 1, \ldots, n - \nu.$$

Let $\mathcal{V}'$ be the sub-bundle of $\mathbb{C}TU'$ that is orthogonal to the bundle generated by $dz_1, \ldots, dz_{m+n-\nu}$. It is easy to see that $(U', \mathcal{V}')$ is a CR structure and for any $L \in \mathcal{V}_p$,

$$L' = L - i \sum_{l=1}^{n-\nu} (Ly_{\nu+l}) \frac{\partial}{\partial x_{m+l}} \in \mathcal{V}'_{p'}.$$

Here for $p \in U$, we write $p' \in U'$ to be any point of the form $p' = (p, x)$. Moreover, the preceding association $L \to L'$ is an isomorphism of $\mathcal{V}_p$ onto $\mathcal{V}_{p'}$. In particular, any solution of $\mathcal{V}$ is also a solution of $\mathcal{V}'$ depending on fewer variables. Characteristic covectors $\sigma \in T_p^0 U$ embed into characteristic covectors $(\sigma, 0) \in T_{p'}^0 U'$ for any $p' = (p, x)$. If $\mathcal{N}$ is a strongly noncharacteristic submanifold of U, then $\mathcal{N}' = \mathcal{N} \times \mathbb{R}^{n-\nu}$ is a strongly noncharacteristic submanifold of U' and if $p \in \mathcal{N}$ and $p' = (p, x) \in \mathcal{N}'$, we have:

$$\mathcal{V}_{p'}^{\mathcal{N}'} = \{L' : L \in \mathcal{V}_p^{\mathcal{N}}\}$$

where L' is determined by L as above. If $\mathcal{W}$ is a wedge with edge $\mathcal{N}$ in U, then $\mathcal{W}' = \mathcal{W} \times \mathbb{R}^{n-\nu}$ is a wedge in U' with edge $\mathcal{N}'$ and

$$\Gamma_{p'}^{\mathcal{V}'}(\mathcal{W}') = \{L' : L \in \Gamma_p^{\mathcal{V}}(\mathcal{W})\}.$$

Finally, if $u \in \mathcal{D}'(\mathcal{N})$, it may be viewed as a distribution in $\mathcal{N}'$ and it is easy to see that

$$WF_{\text{ha}, \, p}^{\mathcal{N}}(u) \times \{0\} \subseteq WF_{\text{ha}, \, p'}^{\mathcal{N}'}(u).$$

We are now ready to present an application of the FBI transform to the hypoanalytic wave front set of a distribution u on a strongly noncharacteristic $\mathcal{N}$ which extends to a solution in a wedge. The result is due to Eastwood and Graham ([**EG1**]).

THEOREM V.6.4. ([**EG1**]) *Let* $(\mathcal{M}, \mathcal{V})$ *be a hypoanalytic structure,* $\mathcal{N}$ *a strongly noncharacteristic submanifold, and let* W *be a wedge in* $\mathcal{M}$ *with edge . Suppose* $f \in \mathcal{D}'(\mathcal{N})$ *is the boundary value of a solution of* $\mathcal{V}$ *on* W. *Then* $WF_{\mathrm{ha}}(f) \subseteq \Gamma^T(W)^0 =$ *the polar of* $\Gamma^T(W)$ *in the duality between* $T\mathcal{N}$ *and* $T^*\mathcal{N}$.

PROOF. Let $p \in \mathcal{N}$ and $\sigma \in T_p^*\mathcal{N}/\{0\}$ satisfy $\sigma \notin \Gamma_p^T(W)^0$. If we embed $\mathcal{M}$ near p into a CR structure as in the preceding discussion, then $\sigma' = (\sigma, 0) \notin (\Gamma_{p'}^T(W'))^0$, and so because of the relation between $WF_{\mathrm{ha},\,p}^{\mathcal{N}}(f)$ and $WF_{\mathrm{ha},\,p'}^{\mathcal{N}'}(f)$, it suffices to prove the theorem under the assumption that $(\mathcal{M}, \mathcal{V})$ is CR. Since $\sigma \notin \Gamma_p^T(W)^0$, there is $L \in \Gamma_p^{\mathcal{V}}(W)$ such that $\langle \sigma, \Re L \rangle < 0$. By Lemma V.6.3, there is a maximally real submanifold $\mathcal{X} \subseteq \mathcal{N}$ with $p \in \mathcal{X}$ and $\Re L \in T_p \mathcal{X}$ (note that the induced structure on $\mathcal{N}$ is CR). Since $\mathcal{X}$ is maximally real and $L \neq 0$, $\Im L \notin T_p \mathcal{X}$. Choose a submanifold $\mathcal{Y}$ of $\mathcal{M}$ such that $\mathcal{X} \subseteq \mathcal{Y}$, and $T_p \mathcal{Y}$ is spanned by $T_p \mathcal{X}$ and $\Im L$. Thus $\mathcal{X}$ is a hypersurface in $\mathcal{Y}$. Since $\mathcal{X}$ is maximally real, $\mathcal{Y}$ inherits a hypoanalytic structure of codimension 1 from $(\mathcal{M}, \mathcal{V})$. This induced structure on $\mathcal{Y}$ is CR near p, is generated by L at p, and $\mathcal{X}$ is a maximally real submanifold of $\mathcal{Y}$. We may assume that near p, $\mathcal{X}$ divides $\mathcal{Y}$ into two components $\mathcal{Y}^+$, $\mathcal{Y}^-$ where $\mathcal{Y}^+$ is the side toward which $\Im L$ points. Since $\Im L \in \Gamma_p(W)$, $\mathcal{Y}^+ \subseteq W$ near p. $\mathcal{Y}^+$ may be regarded as a wedge in $\mathcal{Y}$ with edge $\mathcal{X}$. If F is the solution in W with $bF = f$ on $\mathcal{N}$, then F restricts to $\mathcal{Y}^+$ (since $\mathcal{Y}^+$ is noncharacteristic) and this restriction is a solution for the structure on $\mathcal{Y}$. Moreover, this restriction has a boundary value equal to $f \mid_{\mathcal{X}}$. To prove the theorem, we have to show that $i_{\mathcal{X}}^* \sigma \notin WF_{\mathrm{ha},\,p}(f \mid_{\mathcal{X}})$. Note that we also have $\langle i_{\mathcal{X}}^* \sigma, \Re L \rangle < 0$. Choose local coordinates $x_1, \ldots, x_m, t$ on $\mathcal{Y}$ vanishing at p so that in these coordinates $\mathcal{X}$ is given by $t = 0$ and $L = A + i\frac{\partial}{\partial t}$ where $A = \sum_1^m A_j \frac{\partial}{\partial x_j}$ is a real vector field. We therefore need to show that if $\sigma \in T_0^*\mathbb{R}^m$ and $\langle A, \sigma \rangle < 0$, then $\sigma \notin WF_{\mathrm{ha}} bf$. This will follow from Theorem V.6.9. $\qquad\square$

COROLLARY V.6.5. *Suppose* $\mathcal{X} \subset \mathcal{M}$ *is a maximally real submanifold,* $p \in \mathcal{X}$, *and let* W^+ *and* W^- *be wedges in* $\mathcal{M}$ *with edge* $\mathcal{X}$ *such that* $\Gamma_p(W^+) = -\Gamma_p(W^-)$. *If* $f \in \mathcal{D}'(\mathcal{X})$ *is the boundary value of a solution of* $\mathcal{V}$ *on* W^+ *and also the boundary value of a solution of* $\mathcal{V}$ *on* W^-, *then* $WF_{\mathrm{ha},p}(f) \subset i_{\mathcal{X}}^* T_p^0 \mathcal{M}$.

PROOF. By Theorem V.6.4,

$$WF_{\mathrm{ha},p}(f) \subseteq \Gamma_p^T(W^+)^0 \cap \Gamma_p^T(W^-)^0.$$

Note that since $\Gamma_p(W^+) = -\Gamma_p(W^-)$, $\Gamma_p^T(W^+) = -\Gamma_p^T(W^-)$. Hence if $\sigma \in \Gamma_p^T(W^+)^0 \cap \Gamma_p^T(W^-)^0$, then $\langle \sigma, v \rangle = 0$ for every $v \in \Gamma_p^T(W^+)$. Since $\Gamma_p^T(W^+)$

is an open cone in $\Re V_p \cap T_p \mathcal{X}$, it follows that $\sigma \in \left(\Re V_p \cap T_p \mathcal{X} \right)^{\perp}$. Therefore the corollary follows from the fact that

$$i_{\mathcal{X}}^* T_p^0 = \left(\Re V_p \cap T_p \mathcal{X} \right)^{\perp}.$$

$\square$

COROLLARY V.6.6. (Theorem V.3.1 in [BCT].) *If f is defined in a full neighborhood of p and $p \in \mathcal{N}$ is strongly noncharacteristic, then*

$$WF_{\mathrm{ha}, p}^{\mathcal{N}} f \subset i_{\mathcal{N}}^* T_p^0 \mathcal{M}.$$

COROLLARY V.6.7. (The edge-of-the-wedge theorem.) *If the structure V on $\mathcal{M}$ is an elliptic structure and f is the boundary value of solutions in two wedges $\mathcal{W}^+, \mathcal{W}^-$ with edge a maximally real $\mathcal{X}$ as in Corollary V.6.5, then f extends to a hypoanalytic function in a full neighborhood of p in $\mathcal{M}$.*

Corollary V.6.7 is a generalization of the classical edge-of-the-wedge theorem of several complex variables. The example of the structure in the plane generated by $\frac{\partial}{\partial y}$ for which the x-axis is maximally real shows that the corollary may not be valid when the structure is not elliptic.

REMARK V.6.8. Notice that in general $i_{\mathcal{N}}^* T_p^0 \mathcal{M} \subseteq \Gamma_p^T(\mathcal{W})^0$.

We will next present a result on the hypoanalytic wave front set of the trace of a solution when the vector field in question is locally integrable.

We consider a smooth vector field $L = X + iY$ where X and Y are real vector fields defined in a neighborhood U of the origin. Let Σ be an embedded hypersurface through the origin in U dividing the set U into two regions, U^+ and U^-, where U^+ denotes the region toward which X is pointing. We assume that L is noncharacteristic on Σ, which means (after multiplying L by i if necessary) that X is noncharacteristic. Our considerations will be local and so after an appropriate choice of local coordinates (x, t) and multiplication of L by a nonvanishing factor, the vector field is given by

$$L = \frac{\partial}{\partial t} + \sum_{j=1}^m a_j(x, t) \frac{\partial}{\partial x_j} \qquad (\text{V}.17)$$

and Σ and U^+ are given by $t = 0$ and $t > 0$ respectively. We will need to consider the integral curve $(-\epsilon, \epsilon) \ni s \mapsto \gamma(s)$ of X that passes through the origin, i.e., $\gamma'(s) = X \circ \gamma(s)$, $\gamma(0) = 0$. It is clear that for small $\epsilon > 0$ and $|s| < \epsilon$, $\gamma(s) \in U^+$ if and only if $s > 0$, so $\gamma(-\epsilon, \epsilon) \cap U^+ = \gamma(0, \epsilon)$. To simplify the notation we will simply write γ^+ to denote $\gamma(0, \epsilon)$.

THEOREM V.6.9. *Let* $L = \frac{\partial}{\partial t} + \sum_{j=1}^{m} a_j(x, t) \frac{\partial}{\partial x_j}$ *be locally integrable. Suppose* $f \in \mathcal{D}'(U_+)$ *has a boundary value at* $t = 0$ *and*

$$Lf(x, t) = 0, \qquad (x, t) \in U^+.$$

Assume that there is a sequence $p_k \in \gamma^+$, $p_k \to 0$ *such that for each* $k = 1, 2, \ldots$, $X(p_k)$ *and* $Y(p_k)$ *are linearly independent. Then there exists a unit vector* v *such that*

$$\xi^0 \in \mathbb{R}^n, \quad v \cdot \xi^0 > 0 \Longrightarrow (0, \xi^0) \notin WF_{\mathrm{ha}}(bf).$$

In particular, the hypoanalytic wave front set of bf *at the origin is contained in a closed half-space.*

PROOF. Let $Z_1, \ldots, Z_m$ be a complete set of smooth first integrals of L near the origin in U and choose new local coordinates (x, t) in which the Z_j's may be written as

$$Z_j(x, t) = x_j + i\Phi_j(x, t), \quad k = 1, \ldots, m,$$

with $\Phi(0, 0) = 0$, $\Phi_x(0, 0) = 0$, and $\Phi_{xx}(0, 0) = 0$. For $j = 1, \ldots, m$ let $M_j = \sum_{k=1}^{m} b_{jk}(x, t) \frac{\partial}{\partial x_k}$ be vector fields satisfying

$$M_j Z_k = \delta_j^k, \qquad [M_j, M_k] = 0.$$

It is readily checked that for each $j = 1, \ldots, m$,

$$[M_j, L] = 0. \tag{V.18}$$

For any C^1 function g, the differential may be expressed as

$$dg = Lg \, dt + \sum_{k=1}^{m} M_k g \, dZ_k. \tag{V.19}$$

Using (V.19) we get:

$$d(g dZ_1 \wedge \cdots \wedge dZ_m) = Lg \, dt \wedge dZ_1 \wedge \cdots \wedge dZ_m. \tag{V.20}$$

For $\zeta \in \mathbb{C}^m, z \in \mathbb{C}^m$, let

$$E(z, \zeta, x, t) = i\zeta \cdot (z - Z(x, t)) - \kappa\langle\zeta\rangle[z - Z(x, t)]^2.$$

Let B denote a small ball centered at 0 of radius r in $\mathbb{R}^m$ and $\phi \in C_0^\infty(B)$, $\phi \equiv 1$ for $|x| \leq r/2$, the precise value of r as well as the value of the positive parameter κ in the definition of E will be determined later. We will apply (V.20) to the function

$$g(z, \zeta, x, t) = \phi(x) f(x, t) e^{E(z, \zeta, x, t)}$$

where (z, ζ) are parameters. We get:

$$d(gdZ) = fL\phi e^E dt \wedge dZ, \qquad (V.21)$$

where $dZ = dZ_1 \wedge \cdots \wedge dZ_m$. Next by Stokes' theorem we have, for $t_1 > 0$ small:

$$\int_B g(z, \zeta, x, 0) d_x Z(x, 0) = \int_B g(z, \zeta, x, t_1) d_x Z(x, t_1) + \int_0^{t_1} \int_B d(gdZ).$$
$$(V.22)$$

We will estimate the two integrals on the right in (V.22) and our aim is to show that for z close to the origin in complex space, both decay exponentially as $\zeta \to \infty$ in a conic neighborhood of ξ^0. Write

$$Z = (Z_1, \ldots, Z_m) = x + i\Phi(x, t), \quad \Phi = (\Phi_1, \ldots, \Phi_m).$$

Observe that, assuming without loss of generality that $|\xi^0| = 1$,

$$\Re E(0, \xi^0, x, t) = \Phi(x, t) \cdot \xi^0 - \kappa(|x|^2 - |\Phi(x, t)|^2).$$

Our main task will be to determine convenient values of t_1, κ and r such that for some $\gamma > 0$

(i) $\Re E(0, \xi^0, x, t_1) \leq -\gamma$ for $|x| \leq r$;
(ii) $\Re E(0, \xi^0, x, t) \leq -\gamma$ for $0 \leq t \leq t_1$ and $r/2 \leq |x| \leq r$.

In order to find the vector v mentioned in the statement of the theorem we will need

LEMMA V.6.10. *There exists a sequence $t_k \searrow 0$ such that*

(1) $\Phi(0, t_k) \neq 0$;
(2) $|\Phi(0, t)| \leq |\Phi(0, t_k)|$ *for* $0 \leq t \leq t_k$;
(3) $\lim_{t_k \to 0} \Phi(0, t_k)/|\Phi(0, t_k)| = -v$.

We will postpone the proof of Lemma V.6.10 and continue our reasoning with v given by (3) in Lemma V.6.10. The assumptions on Φ allow us to write

$$\Phi(x, t) = \Phi(0, t) + e(x, t), \quad |e(x, t)| \leq A|xt| + B|x|^2 \qquad (V.23)$$

for some positive constants A and B. Suppose first $\Phi_t(0, 0) \neq 0$, which is the case that is needed for Theorem V.6.4. Then there is $\lambda < 0$ such that $\Phi_t(0, 0) = \lambda v$. Since $\Phi(0, 0) = 0$ and $\Phi_x(0, 0) = 0$, we can write

$$\Phi(x, t) \cdot \xi^0 = \Phi_t(0, 0) \cdot \xi^0 + O(|x|^2 + t^2)$$
$$= \lambda v \cdot \xi^0 + O(|x|^2 + t^2).$$

Hence given $\kappa > 0$, we can find t_1, r and $\gamma > 0$ such that (i) and (ii) above hold. We may therefore assume that $\Phi_t(0,0) = 0$ and so the quotient $|\Phi(0,t)|/t^2 \leq C$ for $(0,t) \in U^+$. We have $\Phi(0,t_k) + |\Phi(0,t_k)| \, v = o(|\Phi(0,t_k)|)$. We recall that by hypothesis $\xi^0 \cdot v > 0$. Hence,

$$
\begin{aligned}
\Phi(0,t_k) \cdot \xi^0 &= -|\Phi(0,t_k)| \, v \cdot \xi^0 + o(|\Phi(0,t_k)|) \\
&< -|\Phi(0,t_k)| \, v \cdot \xi^0/2 = -c|\Phi(0,t_k)|,
\end{aligned}
$$

for t_k small and $0 < c < 1$. We now take $r = \alpha|\Phi(0,t_k)|/t_k$, with α and t_k small to be chosen later. Hence, for $|x| \leq r$ and $0 \leq t \leq t_k$, we can choose α small enough (depending on A, B and C but not on t_k) so that

$$
\begin{aligned}
|e(x,t)| &\leq A\alpha |\Phi(0,t_k)| \frac{t}{t_k} + B\alpha^2 \frac{|\Phi(0,t_k)|}{t_k^2} |\Phi(0,t_k)| \\
&\leq c\frac{|\Phi(0,t_k)|}{2}.
\end{aligned}
\tag{V.24}
$$

This implies that on the support of $\phi(x)$ we have

$$
-(1+c)|\Phi(0,t_k)| \leq \Phi(x,t_k) \cdot \xi^0 \leq -\frac{c}{2}|\Phi(0,t_k)|.
$$

Let $\kappa = \epsilon/|\Phi(0,t_k)|$. A consequence of (V.23), (V.24) and the fact that $|\Phi(0,t)| \leq |\Phi(0,t_k)|$ for $0 \leq t \leq t_k$ is

$$
\begin{aligned}
|\Phi(x,t)| &\leq (1+c)|\Phi(0,t_k)|, \\
|\Phi(x,t)|^2 &\leq (1+c)^2|\Phi(0,t_k)|^2, \\
\kappa|\Phi(x,t)|^2 &\leq \epsilon(1+c)^2|\Phi(0,t_k)|
\end{aligned}
\tag{V.25}
$$

for x in the support of $\phi(x)$ and $0 \leq t \leq t_k$. Choosing $\epsilon = c/4(1+c)^2$ (thus, independent of t_k), we get, on the support of $\phi(x)$,

$$
\begin{aligned}
\Phi(x,t_k) \cdot \xi^0 + \kappa|\Phi(x,t_k)|^2 &\leq -\frac{c}{2}|\Phi(0,t_k)| + \epsilon(1+c)^2|\Phi(0,t_k)| \\
&\leq -\frac{c}{4}|\Phi(0,t_k)|
\end{aligned}
$$

which leads to an exponential decay in the first integral on the right of (V.22) for z complex near 0 and ζ in a complex conic neighborhood of ξ^0, as soon as we replace t_1 by t_k. For the second integral, note that for $0 \leq t \leq t_k$ and x in the support of ϕ, we may invoke again (V.25) to estimate the size of $|\Phi(x,t)|$ and $\kappa|\Phi(x,t)|^2$ which gives, in view of the previous choice of ϵ,

$$
|\Phi(x,t)| + \kappa|\Phi(x,t)|^2 \leq (1+c)|\Phi(0,t_k)| + \frac{c}{4}|\Phi(0,t_k)| \leq (1+2c)|\Phi(0,t_k)|
$$

while on the support of $L\phi$, $|x| \geq r/2 = \alpha|\Phi(0, t_k)|/2t_k$ so

$$\kappa|x|^2 \geq \frac{\epsilon\alpha^2|\Phi(0, t_k)|}{4t_k^2}$$

and

$$\Phi(x, t) \cdot \xi^0 - \kappa(|x|^2 - |\Phi(x, t)|^2) \leq \left(1 + 2c - \frac{\epsilon\alpha^2}{4t_k^2}\right)|\Phi(0, t_k)|.$$

Hence, if t_k is chosen sufficiently small, we also get exponential decay for the second integral on the right-hand side of (V.22) with t_1 replaced by t_k.

We have thus shown that the function

$$F(z, \zeta) = \int_B e^{E(z,\zeta,x,0)} \phi(x) f(x, 0) \, d_x Z(x, 0)$$

satisfies an exponential decay of the form

$$|F(z, \zeta)| \leq Ce^{-R|\zeta|}$$

for z near 0 in $\mathbb{C}^m$ and ζ in a complex conic neighborhood of ξ^0 in $\mathbb{C}^m$. In particular, since $Z(0, 0) = 0$ and $d_x \Phi(0, 0) = 0$, by Theorem V.4.8, $(0, \xi^0) \notin WF_{ha}(bf)$.

We now return to the proof of Lemma V.6.10; it is here that we use the fact that X and Y are linearly independent on a sequence $p_k \in \gamma^+$ that approaches the origin. We will show that $\Phi(0, t)$ cannot vanish identically on any interval $(0, \epsilon')$. Let us write $L = \partial_t + a \cdot \partial_x$, $Z = x + i\Phi$, $Z_x = I + i^t\Phi_x$, and recall that $^t\Phi_x$ has small norm for (x, t) close to 0. Now $LZ = 0$ leads to $a = -i(I + i^t\Phi_x)^{-1} \Phi_t$. If $\Phi(0, t)$ vanishes identically on $[0, \epsilon']$ we will have, for those values of t, that $\Phi_t(0, t) = 0$, $a(0, t) = 0$, and $Y(0, t) = \Im a(0, t) = 0$. Furthermore, $X(0, t) = \partial_t$ for $0 < t < \epsilon'$, showing that $\gamma(s) = (0, \ldots, 0, s)$ for $0 < s < \epsilon'$. Thus, $X(\gamma(s))$ and $Y(\gamma(s))$ are linearly dependent for $0 < s < \epsilon'$, a contradiction. Therefore, there exists a sequence $s_k \searrow 0$ such that $|\Phi(0, s_k)| > 0$ and since $\Phi(0, 0) = 0$ there is another sequence $t_k \searrow 0$ satisfying (1) and (2), which in turn possesses a subsequence that satisfies (1), (2), and (3). □

V.7 Application to the F. and M. Riesz theorem

The classical F. and M. Riesz theorem states that a complex measure μ defined on the boundary $\mathbb{T}$ of the unit disk Δ all of whose negative Fourier coefficients vanish, i.e.,

$$\hat{\mu}(k) = \int_0^{2\pi} \exp(-ik\theta) \, d\mu(\theta) = 0, \qquad k = -1, -2, \ldots, \qquad (V.26)$$

is absolutely continuous with respect to Lebesgue measure $d\theta$.

Observe that condition (V.26) is equivalent to the existence of a holomorphic function $f(z)$ defined on Δ whose weak boundary value is μ. In other words, the theorem asserts that if a holomorphic function f on Δ has a weak boundary value bf that is a measure, then in fact $bf \in L^1(\mathbb{T})$.

The F. and M. Riesz theorem has inspired an extensive generalization in two different directions: (i) generalized analytic function algebras, which has as a starting point the fact that (V.26) means that μ is orthogonal to the algebra of continuous functions f on $\mathbb{T}$ that extend holomorphically to F on Δ with $F(0) = 0$; (ii) ordered groups, which emphasizes instead the role of the group structure of $\mathbb{T}$ in the classical result. We will next briefly describe these two directions.

Let A denote the algebra of continuous functions f on $\mathbb{T}$ which have a holomorphic extension F into Δ. The map $f \longmapsto F(0)$ is a continuous homomorphism ϕ of A and so there is a set M_ϕ of measures on $\mathbb{T}$ each of which represents ϕ. In this case, it is clear that the normalized Lebesgue measure $d\theta$ is the unique element of M_ϕ. The kernel of ϕ is the closure of the linear span A_0 of $\exp(in\theta)$, $n > 0$. Hence the condition $\widehat{\mu}(n) = 0$ for all $n < 0$ is equivalent to $\mu \in A_0^\perp$. Such a μ decomposes as $\mu = \mu_a + \mu_s$, where μ_a (resp. μ_s) is absolutely continuous (resp. singular) with respect to $d\theta$, that is, with respect to every measure in M_ϕ. The classical F. and M. Riesz theorem consists of two parts: $\mu \in A_0^\perp \Rightarrow \mu_s \in A_0^\perp$ and $\mu_s \in A_0^\perp \Rightarrow \mu_s = 0$.

For function algebras A on compact Hausdorff spaces X other than $\mathbb{T}$, one looks at continuous homomorphisms ϕ of A and their sets of representing measures M_ϕ. It is known that any measure μ on X can be decomposed as $\mu = \mu_a + \mu_s$, with μ_a (resp. μ_s) absolutely continuous (resp. singular) with respect to every measure in M_ϕ. Under a variety of hypotheses on A or M_ϕ, the implication $\mu \in A_0^\perp \Rightarrow \mu_s \in A_0^\perp$ has been proved and this kind of result turns out to be a crucial ingredient in the theory of generalized analyticity in the algebra A. For more details on this, we mention the book [**BK**] by Klaus Barbey and Heinz Konig.

In the second direction of generalization, one starts with a locally compact abelian group G. Its dual group $\widehat{G}$, written additively, is assumed to contain an order, that is, a semigroup P which satisfies $P \cup -P = \widehat{G}$. Denote by $M(E)$ the convolution algebra of complex Borel measures on G whose Fourier transforms vanish on the subset E of $\widehat{G}$. Each measure μ decomposes as $\mu_a + \mu_s$ with respect to Haar measure on G. In this set-up, the implication $\mu \in M(P) \Rightarrow \mu_s \in M(P)$ has been proved. Under some conditions on G and P, the implication $\mu \in M(P) \Rightarrow \mu_s = 0$ has also been proved. There are also results for compact groups (see [**K1**] and [**K2**]).

Thus, although absolute continuity with respect to Lebesgue measure is a local property, the generalizations mentioned above involve global objects: function algebras and groups.

In the paper [**B**], Brummelhuis used microlocal analysis to prove generalizations of a local version of the theorem of F. and M. Riesz. Among other things, in [**B**] it is shown that if a CR measure on a hypersurface of $\mathbb{C}^n$ is the boundary value of a holomorphic function defined on a side, then it is absolutely continuous with respect to Lebesgue measure. It is easy to use his methods to get a similar result for CR measures on CR submanifolds of any codimension whenever the measure is the boundary value of a holomorphic function defined in a wedge. Another proof of this result was given by Rosay in [**Ro**]. There are also results when the edge of the wedge has lower regularity ([**CR2**] and [**BH8**]). Another way of stating the F. and M. Riesz theorem is to say that if a holomorphic function $f(z)$ defined on a smoothly bounded domain D of the complex plane has tempered growth at the boundary and its weak boundary value is a measure, then the measure is absolutely continuous with respect to Lebesgue measure.

If we regard holomorphic functions as solutions of the homogeneous equation $\overline{\partial} f = 0$, it is natural to ask for which complex vector fields L it is possible to draw the same conclusion for solutions of the equation $Lf = 0$. We will present here an extension of the F. and M. Riesz theorem to all locally integrable, smooth complex vector fields in the plane for smooth domains at the noncharacteristic part of the boundary. We recall that a nowhere vanishing smooth vector field

$$L = a(x, y)\frac{\partial}{\partial x} + b(x, y)\frac{\partial}{\partial y}$$

is said to be locally integrable in an open set Ω if each $p \in \Omega$ is contained in a neighborhood which admits a smooth function Z with the properties that $LZ = 0$ and the differential $dZ \neq 0$.

THEOREM V.7.1. *Suppose $L = \frac{\partial}{\partial t} + a(x, t)\frac{\partial}{\partial x}$ is smooth in a neighborhood U of the origin in the plane. Let $U_+ = U \cap \mathbb{R}^2_+$, and suppose $f \in C(U_+)$ satisfies $Lf = 0$ in U_+ and for some integer N,*

$$|f(x, t)| = O(t^{-N}) \qquad as\ t \to 0^+.$$

Assume that L is locally integrable in U. If the trace $bf = f(x, 0)$ is a measure, then it is absolutely continuous with respect to Lebesgue measure.

The existence of the trace $bf = f(x, 0)$ under the assumptions on f follows from theorem 1.1 in [**BH1**]. In his work [**B**], the author gives a microlocal criterion for the absolute continuity of a measure analogous to (V.26) based on Uchiyama's deep characterization of BMO ($\mathbb{R}^n$) [**U**]. Similarly, one of the main steps in the generalization of the F. and M. Riesz theorem is Theorem V.6.9, which involves the location of the hypoanalytic wave front set of the trace of a solution of a locally integrable vector field in $\mathbb{R}^n$. On the other hand, while in the classical case and the generalizations in [**B**] the location of the wave front set of the measure under consideration always satisfies a restrictive hypothesis which leads to absolute continuity, this restriction is not fulfilled in general by the trace of a solution of an arbitrary locally integrable vector field even if the solution is smooth (an example concerning a vector field with real-analytic coefficients is shown in example 4.3 of [**BH1**]). Thus, we need to deal as well with points where the wave front set of the measure may contain all directions; at those points, the vector field L exhibits a behavior close to that of a real vector field (in a sense made precise in Lemma V.7.2 below) and absolute continuity may be proved directly.

LEMMA V.7.2. *Let*

$$L = \frac{\partial}{\partial t} + i \sum_{j=1}^{n} b_j(x,t) \frac{\partial}{\partial x_j}$$

be smooth on a neighborhood $U = B(0,a) \times (-T,T)$ of the origin in $\mathbb{R}^{n+1}$ with $B(0,a) = \{x \in \mathbb{R}^n : |x| < a\}$. We will assume that the coefficients $b_j(x,t)$, $j = 1, \ldots, n$ are real and that all of them vanish on $F \times [0,T)$, where $F \subset B(0,a)$ is a closed set. Assume that $f \in C(U^+)$ satisfies $Lf = 0$ on $U^+ = B(0,a) \times (0,T)$, has tempered growth as $t \searrow 0$ and its boundary value $bf(x) = f(x,0)$ is a Radon measure μ. Then the restriction μ_F of μ to F defined on Borel sets $X \subset B(0,a)$ by $\mu_F(X) = \mu(X \cap F)$ is absolutely continuous with respect to Lebesgue measure.

PROOF. If $\tilde{x}$ is an arbitrary point in F we may write

$$b_j(x,t) = \sum_{k=1}^{n} (x_k - \tilde{x}_k) \beta_{jk}(x, \tilde{x}, t)$$

with $\beta_{jk}(x, \tilde{x}, t)$ real and smooth. The proof of theorem 1.1 in [**BH1**] shows that for any $\phi \in C^\infty(-a,a)$ we have

$$\langle \mu, \phi \rangle = \int f(x,T) \Phi^k(x,T) \, ds + \int_0^T \int_{B(0,a)} f(x,t) L^t \Phi^k(x,t) \, dx dt, \quad \text{(V.27)}$$

$$\text{where} \quad \Phi^k(x,t) = \sum_{j=0}^{k} \phi_j(x,t) \frac{t^j}{j!}, \quad \phi_0(x,t) = \phi(x), \quad \text{(V.28)}$$

and $\quad \phi_j(x, t) = -\dfrac{\partial}{\partial t}\phi_{j-1}^{\epsilon}(x, t) - \displaystyle\sum_{s, \ell=1}^{n} \dfrac{\partial}{\partial x_s}(x_\ell - \tilde{x}_\ell)\beta_{j\ell}(x, \tilde{x}, t)\phi_{j-1}(x, t)$

for $j = 1, \ldots, k$, with k a convenient and fixed positive integer. We can write

$$\Phi^k(x, t) = A(x, t, D_x)\phi(x) \tag{V.29}$$

where $A(x, t, D_x) = \sum_{|\alpha|\leq k} a_\alpha(x, t)D_x^{\alpha}$ is a linear differential operator of order k in the x variables with coefficients depending smoothly on t. The coefficients a_α are obtained from the coefficients $b_j(x, t)$ of L by means of algebraic operations and differentiations with respect to x and t. Observe that given any point $\tilde{x} \in F$, $A(x, t, D_x)$ may be written as

$$A(x, t, D_x) = \sum_{|\alpha|\leq k}\sum_{\ell=1}^{n} A_{\alpha\ell}(x, \tilde{x}, t)\left((x_\ell - \tilde{x}_\ell)D_x\right)^{\alpha}. \tag{V.30}$$

Notice that $|A_{\alpha\ell}(x, \tilde{x}, t)| \leq C$, for $x \in B(0, a)$, $\tilde{x} \in F$, $t \in [0, T)$, $|\alpha| \leq k$, and $\ell = 1, \ldots, n$ because the coefficients of L have uniformly bounded derivatives on $B(0, a)$. Hence, we obtain from (V.29) and (V.30) the estimate

$$\left|\int f(x, T)\Phi^k(x, T)\,dx\right| \leq C \sum_{|\alpha|\leq k+1}\int_{B(0,a)} d(x, F)^{|\alpha|}\,|D_x^{\alpha}\phi(x)|\,dx, \tag{V.31}$$

where $d(x, F) = \inf_{\tilde{x}\in F}|x - \tilde{x}|$. We next consider the second integral on the right in (V.27). We will first show that for any j,

$$L^t(\Phi^j) = \dfrac{\phi_{j+1}}{j!}t^j. \tag{V.32}$$

To see this, note first that (V.32) holds for $j = 0$ from the definition of ϕ_1. To proceed by induction, assume (V.32) for $j \leq m$. Then

$$\begin{aligned}
L^t(\Phi^{m+1}) &= L^t(\Phi^m) + L^t\left(\dfrac{\phi_{m+1}}{(m+1)!}t^{m+1}\right) \\
&= \dfrac{\phi_{m+1}}{m!}t^m + L^t\left(\dfrac{\phi_{m+1}}{(m+1)!}t^{m+1}\right) \\
&= \dfrac{L^t(\phi_{m+1})}{(m+1)!}t^{m+1} \\
&= \dfrac{\phi_{m+2}}{(m+1)!}t^{m+1}.
\end{aligned}$$

This proves (V.32). Next we observe that since the coefficients $b_j(x, t)$ vanish on $F \times [0, T]$, each ϕ_j has the form

$$\phi_j(x, t) = \sum_{|\alpha|\leq j} c_\alpha(x, t)D_x^{\alpha}\phi(x) \tag{V.33}$$

where the c_α are smooth and satisfy the estimate

$$|c_\alpha| \le C d(x, F)^{|\alpha|}.$$

The form (V.33) is clearly valid for $\phi_0 = \phi$. Assume it is valid for ϕ_j. Then it will also be valid for ϕ_{j+1} since by definition, $\phi_{j+1} = L^t \phi_j$. If we now choose $k = N + 1$, (V.32) and (V.33) imply that

$$\left| \int_0^T \int_{B(0,a)} f(x, t) L^t \Phi^k(x, t) \,dxdt \right| \le \int_0^T \int_{B(0,a)} |f(x, t)| \frac{\phi_{k+1}(x, t)}{k!} t^k \,dxdt$$

$$\le C \int_0^T \int_{B(0,a)} |\phi_{k+1}(x, t)| \,dxdt \tag{V.34}$$

$$\le C \sum_{|\alpha| \le k+1} \int_{B(0,a)} d(x, F)^{|\alpha|} |D_x^\alpha \phi(x)| \,dx.$$

Thus the second integral on the right-hand side of (V.27) also satisfies an estimate of the kind in (V.31). Consider now a compact subset $K \subset F$ with Lebesgue measure $|K| = 0$ and choose a sequence

$$0 \le \phi_\epsilon(x) \le 1 \in C^\infty(B(0, a)) \quad \epsilon \to 0,$$

such that (i) $\phi_\epsilon(x) = 1$ for all $x \in K$; (ii) $\phi_\epsilon(x) = 0$ if $d(x, K) > \epsilon$; (iii) $|D_x^\alpha \phi_\epsilon(x)| \le C_\alpha \epsilon^{-|\alpha|}$. Note that $\phi_\epsilon(x)$ converges pointwise to the characteristic function of K as $\epsilon \to 0$ while $D^\alpha \phi_\epsilon(x) \to 0$ pointwise if $|\alpha| > 0$. Let $\psi \in C^\infty(B(0, a))$ and use (V.31) and (V.34) with $\phi = \phi_\epsilon \psi$ keeping in mind the trivial estimate $d(x, F) \le d(x, K)$. By the dominated convergence theorem,

$$\langle \mu, \phi_\epsilon \psi \rangle \to \int_K \psi \,d\mu$$

while

$$|d(x, K)^{|\alpha|} D_x^\alpha \phi_\epsilon(x)\|_{L^1} \le \|\epsilon^{|\alpha|} D_x^\alpha \phi_\epsilon(x)\|_{L^1} \to 0$$

as $\epsilon \to 0$ (when $\alpha = 0$ one uses the fact that $|K| = 0$). Thus, (V.31) and (V.34) show that

$$\int_K \psi \,d\mu = 0, \quad \psi \in C^\infty(B(0, a)),$$

which implies that the same conclusion holds for any continuous function ψ on K (first extend ψ to a compactly supported function on $B(0, a)$ and then approximate the extension by test functions). Thus the total variation $|\mu|(K)$ of μ on K is zero and by the regularity of μ it follows that $|\mu|(F') = 0$ whenever $F' \subset F$ is a Borel set with $|F'| = 0$. This proves that μ_F is absolutely continuous with respect to Lebesgue measure. $\qquad\square$

We now consider the set

$$F_0 = \{x \in B(0, a) : \quad \exists \epsilon > 0 : b_j(x, t) = 0, \ \forall t \in [0, \epsilon], \ j = 0, \dots, n\}$$

which is a countable union of the closed sets

$$F_k = \{x \in B(0, a) : \quad b_j(x, t) = 0, \ \forall \ 0 \le t \le \frac{1}{k}, \ j = 0, \dots, n\}$$

to which we can apply Lemma V.7.2 and conclude that μ_{F_k} is absolutely continuous with respect to Lebesgue measure. Thus, μ_{F_0} is also absolutely continuous with respect to Lebesgue measure and the Radon–Nikodym theorem implies that there exists $g \in L^1_{\text{loc}}(B(0, a))$ such that

$$\mu_{F_0}(X) = \int_X g(x) \, dx, \quad X \subset B(0, a) \text{ a Borel set.}$$

Theorem V.6.9 and Lemma V.7.2 imply Theorem V.7.1:

End of the proof of Theorem V.7.1. We may assume that the vector field has the form

$$L = \frac{\partial}{\partial t} + ib(x, t) \frac{\partial}{\partial x}$$

where $b(x, t)$ is real and smooth on a neighborhood of $U = B(-a, a) \times (-T, T)$ of the origin in $\mathbb{R}^2$. Since the trace bf is a measure, by the Radon–Nikodym theorem, we may write

$$bf = g + \mu$$

where g is a locally integrable function and μ is a measure supported on a set E of Lebesgue measure zero. Suppose x_0 is a point for which we can find a sequence t_j converging to 0 with $b(x_0, t_j) \ne 0$. Let $Z(x, t)$ be a first integral satisfying $Z(x_0, 0) = 0$, and $Z_x(x_0, 0) = 1$. If $\Im Z_t(x_0, 0) \ne 0$, then L will be elliptic in a neighborhood of $(x_0, 0)$ and so by the classical F. and M. Riesz theorem, we can conclude that bf is absolutely continuous near $(x_0, 0)$. Otherwise, the proof of Theorem V.6.9 shows that the FBI transform with this Z as a first integral and arbitrarily large κ decays exponentially in a complex conic neighborhood of (x_0, ξ_0), for some nonzero covector. By theorem 2.2 in [**BCT**], it follows that near the point x_0, modulo a smooth nonvanishing multiple, the trace bf is the weak boundary value of a holomorphic function F defined on a side of the curve $x \longmapsto Z(x, 0)$. But then, again by the classical F. and M. Riesz theorem, bf is locally integrable near x_0, that is, $x_0 \notin E$. Hence the set E is contained in the set

$$F_0 = \{x \in B(0, a) : \quad \exists \epsilon > 0 : b_j(x, t) = 0, \forall t \in [0, \epsilon], j = 0, \dots, n\}.$$

But we already observed that the restriction of bf to F_0 is absolutely continuous with respect to Lebesgue measure which implies that μ is zero.

Notes

For a more detailed account of CR manifolds the reader is referred to
the books [**Bog**] and [**BER**]. The book [**T5**] contains a detailed discussion of hypoanalytic manifolds. The characterization of microlocal analyticity (Theorem V.2.14) was proved by Bony. Microlocal analyticity was
generalized to microlocal hypoanalyticity in the work [**BCT**]. Several mathematicians have used the FBI transform to study the regularity of solutions in
involutive structures and higher-order partial differential equations. Some of
these applications can be found in the works [**BCT**], [**BT3**], [**BRT**], [**Hi**] and
[**HaT**], [**Sj1**], and [**EG1**]. Theorem V.5.5 was proved by Chemin [**Che**] by
using para-differential calculus. The main ideas for the proof presented here
are due to Hanges and Treves ([**HaT**]), who proved the analytic version of
Chemin's result. Subsequently, Asano [**A**] used the techniques in [**HaT**] to
give a new proof of Chemin's result. Most of the material in Section V.6 is
taken from a paper of Eastwood and Graham ([**EG1**]). Section V.7 is taken
from [**BH1**]. For a generalization of the F. and M. Riesz theorem to systems
of vector fields, we refer the reader to [**BH7**].

VI

Some boundary properties of solutions

In this chapter we will explore certain boundary properties of the solutions of locally integrable vector fields. In the first section we present a growth condition that ensures the existence of a distribution boundary value for a solution of a locally integrable complex vector field in $\mathbb{R}^N$. This condition extends the well-known tempered growth condition for holomorphic functions which we will recall in Theorem VI.1.1 below. Section VI.2 considers the pointwise convergence of solutions of planar, locally integrable vector fields to their boundary values. Sections VI.3 and VI.4 explore the class of vector fields in the plane for which Hardy space-like properties are valid. The chapter concludes with applications to the boundary regularity of solutions. The boundary variant of the Baouendi–Treves approximation theorem, namely, Theorem II.4.12, will be crucial for the results in Sections VI.2 and VI.4.

VI.1 Existence of a boundary value

Suppose L is a smooth complex vector field,

$$L = \sum_{j=1}^{N} a_j(x) \frac{\partial}{\partial x_j}$$

defined on a domain $\Omega \subseteq \mathbb{R}^N$ and $u \in C(\Omega)$ is such that $Lu = 0$ in Ω. Assume $\partial\Omega$ is smooth. We would like to explore conditions on u that guarantee that u will have a distribution boundary value on $\partial\Omega$. Theorem V.2.6 showed us that when u is holomorphic on a domain $D \subseteq \mathbb{C}^n$, then u has a boundary value if

$$|u(z)| \leq \frac{C}{\text{dist}(z, \partial D)^k} \tag{VI.1}$$

271

for some $C, k > 0$. Conversely, it is well known that if a holomorphic function on Ω has a distribution trace on ∂D, then $u(z)$ has a tempered growth as in (VI.1). For simplicity, we recall here a precise version in the planar case:

THEOREM VI.1.1 (Theorems 3.1.11, 3.1.14 [**H2**].). *Let $A, B > 0$, $Q = (-A, A)$ $\times (0, B)$ and f holomorphic on Q.*

(i) *If for some integer $N \geq 0$ and $C > 0$,*

$$|f(x + iy)| \leq Cy^{-N}, \quad x + iy \in Q,$$

then there exists $bf \in D'(-A, A)$ of order $N + 1$ such that

$$\lim_{y \to 0^+} \int f(x + iy)\psi(x)dx = \langle bf, \psi \rangle \quad \forall \psi \in C_0^{N+1}(-A, A).$$

(ii) *If $\lim_{y \to 0^+} f(\cdot + iy)$ exists in $D'^k(-A, A)$, then for any $0 < A' < A$, and $0 < B' < B$, there exists C' such that*

$$|f(x + iy)| \leq C'y^{-k-1}, \quad x + iy \in (-A', A') \times (0, B').$$

Because of the local equivalence of L^1 and sup norms for solutions in the elliptic (Cauchy–Riemann) case, the preceding theorem asserts that a holomorphic function f on Q has a trace at $y = 0$ if and only if for some integer $N > 0$,

$$\iint_Q |f(x + iy)||y|^N \, dxdy < \infty.$$

It is natural to investigate generalizations of this theorem for nonelliptic vector fields. It turns out that the tempered growth condition (VI.1) is sufficient to ensure the existence of a boundary value for a general nonvanishing vector field that may not be locally integrable. Indeed, we have:

THEOREM VI.1.2 (Theorem 1.1 [**BH4**]). *Let L be a C^∞ complex vector field in a domain $\Omega \subseteq \mathbb{R}^n$, $f \in C(\Omega)$, $Lf = 0$ in Ω. Suppose*

$$|f(x)| \leq C \, \mathrm{dist}(x, \partial\Omega)^{-N}$$

for some $C, N > 0$. If $\Sigma \subseteq \partial\Omega$ is open, smooth and noncharacteristic for L, then f has a distribution boundary value on Σ.

The preceding result suggests that for a locally integrable vector field, in general, one should seek a growth condition that is weaker than a tempered growth expressed in terms of $\mathrm{dist}(x, \partial\Omega)$.

As a motivation, suppose $Z = x + i\varphi(x, y)$ is smooth in a neighborhood of the origin in $\mathbb{R}^2$, φ real-valued. Then Z is a first integral for

$$L = \frac{\partial}{\partial y} - \frac{i\varphi_y}{1 + i\varphi_x} \frac{\partial}{\partial x}.$$

Assume that $\varphi(x, y) > 0$ when $y > 0$ and $\varphi(x, 0) = 0$, for all x. Then for any integer $N > 0$, since the holomorphic function $\frac{1}{(x+iy)^N}$ has a boundary value as $y \to 0^+$, it is not hard to see that

$$u_N(x, y) = \frac{1}{Z(x, y)^N}$$

also has the same boundary value.

Note that $Lu_N = 0$ when $y > 0$, $|u_N(0, y)| = \frac{1}{|\varphi(0, y)|^N}$, while

$$|u_N(x, y)| \le \frac{1}{|\varphi(x, y)|^N} = \frac{1}{|Z(x, y) - Z(x, 0)|^N}.$$

Observe that φ may be chosen so that $u_N(x, y)$ is not bounded by any power of y as $y \to 0^+$. In general, if L is locally integrable, Z is a first integral of L near the origin and $Lu = 0$ in the region $y > 0$, then the growth condition

$$|u(x, y)||Z(x, y) - Z(x, 0)|^N \le C < \infty \tag{VI.2}$$

is sufficient for u to have a distribution boundary value at $y = 0$. When L is real-analytic, (VI.2) is also a necessary condition for the existence of a boundary trace at $y = 0$ (see [BH5]). Before we state the main result of this section, as a motivation for its proof, we review the classical case of holomorphic functions. Consider a holomorphic function f on the rectangle $Q = (-A, A) \times (-B, B)$ satisfying the growth condition

$$|f(x + iy) y^N| \le C < \infty.$$

We wish to show that f has a boundary value at $y = 0$. Let $\psi \in C_0^\infty(-A, A)$. Fix $0 < T < B$. For each integer $m \ge 0$, choose $\psi_m(x, y) \in C^\infty((-A, A) \times [0, B])$ such that

(i) $\psi_m(x, 0) = \psi(x)$ and
(ii) $|\bar{\partial}\psi_m(x, y)| \le Cy^m$

where C depends only on the size of the derivatives of ψ up to order $m + 1$. Indeed, if we define

$$\psi_m(x, y) = \sum_{k=0}^m \frac{\psi^{(k)}(x)}{k!} (iy)^k,$$

then it is easy to see that (i) and (ii) hold. Note that since f is holomorphic, for any $0 < \epsilon < T$, and $g \in C_0^1(-A, A)$, integration by parts gives:

$$\int_{-A}^{A} f(x+i\epsilon)g(x, \epsilon)\,dx = \int_{-A}^{A} f(x+iT)g(x, T)dx$$
$$+ 2i \int_{-A}^{A} \int_{\epsilon}^{T} f(x+iy)\bar{\partial}g(x, y)\,dxdy.$$

Plugging $g(x, y) = \psi_N(x, y - \epsilon)$ in the preceding formula yields

$$\int_{-A}^{A} f(x+i\epsilon)\psi(x)\,dx = \int_{-A}^{A} f(x+iT)\psi_N(x, T - \epsilon)dx$$
$$+ 2i \int_{-A}^{A} \int_{\epsilon}^{T} f(x+iy)e(x, y, \epsilon)\,dxdy$$

where $|e(x, y, \epsilon)| \le C|y - \epsilon|^N$. Since $|f(x+iy)\,y^N| \le C$, as $y \to 0$, the right-hand side in the formula converges. This proves that $f(x+iy)$ has a boundary value at $y = 0$.

We will prove now the sufficiency of (VI.2) in a more general set-up. Let L be a smooth, locally integrable vector field defined near the origin in $\mathbb{R}^{m+1}$. In appropriate coordinates (x, t) we may assume that L possesses m smooth first integrals of the form $Z_j(x, t) = A_j(x, t) + iB_j(x, t), j = 1, \ldots, m$ defined on a neighborhood of the closure of the cylinder $Q = B_r(0) \times (-T, T)$ where $B_r(0)$ is a ball in x space $\mathbb{R}^m$ and $Z_x(0, 0)$ is invertible. Thus, after multiplication by a nonvanishing factor, L may be written as

$$L = \frac{\partial}{\partial t} - \sum_{k=1}^{m} \frac{\partial Z_k}{\partial t} M_k \qquad (\text{VI.3})$$

where the M_k are the vector fields in x space satisfying $M_k Z_j = \delta_{kj}, 1 \le k, j \le m$. The next theorem gives, in particular, a sufficient condition for the existence of a boundary value of a continuous function f when f is a solution of $Lf = 0$.

THEOREM VI.1.3. *Let L be as above and let f be continuous on $Q^+ = B_r(0) \times (0, T)$. Suppose*

(i) *$Lf \in L^1(Q^+)$;*
(ii) *there exists $N \in \mathbb{N}$ such that*

$$\int_0^T \int_{B_r(0)} |Z(x, t) - Z(x, 0)|^N |f(x, t)|dxdt < \infty.$$

Then $\lim_{t \to 0^+} f(x, t) = bf$ exists in $D'(B_r(0))$ and it is a distribution of order $N + 1$.

PROOF. Note first that by taking complex, linear combinations of the Z_j's, we may assume that $Z_x(0, 0) = \text{Id}$, the identity matrix. This will not affect hypothesis (ii) in the theorem. Let $\psi \in C_0^\infty(B_r(0))$. For each integer $k \geq 0$, we will show that there exists $\psi_k(x, t) \in C^\infty(B_r(0) \times [0, T])$ such that

(i) $\psi_k(x, 0) = \psi(x)$ and
(ii) $|L\psi_k(x, t)| \leq C|Z(x, t) - Z(x, 0)|^k$

where C depends only on the size of $D^\alpha \psi(x)$ for $|\alpha| \leq k+1$. To get $\psi_k(x, t)$ with these properties, we will use a smooth function $u_k = u_k(x, y)$ defined near $0 \in \Sigma = \{Z(x, 0)\}$ in $\mathbb{C}^m$ and satisfying:

(a) $u_k(Z(x, 0)) = \psi(x)$ and
(b) $|(\frac{\partial}{\partial x_j} + i\frac{\partial}{\partial y_j})u_k(x, y)| \leq C \operatorname{dist}((x, y), \Sigma)^k$ for $j = 1, \ldots, m$.

Assuming for the moment that such a u_k with these properties exists, we set

$$\psi_k(x, t) = u_k(A(x, t), B(x, t))$$

where

$$A(x, t) = (A_1(x, t), \ldots, A_m(x, t)), \quad B(x, t) = (B_1(x, t), \ldots, B_m(x, t)).$$

Then $\psi_k(x, 0) = \psi(x)$ so that (i) above holds. To check (ii), observe that from the equations

$$L(Z_j) = L(A_j + iB_j) = 0, \quad j = 1, \ldots, m,$$

we have

$$L(\psi_k) = \sum_{j=1}^m \left(\frac{\partial u_k}{\partial x_j} L(A_j) + \frac{\partial u_k}{\partial y_j} L(B_j) \right) = 2 \sum_{j=1}^m L(A_j) \frac{\partial u_k}{\partial \bar{z}_j}. \tag{VI.4}$$

It follows that

$$|L(\psi_k)(x, t)| \leq C_1 |\bar{\partial} u_k(A(x, t), B(x, t))|$$
$$\leq C_2 \operatorname{dist}(A(x, t) + iB(x, t), \Sigma)^k$$
$$\leq C_2 |Z(x, t) - Z(x, 0)|^k.$$

Thus if u_k satisfies (a) and (b), then $\psi_k(x, t)$ will satisfy (i) and (ii). We will next write a formula for the u_k. Since the map $x \mapsto A(x, 0)$ is invertible, there is a smooth map $G = (G_1, \ldots, G_m)$ such that

$$\Im(Z(x, 0)) = B(x, 0) = G(A(x, 0)).$$

This and some of what follows may require decreasing the neighborhood around the origin. Note that since $dB(0, 0) = 0$, and $dA(0, 0) \neq 0$, $dG(0, 0) = 0$.

Let V_j be the vector fields satisfying $V_j(x_s + iG_s(x)) = \delta_{js}$, $1 \leq j, s \leq m$. For each $k = 1, 2, \ldots$ define

$$u_k(x, y) = \sum_{|\alpha| \leq k} \frac{i^{|\alpha|}}{\alpha!} V^\alpha \tilde{\psi}(x)(y - G(x))^\alpha$$

where by definition, $\tilde{\psi}(x) = \psi(A(x, 0)^{-1})$. Clearly, $u_k(Z(x, 0)) = \psi(x)$. We claim that for each $j = 1, \ldots, m$,

$$2\frac{\partial u_k}{\partial \bar{z}_j} = i^k \sum_{|\alpha| = k} \frac{1}{\alpha!} \frac{\partial}{\partial x_j} \left(V^\alpha \tilde{\psi}(x) \right) (y - G(x))^\alpha. \tag{VI.5}$$

In particular, the claim implies property (b) for u_k. Indeed, after contracting the neighborhood of the origin, we may assume that $\Sigma = \{x + iG(x)\}$. Since $dG(0, 0) = 0$, it follows that

$$|y - G(x)| \leq \text{dist}\,((x, y), \Sigma)$$

which gives (b). The claim will be proved by induction. We have:

$$\frac{\partial u_1}{\partial y_j}(x + iy) = iV_j(\tilde{\psi}(x))$$

and

$$\frac{\partial u_1}{\partial x_j}(x + iy) = \frac{\partial \tilde{\psi}}{\partial x_j} - i\sum_{s=1}^m V_s(\tilde{\psi})\frac{\partial G_s}{\partial x_j} + i\sum_{s=1}^m \frac{\partial}{\partial x_j}\left(V_s(\tilde{\psi})\right)(y_s - G_s(x)).$$

Next observe that

$$\frac{\partial}{\partial x_j} = i\sum_{s=1}^m \frac{\partial G_s(x)}{\partial x_j} V_s + V_j \tag{VI.6}$$

which can be seen by applying both sides to the m linearly independent functions $x_1 + iG_1(x), \ldots, x_m + iG_m(x)$. Hence

$$\frac{\partial u_1}{\partial x_j} + i\frac{\partial u_1}{\partial y_j} = i\sum_{s=1}^m \frac{\partial}{\partial x_j}\left(V_s(\tilde{\psi})\right)(y_s - G_s(x))$$

which proves the claim for $k = 1$. Assume next that (VI.5) holds for $k - 1$, $k \geq 1$. We can write

$$u_k(x, y) = u_{k-1}(x, y) + E_k(x, y) \tag{VI.7}$$

where

$$E_k(x, y) = i^k \sum_{|\alpha| = k} \frac{1}{\alpha!}\left(V^\alpha \tilde{\psi}(x)\right)(y - G(x))^\alpha.$$

For any $1 \leq j \leq m$, by the induction assumption, we have

$$2\frac{\partial u_{k-1}}{\partial \overline{z}_j} = i^{k-1} \sum_{|\beta|=k-1} \frac{1}{\beta!} \frac{\partial}{\partial x_j}\left(V^\beta \tilde{\psi}\right)(y-G(x))^\beta. \tag{VI.8}$$

Observe that

$$\frac{\partial E_k}{\partial x_j}(x,y) = i^k \sum_{|\alpha|=k} \frac{1}{\alpha!}\left(\frac{\partial}{\partial x_j}V^\alpha\tilde{\psi} \quad (y-G(x))^\alpha + V^\alpha(\tilde{\psi})\frac{\partial}{\partial x_j}(y-G(x))^\alpha\right) \tag{VI.9}$$

and

$$\frac{\partial E_k}{\partial y_j}(x,y) = i^k \sum_{|\alpha|=k} \frac{V^\alpha(\tilde{\psi})}{\alpha!} \frac{\partial}{\partial y_j}(y-G(x))^\alpha. \tag{VI.10}$$

Using the expression for $\frac{\partial}{\partial x_j}$ from (VI.6), (VI.8) can be written as

$$2\frac{\partial u_{k-1}}{\partial \overline{z}_j} = i^k \sum_{|\beta|=k-1}\sum_{s=1}^{m} \frac{1}{\beta!}\frac{\partial G_s}{\partial x_j}(x)V_s\left(V^\beta\tilde{\psi}\right)(y-G(x))^\beta$$

$$+ i^{k-1} \sum_{|\beta|=k-1} \frac{1}{\beta!}V_j\left(V^\beta\tilde{\psi}\right)(y-G(x))^\beta. \tag{VI.11}$$

From (VI.7), (VI.9), (VI.10) and (VI.11), we get

$$2\frac{\partial u_k}{\partial \overline{z}_j} = i^k \sum_{|\alpha|=k} \frac{1}{\alpha!}\frac{\partial}{\partial x_j}\left(V^\alpha\tilde{\psi}(x)\right)(y-G(x))^\alpha$$

which establishes property (b) for u_k. Hence for each k we have ψ_k which satisfies (i) and (ii) and has the form

$$\psi_k(x,t) = \sum_{|\alpha|\leq k}(P_\alpha(x,t,D_x)\tilde{\psi}(A(x,t)))(B(x,t)-G(A(x,t)))^\alpha \tag{VI.12}$$

where $P_\alpha(x,t,D_x)$ is a differential operator of order $|\alpha|$ involving differentiations only in x. Observe next that if $g(x,t)$ is a C^1 function, the differential of the m form $g(x,t)dZ_1 \wedge \cdots \wedge dZ_m$ where $Z_j = A_j(x,t) + iB_j(x,t)$ is given by

$$d(g\,dZ_1 \wedge \cdots \wedge dZ_m) = Lg\,dt \wedge dZ_1 \wedge \cdots \wedge dZ_m.$$

This observation and integration by parts lead to:

$$\int_{B_r(0)} f(x,\epsilon)\psi_N(x,\epsilon)dZ(x,\epsilon) = \int_{B_r(0)} f(x,T)\psi_N(x,T)\,dZ(x,T)$$

$$+ \int_{B_r(0)}\int_\epsilon^T f(x,t)L\psi_N(x,t)\,dt \wedge dZ \tag{VI.13}$$

$$+ \int_{B_r(0)}\int_\epsilon^T Lf(x,t)\psi_N(x,t)\,dt \wedge dZ$$

where $dZ = dZ_1 \wedge dZ_2 \wedge \cdots \wedge dZ_m$. Now by the hypotheses on $f(x, t)$ and property (ii) of $\psi_N(x, t)$, $|f(x, t)L\psi_N(x, t)| \in L^1$ and so the second integral on the right in (VI.13) has a limit as $\epsilon \to 0$. The third integrand on the right is in L^1 since Lf is. Therefore,

$$\lim_{\epsilon \to 0} \int_{B_r(0)} f(x, \epsilon)\psi_N(x, \epsilon) \, dZ(x, \epsilon) \quad \text{exists.} \tag{VI.14}$$

We can clearly modify ψ_n by dropping the tilde in its definition and use (VI.14) to conclude:

$$\lim_{\epsilon \to 0} \int_{B_r(0)} f(x, \epsilon)\Psi_N(x, \epsilon) \, dZ(x, \epsilon) \quad \text{exists} \tag{VI.15}$$

where for any smooth function $\psi(x)$,

$$\Psi_n(x, t) = \sum_{|\alpha| \le n} (P_\alpha(x, t, D_x)\psi(A(x, t)))(B(x, t) - G(A(x, t)))^\alpha.$$

Let $P(x, t) = B(x, t) - G(A(x, t))$. For $g(x, t) \in C^\infty(B_r(0) \times (-T, T))$ whose x-support is contained in a fixed compact set independent of t, and n a non-negative integer, define

$$T_n g(x, t) = \sum_{|\alpha| \le n} P_\alpha(x, t, D_x)(g(x, t))P(x, t)^\alpha, \quad T_0 g(x, t) = g(x, t). \tag{VI.16}$$

Using (VI.15), we will show next that in fact,

$$\lim_{t \to 0} \int_{B_r(0)} f(x, t)T_N g(x, t) \, dZ(x, t) \quad \text{exists} \tag{VI.17}$$

for any $g = g(x, t)$. To see this, for $\psi = \psi(x)$, we change variables $y = A(x, t)$ in (VI.15) to write

$$\int f(x, t)\Psi_N(x, t) \, dZ(x, t) = \int f(H(y, t), t)Q(y, t, D_y)\psi(y) \, dy$$

where Q is a differential operator (with differentiation only in y) and $y \mapsto H(y, t)$ is the inverse of $x \mapsto A(x, t)$. Since

$$\lim_{t \to 0} \int f(H(y, t), t)Q(y, t, D_y)\psi(y) \, dy \quad \text{exists},$$

it follows that

$$\lim_{t \to 0} \int f(H(y, t), t)Q(y, t, D_y)\psi(y, t) \, dy \quad \text{exists},$$

for any smooth $\psi(y, t)$ with a fixed compact support in y. Going back to the x coordinates, we have shown that

$$\lim_{t \to 0} \int_{B_r(0)} f(x, t)S_N g(x, t) \, dZ(x, t) \quad \text{exists} \tag{VI.18}$$

where by definition

$$S_n g(x, t) = \sum_{|\alpha| \le n} (P_\alpha(x, t, D_x) g(A(x, t), t)) P(x, t)^\alpha$$

for any smooth $g = g(x, t)$. Observe that the integral in (VI.18) can be written in the form

$$\int u(x, t) g(A(x, t), t) \, dx$$

where this latter integral denotes the action of a distribution $u(., t)$ on the smooth function $x \mapsto g(A(x, t), t)$. Now since $(x, t) \mapsto (A(x, t), t)$ is a diffeomorphism near the origin, any function $\psi(x, t)$ is of the form $g(A(x, t), t)$ for some $g = g(x, t)$. We can therefore use (VI.18) to conclude that for any $g(x, t)$,

$$\lim_{t \to 0} \int_{B_r(0)} f(x, t) T_N g(x, t) \, dZ(x, t) \quad \text{exists,} \tag{VI.19}$$

which proves (VI.17). For $\psi(x, t) \in C^\infty(B_r(0) \times (-T, T))$ whose x-support is contained in a fixed compact set and a given multi-index β with $|\beta| = N$, plug $g(x, t) = \psi(x, t) P(x, t)^\beta = \psi(x, t)(B(x, t) - G(A(x, t)))^\beta$ in (VI.19). Note that we may write

$$T_N(\psi P^\beta)(x, t) = \psi P^\beta + \psi \sum_{|\alpha| = N} e_\alpha(x, t) P^\alpha + \sum_{|\gamma| > N} h_\gamma(x, t) P^\gamma \tag{VI.20}$$

where the h_γ and e_α are smooth functions and

$$\lim_{t \to 0} D_x^{\alpha'} e_\alpha(x, t) = 0 \quad \forall \alpha, \alpha'.$$

Observe that for each γ with $|\gamma| > N$,

$$\lim_{t \to 0} \int_{B_r(0)} f(x, t) h_\gamma(x, t) P(x, t)^\gamma \, dZ(x, t) \quad \text{exists.} \tag{VI.21}$$

Indeed, this follows from applying the integration by parts formula (VI.13) to the m-form $f(x, t) h_\gamma(x, t) P(x, t)^\gamma \, dZ_1 \wedge \cdots \wedge dZ_m$, using the hypotheses on f, and the bound $|P(x, t)| \le |Z(x, t) - Z(x, 0)|$. From (VI.19) and (VI.21) we conclude that

$$\lim_{t \to 0} \int_{B_r(0)} f(x, t) \left(\psi P^\beta + \psi \sum_{|\alpha| = N} e_\alpha(x, t) P^\alpha \right) dZ(x, t) \quad \text{exists.} \tag{VI.22}$$

We can plug ψ_β for ψ in (VI.22) and sum over β with $|\beta| = N$ to conclude

$$\lim_{t \to 0} \int_{B_r(0)} f(x, t) \sum_{|\beta| = N} P^\beta \left(\psi_\beta + \left(\sum_{|\alpha| = N} \psi_\alpha \right) E_\beta(x, t) \right) dZ(x, t) \quad \text{exists}$$

$$\tag{VI.23}$$

where all order derivatives of the E_β go to zero as $t \to 0$. Observe that given $\{\psi_\beta\}_{|\beta|=N}$ as above, we can find $\{\phi_\beta\}_{|\beta|=N}$ such that

$$\sum_{|\beta|=N} P^\beta \left(\phi_\beta + \left(\sum_{|\alpha|=N} \phi_\alpha \right) E_\beta \right) = \sum_{|\beta|=N} P^\beta \psi_\beta.$$

It follows that

$$\lim_{t\to 0} \int_{B_r(0)} f(x,t) \sum_{|\beta|=N} \psi_\beta P^\beta \, dZ(x,t) \quad \text{exists} \qquad \text{(VI.24)}$$

whenever the functions $\psi_\beta(x,t) \in C^\infty(B_r(0) \times (-T,T))$ have their x-support contained in a fixed compact set independent of t. We now return to a general $g(x,t) \in C^\infty(B_r(0) \times (-T,T))$ with x-support contained in a fixed compact set independent of t. From (VI.19) and (VI.24) we conclude that

$$\lim_{t\to 0} \int_{B_r(0)} f(x,t) T_{N-1} g(x,t) \, dZ(x,t) \quad \text{exists} \qquad \text{(VI.25)}$$

for any $g(x,t) \in C^\infty(B_r(0) \times (-T,T))$ with x-support contained in a fixed compact set independent of t. We will prove by descending induction that for any such $g(x,t)$ and $0 \le k \le N$,

$$\lim_{t\to 0} \int_{B_r(0)} f(x,t) T_k g(x,t) \, dZ(x,t) \quad \text{exists},$$

which for $k=0$ and $g(x,t) = \psi(x) \in C_c^\infty(B_r(0))$ gives us the desired limit. To proceed by induction, suppose $1 \le k \le N$ and assume that for any multi-index β with $|\beta| = k$, the limits

$$\lim_{t\to 0} \int_{B_r(0)} f(x,t) P^\beta(x,t) g(x,t) \, dZ(x,t) \quad \text{and}$$

$$\lim_{t\to 0} \int_{B_r(0)} f(x,t) T_{k-1} g(x,t) \, dZ(x,t) \qquad \text{(VI.26)}$$

both exist for any $g(x,t) \in C^\infty(B_r(0) \times (-T,T))$ with x-support contained in a fixed compact set independent of t. We have already seen that (VI.26) is true for $k=N$ as follows from (VI.24) and (VI.25). Fix β' with $|\beta'| = k-1$. Plug $g(x,t) = \psi(x,t) P(x,t)^{\beta'}$ in the limit on the right in (VI.26) and observe that $T_{k-1} g$ may be written as

$$T_{k-1} g(x,t) = \psi P^{\beta'} + \psi \sum_{|\alpha|=k-1} e_\alpha(x,t) P^\alpha + \sum_{|\gamma|\ge k} h_\gamma(x,t) P^\gamma \qquad \text{(VI.27)}$$

where the e_α and h_γ are smooth, the x-supports of the $h_\gamma(x,t)$ are contained in a compact set that is independent of t, and all order derivatives of the e_α

go to zero as $t \to 0$. From the existence of the two limits in (VI.26) we derive that

$$\lim_{t \to 0} \int_{B_r(0)} f(x, t) \left(\psi P^{\beta'} + \psi \sum_{|\alpha| = k-1} e_\alpha(x, t) P^\alpha \right) dZ(x, t) \qquad \text{(VI.28)}$$

exists. We now argue as before by replacing ψ by $\psi_{\beta'}$ and summing over $|\beta'| = k - 1$ to conclude that

$$\lim_{t \to 0} \int_{B_r(0)} f(x, t) P(x, t)^\beta \psi(x, t) dZ(x, t) \quad \text{exists} \qquad \text{(VI.29)}$$

for all β with $|\beta| = k - 1$ and $\psi(x, t) \in C^\infty(B_r(0) \times (-B, B))$ with x-support contained in a fixed compact set independent of t. Hence, taking account of (VI.26) and (VI.29) we conclude that

$$\lim_{t \to 0} \int_{B_r(0)} f(x, t) T_{k-2} g(x, t) dZ(x, t) \quad \text{exists.} \qquad \text{(VI.30)}$$

We have thus proved that (VI.26) holds for $k - 1$, completing the inductive step. Therefore,

$$\lim_{\epsilon \to 0} \int_{B_r(0)} f(x, \epsilon) \psi(x) dZ(x, \epsilon) \quad \text{exists} \qquad \text{(VI.31)}$$

and thus $bf = \lim_{t \to 0} f(., t)$ exists. Moreover, since the functions

$$x \longmapsto \psi_N(x, \epsilon) - \psi(x) \quad \text{and} \quad x \longmapsto Z(x, \epsilon) - Z(x, 0)$$

and all their x-derivatives converge to zero as $\epsilon \to 0$, (VI.13), (VI.14), and (VI.31) imply the following formula for bf:

$$\langle Z_x(x, 0) bf, \psi \rangle = \int_{B_r(0)} f(x, T) \psi_N(x, T) dZ \qquad \text{(VI.32)}$$

$$+ \int_{B_r(0)} \int_0^T f(x, t) L \psi_N(x, t) dt \wedge dZ$$

$$+ \int_{B_r(0)} \int_0^T L f(x, t) \psi_N(x, t) dt \wedge dZ.$$

This formula shows that bf is a distribution of order $N + 1$. $\qquad \square$

VI.2 Pointwise convergence to the boundary value

Suppose L is a locally integrable vector field in a planar domain Ω with a smooth boundary. Let $f \in L^1_{loc}(\Omega)$, and assume that f has a weak trace bf which is in $L^1_{loc}(\partial\Omega)$. In this section we will discuss the pointwise convergence of f to bf. It is classical that when L is the Cauchy–Riemann operator, the holomorphic function f converges nontangentially to $bf(p)$ for almost all p in $\partial\Omega$. In general, this approach region cannot be relaxed. Indeed, we recall:

THEOREM VI.2.1. (Theorem 7.44 in [**Zy**].) *Let C_0 be any simply closed curve passing through $z = 1$ situated, except for that point, totally inside the circle $|z| = 1$, and tangent to the circle at that point. Let C_θ be the curve C_0 rotated around $z = 0$ by the angle θ. There is a Blaschke product $B(z)$ which, for almost all θ_0, doesn't tend to any limit as $z \mapsto \exp(i\theta_0)$ inside C_{θ_0}.*

This theorem shows us that for nonelliptic vector fields, we can't expect nontangential convergence. Indeed, by the theorem, if

$$L_k = \frac{\partial}{\partial t} - i(k+1)t^k \frac{\partial}{\partial x} \quad (k = 1, 2, 3, \dots)$$

then for each k, we can get a bounded solution $f_k = F_k(x + it^{k+1})$ of L_k with F_k holomorphic in a semidisk in the upper half-plane, $bf_k(x) = bF_k(x) \in L^1(-1, 1)$, but each $f_k(x, t)$ doesn't converge nontangentially on a subset of $(-1, 1)$ of positive measure. It suffices to take F_k holomorphic and bounded on the semidisk $\{z : |z| < 1, \Im z > 0\}$ such that on a set of full measure in $(-1, 1)$, F_k has no limit in certain appropriate regions. By considering the L_k with k even, we see that nontangential convergence may fail even for vector fields that are C^∞ and analytic hypoelliptic. Note that for each k, and for almost all $p \in (-1, 1)$, there is an open region $\Gamma_k(p)$ with $p \in \overline{\Gamma}_k(p)$ such that $f_k(x, t)$ converges to $bf_k(p)$ in $\Gamma_k(p)$. On the other hand, if we take the real vector field $\frac{\partial}{\partial t}$, and the solution $u(x, t) \equiv bu(x) = \chi$, the characteristic function of a Cantor set C of positive measure in $(-1, 1)$, the only sets of approach for which $u(x, t) \to bu(x)$, $x \in C$, are the vertical segments. Thus for a general locally integrable vector field, we cannot get approach sets for convergence larger than curves. Suppose now $L = X + iY$ is a smooth, locally integrable vector field near the closure of a planar domain Ω. Assume $\Sigma \subseteq \partial\Omega$ is a smooth curve that is noncharacteristic for L, $f \in L^1_{\text{loc}}(\Omega)$, $Lf = 0$ and f has a trace $bf \in L^1(\Sigma)$. Multiplying by i if necessary, we may assume that X is not tangent to Σ anywhere and that it points toward Ω. For each $p \in \Sigma$, let γ_p be the integral curve of X through p and set $\gamma_p^+ = \gamma_p \cap \Omega$. We shall classify the points of Σ into two types:

(I) A point $p \in \Sigma$ is a type I point if the vector fields X and Y are linearly dependent on an arc $\{\gamma_p^+(s) : 0 < s < \epsilon\}$ for some $\epsilon > 0$.

(II) A point $q \in \Sigma$ is a type II point if there is a sequence $q_k \in \gamma_p^+$ converging to q such that L is elliptic at each q_k.

THEOREM VI.2.2. *Let* $Lu = 0$ *in* Ω, $u \in L^1_{loc}(\Omega)$, $bu \in L^1(\Sigma)$, *and* Σ *is noncharacteristic for* L. *Assume* L *is locally integrable in a neighborhood of* Σ. *For each* $p \in \Sigma$, *there is an approach set* $\Gamma(p) \subseteq \Omega$ *such that:*

(i) $p \in \overline{\Gamma}(p)$ *and if* $q \in \Sigma \cap \overline{\Gamma}(p)$, *then* $q = p$;
(ii) $\gamma_p^+ \subseteq \Gamma(p)$;
(iii) *for a.e.* $p \in \Sigma$, $\lim_{\Gamma(p) \ni q \to p} u(q) = u(p)$;
(iv) *if* p *is a type II point,* $\Gamma(p)$ *is an open set, otherwise* $\Gamma(p) = \gamma_p^+$.

PROOF. Since the problem is local, we may assume that we are in coordinates (x, t) where $\Omega = (-1, 1) \times (0, 1)$, $\Sigma = (-1, 1) \times \{0\}$, and $Z(x, t) = x + i\varphi(x, t)$ is a first integral of L with φ real, $\varphi(0, 0) = 0$ and $\varphi_x(0, 0) = 0$. Modulo a nonvanishing factor,

$$L = \frac{\partial}{\partial t} - i\frac{\varphi_t}{1 + i\varphi_x}\frac{\partial}{\partial x}$$

and so

$$X = \frac{\partial}{\partial t} - \left(\frac{\varphi_t \varphi_x}{1 + \varphi_x^2}\right)\frac{\partial}{\partial x}, \quad Y = \frac{-\varphi_t}{1 + \varphi_x^2}\frac{\partial}{\partial x}.$$

Observe that L is elliptic, i.e., X and Y are linearly independent precisely at the points where $\varphi_t \neq 0$. Assume now that $0 \in \Sigma$ is a type II point. Then $t \mapsto \varphi(0, t)$ can't vanish on any interval $[0, \epsilon]$, $\epsilon > 0$. Indeed, otherwise, we would conclude that $L = X$ on $\{0\} \times [0, \epsilon)$—contradicting the hypothesis that 0 is a type II point. For $\delta > 0$ small, define

$$m(x) = \inf_{0 \le t \le \delta} \varphi(x, t), \quad M(x) = \sup_{0 \le t \le \delta} \varphi(x, t).$$

Then since $m(0) < M(0)$, we may choose $A > 0$ so that $m(x) < M(x)$ for $|x| \le A$. After decreasing A and δ, by the boundary version of the Baouendi–Treves approximation theorem in Chapter II (Theorem II.4.12), there is a sequence of entire functions F_k satisfying:

(a) $F_k(Z(x, t)) \to u(x, t)$ pointwise a.e. on $(-A, A) \times (0, \delta)$;
(b) $F_k(Z(x, 0)) \to bu(x)$ a.e. on $(-A, A)$.

Set

$$\Omega_A = \{\zeta = \xi + i\eta : |\xi| < A, m(\xi) < \eta < M(\xi)\}.$$

We may assume that the sequence F_k converges uniformly on compact subsets of Ω_A to a holomorphic function F and $u(x, t) = F(Z(x, t))$ for $(x, t) \in Z^{-1}(\Omega_A)$. Indeed, this is clearly true if $u(x, t)$ is continuous for $t > 0$. In general, we can use the fact that we can express u as Qh where h is a continuous solution and Q is an elliptic differential operator that maps solutions to solutions. The operator Q can be taken to be a convenient power of the operator D defined in Section IV.2. Since 0 is a type II point, by theorem 3.1 in [**BH1**] and [**BCT**] (page 465), for some $0 < A_1 < A$, $0 < \delta_1 < \delta$, there is a holomorphic function G of tempered growth defined on the region $\Omega_1 = \{Z(x, 0) + iZ_x(x, 0)v : |x| < A_1, 0 < v < \delta_1\}$ such that for every $\psi \in C_0^\infty(-A_1, A_1)$,

$$\langle bu, \psi \rangle = \lim_{v \downarrow 0} \int G(Z(x, 0) + iZ_x(x, 0)v)\psi(x)\mathrm{d}x.$$

Since $bu \in L^1$, the holomorphic function $G(z)$ converges nontangentially to $bu(x)$ a.e. in $(-A_1, A_1)$. We may assume that A_1 and δ_1 are small enough so that $\Omega_1 \subseteq \Omega_A$. We will show that $G = F$ on Ω_1. Define the subsets of $[-A_1, A_1]$:

$$E_1 = \{x : \varphi(x, t) = \varphi(x, 0), \ t \in [0, \tau] \quad \text{for some } \tau > 0\},$$
$$E_2 = \{x : \varphi(x, t) \geq \varphi(x, 0), \ t \in [0, \tau] \quad \text{for some } \tau > 0\},$$
$$E_3 = \{x : \varphi(x, t) \leq \varphi(x, 0), \ t \in [0, \tau] \quad \text{for some } \tau > 0\},$$
$$E_4 = \{x : \text{for some} \quad t_j \to 0, \ s_j \to 0, \ \varphi(x, s_j) < \varphi(x, 0) < \varphi(x, t_j)\}.$$

Observe that $[-A_1, A_1] = E_1 \cup E_2 \cup E_3 \cup E_4$. If $x_0 \in E_4$, then by theorem 3.1 in [**BH1**], there is a holomorphic function H defined in a neighborhood of $Z(x_0, 0)$ such that $u(x, t) = H(Z(x, t))$ for (x, t) in a neighborhood of $(x_0, 0)$, $t > 0$. Hence in this case, $F(z)$ has a holomorphic extension to a neighborhood of $Z(x_0, 0)$ and since $u(x, t) = F(Z(x, t))$ for $t > 0$, we have $F(Z(x, 0)) = bu(x) = bG(Z(x, 0))$. Therefore, by theorem 2.2 in [**Du**], $F(z) = G(z)$ on Ω_1. We may therefore assume that $E_4 = \emptyset$. Each of the other three sets E_1, E_2, and E_3 can be written as a countable union of closed sets as follows: $E_1 = \bigcup_{j=1}^\infty E_{1j}$, where $E_{1j} = \{x \in [-A_1, A_1] : \varphi(x, t) = \varphi(x, 0), t \in [0, \frac{1}{j}]\}$; $E_2 = \bigcup_{j=1}^\infty E_{2j}$, where $E_{2j} = \{x \in [-A_1, A_1] : \varphi(x, t) \geq \varphi(x, 0), t \in [0, \frac{1}{j}]\}$; and $E_3 = \bigcup_{j=1}^\infty E_{3j}$, where $E_{3j} = \{x \in [-A_1, A_1] : \varphi(x, t) \leq \varphi(x, 0), t \in [0, \frac{1}{j}]\}$. Thus the interval $[-A_1, A_1]$ is a countable union of the closed sets E_{ij} and hence by Baire's Category Theorem, one of these sets contains an interval with nonempty interior.

Case 1: Suppose $\varphi(x, t) = \varphi(x, 0)$ on $[A_2, A_3] \times [0, T]$ for some $T > 0$, $A_2 < A_3$. Then $L = \frac{\partial}{\partial t}$ on $[A_2, A_3] \times [0, T]$ and so $u(x, t) = bu(x)$ on this rectangle. This implies that $F(z)$ extends as a continuous function in Ω_1 up to the boundary piece $\{Z(x, 0) : A_2 < x < A_3\}$ and therefore $bF(Z(x, 0)) = bu(x)$ for $x \in (A_2, A_3)$. But then $F \equiv G$ in Ω_1.

Case 2: Suppose $\varphi(x, t) \geq \varphi(x, 0)$ on $[A_2, A_3] \times [0, T]$, for some $T > 0$, $A_2 < A_3$. For $\epsilon > 0$ sufficiently small, define

$$u_\epsilon(x, t) = G(Z(x, t) + i\epsilon), \quad (x, t) \in (A_2, A_3) \times (0, T).$$

Observe that $Lu_\epsilon = 0$. Recall that G is holomorphic on the region $\Omega_1 = \{Z(x, 0) + iZ_x(x, 0)v : |x| < A_1, 0 < v < \delta_1\}$. Let $\Omega_2 = \{Z(x, 0) + iZ_x(x, 0)v : |x| < A_1, 0 < v < \delta_2\}$ for some $0 < \delta_2 < \delta_1$, and for each $p = Z(x, 0)$, $|x| < A_1$, define the nontangential approach region

$$\Gamma(p) = \{z \in \Omega_2 : |z - p| < 2\,\mathrm{dist}(z, \partial\Omega_2)\}.$$

Denote by $G^*(x)$ the nontangential maximal function of $G(z)$, that is,

$$G^*(x) = \sup\{|G(z)| : z \in \Gamma(Z(x, 0))\}.$$

We have:

$$|u_\epsilon(x, t)| \leq G^*(x) \in L^1(A_2, A_3).$$

Let

$$w(x, t) = \lim_{\epsilon \to 0} u_\epsilon(x, t) \quad \text{(the pointwise limit)}$$

$$= \begin{cases} G(x + i\varphi(x, t)), & \text{if} \quad \varphi(x, t) > \varphi(x, 0) \\ bu(x), & \text{if} \quad \varphi(x, t) = \varphi(x, 0). \end{cases}$$

Then $u_\epsilon \to w$ in $L^1((A_2, A_3) \times (0, T))$ and so $Lw = 0$ in $(A_2, A_3) \times (0, T)$. Since

$$|G(x + i\varphi(x, t))| \leq G^*(x) \quad \text{and a.e.} \quad G(x + i\varphi(x, t)) \to bu(x) \quad \text{as} \quad t \to 0,$$

we conclude that

$$w(x, t) \to bu(x) \quad \text{in} \quad L^1(A_2, A_3) \quad \text{as} \quad t \to 0.$$

Therefore $u(x, t) = w(x, t)$ in a neighborhood of $(A_2, A_3) \times \{0\}$, $t > 0$. In particular, since we may assume that

$$\{(x, t) \in (A_2, A_3) \times (0, T) : \varphi(x, t) > \varphi(x, 0)\}$$

is not empty (otherwise, we would be placed under Case 1), $F(z) \equiv G(z)$ on Ω_1.

Case 3: Suppose $\varphi(x, t) \leq \varphi(x, 0)$ on $[A_2, A_3] \times [0, T]$, $T > 0$, $A_2 < A_3$. We may assume that there exists $x_0 \in (A_2, A_3)$ and $s_j \to 0$ such that $\varphi(x_0, s_j) < \varphi(x_0, 0)$. Indeed, otherwise, matters will reduce to Case 1. By theorem 3.1

in [**BH1**] and [**BCT**] (page 465), after decreasing $[A_2, A_3] \times [0, T]$, we get a tempered holomorphic function $G_1(z)$ defined on the region

$$\Omega_1' = \{Z(x, 0) + iZ_x(x, 0)v : A_2 < x < A_3, \ -T < v < 0\}$$

such that for every $\psi \in C_0^\infty(A_2, A_3)$,

$$\langle bu, \psi \rangle = \lim_{v \to 0} \int G_1(Z(x, 0) + iZ_x(x, 0)v)\psi(x)\mathrm{d}x.$$

By the edge-of-the-wedge theorem, there is a holomorphic function $v(z)$ defined in a neighborhood of $\{Z(x, 0) : A_2 < x < A_3\}$ that extends G and G_1. Hence $F(z) = G(z)$ in Ω_1. We have thus shown that $F \equiv G$ on Ω_1.

Now for almost every $p \in (-A_1, A_1)$, $G(z)$ converges nontangentially at $Z(p, 0)$ (in Ω_1) to $bu(p)$. Pick such a point p and let $\tilde{\Gamma}(p)$ be a nontangential approach region for $G(z)$ at $Z(p, 0)$. Define $\Gamma(p) = Z^{-1}(\tilde{\Gamma}(p))$. Then

$$\lim_{\Gamma(p)\ni(x,t)\to p} u(x, t) = \lim_{\Gamma(p)\ni(x,t)\to p} F(Z(x, t))$$

$$= \lim_{\tilde{\Gamma}(p)\ni z} G(z) = bu(p).$$

We have thus shown that if p is a type II point, then there is an interval around it such that a.e. in the interval, pointwise convergence holds as asserted. Consider now a type I point $(x_0, 0)$. Then $Z(x_0, t) \equiv Z(x_0, 0)$ for t in some interval $[0, \epsilon]$. This implies that $F_k(Z(x_0, t)) \equiv F_k(Z(x_0, 0))$ for $t \in [0, \epsilon]$, and so because of the a.e. convergence stated in (a) and (b), we conclude that for almost every type I point x, $u(x, t) \to bu(x)$ as $t \to 0$. $\qquad\square$

VI.3 One-sided local solvability in the plane

In Section VI.4 we will explore the boundary regularity of solutions of the inhomogeneous equation $Lf = g$ where

$$L = A(x, t)\frac{\partial}{\partial t} + B(x, t)\frac{\partial}{\partial x}$$

is a smooth, locally integrable complex vector field defined on a subdomain Ω of $\mathbb{R}^2$.

If $Lf = g$ in Ω, and f has a trace bf on $\partial\Omega$ with a certain degree of regularity, we will investigate whether the regularity persists near $\partial\Omega$ under some smoothness assumption on g. As usual, the motivation comes from what is known in the elliptic case. Suppose $h(z)$ is a holomorphic function of one variable defined on the rectangle $Q = (-A, A) \times (0, T)$ with a weak trace bh

at $y = 0$. From the local version of the classical Hardy space (H^p) theory for holomorphic functions in the unit disk, we have:

(i) if $bh \in C^\infty(-A, A)$, then h is C^∞ up to $y = 0$;
(ii) if $bh \in L^p(-A, A)$ $(1 \le p \le \infty)$, then for any $B < A$, the norms of the traces $h(\cdot, y)$ in $L^p(-B, B)$ are uniformly bounded as $y \to 0^+$.

The main results of Section VI.4 will extend (i) and (ii) above to solutions of complex vector fields that satisfy a one-sided solvability condition. In the elliptic case, property (i) follows easily from part (ii) of Theorem VI.1.1. We will show in Section VI.4 that in general, property (i) follows from property (ii) above and a boundary solvability condition. When a vector field exhibits property (ii), we will say that it has the H^p property. To describe the class of vector fields with the H^p property, consider a curve Σ in Ω such that $\Omega \backslash \Sigma$ has two connected components, $\Omega \backslash \Sigma = \Omega^+ \cup \Omega^-$. It turns out that the local solutions of the equation $Lu = 0$ on Ω^+ possess the (H^p) property at $q \in \Sigma$ if and only if there is a neighborhood U of q such that L satisfies the solvability condition $(\mathcal{P})$ of Nirenberg and Treves ([**NT**]) on $U \cap \Omega^+$. This leads to a one-sided version of $(\mathcal{P})$ that we denote by $(\mathcal{P}^+)$ (or $(\mathcal{P}^-)$ if Ω^+ is replaced by Ω^-) to indicate the side where it holds. If $(\mathcal{P})$ holds at q, then both $(\mathcal{P}^+)$ and $(\mathcal{P}^-)$ hold at q. However, $(\mathcal{P}^+)$ and $(\mathcal{P}^-)$ may hold at $q \in \Sigma$ and yet $(\mathcal{P})$ may not hold in a neighborhood of q. The Mizohata vector field provides an example illustrating this. Write $L = X + iY$ with X and Y real. Let $\mathcal{O} \subset U$ be a two-dimensional orbit of L in U and consider $X \wedge Y \in C^\infty(U; \bigwedge^2(T(U)))$. Since $\bigwedge^2(T(U))$ has a global nonvanishing section $e_1 \wedge e_2$, $X \wedge Y$ is a real multiple of $e_1 \wedge e_2$ and this gives a meaning to the requirement that $X \wedge Y$ does not change sign on any two-dimensional orbit $\mathcal{O}$ of $\{X, Y\}$ in U. Recall from Chapter IV that the vector field L satisfies condition $(\mathcal{P})$ at $p \in \Sigma$ if there is a disk $U \subseteq \Omega$ centered at p such that $X \wedge Y$ does not change sign on any two-dimensional orbit of L in U.

DEFINITION VI.3.1. *We say that L satisfies condition $(\mathcal{P}^+)$ at $p \in \Sigma$ if there is a disk $U \subseteq \Omega$ centered at p such that $X \wedge Y$ does not change sign on any two-dimensional orbit of L in $U^+ = U \cap \Omega^+$.*

DEFINITION VI.3.2. *We say that L is one-sided locally solvable in L^p, $1 < p < \infty$ (resp. in C^∞) at $q \in \Sigma$ if there is a neighborhood $U \subseteq \Omega$ of q such that—after interchanging Ω^+ and Ω^- if necessary—for every $f \in L^p(U)$ (resp. $f \in C^\infty(U \cap \Omega^+)$) there exists $u \in L^p(U)$ (resp. $u \in C^\infty(U \cap \Omega^+)$) such that $Lu = f$ on $U^+ = U \cap \Omega^+$.*

DEFINITION VI.3.3. *We say that L is one-sided locally integrable at $p \in \Sigma$ if there is a disk $U \subset \Omega$ centered at p such that—after interchanging Ω^+ and Ω^- if necessary—there exists $Z \in C^\infty(U)$ such that:*

(1) *LZ vanishes identically on $U^+ = U \cap \Omega^+$;*
(2) $dZ(p) \neq 0$.

Let us assume that L is one-sided locally integrable at $p \in \Sigma$ and let Z satisfy (1) and (2) of Definition VI.3.3. Replacing Z by iZ if necessary and decreasing U we may choose local coordinates (x, t) such that $x(p) = t(p) = 0$,

$$Z(x, t) = x + i\varphi(x, t) \tag{VI.33}$$

with φ real, U is the rectangle $U = (-a, a) \times (-T, T)$, $\Sigma \cap U = \{(x, 0) : |x| < a\}$ and $U^+ = (-a, a) \times (0, T)$. Thus, modulo a nonvanishing multiple, we may assume that

$$L = \frac{\partial}{\partial t} - i \frac{\varphi_t(x, t)}{1 + i\varphi_x(x, t)} \frac{\partial}{\partial x}, \tag{VI.34}$$

$$X = \frac{\partial}{\partial t} + \frac{\varphi_t \varphi_x}{1 + \varphi_x^2} \frac{\partial}{\partial x}, \qquad Y = -\frac{\varphi_t}{1 + \varphi_x^2} \frac{\partial}{\partial x},$$

and so

$$X \wedge Y = \frac{\varphi_t(x, y)}{1 + \varphi_x^2} \frac{\partial}{\partial x} \wedge \frac{\partial}{\partial t}.$$

The proof of the following lemma is essentially the same as the one for Lemma IV.2.2.

LEMMA VI.3.4. *Let $Z(x, t)$ and L be given by (VI.33) and (VI.34) respectively. Then, L satisfies $(\mathcal{P}^+)$ at the origin if and only there exist $T, a > 0$ such that $(0, T) \ni t \mapsto \varphi(x, t)$ is monotone for every $x \in (-a, a)$.*

We now recall from [**BH6**] the local equivalence between $(\mathcal{P}^+)$ and one-sided solvability. More precisely,

THEOREM VI.3.5. *Let $Z(x, t)$ and L be given by (VI.33) and (VI.34) respectively. The following properties are equivalent:*

(1) *L satisfies $(\mathcal{P}^+)$ (or $(\mathcal{P}^-)$) at the origin;*
(2) *L is one-sided locally solvable in L^p, $1 < p < \infty$, at the origin;*
(3) *L is one-sided locally solvable in C^∞ at the origin.*

The following proposition is concerned with continuous solvability up to the boundary and will be useful in the applications to boundary regularity in Section VI.4.

PROPOSITION VI.3.6. *Let $Z(x, t)$ and L be given by (VI.33) and (VI.34) respectively and assume that L satisfies $(\mathcal{P}^+)$ at the origin, i.e., for some $U^+ = (-r, r) \times (0, T)$, the function $(0, T) \ni t \mapsto \varphi(x, t)$ is monotone for $|x| < r$. If $f(x, t) \in \mathrm{Lip}(U)$ there exists $u \in \bigcap_{0 < \alpha < 1} C^\alpha((-r, r) \times [0, T))$ such that $Lu = f$ in U^+.*

The proof of the proposition is based on the following lemma.

LEMMA VI.3.7. *Let $F(\zeta) \in L_c^\infty(\mathbb{C})$ and let $f(x, t) = F \circ Z(x, t)$. There exists $v \in \bigcap_{0 < \alpha < 1} C^\alpha((-r, r) \times (-T, T))$ such that $Lv = 2i\varphi_t Z_x^{-1} f$ on $Q = (-r, r) \times (-T, T)$.*

PROOF. Let $E = 1/(\pi\zeta)$ be the fundamental solution of $\partial/\partial\bar{\zeta}$ and set $V = E * F$. Then $V \in \bigcap_{0 < \alpha < 1} C^\alpha$ locally and $\bar{\partial}_\zeta V = F$ in the sense of distributions. If we set $v = V \circ Z$ it follows that v is in $\bigcap_{0 < \alpha < 1} C^\alpha((-r, r) \times [0, T))$ and the chain rule gives $Lv = -2i\varphi_t Z_x^{-1} (\bar{\partial}_\zeta V) \circ Z = -2i\varphi_t Z_x^{-1} f$. $\square$

PROOF OF PROPOSITION VI.3.6. Let $f \in \mathrm{Lip}(U)$. Set $u_0(x, t) = \int_0^t f(x, s) \, ds$. Then, $u_0 \in \mathrm{Lip}(U)$ and $Lu_0 - f = -i\varphi_t Z_x^{-1} \int_0^t \partial_x f \, ds = 2i\varphi_t Z_x^{-1} f_1$ where f_1 is bounded. It is clear that we will be able to solve $Lu = f$ on Q^+ if we can solve

$$Lu_1 = 2i\varphi_t Z_x^{-1} f_1 \quad \text{on } Q^+ \tag{VI.35}$$

by setting $u = u_0 - u_1$. In view of Lemma VI.3.7 we wish to write $f_1 = F_1 \circ Z(x, t)$ and the obstruction to doing so is the fact that f_1 may not be constant on the fibers $Z^{-1}(\zeta)$, $\zeta \in Z(Q^+)$. However, we are free to modify arbitrarily f_1 on the set $\{\varphi_t = 0\} \cup \{t \le 0\}$ without modifying the right-hand side of (VI.35). Hence, we declare that f_1 vanishes on $\{\varphi_t = 0\}$ as well as on $t \le 0$. Since Z is a diffeomorphism on $Q^+ \setminus \{\varphi_t = 0\}$, we may write $f_1 = F_1 \circ Z(x, t)$ with F_1 bounded on $Z(Q^+)$ and extend F_1 as zero outside $Z(Q^+)$, so $F_1 \in L_c^\infty(\mathbb{C})$. An application of Lemma VI.3.7 shows that there exists a function u_1 of class $C^\alpha(U)$ for any $0 < \alpha < 1$ whose restriction to U^+ satisfies (VI.35). Then $u = u_0 - u_1 \in C^\alpha(U^+) = C^\alpha(\overline{U^+})$. $\square$

VI.4 The H^p property for vector fields

Consider a one-sided locally integrable smooth vector field

$$L = \frac{\partial}{\partial t} + a(x, t) \frac{\partial}{\partial x}$$

defined on a neighborhood $Q = (-A, A) \times (-B, B)$ of the origin with a one-sided first integral $Z(x, t) = x + i\varphi(x, t)$ defined on Q satisfying $LZ = 0$

for $t \geq 0$. In this section we will assume that L satisfies condition $(\mathcal{P}^+)$ at the origin in $\Sigma = (-A, A) \times \{0\}$. We may clearly assume that $\varphi(0, 0) = \varphi_x(0, 0) = 0$ and

$$|\varphi_x(x, t)| < \frac{1}{2} \quad \text{on a neighborhood of } Q.$$

After a further contraction of Q about the origin, Lemma VI.3.4 shows that

for every $x \in (-A, A)$, the map $(0, B) \ni t \mapsto \varphi(x, t)$ is monotone.

The main result of this section is as follows:

THEOREM VI.4.1. *Suppose f is a distribution solution of $Lf = 0$ in the rectangle $Q = (-A, A) \times (0, B)$. Assume f has a weak boundary value $bf = f(x, 0)$ at $y = 0$. Then there exist $A_0 > 0$ and $T_0 > 0$ such that for any $0 < T \leq T_0$ and $0 < a < A_0$, if $f(., 0)$ and $f(., T) \in L^p(-A_0, A_0)$, $f(., t) \in L^p(-a, a)$ for any $0 < t < T$ and for almost all $0 < a < A_0$, there exists $C = C(a, T)$ such that*

(i) *if $1 \leq p < \infty$, then*

$$\int_{-a}^{a} |f(x, t)|^p \, dx \leq C \left(\int_{-a}^{a} |f(x, 0)|^p \, dx + \int_{-a}^{a} |f(x, T)|^p \, dx \right.$$

$$+ \int_0^T |f(a, s)|^p |\varphi_s(a, s)| \, ds$$

$$\left. + \int_0^T |f(-a, s)|^p |\varphi_s(-a, s)| \, ds \right);$$

(ii) *if $p = \infty$, then $f \in L^\infty((-a, a) \times (0, T))$.*

Before proving Theorem VI.4.1, we will need to recall some concepts and results from the classical theory of Hardy spaces for bounded, simply connected domains in the complex plane. Let D be a such a domain with rectifiable boundary. There are several definitions of a Hardy space for such a domain (see [L] and [Du]). For our purpose here, we need to recall two of the definitions:

DEFINITION VI.4.2. [Du] *For $1 \leq p < \infty$, a holomorphic function g on a bounded domain D with rectifiable boundary is said to be in $E^p(D)$ if there exists a sequence of rectifiable curves C_j in D tending to bD in the sense that the C_j eventually surround each compact subdomain of D, such that*

$$\int_{C_n} |g(z)|^p |dz| \leq M < \infty.$$

The norm of $g \in E^p(D)$ is defined as

$$\|g\|_{E^p(D)}^p = \inf_j \sup \int_{C_j} |g(z)|^p |dz|$$

where the inf is taken over all sequences of rectifiable curves C_j in D tending to ∂D.

DEFINITION VI.4.3. *Suppose for a bounded region $\Omega \subseteq \mathbb{C}$ there is $\alpha = \alpha(\Omega) > 0$ with the property that almost every point p in the boundary admits a nonempty nontangential approach subregion*

$$\Gamma_\alpha(p) = \{z \in \Omega : |z - p| \leq (1 + \alpha)\text{dist}(z, \partial\Omega)\}$$

that is, for a.e. $p \in \partial\Omega$, $\Gamma_\alpha(p)$ is open and p is in the closure of $\Gamma_\alpha(p)$. Let u be a function defined on Ω. The nontangential maximal function of u, u^*, and the nontangential limit of u, u^+, are defined as follows:

$$u^*(p) = \sup_{\zeta \in \Gamma_\alpha(p)} |u(\zeta)|, \quad a.e. \quad p \in \partial\Omega,$$

$$u^+(p) = \lim_{\zeta \in \Gamma_\alpha(p)} u(\zeta), \quad a.e. \quad p \in \partial\Omega.$$

DEFINITION VI.4.4. *For $1 \leq p < \infty$ the Hardy space $H^p(\Omega)$ is defined by*

$$H^p(\Omega) = \{G \in \mathbb{O}(\Omega) : G^* \in L^p(\partial\Omega)\}$$

where $\mathbb{O}(\Omega)$ denotes the holomorphic functions on Ω and G^* denotes the nontangential maximal function defined using the $\Gamma_\alpha(p)$ as in the definition above.

When Ω is the unit disk, it is a classical fact that both definitions of Hardy spaces agree ([**Du**]). By the Riemann mapping theorem, this is also true for any bounded, simply connected domain with a smooth boundary. In the work [**L**], it is shown that when $1 < p < \infty$, these spaces agree if Ω is bounded, simply connected with a Lipschitz boundary.

DEFINITION VI.4.5. *For $1 < q < \infty$, the maximal operator T_* on $L^q(\partial\Omega)$ is defined by*

$$T_*u(p) = \sup_{\epsilon > 0} \left| \int_{|\zeta - p| > \epsilon} \frac{1}{\zeta - p} u(\zeta)\, d\zeta \right|, \quad a.e. \quad p \in \partial\Omega.$$

Let us denote the Cauchy integral of a function u by Cu. We will be interested in the L^p boundedness of the nontangential maximal operator $(Cu)^*$ on certain kinds of domains which we now describe:

DEFINITION VI.4.6. *A bounded, simply connected domain Ω is called Ahlfors-regular if there is a constant $c > 0$ such that for every $q \in \partial\Omega$, and for every $r > 0$, the arclength measure of the portion of the boundary contained in the disk of radius r centered at q is less than cr.*

We note that examples of Ahlfors-regular domains include simply connected domains with Lipschitz boundary. Ahlfors-regular domains admit nontangential approach regions $\Gamma_\alpha(p)$ as in Definition VI.4.3. The study of the boundedness of the operator T_* on domains with Lipschitz boundary was initiated by A. Calderón in the 1970s. He proved that T_* is well-defined and bounded on $L^q(\partial\Omega)$ $(1 < q < \infty)$ provided the Lipschitz character of Ω is smaller than an absolute constant. Later, R. Coifman, A. McIntosh and Y. Meyer extended this result to the entire Lipschitz class. G. David has shown that the Ahlfors-regular domains are the largest rectifiable domains on which T_* is bounded. More precisely, he proved:

THEOREM VI.4.7. **[D]** *Let $\Omega \subseteq \mathbb{C}$ be a bounded, simply connected domain with rectifiable boundary. Then T_* is bounded on $L^q(\partial\Omega)$, $1 < q < \infty$, if and only if Ω is an Ahlfors-regular domain.*

The Hardy–Littlewood maximal function Mu on $\partial\Omega$ is defined by

$$Mu(z) = \sup \frac{1}{|I|} \int_I |u(\zeta)| \, |d\zeta|$$

where the sup is taken over all subarcs $I \subseteq \partial\Omega$ that contain z and $|I|$ denotes the arclength of I. It is well known that the Hardy–Littlewood maximal function of $\partial\Omega$ is L^p bounded $(1 < p < \infty)$ for a class of domains that includes the Ahlfors-regular domains (**[D]**). The following lemma therefore reduces the boundedness of $(Cu)^*$ to that of T_*.

LEMMA VI.4.8. *Let $\Omega \subseteq \mathbb{C}$ be an Ahlfors-regular domain. The following inequality holds for every $u \in L^q(\partial\Omega)$, $1 < q < \infty$, and every $p \in \partial\Omega$:*

$$(Cu)^*(p) \leq T_*u(p) + c(\alpha)Mu(p), \tag{VI.36}$$

where $(Cu)^*$ denotes the nontangential maximal function of the Cauchy integral of u and $c(\alpha)$ is a positive constant depending exclusively on the aperture of the cone $\Gamma_\alpha(p)$.

PROOF. For $p \in \partial\Omega$ arbitrary, it suffices to show that

$$|Cu(x)| \leq T_*u(p) + c(\alpha)Mu(p) \quad \text{for every } x \in \Gamma_\alpha(p).$$

Let $r := |x - p|$. We have

$$2\pi i C u(x) = \int_{|\zeta - p| > 2r} \frac{u(\zeta)}{\zeta - p} \, d\zeta$$

$$+ \int_{|\zeta - p| > 2r} \left(\frac{u(\zeta)}{\zeta - x} - \frac{u(\zeta)}{\zeta - p} \right) d\zeta$$

$$+ \int_{|\zeta - p| < 2r} \frac{u(\zeta)}{\zeta - x} \, d\zeta$$

$= I_1 + I_2 + I_3$. We will now proceed to estimate $|I_i|$, $i = 1, 2, 3$. Clearly, $|I_1| \le T_* u(p)$.

To estimate I_2 observe that

$$\left| \frac{1}{\zeta - x} - \frac{1}{\zeta - p} \right| = \frac{r}{|\zeta - x| \, |\zeta - p|}. \tag{VI.37}$$

But $|\zeta - p| \le |\zeta - x| + |x - p|$ and since $x \in \Gamma_\alpha(p)$, we have: $|\zeta - p| \le (2 + \alpha)|x - \zeta|$. Hence (VI.37) becomes

$$\left| \frac{1}{\zeta - x} - \frac{1}{\zeta - p} \right| \le \frac{(2 + \alpha)r}{|\zeta - p|^2}.$$

I_2 can thus be estimated as follows:

$$|I_2| \le (2 + \alpha) \int_{|\zeta - p| > 2r} \frac{r}{|p - \zeta|^2} |u(\zeta)| \, d\sigma(\zeta)$$

$$\le (2 + \alpha) \sum_{j=1}^{\infty} \int_{2^j r < |p - \zeta| < 2^{j+1} r} \frac{r}{(2^j r)^2} |u(\zeta)| \, d\sigma(\zeta)$$

$$\le 2(2 + \alpha) \sum_{j=1}^{\infty} \frac{1}{2^j} \left(\frac{1}{2^{j+1} r} \int_{|p - \zeta| < 2^{j+1} r} |u(\zeta)| \, d\sigma(\zeta) \right)$$

$$\le c(\alpha) M u(p).$$

Finally, in order to estimate I_3 we observe that $x \in \Gamma_\alpha(p)$ and $\zeta \in \partial\Omega$ imply

$$\frac{1}{|\zeta - x|} \le \frac{1 + \alpha}{r}.$$

Using the latter estimate we obtain:

$$|I_3| \le \frac{(1 + \alpha)}{2\pi r} \int_{|p - \zeta| < 2r} |u(\zeta)| \, d\sigma(\zeta) \le c(\alpha) M u(p). \qquad \square$$

Our next aim is to prove that $E^p(\Omega) = H^p(\Omega)$ for a particular class of domains Ω that includes the domains U_k that will appear in the proof of Theorem VI.4.1. We consider smooth regions U that are bounded by two smooth curves C_1 and C_2 that cross each other at two points A and B where they meet at angles $0 \le \theta(A),\ \theta(B) < \pi$. If $\theta(A), \theta(B) > 0$ then U has a Lipschitz boundary and by the result in [L] we know that $E^p(U) = H^p(U)$ for $p > 1$. Our methods will show that this equivalence still holds when the values $\theta(A) = 0,\ \theta(B) = 0$, and $p = 1$ are allowed. By a conformal map argument we may assume that

(1) $A = 0$ and $B = 1$;
(2) the part C_1 in the boundary of U is given by $[0, 1] \ni t \mapsto t$;
(3) the part C_2 in the boundary of U is given by $[0, 1] \ni t \mapsto x(t) + iy(t)$ where $x(t), y(t)$ are smooth real functions such that $x(0) = y(0) = y(1) = 0$, $x(1) = 1$.

We first prove that $H^p(U) \subseteq E^p(U)$. We construct for a large integer j a curve C_j as follows. To every point $z \in C_2 \cap \partial U$ we assign the point $\gamma_{j,2}(z) = z + j^{-1}\mathbf{n}(z)$ where $\mathbf{n}(z)$ is the inward unit normal to C_2 at z. For large j, $C_2 \ni z \mapsto \gamma_{j,2}(z)$ is a diffeomorphism and

$$\mathrm{dist}\,(\gamma_{j,2}(z), C_2) = |\gamma_{j,2}(z) - z| = \frac{1}{j}. \tag{VI.38}$$

Observe that the set

$$D_j = \left\{ z:\ \mathrm{dist}(z, [0, 1] \times \{0\}) \le \frac{1}{j} \right\}$$

has a C^1 boundary ∂D_j formed by two straight segments and two circular arcs. Fix a point $z_0 \in C_2$, choose j such that $z_0 \notin D_j$ and consider the connected component of

$$\left\{ z:\ \mathrm{dist}(\gamma_{j,2}(z), D_j) \ge \frac{1}{j} \right\}$$

that contains z_0. Thus, we obtain a curve $C_{j,2}$ given by $[0, 1] \supseteq [a_j, b_j] \ni t \mapsto \gamma_{j,2}(x(t) + iy(t)) \subset U$ that meets ∂D_j at its endpoints A_j, B_j and remains off D_j for $a_j < t < b_j$. Hence, we obtain a closed curve C_j completing the curve $C_{j,2}$ with the portion $C_{j,1}$ of ∂D_j contained in U that joins A_j to B_j. Because we are assuming that $\theta(A), \theta(B) < \pi$ we see that, for large j, $C_{j,1}$ is a horizontal segment at height $1/j$. It is clear that all points in C_j have distance $1/j$ to the boundary. Furthermore, if $q \in C_{j,2}$, $q \ne A_j$, and $q \ne B_j$ then $\mathrm{dist}(q, \partial U) = \mathrm{dist}(q, C_2) = 1/j$ because of (VI.38) and the fact that $\mathrm{dist}(q, [0, 1] \times \{0\}) > 1/j$. Similarly, if $q \in C_{j,1}$, $q \ne A_j$, and $q \ne B_j$ then $\mathrm{dist}(q, \partial U) = \mathrm{dist}(q, C_1) = 1/j$. Thus, every point $q \in C_j$ is at a distance $1/j$ of ∂U, we can always find $z \in \partial U$ such that $|q - z| = \mathrm{dist}(q, \partial U)$, and z is uniquely determined by q except when $q = A_j$ or $q = B_j$ (in which case the distance may be attained at two distinct boundary points). In particular, whatever the value of $\alpha > 0$, $q \in \Gamma_\alpha(z)$ for all $q \in C_j$ and $|g(q)| \leq g^*(z)$ for any function g defined on U. Given $g \in H^p(U)$ we must show that

$$\sup_j \int_{C_j} |g(z)|^p |dz| \leq M < \infty. \tag{VI.39}$$

We have

$$\int_{C_{j,2}} |g(q)|^p |dq| = \int_{\gamma_{j,2}^{-1}(C_j)} |g(\gamma_{j,2}(z))|^p \, |\gamma'_{j,2}(z)| \, |dz|$$

$$\leq \int_{\gamma_{j,2}^{-1}(C_j)} |g^*(z)|^p \, |\gamma'_{j,2}(z)| \, |dz|$$

$$\leq C \int_{C_2} |g^*(z)|^p \, |dz|. \tag{VI.40}$$

Similarly, using the map $\gamma_{j,1}(x) = x + i(1/j) \in C_{j,1}$, we get

$$\int_{C_{j,1}} |g(q)|^p |dq| \leq C \int_{C_1} |g^*(z)|^p \, |dz|, \tag{VI.41}$$

so adding (VI.40) and (VI.41) we obtain

$$\int_{C_j} |g(q)|^p |dq| \leq C \int_{\partial U} |g^*(z)|^p \, |dz|$$

which implies (VI.39) with $M = C \, \|g\|_{H^p}^p$.

To prove the other inclusion we first assume that $p = 2$. Given $f \in E^2(U) \subseteq E^1(U)$ it has an a.e defined boundary value $f^+ = bf \in L^2(\partial U)$ and the Cauchy integral representation

$$f(z) = \frac{1}{2\pi i} \int_{\partial U} \frac{bf(\zeta)}{\zeta - z} \, d\zeta, \quad z \in U$$

is valid ([**Du**], theorem 10.4). Furthermore, $\|f\|_{E^p(U)} \simeq \|f^+\|_{L^p(\partial U)}$. Next we recall Lemma VI.4.8 that gives the estimate

$$f^*(z) \leq T_* f^+(z) + CMf^+(z), \quad z \in \partial U \backslash \{A, B\}. \tag{VI.42}$$

It is well known that M is bounded in $L^2(\partial U)$. Furthermore, T_* is also bounded in $L^2(\partial U)$ by Theorem VI.4.7. Therefore (VI.42) implies that

$$\|f\|_{H^2(U)} = \|f^*\|_{L^2(\partial U)} \leq C\|f^+\|_{L^2(\partial U)} \leq C'\|f\|_{E^2(U)}.$$

The same technique leads to the inclusion $E^p(U) \subset H^p(U)$ for $p > 1$ because T_* and M are bounded as well in $L^p(\partial U)$ for $1 < p < \infty$ but the method breaks down for $p = 1$. This case will be handled in the proof of Theorem VI.4.1 using the fact that if $f \in E^p(U)$, $1 \leq p < \infty$, f has a canonical factorization $f = FB$ where F has no zeros, and $|B| \leq 1$. This is classical for the unit disk

Δ, where B is obtained as a Blaschke product and the general case is obtained from the classical result.

We are now ready to present the proof of Theorem VI.4.1. We begin by defining

$$m(x) = \min_{0 \le y \le B} \varphi(x, y), \qquad M(x) = \max_{0 \le y \le B} \varphi(x, y), \qquad -A \le x \le A.$$

The function $Z(x, y)$ takes the rectangle $Q = [-A, A] \times [0, B]$ onto

$$Z(Q) = \{\xi + i\eta : \quad -A \le \xi \le A, \quad m(\xi) \le \eta \le M(\xi)\}.$$

The interior of $Z(Q)$ is

$$\{\xi + i\eta : \quad -A < \xi < A, \quad m(\xi) < \eta < M(\xi)\}.$$

We will consider three essential cases, in each of which we will show that the assertions of the theorem are valid on a half-interval $[0, a]$. Since the same arguments also apply to the half-intervals $[-a, 0]$, the theorem will follow.

Case 1: Assume that $M(0) = m(0)$ and $M(a) = m(a)$ for some $a > 0$. In this case we will first assume that the solution f is smooth on $\overline{Q}$. If $M(x) = m(x)$ for every $x \in [0, a]$, then L would be $\frac{\partial}{\partial t}$ in $[0, a]$ and $f(x, t) = f(x, 0)$ for all $t \in [0, B]$, which trivially leads to the inequality we seek on the half-interval $[0, a]$. Hence we may assume that there is $x \in (0, a)$ for which $m(x) < M(x)$. Then the set $Z((0, a) \times (0, B))$ has nonempty interior. Every component of the interior of this set has the form

$$\{\xi + i\eta : \alpha < \xi < \beta, m(\xi) < \eta < M(\xi)\}$$

where (α, β) is a component of the open set $\{x \in (0, a) : M(x) > m(x)\}$. Let

$$\{x \in (0, a) : M(x) > m(x)\} = \bigcup_k (\alpha_k, \beta_k)$$

be a decomposition into components. Fix k and consider one of these components (α_k, β_k). Note that $m(\alpha_k) = M(\alpha_k)$ and $m(\beta_k) = M(\beta_k)$. Since for each x, the function

$$t \longmapsto \varphi(x, t) \quad \text{is monotonic,}$$

either $m(x) = \varphi(x, 0)$ and $M(x) = \varphi(x, B)$ or $m(x) = \varphi(x, B)$ and $M(x) = \varphi(x, 0)$ on (α_k, β_k). Without loss of generality, we may assume that $m(x) = \varphi(x, 0)$ and $M(x) = \varphi(x, B)$ for every $x \in (\alpha_k, \beta_k)$. Let $U_k =$ the interior of $Z((\alpha_k, \beta_k) \times (0, B))$. Thus

$$U_k = \{x + iy : \alpha_k < x < \beta_k, \ \varphi(x, 0) < y < \varphi(x, B)\}.$$

Since the solution f is assumed smooth on $\overline{Q}$ in the case under consideration, by the Baouendi–Treves approximation theorem, there exists $F_k \in C^\infty(\overline{U_k})$, holomorphic in U_k such that

$$f(x, y) = F_k(Z(x, y)) \quad \forall (x, y) \in [\alpha_k, \beta_k] \times [0, B].$$

Note that U_k is a bounded, simply connected region lying between two smooth graphs and its boundary ∂U_k is smooth except at the two end points $(\alpha_k, M(\alpha_k))$ and $(\beta_k, M(\beta_k))$. Note also that U_k has a rectifiable boundary of length bounded by

$$|\partial U_k| \leq \int_{\alpha_k}^{\beta_k} \sqrt{1 + \varphi_x^2(x, B)} \, dx + \int_{\alpha_k}^{\beta_k} \sqrt{1 + \varphi_x^2(x, 0)} \, dx$$

$$\leq 2(\beta_k - \alpha_k) \sqrt{1 + \sup_Q |\nabla \varphi|^2} = K(\beta_k - \alpha_k)$$

where the constant K is independent of k. For each $p \in \partial U_k$, and $p \notin \{(\alpha_k, M(\alpha_k)), (\beta_k, M(\beta_k))\}$, define the approach region

$$\Gamma_p = \{z \in U_k : |z - p| \leq 2 \operatorname{dist}(z, \partial U_k)\}.$$

Define the maximal functions F_k^* and $T_* F_k$ on ∂U_k (except at the two cusps) by

$$F_k^*(p) = \sup_{\zeta \in \Gamma_p} |F_k(\zeta)|$$

and

$$T_* F_k(z) = \sup_{\epsilon > 0} \left| \int_{\{\zeta \in \partial U_k : |\zeta - z| > \epsilon\}} \frac{1}{\zeta - z} F_k(\zeta) \, d\zeta \right|, \quad z \in \partial U_k.$$

Recall the Hardy–Littlewood maximal function

$$MF_k(z) = \sup \frac{1}{|I|} \int_I |f^+(\zeta)| \, |d\zeta|, \quad z \neq \alpha_k + iM(\alpha_k), \beta_k + iM(\beta_k)$$

where the sup is taken over all subarcs $I \subseteq \partial U_k$ that contain z and $|I|$ denotes the arclength of I. Next, since each U_k is Ahlfors-regular, Lemma VI.4.8 gives the estimate

$$F_k^*(z) \leq T_* F_k(z) + CMF_k(z), \quad z \in \partial U_k \backslash \{\alpha_k + iM(\alpha_k), \beta_k + iM(\beta_k)\}.$$
$$\text{(VI.43)}$$

The constant C in (VI.43) is independent of k because the aperture of the Γ_p is independent of k. Next we will show that any $z \in U_k$ lies in Γ_p for some $p \in \partial U_k$. Let $z \in U_k$. Then for some $(x, t) \in (\alpha_k, \beta_k) \times (0, B)$, $z = x + i\varphi(x, t)$

and $\varphi(x, 0) < \varphi(x, t) < \varphi(x, B)$. Let $p = x + i\varphi(x, B)$ and $q = x + i\varphi(x, 0)$. We claim that $z \in \Gamma_p \cup \Gamma_q$. Indeed suppose first

$$|\varphi(x, B) - \varphi(x, t)| \le |\varphi(x, t) - \varphi(x, 0)|. \tag{VI.44}$$

Then for any y:

$$|x + i\varphi(x, t) - y - i\varphi(y, B)| \ge \frac{1}{2}(|x - y| + |\varphi(x, t) - \varphi(y, B)|)$$

$$\ge \frac{1}{2}(|x - y| + |\varphi(x, t) - \varphi(x, B)| - |\varphi(x, B) - \varphi(y, B)|)$$

$$\ge \frac{1}{2}(|\varphi(x, t) - \varphi(x, B)| \quad \text{since} |\varphi_x| \le \frac{1}{2}$$

$$= \frac{1}{2}|z - p|. \tag{VI.45}$$

We also have:

$$|x + i\varphi(x, t) - y - i\varphi(y, 0)| \ge \frac{1}{2}(|x - y| + |\varphi(x, t) - \varphi(y, 0)|)$$

$$\ge \frac{1}{2}(|\varphi(x, t) - \varphi(x, 0)|$$

$$\ge \frac{1}{2}(|\varphi(x, B) - \varphi(x, t)| \quad \text{by (VI.44)}$$

$$= \frac{1}{2}|z - p|. \tag{VI.46}$$

From (VI.45) and (VI.46) we see that if (VI.44) holds, then $z \in \Gamma_p$. By a similar reasoning, if (VI.44) does not hold, then $z \in \Gamma_q$. We have thus shown that

$$U_k \subseteq \bigcup_{p \in \partial U_k} \Gamma_p. \tag{VI.47}$$

Next fix $(x, t) \in (\alpha_k, \beta_k) \times (0, B)$. If $x + i\varphi(x, t) \in U_k$, i.e., if $\varphi(x, 0) < \varphi(x, t) < \varphi(x, B)$, then by (VI.47),

$$|F_k(x + i\varphi(x, t))| \le F_k^*(x + i\varphi(x, 0)) + F_k^*(x + i\varphi(x, B)). \tag{VI.48}$$

On the other hand, if $\varphi(x, t) = \varphi(x, 0)$, then since $\varphi(x, 0) < \varphi(x, B)$, there exists $t \le y < B$ such that $\varphi(x, y) = \varphi(x, 0) = \varphi(x, t)$ and y is the maximum such. Let $y_m \to y$, $y_m > y$. Then by (VI.48),

$$|F_k(x + i\varphi(x, y_m))| \le F_k^*(x + i\varphi(x, 0)) + F_k^*(x + i\varphi(x, B)).$$

Letting $m \to \infty$, we get

$$|F_k(x + i\varphi(x, t))| = |F_k(x + i\varphi(x, y))|$$

$$\le F_k^*(x + i\varphi(x, 0)) + F_k^*(x + i\varphi(x, B)).$$

Thus for any $(x, t) \in (\alpha_k, \beta_k) \times (0, B)$, we have:

$$|f(x, t)| = |F_k(x + i\varphi(x, t))| \le F_k^*(x + i\varphi(x, 0)) \tag{VI.49}$$
$$+ F_k^*(x + i\varphi(x, B)).$$

From (VI.43) and (VI.49), for any $(x, t) \in (\alpha_k, \beta_k) \times (0, B)$, we have:

$$|f(x, t)| \le T_* F_k(x + i\varphi(x, 0)) + T_* F_k(x + i\varphi(x, B))$$
$$+ C(MF_k(x + i\varphi(x, 0)) + MF_k(x + i\varphi(x, B))), \tag{VI.50}$$

where we recall that the constant C is independent of k. Let $1 < p < \infty$. The cases $p = 1, \infty$ will be treated separately at the end. Since U_k is an Ahlfors-regular domain, both T_* and M are bounded in $L^p(\partial U_k)$ ([**D**]) and so (VI.50) leads to

$$\int_{\alpha_k}^{\beta_k} |f(x, t)|^p \, dx \le C \int_{\partial U_k} |F_k(z)|^p \, |dz| \quad \text{for any } 0 < t < B. \tag{VI.51}$$

Since $f(x, t) = F_k(Z(x, t))$ on $[\alpha_k, \beta_k] \times [0, B]$, we conclude that for any $0 < t < B$:

$$\int_{\alpha_k}^{\beta_k} |f(x, t)|^p \, dx \le C \left(\int_{\alpha_k}^{\beta_k} |f(x, 0)|^p \, dx + \int_{\alpha_k}^{\beta_k} |f(x, B)|^p \, dx \right) \tag{VI.52}$$

where C is independent of k. We can write

$$(0, a) = \left(\bigcup_k (\alpha_k, \beta_k) \right) \bigcup S$$

where $S = \{x \in (0, a) : \varphi(x, 0) = \varphi(x, B)\}$. Observe that for $x \in S$, the function $t \longmapsto f(x, t)$ is constant since $L = \frac{\partial}{\partial t}$ on $\{x\} \times (0, B)$. Hence for any $0 \le t \le B$,

$$\int_S |f(x, t)|^p \, dx = \int_S |f(x, B)|^p \, dx. \tag{VI.53}$$

Using (VI.53) and summing up over k in (VI.52), we conclude:

$$\int_0^a |f(x, t)|^p \, dx \le C \left(\int_0^a |f(x, 0)|^p \, dx + \int_0^a |f(x, B)|^p \, dx \right) \tag{VI.54}$$

for any $0 < t < B$. Finally, we use a refinement of the approximation theorem as in Theorem II.4.12 to remove the smoothness of f.

Case 2: Assume that $M(0) = m(0)$ and $M(x) > m(x)$ for every $0 < x \le A$. We will need to use the boundary version of the Baouendi–Treves approximation formula. Let $h(x) \in C_0^\infty(-A, A)$, $h(x) \equiv 1$ in a neighborhood of 0. For $\tau > 0$, define

$$E_\tau f(x, t) = (\tau/\pi)^{1/2} \int_{\mathbb{R}} e^{-\tau [Z(x, t) - Z(x', 0)]^2} f(x', 0) h(x') Z_x(x', 0) \, dx'$$

and

$$G_\tau f(x, t) = (\tau/\pi)^{1/2} \int_{\mathbb{R}} e^{-\tau[Z(x,t)-Z(x',t)]^2} f(x', t) h(x') Z_x(x', t) \, dx'$$

where $f(x', t)$ is the distribution trace of f at $t \geq 0$. Let

$$R_\tau f(x, t) = E_\tau f(x, t) - G_\tau f(x, t).$$

The Baouendi–Treves approximation theorem asserts that after decreasing A and B, $E_\tau f(x, t)$ converges to $f(x, t)$ in the sense of distributions in the open set $(-A, A) \times (0, B)$. However, here we need the refined boundary result in Chapter II (Theorem II.4.12) which guarantees convergence up to $t = 0$ in appropriate function spaces. More precisely, according to the result, there exist $a, b > 0$ such that

$$R_\tau f(x, t) \to 0 \quad \text{in } C^\infty([-a, a] \times [0, b]).$$

Since it is clear that $G_\tau f(x, t) \to f(x, t)$ in $L^p(-a, a)$ whenever $f(., t) \in L^p(-a, a)$, it follows that

$$E_\tau f(x, t) \to f(x, t) \quad \text{in } L^p([-a, a]), \quad \text{if } f(., t) \in L^p(-a, a). \tag{VI.55}$$

Let $F_\tau(z)$ be the entire function satisfying $F_\tau(Z(x, t)) = E_\tau f(x, t)$. Let $U_a =$ the interior of $Z((0, a) \times (0, b))$. Recall that $m(0) = M(0)$ but $m(x) < M(x)$ for any $0 < x \leq A$. The domain U_a is also an Ahlfors-regular domain. Therefore, we can apply the arguments in Case 1 to the smooth functions $E_\tau f$ to arrive at:

$$\int_0^a |E_\tau f(x, t)|^p \, dx \leq C \int_{\partial U_a} |F_\tau f(z)|^p \, |dz|. \tag{VI.56}$$

Note that this time ∂U_a has three pieces and so (VI.56) leads to:

$$\int_0^a |E_\tau f(x, t)|^p \, dx \leq C \left(\int_0^a |E_\tau f(x, 0)|^p \, dx + \int_0^a |E_\tau f(x, b)|^p \, dx \right.$$
$$\left. + \int_0^b |E_\tau f(a, s)|^p |\varphi_s(a, s)| \, ds \right), \quad 0 < t < b. \tag{VI.57}$$

We now wish to let $\tau \to \infty$ in (VI.57). From (VI.55) we know that if $f(., 0)$ and $f(., b)$ are in $L^p(-a, a)$, then

$$\int_0^a |E_\tau f(x, 0)|^p \, dx \to \int_0^a |f(x, 0)|^p \, dx \quad \text{and}$$
$$\int_0^a |E_\tau f(x, b)|^p \, dx \to \int_0^a |f(x, b)|^p \, dx.$$

We thus need only compute the limit of the s integral in (VI.57). We will show that for almost all a',

$$\int_0^b |E_\tau f(a', s)|^p |\varphi_s(a', s)| \, ds \to \int_0^b |f(a', s)|^p |\varphi_s(a', s)| \, ds. \qquad (VI.58)$$

We know that $M(x) > m(x)$ for every $0 < x \le A$. We may also assume that $\varphi(x, t) > \varphi(x, 0)$ for every $x \in (0, A]$, $t \in (0, b]$. Indeed, otherwise, we will be placed in the context of Case 1. The approximation theorem then implies that for each $x > 0$, f is continuous at (x, t) for $t > 0$ small. Since $R_\tau f(x, t) \to 0$ uniformly in $[0, a] \times [0, b]$, (VI.58) will follow if we show that for almost all a',

$$\int_0^b |G_\tau f(a', s)|^p |\varphi_s(a', s)| \, ds \to \int_0^b |f(a', s)|^p |\varphi_s(a', s)| \, ds. \qquad (VI.59)$$

Choose two numbers a_1, a_2 such that $0 < a_1 < a < a_2 \le A$. By the approximation theorem, after decreasing b, since f is continuous at (x, t) for $t = t(x) > 0$ small, there exists F continuous in $Z((a_1, a_2) \times (0, b))$, holomorphic in $W = $ the interior of $Z((a_1, a_2) \times (0, b))$ such that $F(Z(x, t)) = f(x, t)$. Observe that

$$W = \{x + iy : x \in (a_1, a_2), \ \varphi(x, 0) < y < \varphi(x, b)\}$$

and F has a distributional boundary value $= f(x, 0)$ on the curve $\{x + i\varphi(x, 0) : a_1 < x < a_2\}$. For $x \in (a_1, a_2)$, define

$$F^*(x) = \sup_{0 < t < b} |F(x + i\varphi(x, t))|.$$

Since F has an L^p boundary value, it is well known (see, for example, [**Ro**]) that $F^* \in L^p_{\text{loc}}(a_1, a_2)$. Let $\psi \in C_0^\infty(a_1, a_2)$, $\psi \ge 0$, $\psi(x) \equiv 1$ near a. Write $G_\tau f(x, t) = G_\tau^1 f(x, t) + G_\tau^2 f(x, t)$, where

$$G_\tau^1 f(x, t) = (\tau/\pi)^{1/2} \int_{\mathbb{R}} e^{-\tau[Z(x,t) - Z(x',t)]^2} \psi(x') f(x', t) h(x') Z_x(x', t) \, dx'$$

and $G_\tau^2 f(x, t) = G_\tau f(x, t) - G_\tau^1 f(x, t)$. Consider first $G_\tau^2 f(x, t)$ for x near a. Observe that the integrand is zero for x' near a and hence for x near a and $t \in [0, b]$,

$$G_\tau^2 f(x, t) \to 0 \quad \text{uniformly.} \qquad (VI.60)$$

In the integrand of $G_\tau^1 f(x, t)$, $f(x', t)$ can be replaced by $F(Z(x', t)) = F(x' + i\varphi(x', t))$ and hence we have:

$$|G_\tau^1 f(x, t)| \le C(\tau/\pi)^{1/2} \int_{\mathbb{R}} e^{-\frac{1}{2}\tau|x - x'|^2} \psi(x') F^*(x') \, dx' \qquad (VI.61)$$

where C is independent of τ. Thus if we define

$\eta(x) = \pi^{-1/2}e^{-\frac{x^2}{2}}$, and $\eta_\tau(x) = \tau^{1/2}\eta(\tau^{1/2}x)$, then (VI.61) says that

$$|G_\tau^1 f(x,t)| \le C(\eta_\tau * \psi F^*)(x) \quad \forall t \in [0,b]. \tag{VI.62}$$

Since $\psi F^* \in L^p(-\infty, \infty)$ and η is a radial decreasing function in $|x|$, by a proposition in [**S2**, page 57],

$$\sup_{\tau > 0} \eta_\tau * \psi F^*(x) \quad \text{is finite a.e.}$$

Pick a point x_0 where this supremum is finite and where $F^*(x_0) < \infty$. Then at such a point, the functions $|G_\tau^1 f(x_0, t)|$ are bounded on $[0,b]$. Since pointwise,

$$G_\tau^1 f(x_0, t) \to f(x_0, t) \quad \forall t \in [0,b],$$

it follows that

$$\int_0^b |G_\tau f(x_0, s)|^p |\varphi_s(x_0, s)| \, ds \to \int_0^b |f(x_0, s)|^p |\varphi_s(x_0, s)| \, ds. \tag{VI.63}$$

From (VI.59) and (VI.63), we conclude that

$$\int_0^b |E_\tau f(a', s)|^p |\varphi_s(a', s)| \, ds \to \int_0^b |f(a', s)|^p |\varphi_s(a', s)| \, ds \tag{VI.64}$$

for almost all a'. We can therefore let $\tau \to \infty$ in (VI.57) and conclude that for almost all a:

$$\int_0^a |f(x,t)|^p \, dx \le C \left(\int_0^a |f(x,0)|^p \, dx + \int_0^a |f(x,b)|^p \, dx \right.$$

$$\left. + \int_0^b |f(a,s)|^p |\varphi_s(a,s)| \, ds \right), \quad 0 < t < b. \tag{VI.65}$$

Case 3: Assume $M(0) > m(0)$. Let $a > 0$ such that $M(x) > m(x)$ for every $x \in (-a, a)$. If $W_a = Z((-a, a) \times (0, B))$, there is a function F holomorphic on the interior of W_a such that $f(x,y) = F(Z(x,y))$. This time the boundary of W_a has four pieces. One can then reason as in the previous case to get the required estimate on the interval $(-a, a)$. Finally, observe that estimates on the interval of the form $[-a, 0]$ are also valid under Cases 1 and 2. The theorem for $1 < p < \infty$ follows from these three cases.

We consider next the case when $p = 1$.

Assume we are in the situation of Case 1 where $M(0) = m(0)$ and $M(a) = m(a)$ for some $a > 0$. As before we assume first that $f(x,t)$ is smooth on $\overline{Q^+}$, $F_k \in C^\infty(\overline{U_k})$, holomorphic in U_k and $f(x,y) = F_k(Z(x,y))$ on $[\alpha_k, \beta_k] \times [0, B]$. Since U_k is simply connected, by a classical result (see the corollary of theorem 10.1 in [**Du**]), F_k has a factorization $F_k = G_k B_k$ where each factor is holomorphic in U_k, G_k has no zeros, $G_k \in E^1(U_k)$, $|B_k(z)| \leq 1$, and $|B_k(z)| = 1$ on ∂U_k. The fact that $G_k \in E^1(U_k)$ implies (see theorem 10.4 in [**Du**]) that it has a nontangential limit bG_k a.e. on ∂U_k, and G_k equals the Cauchy transform of bG_k. Observe that since $|B_k(z)| = 1$ on ∂U_k, $|bG_k(z)| = |F_k(z)|$ on ∂U_k. Since G_k has no zeros on the simply connected region U_k, it has a holomorphic square root H_k. Note that $H_k \in E^2(U_k) = H^2(U_k)$ (by the discussion preceding this proof). We have

$$H_k^*(z) \leq T_*(bH_k)(z) + CM(bH_k)(z). \qquad (\text{VI.66})$$

Using (VI.66) and the equality $|G_k| = |F_k|$ on ∂U_k we get:

$$\int_{\alpha_k}^{\beta_k} |f(x,t)|\,dx = \int_{\alpha_k}^{\beta_k} |F_k(x + i\varphi(x,t))|\,dx$$

$$\leq \int_{\alpha_k}^{\beta_k} |G_k(x + i\varphi(x,t))|\,dx = \int_{\alpha_k}^{\beta_k} |H_k(x + i\varphi(x,t))|^2\,dx$$

$$\leq \int_{\partial U_k} |H_k^*(z)|^2\,|dz|$$

$$\leq C \int_{\partial U_k} |bH_k(z)|^2\,|dz| \quad \text{by the } L^2 \text{ boundedness of } T_* \text{ and } M$$

$$= C \left(\int_{\alpha_k}^{\beta_k} |f(x,0)|\,dx + \int_{\alpha_k}^{\beta_k} |f(x,B)|\,dx \right) \text{ for any } 0 < t < B.$$

$$(\text{VI.67})$$

Summing up over k and adding the contributions from the set $S = (0, a) \setminus \bigcup_k (\alpha_k, \beta_k)$, we get:

$$\int_0^a |f(x,t)|\,dx \leq C \left(\int_0^a |f(x,0)|\,dx + \int_0^a |f(x,B)|\,dx \right) \qquad (\text{VI.68})$$

$$\text{for } 0 < t < B,$$

whenever f is a solution and $f \in C^\infty(\overline{Q^+})$. In general, for $f \in \mathcal{D}'(Q^+)$ satisfying the hypotheses of Theorem VI.4.1, let $\{f_m(x,t)\}$ be a sequence of C^∞ solutions on $\overline{Q^+}$ satisfying:

(i) for each $0 \leq t \leq B$, $f_m(.,t) \to f(.,t)$ in $\mathcal{D}'(-a,a)$;
(ii) $f_m(x,0) \to f(x,0)$ and $f_m(x,B) \to f(x,B)$ in $L^1(-a,a)$.

We now apply inequality (VI.68) to $f_m - f_n$, let m and n tend to ∞, and use (i) and (ii) above to conclude that (VI.68) also holds for f. Cases 2 and 3 are also treated in a similar fashion. Finally we consider the case where $p = \infty$. Suppose we are in the situation of Case 1 where $M(0) = m(0)$ and

$M(a) = m(a)$ for some $a > 0$. Assume first that $f(x, t) \in C^\infty(\overline{Q})$ and for k fixed as before, let

$$U_k = \{x + iy : \alpha_k < x < \beta_k, \; \varphi(x, 0) < y < \varphi(x, B)\}$$

and $f(x, y) = F_k(Z(x, y))$ on $[\alpha_k, \beta_k] \times [0, B]$, F_k holomorphic on U_k and continuous on the closure. We apply the maximum modulus principle to F_k and use the constancy of f on the vertical segments $x = \alpha_k$ and $x = \beta_k$ to conclude that

$$|f(x, y)| \le ||f(., 0)||_{L^\infty(0,a)} + ||f(., B)||_{L^\infty(0,a)} \quad \forall (x, y) \in [\alpha_k, \beta_k] \times [0, B].$$

If S is the set as before with

$$(0, a) = \left(\bigcup_k (\alpha_k, \beta_k)\right) \bigcup S,$$

then $f(x, y) = f(x, B) \quad \forall (x, y) \in S \times (0, B)$, and so we conclude that

$$|f(x, y)| \le ||f(., 0)||_{L^\infty(0,a)} + ||f(., B)||_{L^\infty(0,a)} \qquad (VI.69)$$
$$\forall (x, y) \in (0, a) \times (0, B).$$

For a solution $f \in \mathcal{D}'(Q^+)$ satisfying $f(., 0)$ and $f(., B) \in L^\infty(-A, A)$, we use the refinement of the approximation theorem in Chapter II according to which

$$f(x, y) = \lim_{\tau \to \infty} E_\tau f(x, y) \quad \text{a.e. in} \quad (0, a) \times (0, B), \qquad (VI.70)$$

provided that A and B are small enough. Moreover,

$$|G_\tau f(x, B)| \le c_1 \tau^{\frac{1}{2}} \int e^{-c_2 \tau |x - x'|^2} |f(x', B)| |h(x')| \, dx'$$
$$\le c_3 ||f(., B)||_{L^\infty} \quad \forall \tau > 0 \qquad (VI.71)$$

and likewise,

$$|G_\tau f(x, 0)| \le c ||f(., 0)||_{L^\infty}. \qquad (VI.72)$$

Letting $\tau \to \infty$, and recalling that $R_\tau f \to 0$ uniformly, we get

$$\overline{\lim_{\tau \to \infty}} |E_\tau f(x, 0)| \le C ||f(., 0)||_{L^\infty} \quad \text{and}$$
$$\overline{\lim_{\tau \to \infty}} |E_\tau f(x, B)| \le C ||f(., B)||_{L^\infty} \qquad (VI.73)$$

for some $C > 0$. From (VI.69) (applied to $E_\tau f$), (VI.70) and (VI.73), we conclude that for every $(x, y) \in (0, a) \times (0, B)$,

$$|f(x, y)| \le C \left(||f(., 0)||_{L^\infty(0,a)} + ||f(., B)||_{L^\infty(0,a)} \right). \qquad (VI.74)$$

Next we consider Case 2 where $M(0) = m(0)$ and $M(x) > m(x)$ for every $0 < x \le A$. As before, let $a, b > 0$ such that

$$E_\tau f(x, t) \to f(x, t) \quad \text{a.e. in } [-a, a] \times [0, b]. \tag{VI.75}$$

Let $U_a = Z((0, a) \times (0, b))$ and consider the holomorphic function F_τ such that $F_\tau(Z(x, t)) = E_\tau f(x, t)$. The maximum principle applied to F_τ on U_a leads to

$$|E_\tau f(x, y)| \le ||E_\tau f(., 0)||_{L^\infty(0,a)} + ||E_\tau f(., b)||_{L^\infty(0,a)}$$
$$+ ||E_\tau f(a, .)||_{L^\infty(0,b)} \quad \forall (x, y) \in [0, a] \times [0, b]. \tag{VI.76}$$

As observed already, the terms $||E_\tau f(., 0)||_{L^\infty(0,a)}$ and $||E_\tau f(., b)||_{L^\infty(0,a)}$ are dominated by a constant multiple of

$$||f(., 0)||_{L^\infty(0,a)} + ||f(., b)||_{L^\infty(0,a)}.$$

We therefore only need to estimate the term $||E_\tau f(a, .)||_{L^\infty(0,b)}$ for which it suffices to estimate $||G_\tau f(a, .)||_{L^\infty(0,b)}$. Let $0 < a_1 < a < a_2 < A$ be as before, F holomorphic such that

$$f(x, y) = F(x + i\varphi(x, y)) \quad \text{on} \quad [a_1, a_2] \times (0, b].$$

Since $bF = bf \in L^\infty(a_1, a_2)$, by the generalized maximum principle applied to F there exists $M > 0$ such that

$$|F(x + i\varphi(x, y))| = |f(x, y)| \le M \quad \text{on} \quad [a_1', a_2'] \times (0, b],$$

for some $a_1 < a_1' < a < a_2' < a_2$. We write $G_\tau f = G_\tau^1 f + G_\tau^2 f$ as before, except that this time ψ is supported in (a_1', a_2'). Recall that $G_\tau^2 f \to 0$ uniformly while

$$|G_\tau^1 f(x, t)| \le C \sup |\psi(x')f(x', t)| \le CM.$$

Hence for some $C > 0$,

$$||E_\tau f(a, .)||_{L^\infty(0,b)} \le C \quad \forall \tau > 0.$$

We have shown that $f \in L^\infty((0, a) \times (0, b))$ in this case. Case 3 is treated likewise. We conclude that f is bounded. Theorem VI.4.1 has now been proved.

COROLLARY VI.4.9. *Suppose f is a distribution solution of $Lf = g$ in the rectangle $Q = (-A, A) \times (0, B)$. Suppose f has a weak boundary value $bf = f(x, 0)$ at $y = 0$ and that g is a Lipschitz function. Then there exist $A_0 > 0$ and $T_0 > 0$ such that for any $0 < T \le T_0$ and $0 < a < A_0$, if $f(., 0)$ and $f(., T) \in L^p(-A_0, A_0)$, $f(., t) \in L^p(-a, a)$ for any $0 < t < T$.*

PROOF. Using Proposition VI.3.6 we may find a function f_0, uniformly continuous on Q, such that $Lf_0 = g$. Then, $f_1 = f - f_0$ satisfies the hypothesis of
Theorem VI.4.1. It follows that (i) holds for f_1 if $1 \leq p < \infty$ or (ii) if $p = \infty$
and the same conclusion applies to $f = f_0 + f_1$ because f_0 is continuous up
to the boundary. □

COROLLARY VI.4.10. *Let L be as above, $f \in \mathcal{D}'(Q^+)$, $Lf = g$ in Q^+ where
$g \in C^\infty(\overline{Q^+})$. Let A_0 and T_0 be as in Theorem VI.4.1. If f has a weak trace
$f(x, 0) \in C^\infty(-A_0, A_0)$ and $f(., T_0)$ is in $C^\infty(-A_0, A_0)$, then for all $0 < a < A_0$
and $0 < T < T_0$, $f \in C^\infty([-a, a] \times [0, T])$. In particular, f is smooth up to
the boundary $t = 0$.*

PROOF. By Proposition VI.3.6, we can get $u \in C^0((-A, A) \times [0, B))$ that
solves $Lu = g$ in Q^+. Hence $L(u - f) = 0$ in Q^+ and so by Theorem VI.4.1
and the continuity of u up to the boundary, for any $0 < a < A_0$ and $0 < t \leq T_0$
there is a constant $C > 0$ such that

$$\int_{-a}^{a} |f(x, t)|^2 \, dx \leq C \quad \forall t \in [0, T_0]. \tag{VI.77}$$

Define the vector field $M = \frac{1}{Z_x(x,t)} \frac{\partial}{\partial x}$. Since the bracket $[L, M] = 0$ and $Lf = g$,
the distribution Mf is also a solution of $L(Mf) = Mg$ in Q^+. Moreover,
since the traces $Mf(., T_0)$ and $Mf(., 0)$ are smooth, by repeating the same
arguments, for any $0 < a < A_0$ and $0 < T < T_0$ there is a constant $C > 0$ such
that

$$\int_{-a}^{a} |Mf(x, t)|^2 \, dx \leq C \quad \forall t \in [0, T]. \tag{VI.78}$$

Since $\frac{\partial f}{\partial t} = -a(x, t) \frac{\partial f}{\partial x} + g(x, t)$, (VI.78) implies that for some constant C',

$$\int_{-a}^{a} \left| \frac{\partial f}{\partial t}(x, t) \right|^2 \, dx \leq C' \quad \forall t \in [0, T].$$

By iterating this argument, we derive that for every $m, n = 1, 2, \ldots$, there
exists $C = C(m, n) > 0$ such that

$$\int_{-a}^{a} |D_x^m D_t^n f(x, t)|^2 \, dx \leq C \quad \forall t \in [0, T]. \tag{VI.79}$$

From (VI.79) we conclude that $f \in C^\infty([-a, a] \times (0, T])$. Smoothness up to
the boundary now follows from the case $p = \infty$ in Theorem VI.4.1. □

REMARK VI.4.11. Conversely, if a locally integrable vector field L shares the
H^p property as in Theorem VI.4.1, then L has to satisfy condition $(\mathcal{P}^+)$ at
the origin in $\Sigma = (-A, A) \times \{0\}$. See [**BH6**] for the proof.

COROLLARY VI.4.12. *Let L satisfy $(\mathcal{P}^+)$ at the origin as above. Suppose $Lf = g$ in Q^+, $g \in C^\infty(\overline{Q^+})$, and $f \in C^\infty(Q^+)$. If the trace $bf = f(x, 0)$ exists and $f(x, 0) \in C^\infty(-A, A)$, then f is C^∞ up to the boundary $t = 0$.*

Example 4.3 in [BH6] provides a real-analytic vector field L for which Corollary VI.4.12 is not valid even for a solution of the homogeneous equation $Lf = 0$. Example 4.4 in the same paper shows that in Theorem VI.4.1, one needs to assume the integrability of two traces. That is, if we only assume that $bf = f(x, 0) \in L^1$, the traces $f(., t)$ may not be in L^1.

Notes

The results of this chapter in the holomorphic case are classical. For a discussion of the conditions that guarantee the existence of a boundary value we refer to the books [BER] and [H2]. The basic theory of Hardy spaces for bounded, simply connected domains in the complex plane is exposed in [Du] (see also [Po]). The paper [L] and the references in it contain more recent developments on the subject. The planar case of Theorem VI.1.3 as well as the necessity in the real-analytic, planar situation was proved in [BH5]. Lemma VI.4.8 is taken from [L]. Theorem VI.4.1 and its corollaries appeared in [BH6]. The work [HH] extends Theorem VI.4.1 to the case $0 < p < 1$ for vector fields with real-analytic coefficients.

VII

The differential complex associated with a formally integrable structure

In this chapter we shall introduce the differential complex associated with a formally integrable structure and discuss several aspects of its exactness.

VII.1 The exterior derivative

Let Ω be a differentiable manifold of dimension N. As in Chapter I, we shall denote by $\mathfrak{X}(\Omega)$ the space of all complex vector fields over Ω. We then set $\mathfrak{N}_0(\Omega) = C^\infty(\Omega)$ and if $q \geq 1$ is an integer we shall denote by $\mathfrak{N}_q(\Omega)$ the space of all $C^\infty(\Omega)$-multilinear, alternating forms

$$\omega : \underbrace{\mathfrak{X}(\Omega) \times \ldots \times \mathfrak{X}(\Omega)}_{q} \longrightarrow C^\infty(\Omega).$$

Notice that, according to Section I.4, we have $\mathfrak{N}_1(\Omega) = \mathfrak{N}(\Omega)$; notice also that $\mathfrak{N}_q(\Omega)$ has, for each q, the structure of a $C^\infty(\Omega)$-module. We then generalize the concept of one-forms introduced in Section I.4 and call the elements of the direct sum $\oplus_{q=0}^\infty \mathfrak{N}_q(\Omega)$ *differential forms over* Ω. If $\omega \in \mathfrak{N}_q(\Omega)$ we shall say that ω is a *differential form of degree* q (or *q-form* for short). The *exterior product* between $\omega \in \mathfrak{N}_q(\Omega)$ and $\theta \in \mathfrak{N}_r(\Omega)$ is the $(q+r)$-form $\omega \wedge \theta \in \mathfrak{N}_{q+r}(\Omega)$ defined by the formula

$$(\omega \wedge \theta)(X_1, \ldots, X_{q+r}) = \sum_{(A,B)} (\mathrm{sg}\,\sigma)\, \omega(X_{\sigma(1)}, \ldots, X_{\sigma(q)}) \theta(X_{\sigma(q+1)}, \ldots, X_{\sigma(q+r)}),$$

(VII.1)

where $X_j \in \mathfrak{X}(\Omega)$ and the summation is over all partitions (A, B) of $\{1, \ldots, q+r\}$ with $|A| = q$, $|B| = r$ and $\sigma \in S^{q+r}$ is such that $\sigma\{1, \ldots, q\} = A$, $\sigma\{q+1, \ldots, q+r\} = B$. It is easy to see that (VII.1) defines indeed a $(q+r)$-form, that the map

$$(\omega, \theta) \mapsto \omega \wedge \theta$$

is $C^\infty(\Omega)$-bilinear, and that the operation so defined is associative. It follows that $\oplus_{q=0}^\infty \mathfrak{N}_q(\Omega)$ has a structure of a graded $C^\infty(\Omega)$-algebra. We also remark that

$$\omega \wedge \theta = (-1)^{qr} \theta \wedge \omega, \quad \omega \in \mathfrak{N}_q(\Omega), \ \theta \in \mathfrak{N}_r(\Omega). \tag{VII.2}$$

The *exterior differentiation operator* is a $\mathbb{C}$-linear map

$$d : \oplus_{q=0}^\infty \mathfrak{N}_q(\Omega) \to \oplus_{q=0}^\infty \mathfrak{N}_q(\Omega)$$

whose restriction to $\mathfrak{N}_0(\Omega) = C^\infty(\Omega)$ coincides with the operator introduced in Definition I.1.6 [that is, $df(X) = X(f)$ if $f \in C^\infty(\Omega)$ and $X \in \mathfrak{X}(\Omega)$] and is characterized by the following additional properties:

(d₁) $d\mathfrak{N}_q(\Omega) \subset \mathfrak{N}_{q+1}(\Omega)$ for every $q \geq 0$;
(d₂) $d \circ d = 0$;
(d₃) if $\omega \in \mathfrak{N}_q(\Omega)$ and $\theta \in \mathfrak{N}_r(\Omega)$ then

$$d(\omega \wedge \theta) = d\omega \wedge \theta + (-1)^q \omega \wedge d\theta. \tag{VII.3}$$

The only operator d which satisfies these properties can be defined by the expression:

$$d\omega(X_1, \ldots, X_{q+1}) = \sum_{j=1}^{q+1} (-1)^{j+1} X_j \left\{ \omega(X_1, \ldots, \hat{X}_j, \ldots, X_{q+1}) \right\}$$
$$+ \sum_{j<k} (-1)^{j+k} \omega([X_j, X_k], X_1, \ldots, \hat{X}_j, \ldots, \hat{X}_k, \ldots, X_{q+1}),$$

$$\tag{VII.4}$$

where $\omega \in \mathfrak{N}_q(\Omega)$ and $X_j \in \mathfrak{X}(\Omega)$. (Recall that the sign ˆ over a letter means that the letter is missing.)

VII.2 The local representation of the exterior derivative

If $\omega \in \mathfrak{N}_q(\Omega)$ then $\omega(X_1, X_2, \ldots, X_q) = 0$ at p if the vector fields $X_1, \ldots, X_q$ are linearly dependent at p. Indeed if we have, say, $X_1 = \sum_{j=2}^q \alpha_j X_j$ at p and if take $g_j \in C^\infty(\Omega)$ with $g_j(p) = \alpha_j$ then

$$\omega(X_1, X_2, \ldots, X_q) = \omega \left(X_1 - \sum_{j=2}^q g_j X_j, X_2, \ldots, X_q \right)$$

and our claim follows immediately from Lemma I.4.1 applied to the one-form $X \mapsto \omega(X, X_2, \ldots, X_q)$. In particular, we can restrict a q-form over Ω to an open set $W \subset \Omega$, that is, given $\omega \in \mathfrak{N}_q(\Omega)$ there is $\omega|_W \in \mathfrak{N}_q(W)$ which makes the diagram

$$\begin{array}{ccc} \mathfrak{X}(\Omega) \times \ldots \times \mathfrak{X}(\Omega) & \xrightarrow{\omega} & C^\infty(\Omega) \\ \downarrow & & \downarrow \\ \mathfrak{X}(W) \times \ldots \times \mathfrak{X}(W) & \xrightarrow{\omega|_W} & C^\infty(W) \end{array}$$

commutative, where the vertical arrows denote the restriction homomorphisms. Moreover, from (VII.4) it follows easily that the operator d commutes with restrictions.

Let $(U, \mathbf{x})$ be a local chart in Ω. The $C^\infty(U)$-module $\mathfrak{N}_q(U)$ is spanned by the q-forms dx_J, where $J : j_1 < j_2 < \ldots < j_q$ is an ordered multi-index of length q, $j_\ell \in \{1, \ldots, N\}$, and

$$dx_J = dx_{j_1} \wedge \ldots \wedge dx_{j_q}.$$

Every $\omega \in \mathfrak{N}_q(U)$ can be represented as

$$\omega = \sum_{|J|=q} f_J(x) dx_J, \quad f_J \in C^\infty(U) \tag{VII.5}$$

and the properties that characterize d allow us to write

$$d\omega = \sum_{|J|=q} df_J \wedge dx_J. \tag{VII.6}$$

REMARK VII.2.1. The analysis presented at the beginning of this section allows one to extend the notion of pullback for one-forms introduced in Section I.14. If $\mathcal{M}$ is a submanifold of Ω we have well-defined *pullback homomorphisms* $(\iota_\mathcal{M})^* : \mathfrak{N}_q(\Omega) \to \mathfrak{N}_q(\mathcal{M})$ defined by

$$(\iota_\mathcal{M})^* \omega (X_1, \ldots, X_q)(p) = \omega(\tilde{X}_1, \ldots, \tilde{X}_q)(p), \quad \omega \in \mathfrak{N}_q(\Omega), \tag{VII.7}$$

where $p \in \mathcal{M}$, $X_1, \ldots, X_q \in \mathfrak{X}(\mathcal{M})$, and $\tilde{X}_1, \ldots, \tilde{X}_q \in \mathfrak{X}(\Omega)$ are such that $(\iota_\mathcal{M})_* X_j|_p = \tilde{X}_j|_p$ for every $j = 1, \ldots, q$. The pullback homomorphisms commute with the exterior derivative, that is

$$(\iota_\mathcal{M})^* d\omega = d_\mathcal{M}(\iota_\mathcal{M})^* \omega, \quad \omega \in \mathfrak{N}_q(\Omega), \tag{VII.8}$$

where we have denoted by $d_\mathcal{M}$ the exterior derivative operator on the manifold $\mathcal{M}$.

VII.3 The Poincaré Lemma

Let $D \subset \mathbb{R}^N$ be open and convex and let $\mathcal{U}$ be an open subset of $\mathbb{R}^p$. Denote by $\mathfrak{N}_q^\bullet(D \times \mathcal{U})$ the space of all q-forms

$$f = \sum_{|J|=q} f_J(x, y)\mathrm{d}x_J \tag{VII.9}$$

where $f_J \in C^\infty(D \times \mathcal{U})$.

Fix $x^0 \in D$ and set, for $J = (j_1, \ldots, j_q)$,

$$\beta_J(x, x^0) = \sum_{r=1}^q (-1)^{r-1}(x_{j_r} - x_{j_r}^0)\mathrm{d}x_{j_1} \wedge \ldots \wedge \widehat{\mathrm{d}x_{j_r}} \wedge \ldots \wedge \mathrm{d}x_{j_q}.$$

Next we introduce the operators, for $q \geq 1$,

$$G : \mathfrak{N}_q^\bullet(D \times \mathcal{U}) \to \mathfrak{N}_{q-1}^\bullet(D \times \mathcal{U})$$

defined in the following way: if f is as in (VII.9) we set

$$G(f) = \sum_{|J|=q} \left\{ \int_0^1 f_J\left(x^0 + \tau[x - x^0], y\right) \tau^{q-1}\mathrm{d}\tau \right\} \beta_J(x, x^0). \tag{VII.10}$$

The standard Poincaré Lemma states that

$$\mathrm{d}_x G(f) + G(\mathrm{d}_x f) = f \quad \text{if } q \geq 1; \tag{VII.11}$$

$$G(\mathrm{d}_x f) = f - f(x_0, \cdot) \quad \text{if } q = 0, \tag{VII.12}$$

which are formulae that can be proved by direct computation, using (VII.6). In particular we derive, if $q \geq 1$,

$$\mathrm{d}_x G(f) = f \quad \text{if} \quad \mathrm{d}_x f = 0.$$

VII.4 The differential complex associated with a formally integrable structure

Let $\mathcal{V} \subset \mathbb{C}T\Omega$ be a formally integrable structure over Ω. For each $q \geq 1$ we denote by $\mathfrak{N}_q^\mathcal{V}(\Omega)$ the $C^\infty(\Omega)$-submodule of $\mathfrak{N}_q(\Omega)$ defined by all $\omega \in \mathfrak{N}_q(\Omega)$ for which $\omega(X_1, \ldots, X_q) = 0$ if $X_1, \ldots, X_q$ are sections of $\mathcal{V}$ over Ω. Observe that $\mathfrak{N}_q^\mathcal{V}(\Omega) = \mathfrak{N}_q(\Omega)$ if $q > n$ for the sections of $\mathcal{V}$ form, locally, a free C^∞-module of rank n.

Since $\mathcal{V}$ satisfies, by definition, the Frobenius condition it follows immediately from (VII.4) that

$$d\mathfrak{N}_q^{\mathcal{V}}(\Omega) \subset \mathfrak{N}_{q+1}^{\mathcal{V}}(\Omega) \qquad \text{(VII.13)}$$

for every $q \geq 1$. Finally we set

$$\mathfrak{U}_q^{\mathcal{V}}(\Omega) = \mathfrak{N}_q(\Omega)/\mathfrak{N}_q^{\mathcal{V}}(\Omega), \quad q \geq 1. \qquad \text{(VII.14)}$$

Thanks to (VII.13) the exterior derivative defines a complex of $\mathbb{C}$-linear
mappings

$$C^\infty(\Omega) \xrightarrow{\mathrm{d}'} \mathfrak{U}_1^{\mathcal{V}}(\Omega) \xrightarrow{\mathrm{d}'} \ldots \xrightarrow{\mathrm{d}'} \mathfrak{U}_q^{\mathcal{V}}(\Omega) \xrightarrow{\mathrm{d}'} \mathfrak{U}_{q+1}^{\mathcal{V}}(\Omega) \xrightarrow{\mathrm{d}'} \ldots, \qquad \text{(VII.15)}$$

which we shall refer to as the *complex associated with $\mathcal{V}$ over Ω*.

VII.5 Localization

If $W \subset \Omega$ is open there is a well-defined complex homomorphism

$$\left(\mathfrak{U}_q^{\mathcal{V}}(\Omega), \mathrm{d}'\right) \longrightarrow \left(\mathfrak{U}_q^{\mathcal{V}}(W), \mathrm{d}'\right)$$

which is induced by restriction.

Let $p \in \Omega$ and consider an open neighborhood W of p over which there
are defined m differential forms $\omega_1, \ldots, \omega_m \in \mathfrak{N}(W)$ that span $T'|_W$ at every
point. After contracting W around p and a linear change on $\omega_1, \ldots, \omega_m$,
we can obtain a coordinate system $(x_1, \ldots, x_m, t_1, \ldots, t_n)$ defined on W and
centered at p in such a way that

$$\omega_k = \mathrm{d}x_k - \sum_{j=1}^n b_{jk}(x,t)\,\mathrm{d}t_j, \quad k = 1, \ldots, m,$$

with $b_{jk} \in C^\infty(W)$. Next we introduce the linearly independent vector fields
over W

$$L_j = \frac{\partial}{\partial t_j} + \sum_{k=1}^m b_{jk}(x,t)\frac{\partial}{\partial x_k}.$$

Since $\omega_k(L_j) = 0$ for all $j = 1, \ldots, n$ and $k = 1, \ldots, m$ it follows that
$L_1, \ldots, L_n$ span $\mathcal{V}|_W$ at each point.

Next the $C^\infty(W)$-module $\mathfrak{N}_q(W)$ is spanned by the q-forms

$$\omega_J \wedge \mathrm{d}t_K, \quad |J| + |K| = q$$

and since

$$|J| > 0 \Longrightarrow \omega_J \wedge \mathrm{d}t_K \in \mathfrak{N}_q^{\mathcal{V}}(W)$$

it follows that $\mathfrak{U}_q^{\mathcal{V}}(W)$ can be identified with the submodule of $\mathfrak{N}_q(W)$ spanned by $\{dt_K; |K| = q\}$.

If $f \in C^\infty(W)$ then it is plain that

$$df = \sum_{j=1}^{n} (L_j f) dt_j \quad \mathrm{mod}\,[\omega_1, \ldots, \omega_m]$$

since $dt_j(L_{j'}) = \delta_{jj'}$. From this we obtain the representation of the operator d' under the preceding identification: if $f = \sum_J f_J\, dt_J \in \mathfrak{U}_q^{\mathcal{V}}(W)$ then

$$d'f = \sum_{|J|=q} \sum_{j=1}^{n} (L_j f_J) dt_j \wedge dt_J. \tag{VII.16}$$

REMARK VII.5.1. Since $\mathcal{V}$ satisfies the Frobenius condition and since furthermore the vector fields $[L_j, L_{j'}]$ do not involve any differentiation in the t-variables it follows that $[L_j, L_{j'}] = 0$ for every $j, j' = 1, \ldots, n$. Now it is easily seen that this condition is equivalent to the fact that formula (VII.16) defines a differential complex, i.e., that $d' \circ d' = 0$.

VII.6 Germ solvability

In this section we pause to apply some standard functional analytic methods in order to discuss the notion of *exactness in the sense of germs*. The important conclusion is that such a weak notion indeed implies solvability in fixed neighborhoods, and with a bound on the order of the distribution solutions when we are willing to allow even the existence of weak solutions.

Although this is a preparation for all the discussion that will follow, we allow quite general systems of operators.

Let then Ω now denote an open subset of $\mathbb{R}^N$ and let

$$P(x, D) = \{P_{jk}(x, D)\}, \quad Q(x, D) = \{Q_{\ell j}(x, D)\}$$

be matrices of linear partial differential operators (with smooth coefficients) in Ω. We assume $j = 1, \ldots, \beta,\ k = 1, \ldots, \alpha,\ \ell = 1, \ldots, \gamma$ and that

$$C^\infty(\Omega, \mathbb{C}^\alpha) \xrightarrow{P(x,D)} C^\infty(\Omega, \mathbb{C}^\beta) \xrightarrow{Q(x,D)} C^\infty(\Omega, \mathbb{C}^\gamma) \tag{VII.17}$$

defines a differential complex, that is, $Q(x, D)P(x, D) = 0$.

Let $x_0 \in \Omega$. We shall say that (VII.17) *is exact at x_0 (in the sense of germs)* if for every $f \in C^\infty(\Omega, \mathbb{C}^\beta)$ satisfying $Q(x, D)f = 0$ in a neighborhood of x_0 there is $u \in C^\infty(\Omega, \mathbb{C}^\alpha)$ solving $P(x, D)u = f$ in a neighborhood of x_0.

THEOREM VII.6.1. *Suppose that (VII.17) is exact at x_0. Then:*

- *for every open neighborhood U_0 of x_0 in Ω there is another such neighborhood $U_1 \subset\subset U_0$ such that given $f \in C^\infty(U_0, \mathbb{C}^\beta)$ satisfying $Q(x, D)f = 0$ in U_0 there is $u \in C^\infty(U_1, \mathbb{C}^\alpha)$ solving $P(x, D)u = f$ in U_1.*

PROOF. The proof is a well-known category argument due to A. Grothendieck. We fix U_0 and select a fundamental system $\{V_\nu\}_{\nu \in \mathbb{N}}$ of open neighborhoods of x_0, each of them with compact closure in U_0. Set

$$E_\nu = \{(f, v) \in C^\infty(U_0, \mathbb{C}^\beta) \times C^\infty(V_\nu, \mathbb{C}^\alpha) : Q(x, D)f = 0, \ Pv = f \text{ in } V_\nu\}.$$

Each E_ν is a Fréchet space and the linear maps

$$\lambda_\nu : E_\nu \to \{f \in C^\infty(U_0, \mathbb{C}^\beta) : Q(x, D)f = 0\}, \quad \lambda_\nu(f, v) = f$$

are continuous. Now the fact that (VII.17) is exact at x_0 means that

$$\{f \in C^\infty(U_0, \mathbb{C}^\beta) : Q(x, D)f = 0\} = \bigcup_{\nu \in \mathbb{N}} \lambda_\nu (E_\nu).$$

By Baire's category theorem there is ν_0 such that $\lambda_{\nu_0}(E_{\nu_0})$ is of second category in $\{f \in C^\infty(U_0, \mathbb{C}^\beta) : Q(x, D)f = 0\}$ and the open mapping theorem implies that λ_{ν_0} is indeed surjective. This proves the theorem. $\quad\square$

The same argument gives a version of Theorem VII.6.1 where the solutions are now allowed to be distributions.

THEOREM VII.6.2. *Assume that for every $f \in C^\infty(\Omega, \mathbb{C}^\beta)$ satisfying $Q(x, D)f = 0$ in a neighborhood of x_0 there is $u \in \mathcal{D}'(\Omega, \mathbb{C}^\alpha)$ solving $P(x, D)u = f$ in a neighborhood of x_0. Then the following holds:*

- *for every open neighborhood U_0 of x_0 in Ω there are another such neighborhood $U_1 \subset\subset U_0$ and $p \in \mathbb{N}$ such that given $f \in C^\infty(U_0, \mathbb{C}^\beta)$ satisfying $Q(x, D)f = 0$ in U_0 there is $u \in L^2_{-p}(U_1, \mathbb{C}^\alpha)$ solving $P(x, D)u = f$ in U_1.*

PROOF. It suffices to repeat the argument in the proof of Theorem VII.6.1 with

$$E'_\nu = \{(f, v) \in C^\infty(U_0, \mathbb{C}^\beta) \times L^2_{-\nu}(V_\nu, \mathbb{C}^\alpha) : Q(x, D)f = 0, \ Pv = f \text{ in } V_\nu\}$$

in the place of E_ν. $\quad\square$

VII.7 $\mathcal{V}$-cohomology and local solvability

We now return to our original situation where we are given a formally integrable structure $\mathcal{V} \subset \mathbb{C}T\Omega$ over a smooth manifold Ω.

Given $W \subset \Omega$ open we shall denote by $H^q(W; \mathcal{V})$, $q = 0, 1, \ldots, n$, the cohomology spaces of the complex (VII.15). In other words, we have

$$H^0(W; \mathcal{V}) \doteq \mathrm{Ker}\{C^\infty(W) \xrightarrow{\mathrm{d}'} \mathfrak{U}_1^{\mathcal{V}}(W)\} \tag{VII.18}$$

$$H^q(W; \mathcal{V}) \doteq \frac{\mathrm{Ker}\{\mathfrak{U}_q^{\mathcal{V}}(W) \xrightarrow{\mathrm{d}'} \mathfrak{U}_{q+1}^{\mathcal{V}}(W)\}}{\mathrm{Im}\{\mathfrak{U}_{q-1}^{\mathcal{V}}(W) \xrightarrow{\mathrm{d}'} \mathfrak{U}_q^{\mathcal{V}}(W)\}}, \quad q \geq 1. \tag{VII.19}$$

Notice that $H^0(W, \mathcal{V})$ is the space of all smooth functions u on W such that $\mathrm{d}u$ is a section of $T'|_W$.

Given a point $p \in \Omega$ we shall also introduce the direct limits[1]

$$H^q(p; \mathcal{V}) \doteq \lim_{W \to \{p\}} H^q(W; \mathcal{V}), \quad q \geq 0 \tag{VII.20}$$

and the related definition:

DEFINITION VII.7.1. *We shall say that* d' *is solvable in* $W \subset \Omega$ *open in degree* $q \geq 1$ *if* $H^q(W, \mathcal{V}) = 0$. *We shall further say that* d' *is solvable near* $p \in \Omega$ *in degree* $q \geq 1$ *if* $H^q(p; \mathcal{V}) = 0$.

Take an open neighborhood W of p as in Section VII.5. With the identification described there we see that the spaces $\mathfrak{U}_q^{\mathcal{V}}(U)$, $U \subset W$ open, carry natural topologies of Fréchet spaces. As an immediate consequence of Theorem VII.6.1 we derive:

PROPOSITION VII.7.2. *The operator* d' *is solvable near* $p \in \Omega$ *in degree* $q \geq 1$ *if and only if the following holds:*

- *given an open neighborhood* $U \subset W$ *of* p *there is another such neighborhood* $V \subset U$ *such that for every* $f \in \mathfrak{U}_q^{\mathcal{V}}(U)$ *satisfying* $\mathrm{d}'f = 0$ *there is* $u \in \mathfrak{U}_{q-1}^{\mathcal{V}}(V)$ *satisfying* $\mathrm{d}'u = f$ *in* V.

---

[1] We recall that for a sheaf of $\mathbb{C}$-vector spaces $U \mapsto F(U)$ over a topological space X, the *direct limit*

$$\lim_{W \to \{p\}} F(W)$$

at $p \in X$ is the space of all pairs (W, f), with W an open subset of X that contains p and $f \in F(W)$, modulo the following equivalence relation: $(W_1, f_1) \sim (W_2, f_2)$ if there is an open neighborhood W of p, $W \subset W_1 \cap W_2$, such that $f_1|_W = f_2|_W$.

VII.8 The Approximate Poincaré Lemma

Now we assume given a locally integrable structure $\mathcal{V}$ over a smooth manifold Ω. Under this stronger hypothesis a richer description of the differential complex associated with $\mathcal{V}$ can be given. Let $p \in \Omega$ and apply Corollary I.10.2. There is a coordinate system

$$(x_1, \ldots, x_m, t_1, \ldots, t_n)$$

centered at p and there are smooth, real-valued functions $\phi_1, \ldots, \phi_m$ defined in a neighborhood of the origin of $\mathbb{R}^{m+n}$ and satisfying

$$\phi_k(0,0) = 0, \quad d_x\phi_k(0,0) = 0, \quad k = 1, \ldots, m \qquad (\text{VII.21})$$

such that the differentials of the functions

$$Z_k(x,t) = x_k + i\phi_k(x,t), \quad k = 1, \ldots, m \qquad (\text{VII.22})$$

span T' near $p = (0,0)$. We shall set

$$Z = (Z_1, \ldots, Z_m), \qquad \phi = (\phi_1, \ldots, \phi_m).$$

Thus we can write

$$Z(x,t) = x + i\phi(x,t),$$

which we assume defined in an open neighborhood of the closure of $B_0 \times \Theta_0$, where $B_0 \subset \mathbb{R}^m$ and $\Theta_0 \subset \mathbb{R}^n$ are open balls centered at the corresponding origins. Thanks to (VII.21) we can assume that

$$|\phi(x,t) - \phi(x',t)| \le \frac{1}{2}|x - x'|, \quad x, x' \in B_0, \ t \in \Theta_0. \qquad (\text{VII.23})$$

Also recall that $\mathcal{V}$ is spanned, in an open set that contains the closure of $B_0 \times \Theta_0$, by the linearly independent, pairwise commuting vector fields (*cf.* (I.37))

$$L_j = \frac{\partial}{\partial t_j} - i\sum_{k=1}^{m} \frac{\partial \phi_k}{\partial t_j}(x,t)M_k, \quad j = 1, \ldots, n, \qquad (\text{VII.24})$$

where the vector fields

$$M_k = \sum_{\ell=1}^{m} \mu_{k\ell}(x,t)\frac{\partial}{\partial x_\ell}, \quad k = 1, \ldots, m \qquad (\text{VII.25})$$

are characterized by the relations $M_k Z_\ell = \delta_{k\ell}$ (*cf.* (I.35) and (I.36)).

LEMMA VII.8.1. *Let the x-projection of the support of $u \in C^\infty(B_0 \times \Theta_0)$ be a compact subset of B_0. Then, for each $j = 1, \ldots, n$,*

$$\frac{\partial}{\partial t_j}\int u(y,t)\det Z_y(y,t)\,dy = \int (L_ju)(y,t)\det Z_y(y,t)\,dy. \qquad (\text{VII.26})$$

PROOF. In order to prove (VII.26) it suffices to show that, for an arbitrary $\varphi \in C_c^\infty(\Theta_0)$,

$$-\int_{\Theta_0} \left\{ \int u(y,t) \det Z_y(y,t)\, dy \right\} \frac{\partial \varphi}{\partial t_j}(t)\, dt =$$

$$\int_{\Theta_0} \left\{ \int (L_j u)(y,t) \det Z_y(y,t)\, dy \right\} \varphi(t)\, dt.$$

We have $dZ_1(y,t) \wedge \ldots \wedge dZ_m(y,t) \wedge dt = \det Z_y(y,t) dy \wedge dt$. Hence

$$\int_{\Theta_0} \left\{ \int u(y,t) \det Z_y(y,t)\, dy \right\} \frac{\partial \varphi}{\partial t_j}\, dt = \int_{B_0 \times \Theta_0} \frac{\partial \varphi}{\partial t_j}(t) u(y,t)$$

$$dZ_1(y,t) \wedge \ldots \wedge dZ_m(y,t) \wedge dt.$$

Using now the Leibniz rule

$$\frac{\partial \varphi}{\partial t_j} u = L_j(\varphi u) - \varphi L_j u$$

the lemma will be proved if we observe that

$$\int_{B_0 \times \Theta_0} \left[L_j(\varphi u) \right](y,t)\, dZ_1(y,t) \wedge \ldots \wedge dZ_m(y,t) \wedge dt = 0,$$

a fact that follows from Stokes' theorem in conjunction with the identity

$$d \left\{ \varphi(t) u(y,t) dZ_1(y,t) \wedge \ldots \wedge dZ_m(y,t) \wedge dt_1 \wedge \ldots \wedge \widehat{dt_j} \wedge \ldots \wedge dt_n \right\}$$

$$= (-1)^{m+j-1} \left[L_j(\varphi u) \right](y,t)\, dZ_1(y,t) \wedge \ldots \wedge dZ_m(y,t) \wedge dt. \qquad \square$$

We now let

$$f(x,t) = \sum_{|J|=q} f_J(x,t) dt_J \in \mathfrak{U}_q^y(B_0 \times \Theta_0) \qquad (\text{VII.27})$$

satisfy $d'f = 0$. Take $\Psi(x) \in C_c^\infty(B_0)$, $\Psi = 1$ in an open ball $B \subset\subset B_0$ also centered at the origin of $\mathbb{R}^m$ and form

$$F_\nu(z,t) = \left(\frac{\nu}{\pi}\right)^{\frac{m}{2}} \int e^{-\nu[z-Z(y,t)]^2} \Psi(y) f(y,t) \det Z_y(y,t)\, dy. \qquad (\text{VII.28})$$

Notice that F_ν is defined in $\mathbb{C}^m \times \Theta_0$ and is holomorphic in the first variable. Applying Lemma VII.8.1 gives

$$d_t F_\nu(z,t) = \qquad (\text{VII.29})$$

$$\left(\frac{\nu}{\pi}\right)^{\frac{m}{2}} \int e^{-\nu[z-Z(y,t)]^2} (d_0'\Psi)(y,t) \wedge f(y,t) \det Z_y(y,t)\, dy.$$

In (VII.29) the integral is over $B_0 \setminus B$, since Ψ is identically equal to one over B. On the other hand, the real part of the exponent equals $-\nu \mathcal{Q}(z, y, t)$, where

$$\mathcal{Q}(z, y, t) = |\Re z - y|^2 - |\Im z - \phi(y, t)|^2.$$

Now, thanks to (VII.23) we have

$$|\phi(y, t)| \le |\phi(0, t)| + \frac{1}{2}|y|$$

and then

$$|\phi(y, t)|^2 \le 2|\phi(0, t)|^2 + \frac{1}{2}|y|^2.$$

Denote by $b > 0$ the radius of B and use the fact that $\phi(0, 0) = 0$: there is an open ball $\Theta \subset\subset \Theta_1$, centered at the origin in $\mathbb{R}^n$, such that

$$2|\phi(0, t)|^2 \le \frac{1}{4}b^2, \quad t \in \Theta.$$

If $y \in B_0 \setminus B$ and $t \in \Theta$ then we obtain

$$\mathcal{Q}(0, y, t) \ge \frac{1}{2}|y|^2 - 2|\phi(0, t)|^2 \ge \frac{1}{4}b^2$$

and consequently, by continuity we conclude that there are $r > 0$ and $\lambda > 0$ such that

$$(y, t) \in (B_0 \setminus B) \times \Theta, \quad |z| < r \implies \mathcal{Q}(z, y, t) \ge \lambda.$$

We can state:

LEMMA VII.8.2. *Given $\alpha \in \mathbb{Z}_+^m$, $\beta \in \mathbb{Z}_+^n$ there is a constant $C_{\alpha, \beta} > 0$ such that*

$$|\partial_z^\alpha \partial_t^\beta \mathrm{d}_t F_\nu(z, t)| \le C_{\alpha, \beta}\, \mathrm{e}^{-\lambda \nu}, \quad |z| < r, \quad t \in \Theta. \tag{VII.30}$$

Next we apply the Poincaré Lemma, more precisely the homotopy formula (VII.11) in t-space, with base point $t_0 = 0$, considering z as a parameter:

$$F_\nu(z, t) = \mathrm{d}_t \mathrm{G}(F_\nu)(z, t) + \mathrm{G}(\mathrm{d}_t F_\nu)(z, t). \tag{VII.31}$$

If we use Lemma VII.8.2, a close inspection of the formula that defines the operator G (*cf.* (VII.10)) allows us to state:

LEMMA VII.8.3. *Let $\mathcal{U} = \{(z, t) : |z| < r, t \in \Theta\}$. Then, for every $\alpha \in \mathbb{Z}_+^m$, $\beta \in \mathbb{Z}_+^n$*

$$\sup_{(z,t) \in \mathcal{U}} |\partial_z^\alpha \partial_t^\beta \mathrm{G}(\mathrm{d}_t F_\nu)(z, t)| \longrightarrow 0 \quad as \quad \nu \to \infty. \tag{VII.32}$$

$\square$

Taking into account the fact that

$$F_\nu(Z(x, t), t) \longrightarrow \Psi(x) f(x, t) \quad \text{as} \quad \nu \to \infty$$

in the topology of $\mathfrak{U}_q^\nu(B_0 \times \Theta_0)$, we obtain from (VII.31) and (VII.32) the following result:

THEOREM VII.8.4. *Given $B_0 \times \Theta_0$ as above, there is $B \times \Theta \subset\subset B_0 \times \Theta_0$, where $B \subset B_0$ and $\Theta \subset \Theta_0$ are also open balls centered at the origin in $\mathbb{R}^m$ and $\mathbb{R}^n$ respectively, such that if f is as in (VII.27) and satisfies $d'f = 0$ then*

$$\{d_t G(F_\nu)(z, t)\}_{z=Z(x,t)} \longrightarrow f(x, t) \tag{VII.33}$$

in the topology of $\mathfrak{U}_q^\nu(B \times \Theta)$. □

We observe that we can write

$$Q_\nu(z, t) \doteq G(F_\nu)(z, t) = \sum_{|J|=q-1} Q_{J,\nu}(z, t) \, dt_J,$$

where the coefficients are entire holomorphic in $z \in \mathbb{C}^m$ and smooth in $\mathbb{C}^m \times \Theta_0$; moreover (VII.33) gives

$$d' \{Q_\nu(Z(x, t), t)\} = (d_t Q_\nu)(Z(x, t), t) \longrightarrow f(x, t)$$

in the topology of $\mathfrak{U}_q^\nu(B \times \Theta)$. This justifies us referring to the result stated in Theorem VII.8.4 as the *Approximate Poincaré Lemma for the differential complex d'.*

VII.9 One-sided solvability

Let $\mathcal{V}$ be a formally integrable structure over an N-dimensional smooth manifold Ω and let $\Sigma \subset \Omega$ be an embedded submanifold of dimension $N - 1$. We assume that Σ is *noncharacteristic with respect to $\mathcal{V}$*, that is

$$T_p^0 \cap N^*\Sigma_p = 0, \quad \forall p \in \Sigma. \tag{VII.34}$$

Notice that (VII.34) is equivalent, in this particular situation, to

$$T_p' \cap \mathbb{C}N^*\Sigma_p = 0, \quad \forall p \in \Sigma. \tag{VII.35}$$

Indeed it is clear that (VII.35)$\Rightarrow$(VII.34); on the other hand, suppose that for some $p \in \Sigma$ there is $0 \neq \zeta \in T_p' \cap \mathbb{C}N^*\Sigma_p$. Since Σ is one-codimensional it follows that $\mathbb{C}N^*\Sigma_p$ is spanned by one of its nonzero real elements. In particular there are $0 \neq \xi \in N^*\Sigma_p$ and $z \in \mathbb{C}$ such that $\zeta = z\xi$ and thus $\xi = z^{-1}\zeta \in N^*\Sigma_p \cap (T_p' \cap T_p^*\Omega)$, which contradicts (VII.34).

Thanks to (VII.35) it follows that $(\iota_p)^*$ restricted to T'_p is injective and consequently we obtain isomorphisms

$$(\iota_p)^*|_{T'_p} : T'_p \longrightarrow T'\Sigma_p.$$

In particular it follows that $\dim \mathcal{V}(\Sigma)_p = n - 1$ for every $p \in \Sigma$. By Proposition I.14.2, we conclude that Σ is compatible with $\mathcal{V}$; thus $\mathcal{V}(\Sigma)$ defines a formally integrable structure over Σ of rank $n - 1$. One important situation occurs when Σ is the boundary of a *regular* open subset $\Omega_\bullet \subset \Omega$: this means that the topological boundary of $\Omega_\bullet$ equals Σ and that for each $p \in \Sigma$ there is a coordinate system $(U, \mathbf{x})$ centered at p such that $\mathbf{x}(U \cap \Omega_\bullet) = \mathbf{x}(U) \cap \{x = (x_1, \dots, x_N) \in \mathbb{R}^N : x_N > 0\}$. Notice that *a fortiori* $\Omega \backslash \overline{\Omega_\bullet}$ is also regular with boundary Σ.

Let $U \subset \Omega$ be an open set such that $U \cap \Sigma \neq \emptyset$. For each $q = 0, 1, \dots, n$ we shall set

$$\mathfrak{U}_q^\mathcal{V}(U \cap \overline{\Omega_\bullet}) = \{f \in \mathfrak{U}_q^\mathcal{V}(U \cap \Omega_\bullet) : \exists \tilde{f} \in \mathfrak{U}_q^\mathcal{V}(U),\ \tilde{f}|_{U \cap \Omega_\bullet} = f\}.$$

The operator d' induces a differential complex

$$C^\infty(U \cap \overline{\Omega_\bullet}) \overset{d'}{\longrightarrow} \mathfrak{U}_1^\mathcal{V}(U \cap \overline{\Omega_\bullet}) \overset{d'}{\longrightarrow} \dots \overset{d'}{\longrightarrow} \mathfrak{U}_q^\mathcal{V}(U \cap \overline{\Omega_\bullet}) \tag{VII.36}$$

$$\overset{d'}{\longrightarrow} \mathfrak{U}_{q+1}^\mathcal{V}(U \cap \overline{\Omega_\bullet}) \overset{d'}{\longrightarrow} \dots$$

whose cohomology will be denoted by $H^q(U \cap \overline{\Omega_\bullet}; \mathcal{V})$, $q = 0, 1, \dots, n$. If $p \in \Sigma$ we shall set

$$H^q(p, \Omega_\bullet; \mathcal{V}) \doteq \lim_{U \to \{p\}} H^q(U \cap \Omega_\bullet, \mathcal{V}) \tag{VII.37}$$

$$H^q(p, \overline{\Omega_\bullet}; \mathcal{V}) \doteq \lim_{U \to \{p\}} H^q(U \cap \overline{\Omega_\bullet}; \mathcal{V}). \tag{VII.38}$$

DEFINITION VII.9.1. *Let $1 \leq q \leq n$. We say that d' is solvable near $p \in \Sigma$ in degree q with respect to $\Omega_\bullet$ if $H^q(p, \Omega_\bullet; \mathcal{V}) = 0$. We further say that d' is solvable near $p \in \Sigma$ in degree q with respect to $\overline{\Omega_\bullet}$ if $H^q(p, \overline{\Omega_\bullet}; \mathcal{V}) = 0$.*

The following result is an immediate consequence of the arguments in Section VII.6:

PROPOSITION VII.9.2. *Let $1 \leq q \leq n$ and assume that d' is solvable near $p \in \Sigma$ in degree q with respect to $\Omega_\bullet$ (resp. with respect to $\overline{\Omega_\bullet}$). Then to every open neighborhood U of p in Ω there is another such neighborhood $U' \subset U$ such that the natural homomorphism $H^q(U \cap \Omega_\bullet; \mathcal{V}) \to H^q(U' \cap \Omega_\bullet; \mathcal{V})$ (resp. $H^q(U \cap \overline{\Omega_\bullet}; \mathcal{V}) \to H^q(U' \cap \overline{\Omega_\bullet}; \mathcal{V}))$ is trivial.* $\qquad\square$

VII.10 Localization near a point at the boundary

Let $p \in \Sigma$, the boundary of a regular open set $\Omega_\bullet \subset \Omega$. We assume that Σ is noncharacteristic with respect to a *locally integrable* structure $\mathcal{V}$ over Ω of rank n. There is a coordinate system $(y_1, \ldots, y_N)$ defined on an open neighborhood U of p and centered at p such that

$$\Sigma \cap U = \{(y_1, \ldots, y_N) : y_N = 0\};$$

$$\Omega_\bullet \cap U = \{(y_1, \ldots, y_N) : y_N > 0\}.$$

Next, after a possible contraction of U around p, we can select first integrals $Z_1^\flat, \ldots, Z_m^\flat \in C^\infty(U)$ whose differentials span $T'|_U$. Thanks to (VII.35) the forms $dZ_1^\flat(0, 0), \ldots, dZ_m^\flat(0, 0), dy_N$ are linearly independent and consequently, after relabeling, we can assume that

$$A \doteq \frac{\partial(Z_1^\flat, \ldots, Z_m^\flat)}{\partial(y_1, \ldots, y_m)}(0)$$

is nonsingular. We then set

$$Z_k = \sum_{r=1}^m A^{kr}\left[Z_r^\flat - Z_r^\flat(0)\right], \quad k = 1, \ldots, m,$$

where (A^{kr}) denotes the inverse of A. We define

$$x_k = \Re Z_k(y), \quad t_j = y_{m+j}, \quad k = 1, \ldots, m, \ j = 1, \ldots, n. \qquad \text{(VII.39)}$$

Notice that

$$\frac{\partial Z_k}{\partial y_r}(0) = \delta_{kr}; \qquad \text{(VII.40)}$$

in particular it follows from (VII.40) that (VII.39) defines a local diffeomorphism in a neighborhood of the origin. In the new variables $(x_1, \ldots, x_m, t_1, \ldots, t_n)$ we have

$$Z_k(x, t) = x_k + i\phi_k(x, t),$$

where the functions ϕ_k are smooth, real-valued and vanish at the origin. Furthermore, we have

$$\phi_k(x, t) = \Im Z_k(y(x, t))$$

and consequently

$$\frac{\partial \phi_k}{\partial x_s} = \sum_{\ell=1}^N \frac{\partial(\Im Z_k)}{\partial y_\ell}(y(x, t)) \frac{\partial y_\ell}{\partial x_s}(x, t)$$

which, thanks to (VII.40), implies

$$\frac{\partial \phi_k}{\partial x_s}(0,0) = 0 \quad k, s = 1, \ldots, m.$$

We summarize:

PROPOSITION VII.10.1. *Let* $\Omega_\bullet \subset \Omega$, $p \in \Sigma$ *and* $\mathcal{V}$ *as in the beginning of this section. Then there is a coordinate system* $(x_1, \ldots, x_m, t_1, \ldots, t_n)$ *centered at* p *and defined in* $B_0 \times \Theta_0$, *where* $B_0 \subset \mathbb{R}^m$ *(resp.* $\Theta_0 \subset \mathbb{R}^n$*) is an open ball centered at the origin of* $\mathbb{R}^m$ *(resp.* $\mathbb{R}^n$*) such that*

$$\Sigma \cap (B_0 \times \Theta_0) = \{(x, t) \in B_0 \times \Theta_0 : t_n = 0\};$$

$$\Omega_\bullet \cap (B_0 \times \Theta_0) = \{(x, t) \in B_0 \times \Theta_0 : t_n > 0\}$$

and there are smooth, real-valued functions $\phi_1, \ldots, \phi_m$ *defined in* $B_0 \times \Theta_0$ *satisfying (VII.21) in such a way that the differential of the functions (VII.22) span* T' *over* $B_0 \times \Theta_0$.

VII.11 One-sided approximation

We continue the analysis within the set-up of the last section; in particular we apply the conclusions obtained in Proposition VII.10.1. As usual we shall set $Z = (Z_1, \ldots, Z_m)$, $\phi = (\phi_1, \ldots, \phi_m)$ and thus we can write

$$Z(x, t) = x + i\phi(x, t).$$

After contracting $B_0 \times \Theta_0$ we can assume that ϕ is smooth in an open neighborhood of the closure of $B_0 \times \Theta_0$ and also that (VII.23) holds. The vector fields (VII.24) span $\mathcal{V}|_{B_0 \times \Theta_0}$ and $L_1, \ldots, L_{n-1}$ are tangent to $\Sigma \cap (B_0 \times \Theta_0)$ whereas L_n is transversal to it. Clearly $\mathcal{V}(\Sigma)|_{\Sigma \cap (B_0 \times \Theta_0)}$ is spanned by the restriction of the vector fields $L_1, \ldots, L_{n-1}$ to $\Sigma \cap (B_0 \times \Theta_0)$. We now write $\Theta_0^+ = \{t \in \Theta_0 : t_n > 0\}$ and assume given

$$f(x, t) = \sum_{|J|=q} f_J(x, t) \, dt_J \in \mathfrak{U}_q^\mathcal{V}(B_0 \times \Theta_0^+), \tag{VII.41}$$

where $q \in \{0, 1, \ldots, n\}$ and $d'f = 0$. We repeat the analysis presented in Section VII.10. We choose Ψ in the same way and define $F_\nu(z, t)$ by formula (VII.28). Notice that now F_ν is defined in $\mathbb{C}^m \times \Theta_0^+$ and holomorphic in the first variable. If we follow with absolutely no changes the argument that precedes Lemma VII.8.2, we reach the following conclusion:

LEMMA VII.11.1. *Given* $\alpha \in \mathbb{Z}_+^m$, $\beta \in \mathbb{Z}_+^n$ *and* $\varepsilon > 0$ *there is a constant* $C_{\alpha,\beta,\varepsilon} > 0$ *such that*

$$|\partial_z^\alpha \partial_t^\beta \mathrm{d}_t F_\nu(z, t)| \le C_{\alpha,\beta,\varepsilon} \mathrm{e}^{-\lambda\nu}, \quad |z| < r, \quad t \in \Theta \cap \Theta_0^+, \quad \varepsilon \le t_n. \quad \text{(VII.42)}$$

We now fix $t_0 \in \Theta \cap \Theta_0^+$ and consider the homotopy formulae (VII.11), (VII.12) in *t*-space, with base point t_0, considering z as a parameter:

$$F_\nu(z, t) = \mathrm{d}_t G(F_\nu)(z, t) + G(\mathrm{d}_t F_\nu)(z, t); \quad \text{(VII.43)}$$

$$F_\nu(z, t) - F_\nu(z, t_0) = G(\mathrm{d}_t F_\nu)(z, t). \quad \text{(VII.44)}$$

From Lemma VII.11.1 we derive

$$|\partial_z^\alpha \partial_t^\beta G(\mathrm{d}_t F_\nu)(z, t)| \le C_{\alpha,\beta,\varepsilon} \mathrm{e}^{-\lambda\nu}, \quad |z| < r, \quad t \in \Theta \cap \Theta_0^+, \quad \varepsilon \le t_n.$$

Since moreover

$$F_\nu(Z(x, t), t) \longrightarrow \Psi(x) f(x, t) \quad \text{as} \quad \nu \to \infty$$

in the topology of $\mathfrak{U}_q^\nu(B_0 \times \Theta_0^+)$ we obtain, as before:

THEOREM VII.11.2. *Given* $B_0 \times \Theta_0$ *as above there is* $B \times \Theta \subset\subset B_0 \times \Theta_0$, *where* $B \subset B_0$ *and* $\Theta \subset \Theta_0$ *are also open balls centered at the origin in* $\mathbb{R}^m$ *and* $\mathbb{R}^n$ *respectively, such that if f is as in (VII.41) and satisfies* $\mathrm{d}' f = 0$ *then*

$$\{\mathrm{d}_t G(F_\nu)(z, t)\}_{z=Z(x,t)} \longrightarrow f(x, t) \quad \text{if } q \ge 1; \quad \text{(VII.45)}$$

$$F_\nu(Z(x, t), t_0) \longrightarrow f(x, t) \quad \text{if } q = 0; \quad \text{(VII.46)}$$

in the topology of $\mathfrak{U}_q^\nu(B \times (\Theta \cap \Theta_0^+))$. *Moreover, if* $f \in \mathfrak{U}_q^\nu(\overline{\Omega_\bullet} \cap (B_0 \times \Theta_0))$ *then the convergence in (VII.45) and (VII.46) occurs in* $\mathfrak{U}_q^\nu(\overline{\Omega_\bullet} \cap (B_0 \times \Theta_0))$.

The only point that remains to verify in the statement of Theorem VII.11.2 is the very last one, and this follows again from an inspection of the argument, observing that the estimates can be obtained uniformly up to $t_n = 0$. Notice also that in this case the base point t_0 can be chosen to be the origin in *t*-space.

VII.12 A Mayer–Vietoris argument

We continue to work under the following set-up: $\mathcal{V}$ is a locally integrable structure over the smooth manifold Ω, $\Omega_\bullet \subset \Omega$ is a regular open subset of Ω, and the boundary Σ of $\Omega_\bullet$ is noncharacteristic with respect to $\mathcal{V}$. The differential complex on Σ associated with $\mathcal{V}(\Sigma)$ will be denoted by d_Σ'. The next result is one of the main reasons why we introduce such a scheme:

THEOREM VII.12.1. *Let $p \in \Sigma$ and $1 \leq q \leq n-1$.*

(a) *Assume that d' is solvable near p in degree $q+1$. If d' is solvable near p in degree q with respect to $\overline{\Omega_\bullet}$ and with respect to $\Omega \backslash \Omega_\bullet$, then d'_Σ is solvable near p in degree q.*

(b) *Assume that d' is solvable near p in degree q. If d'_Σ is solvable near p in degree q then d' is solvable near p in degree q with respect to $\overline{\Omega_\bullet}$ and with respect to $\Omega \backslash \Omega_\bullet$.*

For the proof we shall first establish some lemmas. We return to the local coordinates and conclusions provided by Proposition VII.10.1. We call attention, in particular, to the properties of the vector fields $L_1, \ldots, L_n$ as described at the beginning of Section VII.11. Recall that $L_1, \ldots, L_{n-1}$ are tangent to Σ and so they have well-defined restrictions to $\Sigma \cap (B_0 \times \Theta_0)$:

$$L_j^0 \doteq L_j|_{t_n=0}.$$

We shall work in an open set of the form $W_0 = B_0 \times (\Theta'_0 \times J_0) \subset B_0 \times \Theta_0$, where Θ'_0 (resp. J_0) is an open ball (resp. open interval) centered at the origin in $\mathbb{R}^{n-1}$ (resp. $\mathbb{R}$).

Given a smooth function (or even a differential form) g on W_0 the notation $g \sim_\Sigma 0$ will indicate that g vanishes to infinite order on $\Sigma \cap W_0$.

LEMMA VII.12.2. *Given $f \in C^\infty(W_0)$ and $u_0 \in C^\infty(\Sigma \cap W_0)$ there is a solution $u \in C^\infty(W_0)$ to the approximate Cauchy problem*

$$\begin{cases} L_n u - f \sim_\Sigma 0 \\ u|_{t_n=0} = u_0. \end{cases} \tag{VII.47}$$

If moreover v is another solution to (VII.47) then $u - v \sim_\Sigma 0$.

PROOF. By the formal Cauchy–Kowalevsky theorem it is possible to solve the Cauchy problem $L_n u = f$, $u|_{t_n=0} = u_0$ uniquely in the ring of formal power series in t_n with coefficients in $C^\infty(B_0 \times \Theta'_0)$. If

$$\sum_{j=0}^{\infty} u_j(x, t') t_n^j$$

is such a formal solution we can obtain a solution to (VII.47) by taking

$$u(x, t) = \sum_{j=0}^{\infty} \zeta(\theta_j t_n) u_j(x, t') t_n^j,$$

where $\zeta \in C_c^\infty(\mathbb{R})$, $\zeta(s) = 1$ for $|s| < 1$, $\zeta(s) = 0$ for $|s| > 2$, and (θ_j) is a suitably chosen sequence of real numbers satisfying $\theta_j < \theta_{j+1}$, $\theta_j \to \infty$.

For the uniqueness it suffices to observe that u and v must *a fortiori* have identical formal power series expansions in t_n, whence the assertion. $\square$

LEMMA VII.12.3. *Given $q \geq 0$ and $g \in \mathfrak{U}_q^{V(\Sigma)}(\Sigma \cap W_0)$ satisfying $d'_\Sigma g = 0$ there is $G \in \mathfrak{U}_q^V(W_0)$ satisfying $(\iota_\Sigma)^*(G) = g$ and $d'G \sim_\Sigma 0$.*

PROOF. We write

$$g = \sum_{|I|=q, I \subset \{1,\dots,n-1\}} g_I(x, t') dt_I$$

and apply Lemma VII.12.2 in order to solve, for each I, the approximate Cauchy problems

$$\begin{cases} L_n G_I \sim_\Sigma 0 \\ G_I|_{t_n=0} = g_I. \end{cases} \tag{VII.48}$$

If we set

$$G = \sum_{|I|=q} G_I(x, t) \, dt_I \in \mathfrak{U}_q^V(W_0)$$

it is clear that, thanks to (VII.48), $(\iota_\Sigma)^* G = g$. We also obtain

$$d'G = \sum_{|I|=q} \sum_{j=1}^{n-1} L_j G_I(x, t) dt_j \wedge dt_I + \sum_{|I|=q} L_n G_I(x, t) dt_n \wedge dt_I. \tag{VII.49}$$

The second term on the right in (VII.49) vanishes to infinite order at $\Sigma \cap W_0$ thanks again to (VII.48). On the other hand, since the vector fields L_j are pairwise commuting, we obtain

$$\begin{cases} L_n L_j G_I \sim_\Sigma 0 \\ L_j^0 \left(G_I|_{t_n=0} \right) = L_j^0 g_I \end{cases}$$

for each $j = 1, \dots, n-1$ and each I. From the uniqueness part in Lemma VII.12.2 together with the fact that $d'_\Sigma g = 0$ it follows that the first term on the right of (VII.49) also vanishes to infinite order on $\Sigma \cap W_0$. This completes the proof. $\square$

LEMMA VII.12.4. *Let $G \in \mathfrak{U}_q^V(W_0)$ satisfy $(\iota_\Sigma)^*(G) = 0$ and $d'G \sim_\Sigma 0$. Then:*

(a) If $q = 0$ then $G \sim_\Sigma 0$.

(b) If $q \geq 1$ then there exists $u \in \mathfrak{U}_{q-1}^V(W_0)$ such that $G - d'u \sim_\Sigma 0$.

PROOF. If $G \in \mathfrak{U}_0^V(W_0)$ is such that $G|_{t_n=0} = 0$ and $d'G \sim_\Sigma 0$ then in particular we have $L_n G \sim_\Sigma 0$. Consequently we obtain $G \sim_\Sigma 0$ thanks to Lemma VII.12.2.

We now prove (b), whose proof is more involved. We assume $q \geq 1$ and write

$$G = \sum_{|I|=q} G_I(x,t)\, dt_I = dt_n \wedge u_1 + \beta_1 = d'(t_n u_1) + \beta_1 - t_n d' u_1,$$

where $u_1 \in \mathfrak{U}^\nu_{q-1}(W_0)$ and $\beta_1 \in \mathfrak{U}^\nu_q(W_0)$ do not involve dt_n. Since $(\iota_\Sigma)^* G = 0$ we have $G_I|_{t_n=0} = 0$ if $n \notin I$. Consequently, all the coefficients of β_1 vanish when $t_n = 0$ and then we can further write

$$G = d'(t_n u_1) + t_n h_1, \tag{VII.50}$$

where $h_1 \in \mathfrak{U}^\nu_q(W_0)$.

We shall construct inductively two sequences $(u_\nu) \subset \mathfrak{U}^\nu_{q-1}(W_0)$, $(h_\nu) \subset \mathfrak{U}^\nu_q(W_0)$, where u_ν do not involve dt_n, such that

$$G = d'\left[\sum_{j=1}^{\nu} \frac{t_n^j}{j} u_j \right] + t_n^\nu h_\nu. \tag{VII.51}$$

Indeed we first observe that (VII.50) gives (VII.51) for $\nu = 1$. We assume then that $u_0, \ldots, u_\nu, h_0, \ldots, h_\nu$ have already been constructed with the required properties and we apply the operator d' to both sides of (VII.51). We obtain

$$d'G = \nu t_n^{\nu-1} dt_n \wedge h_\nu + t_n^\nu d' h_\nu \tag{VII.52}$$

and then, since $d'G$ vanishes to infinite order at $t_n = 0$, we conclude that all the coefficients of $dt_n \wedge h_n u$ vanish at $t_n = 0$. Hence we can write

$$h_\nu = dt_n \wedge u_{\nu+1} + t_n g_\nu, \tag{VII.53}$$

where $u_{\nu+1} \in \mathfrak{U}^\nu_{q-1}(W_0)$ and $g_\nu \in \mathfrak{U}^\nu_q(W_0)$ do not involve dt_n.

Then

$$
\begin{aligned}
G - d'\left[\sum_{j=1}^{\nu+1} \frac{t_n^j}{j} u_\ell \right]
&= t_n^\nu h_\nu - d'\left(\frac{t_n^{\nu+1}}{\nu+1} u_{\nu+1} \right) \\
&= t_n^\nu h_\nu - t_n^\nu dt_n \wedge u_{\nu+1} - \frac{t_n^{\nu+1}}{\nu+1} d' u_{\nu+1} \\
&= t_n^\nu h_\nu + t_n^\nu (t_n g_\nu - h_\nu) - \frac{t_n^{\nu+1}}{\nu+1} d' u_{\nu+1} \\
&= t_n^{\nu+1}\left(g_\nu - \frac{1}{\nu+1} d' u_{\nu+1} \right).
\end{aligned}
$$

Defining $h_{\nu+1} = g_\nu - d' u_{\nu+1}/(\nu+1)$ completes the proof of the inductive argument.

Next we observe that any element $v \in \mathfrak{U}_{q-1}^{\nu}(W_0)$ which does not involve dt_n can be written as

$$v = v|_{t_n=0} + t_n v_1,$$

where $v_1 \in \mathfrak{U}_{q-1}^{\nu}(W_0)$ also does not involve dt_n. Reasoning again by induction we then obtain, from (VII.51), a sequence $(v_\nu) \subset \mathfrak{U}^{\nu(\Sigma)}(W_0)$ such that, for every ν,

$$G = d' \left[\sum_{j=1}^{\nu} v_j t_n^j \right] + O(t_n^\nu). \tag{VII.54}$$

Finally we select $\zeta(s)$ and (θ_j) as in the proof of Lemma VII.12.2 in such a way that

$$\sum_{j=0}^{\infty} \zeta(\theta_j t_n) v_j t_n^j$$

converges in $\mathfrak{U}_{q-1}^{\nu}(W_0)$. Call $u \in \mathfrak{U}_{q-1}^{\nu}(W_0)$ this sum: for every ν, (VII.54) gives

$$G - d'u = d' \left[\sum_{j=1}^{\nu} v_j t_n^j - u \right] + O(t_n^\nu) = O(t_n^\nu),$$

which completes the proof. $\qquad \square$

Proof of Theorem VII.12.1. In order to shorten the notation it is convenient to work with germs of forms at p. Thus we shall introduce the spaces

$$\mathfrak{U}_q^{\nu}(p) = \lim_{U \to \{p\}} \mathfrak{U}_q^{\nu}(U);$$

$$\mathfrak{U}_q^{\nu(\Sigma)}(p) = \lim_{U \to \{p\}} \mathfrak{U}_q^{\nu}(U \cap \Sigma);$$

$$\mathfrak{U}_q^{\nu}(p, \overline{\Omega^\bullet}) = \lim_{U \to \{p\}} \mathfrak{U}_q^{\nu}(U \cap \overline{\Omega^\bullet});$$

$$\mathfrak{U}_q^{\nu}(p, \Omega \backslash \Omega^\bullet) = \lim_{U \to \{p\}} \mathfrak{U}_q^{\nu}(U \cap (\Omega \backslash \Omega^\bullet)).$$

We start by proving (a). Let $g \in \mathfrak{U}_q^{\nu(\Sigma)}(p)$ satisfy $d_{\Sigma}'g = 0$. By Lemma VII.12.3 there is $\underline{f} \in \mathfrak{U}_q^{\nu}(p)$ satisfying $(\iota_{\Sigma})^* \underline{f} = g$, $d'\underline{f} \sim_{\Sigma} 0$. We then define $\mathbf{F} \in \mathfrak{U}_{q+1}^{\nu}(p)$ by the rule

$$\mathbf{F} = \begin{cases} d'\underline{f} & \text{in } (\overline{\Omega^\bullet}, p) \\ -d'\underline{f} & \text{in } (\Omega \backslash \Omega^\bullet, p). \end{cases}$$

Then $\mathbf{F} \in \mathfrak{U}_{q+1}^{\nu}(p)$ and $d'\mathbf{F} = 0$. We now apply our hypothesis: we can find $\underline{f}^* \in \mathfrak{U}_q^{\nu}(p)$ solving $d'\underline{f}^* = \mathbf{F}$ and also $\mathbf{u}^\bullet \in \mathfrak{U}_{q-1}^{\nu}(p, \overline{\Omega^\bullet})$, $\mathbf{u}^{\bullet\bullet} \in \mathfrak{U}_{q-1}^{\nu}(p, \Omega \backslash \Omega^\bullet)$ solving $d'\mathbf{u}^\bullet = \underline{f} - \underline{f}^*$ in $(\overline{\Omega^\bullet}, p)$, $d'\mathbf{u}^{\bullet\bullet} = \underline{f} + \underline{f}^*$ in $(\Omega \backslash \Omega^\bullet, p)$. We then set

$$\mathbf{u} \doteq \frac{1}{2}\left[(\iota_\Sigma)^*\mathbf{u}^\bullet + (\iota_\Sigma)^*\mathbf{u}^{\bullet\bullet}\right].$$

We obtain

$$\begin{aligned}
\mathrm{d}'_\Sigma\mathbf{u} &= \frac{1}{2}\left[(\iota_\Sigma)^*\mathrm{d}'\mathbf{u}^\bullet + (\iota_\Sigma)^*\mathrm{d}'\mathbf{u}^{\bullet\bullet}\right] \\
&= \frac{1}{2}\left[(\iota_\Sigma)^*(\underline{f} - \underline{f}^*) + (\iota_\Sigma)^*(\underline{f} + \underline{f}^*)\right] \\
&= (\iota_\Sigma)^*\underline{f} \\
&= \underline{g}.
\end{aligned}$$

Next we prove (b). We shall prove that d' is solvable near p in degree q with respect to $\Omega_\bullet$, the other case being analogous. Let then $f \in \mathfrak{U}_q^\nu(p, \overline{\Omega^\bullet})$ satisfy $\mathrm{d}'f = 0$. We can of course assume that f has been extended to a germ in $\mathfrak{U}_q^\nu(p)$ (which in general will no longer be d'-closed). Let $\mathbf{v} \in \mathfrak{U}_{q-1}^{\nu(\Sigma)}(p)$ solve $\mathrm{d}'_\Sigma\mathbf{v} = (\iota_\Sigma)^*f$. If $\mathbf{V} \in \mathfrak{U}_{q-1}^\nu(p)$ is such that $(\iota_\Sigma)^*\mathbf{V} = \mathbf{v}$ then $\mathbf{F} \doteq f - \mathrm{d}'\mathbf{V}$ satisfies $(\iota_\Sigma)^*\mathbf{F} = \overline{0}$ and $\mathrm{d}'\mathbf{F} \sim_\Sigma 0$. By Lemma VII.12.4 there is $\mathbf{u} \in \overline{\mathfrak{U}}_{q-1}^\nu(p)$ such that $\mathbf{F} - \mathrm{d}'\mathbf{u} \sim_\Sigma 0$, that is

$$f - \mathrm{d}'(\mathbf{u} + \mathbf{V}) \sim_\Sigma 0. \tag{VII.55}$$

Define

$$\mathbf{G} = \begin{cases} \underline{f} - \mathrm{d}'(\mathbf{u} + \mathbf{V}) & \text{in } (\overline{\Omega^\bullet}, p) \\ 0 & \text{in } (\Omega\backslash\Omega^\bullet, p). \end{cases}$$

Then $\mathbf{G} \in \mathfrak{U}_q^\nu(p)$ thanks to (VII.55) and $\mathrm{d}'\mathbf{G} = 0$. By hypothesis we can solve $\mathrm{d}'\mathbf{h} = \mathbf{G}$ for some $\mathbf{h} \in \mathfrak{U}_{q-1}^\nu(p)$. It follows finally that

$$\mathrm{d}'[\mathbf{u} + \mathbf{V} + \mathbf{h}] = \underline{f} \quad \text{in} \quad (\overline{\Omega^\bullet}, p).$$

The proof of Theorem VII.12.1 is complete. $\qquad\square$

VII.13 Local solvability versus local integrability

We conclude the chapter by presenting a natural generalization of Proposition I.13.6 for locally integrable structures of corank one. Thus we assume that $\mathcal{V}$ is a formally integrable structure of rank n over a smooth manifold of dimension $n + 1$. Fix $p \in \Omega$ and take an open neighborhood W of p and $\omega \in \mathfrak{N}(W)$ which spans $T'|_W$. As in Section VII.4 we can assume that W is the domain of a coordinate system $(x, t_1, \ldots, t_n)$ centered at p and that ω can be written as

$$\omega = dx - \sum_{j=1}^{n} b_j(x, t)\, dt_j,$$

where $b_j \in C^{\infty}(W)$. The linearly independent vector fields

$$L_j = \frac{\partial}{\partial t_j} + b_j(x, t)\frac{\partial}{\partial x}$$

span $\mathcal{V}|_W$ and are pairwise commuting. Since furthermore

$$[L_j, L_{j'}] = \{L_j b_{j'} - L_{j'} b_j\}\frac{\partial}{\partial x}$$

it follows that

$$0 = \frac{\partial}{\partial x}\{L_j b_{j'} - L_{j'} b_j\} = L_j\left\{\frac{\partial b_{j'}}{\partial x}\right\} - L_{j'}\left\{\frac{\partial b_j}{\partial x}\right\}$$

and consequently

$$f_0 \doteq -\sum_{j=1}^{n}\frac{\partial b_j}{\partial x}(x, t)\, dt_j \in \mathfrak{U}_1^{\mathcal{V}}(W) \tag{VII.56}$$

is d'-closed.

THEOREM VII.13.1. *The following properties are equivalent:*

(i) *There is $Z \in C^{\infty}$ in some neighborhood of the origin solving $d'Z = 0$ and satisfying $Z_x \neq 0$.*

(ii) *There is $u \in C^{\infty}$ in some neighborhood of the origin solving $d'u = f_0$.*

In other words, the structure will be locally integrable near p if the class of f_0 in $H^1(p, \mathcal{V})$ vanishes.

PROOF. Assume that (i) holds. If we differentiate the identity

$$Z_{t_j} + b_j Z_x = 0$$

with respect to x we obtain

$$(Z_x)_{t_j} + b_j(Z_x)_x = -(b_j)_x Z_x$$

which gives

$$L_j\{\log Z_x\} = -(b_j)_x.$$

Thus

$$d'\{-\log Z_x\} = f_0,$$

which proves (ii).

Conversely, given u as in (ii) we take

$$G(x, t) = \int_0^x e^{u(y,t)} \, dy.$$

Then

$$L_j G(x, t) = b_j(x, t)e^{u(x,t)} + \int_0^x u_{t_j}(y, t)e^{u(y,t)} dy$$

$$= b_j(x, t)e^{u(x,t)} - \int_0^x [b_j(y, t)u_y(y, t) + (b_j)_y(y, t)]e^{u(y,t)} dy$$

$$= b_j(x, t)e^{u(x,t)} - \int_0^x [b_j(y, t)e^{u(y,t)}]_y dy$$

$$= b_j(0, t)e^{u(0,t)}.$$

If we set

$$B(t) = \sum_{j=1}^{n} b_j(0, t)e^{u(0,t)} dt_j$$

then $d'G = B$ and consequently, in particular, $d'B = d_t B = 0$. By the Poincaré Lemma we can find $F(t)$ smooth near the origin such that $d_t F = B$ and then if we set $Z(x, t) = G(x, t) - F(t)$ we obtain $Z_x = \exp\{u\} \neq 0$ and $d'Z = 0$. □

Notes

The differential complex associated with a formally integrable structure, first presented in [**T4**], is the natural generalization of the de Rham, Dolbeault, and tangential Cauchy–Riemann complexes, associated respectively with real, complex, and CR structures.

Much of the material of this chapter is preparatory for Chapter VIII, and we should just point out that the Approximate Poincaré Lemma is due to Treves ([**T4**]), whereas Theorem VII.12.1 is a consequence of the existence of a natural Mayer–Vietoris sequence, whose existence for hypersurfaces in the complex space was first proved in [**AH1**].

VIII

Local solvability in locally integrable structures

Throughout this chapter we will work with a locally integrable structure $\mathcal{V}$ over a smooth manifold Ω. Our analysis will be for most of the chapter strictly local, and thus, we shall work in a neighborhood of a fixed point $p \in \Omega$. By Corollary I.10.2 there is a coordinate system $(x_1, \ldots, x_m, t_1, \ldots, t_n)$ centered at p and there are smooth, real-valued functions $\phi_1, \ldots, \phi_m$ defined in a neighborhood of the origin of $\mathbb{R}^{m+n}$ and satisfying

$$\phi_k(0,0) = 0, \quad d_x\phi_k(0,0) = 0, \quad k = 1, \ldots, m, \tag{VIII.1}$$

such that the differentials of the functions

$$Z_k(x, t) = x_k + i\phi_k(x, t), \quad k = 1, \ldots, m \tag{VIII.2}$$

span T' near $p = (0, 0)$.

We shall set $Z = (Z_1, \ldots, Z_m)$, $\phi = (\phi_1, \ldots, \phi_m)$. Thus we can write

$$Z(x, t) = x + i\phi(x, t),$$

which we assume is defined in an open neighborhood of the closure of $B_0 \times \Theta_0$, where $B_0 \subset \mathbb{R}^m$ and $\Theta_0 \subset \mathbb{R}^n$ are open balls centered at the corresponding origins. Thanks to (VIII.1) we can assume that

$$|\phi(x, t) - \phi(x', t)| \leq \frac{1}{2}|x - x'|, \quad x, x' \in B_0, \ t \in \Theta_0. \tag{VIII.3}$$

Also recall that $\mathcal{V}$ is spanned, in an open set that contains the closure of $B_0 \times \Theta_0$, by the linearly independent, pairwise commuting vector fields (*cf.* (I.37))

$$L_j = \frac{\partial}{\partial t_j} - i \sum_{k=1}^{m} \frac{\partial \phi_k}{\partial t_j}(x, t) M_k, \quad j = 1, \ldots, n, \tag{VIII.4}$$

331

where the vector fields

$$M_k = \sum_{\ell=1}^{m} \mu_{k\ell}(x, t) \frac{\partial}{\partial x_\ell}, \quad k = 1, \ldots, m \qquad \text{(VIII.5)}$$

are characterized by the relations $M_k Z_\ell = \delta_{k\ell}$ (*cf.* (I.35) and (I.36)).

VIII.1 Local solvability in essentially real structures

If $\mathcal{V}$ defines an essentially real structure over Ω of rank n then the functions ϕ_j can be taken identically zero (Theorem I.9.1). Hence $L_j = \partial/\partial t_j$, $j = 1, \ldots, n$ and the operator d' equals the partial exterior derivative

$$d_t f = \sum_{j=1}^{n} \sum_{|J|=q} \frac{\partial f_J}{\partial t_j} dt_j \wedge dt_J, \qquad \text{(VIII.6)}$$

that is, the d'-complex is nothing other than the standard de Rham complex along the leaves of the foliation defined by $\mathcal{V}$. In particular, if we apply the Poincaré Lemma (Section VII.3) we conclude that *local solvability holds for an essentially real structure near any point and at any degree.*

VIII.2 Local solvability in the analytic category

Now we assume that the manifold Ω and the given locally integrable structure $\mathcal{V}$ are real-analytic. In this case Corollary I.11.1 asserts that the coordinates, functions, and vector fields described at the beginning of the chapter can all be taken real-analytic. We shall show:

PROPOSITION VIII.2.1. *Let $\underline{f} \in \mathfrak{U}_q^{\mathcal{V}}(p)$ have analytic coefficients and satisfy $d'\underline{f} = 0$. If $q \geq 1$ then there is $\mathbf{u} \in \mathfrak{U}_{q-1}^{\mathcal{V}}(p)$, also with analytic coefficients, solving $d'\mathbf{u} = \underline{f}$.*

PROOF. We let

$$f = \sum_{|I|=q} f_I(x, t) dt_I$$

represent $\underline{f}$; the functions f_I are thus real-analytic in a neighborhood of the origin and

$$d'f = \sum_{j=1}^{n} \sum_{|I|=q} L_j f_I(x, t) dt_j \wedge dt_I = 0. \qquad \text{(VIII.7)}$$

Let $1 \leq r \leq n$ be an integer such that f only involves $dt_1, \ldots, dt_r$. Hence we can write $f = f_1 + f_2$, where

$$f_1 = \sum_{|I|=q, I \subset \{1, \ldots, r-1\}} f_I(x, t) \, dt_I$$

and

$$f_2 = \sum_{|J|=q-1, J \subset \{1, \ldots, r-1\}} (-1)^{q-1} f_{J \cup \{r\}} dt_r \wedge dt_J .$$

Notice in particular that (VIII.7) implies

$$L_s f_{J \cup \{r\}} = 0, \quad s > r. \tag{VIII.8}$$

We then apply the Cauchy–Kowalevsky theorem in order to solve, in a neighborhood of the origin, the problems

$$\begin{cases} L_r u_J &= (-1)^{q-1} f_{J \cup \{r\}} \\ u_J|_{t_r=0} &= 0. \end{cases} \tag{VIII.9}$$

Since the vector fields L_j are pairwise commuting, (VIII.8) implies

$$L_r L_s u_J = 0, \quad s > r.$$

Since moreover $L_s u_J = 0$ when $t_r = 0$ it follows from the uniqueness part in the Cauchy–Kowalevsky theorem that $L_s u_J = 0$ for all $s > r$ and all J. Consequently, if we set $u = \sum_J u_J \, dt_J$ then

$$d'u = \sum_{|J|=q-1, J \subset \{1, \ldots, r-1\}} \sum_{j=1}^{r-1} L_j u_J \, dt_j \wedge dt_J +$$
$$\sum_{|J|=q-1, J \subset \{1, \ldots, r-1\}} L_r u_J \, dt_r \wedge dt_J$$

and hence (VIII.9) implies that the d'-closed form $d'u - f$ only involves $dt_1, \ldots, dt_{r-1}$. The proof can then be concluded by an elementary inductive argument, whose details are left to the reader. $\qquad\square$

VIII.3 Elliptic structures

When the structure $\mathcal{V}$ is elliptic the discussion presented at the end of Section I.12 shows that the differential complex associated with $\mathcal{V}$ can be locally realized as the standard elliptic complex in $\mathbb{C}^m \times \mathbb{R}^{n'}$, $n' = n - m$, which we now describe and study in some detail.

Let $m \in \mathbb{Z}_+$ and write the variables in $\mathbb{C}^m \times \mathbb{R}^{n'}$ as

$$(z, t) = (z_1, \ldots, z_m, t_1, \ldots, t_{n'}).$$

We shall also write $z_j = x_j + iy_j$, $j = 1, \ldots, m$, and $n = m + n'$.

The elliptic complex on $\mathbb{C}^m \times \mathbb{R}^{n'}$ is defined as follows: given $\Omega \subset \mathbb{C}^m \times \mathbb{R}^{n'}$ open and $0 \leq q \leq n$ we set $C^\infty(\Omega, \underline{\Lambda}^q)$ as being the space of all smooth differential forms of the kind

$$f = \sum_{|J|+|K|=q} f_{JK} \, d\bar{z}_J \wedge dt_K, \qquad f_{JK} \in C^\infty(\Omega). \tag{VIII.10}$$

We define the differential operator

$$D_q : C^\infty(\Omega, \underline{\Lambda}^q) \longrightarrow C^\infty(\Omega, \underline{\Lambda}^{q+1}) \tag{VIII.11}$$

by the formula

$$D_0 u = \sum_{j=1}^{m} \frac{\partial u}{\partial \bar{z}_j} \, d\bar{z}_j + \sum_{k=1}^{n'} \frac{\partial u}{\partial t_k} \, dt_k \tag{VIII.12}$$

if $u \in C^\infty(\Omega) = C^\infty(\Omega, \underline{\Lambda}^0)$, and by

$$D_q f = \sum_{|J|+|K|=q} D_0 f_{JK} \wedge d\bar{z}_J \wedge dt_K \tag{VIII.13}$$

if f is as in (VIII.10). In particular, when $m = 0$, we have $D_q = d_q$, the exterior derivative acting on q-forms.

It is clear that $D_{q+1} \circ D_q = 0$ and consequently (VIII.11) defines a complex D of differential operators, whose cohomology will be denoted by $\{H_D^q(\Omega) : q = 0, \ldots, n\}$. In particular, $H_D^0(\Omega)$ is the space of all solutions $u \in C^\infty(\Omega)$ of the system $D_0(u) = 0$, that is, the space of all smooth functions on Ω that are holomorphic in z and locally constant in t. Furthermore, when $m = 0$, there are isomorphisms between $H_D^q(\Omega)$ and $H^q(\Omega, \mathbb{C})$, the cohomology groups of Ω with complex coefficients (de Rham's theorem).

THEOREM VIII.3.1. *Let $U \subset \mathbb{C}^m$ be open and pseudo-convex and let $\Theta \subset \mathbb{R}^{n'}$ be open and convex. Then D is solvable in $U \times \Theta$ in degree q, for every $q \geq 1$.*

PROOF. For the proof it is convenient to introduce the natural decomposition

$$C^\infty(U \times \Theta, \underline{\Lambda}^q) = \bigoplus_{r+s=q} C^\infty(U \times \Theta, \underline{\Lambda}^{r,s}),$$

where $C^\infty(U \times \Theta, \underline{\Lambda}^{r,s})$ is the space of forms of the kind

$$f = \sum_{|J|=r, |K|=s} f_{JK} \, d\bar{z}_J \wedge dt_K.$$

Notice that $C^\infty(U \times \Theta, \underline{\Lambda}^{r,s}) = 0$ if either $r > m$ or $s > n'$. We also observe that we have homomorphisms

$$\bar{\partial}_z : C^\infty(U \times \Theta, \underline{\Lambda}^{r,s}) \longrightarrow C^\infty(U \times \Theta, \underline{\Lambda}^{r+1,s}),$$

$$d_t : C^\infty(U \times \Theta, \underline{\Lambda}^{r,s}) \longrightarrow C^\infty(U \times \Theta, \underline{\Lambda}^{r,s+1})$$

such that $D = \bar{\partial}_z + d_t$.

Let $f \in C^\infty(U \times \Theta, \underline{\Lambda}^q)$ satisfy $D_q f = 0$ and decompose $f = \sum_{r,s} f_{r,s}$, where $f_{r,s} \in C^\infty(U \times \Theta, \underline{\Lambda}^{r,s})$ and the sum runs over the pairs (r, s) such that $r + s = q$, $r \le m$, $s \le n'$. Consider, in this decomposition, the term $f_{r,s}$ whose value of s is maximum. From the fact that $Df = 0$ it follows that $d_t f_{r,s} = 0$ and consequently we can apply the Poincaré Lemma (Section VII.3) in order to find $h \in C^\infty(U \times \Theta, \underline{\Lambda}^{r,s-1})$ such that $d_t h = f_{r,s}$. If we set $f^\bullet = f - D_{q-1} h$ it follows that in the analogous decomposition $f^\bullet = \sum_{r,s} f_{r,s}^\bullet$ the maximum value of s that occurs has dropped by one and $D_q f^\bullet = 0$.

If we iterate the argument we will, after a finite number of steps, either solve the equation $D_{q-1} u = f$ or at least find $v \in C^\infty(U \times \Theta, \underline{\Lambda}^{q-1})$ such that $g \doteq f - D_{q-1} v$ does not involve $dt_1, \ldots, dt_{n'}$. If this is the case we can write

$$g = \sum_{|J|=q} g_J(z, t) d\bar{z}_J$$

and the fact that $D_q g = 0$ gives in particular that $d_t g_J = 0$ for all J, that is, g_J are independent of t. Hence g defines a Dolbeault class in U and by the standard complex analysis theory we can determine $w = \sum_{|L|=q-1} w_L(z) d\bar{z}_L$ solving $\bar{\partial}_z w = g$. If we set $u \doteq v + w$ we obtain $D_{q-1} u = f$, which completes the proof. $\qquad\square$

Likewise we can introduce the spaces $\mathcal{D}'(\Omega, \underline{\Lambda}^q)$, which are the spaces of all currents of the form (VIII.11) where now the coefficients are allowed to be elements of $\mathcal{D}'(\Omega)$. By the same expressions (VIII.12) and (VIII.13) we obtain new differential complexes

$$D_q : \mathcal{D}'(\Omega, \underline{\Lambda}^q) \longrightarrow \mathcal{D}'(\Omega, \underline{\Lambda}^{q+1}) \qquad\qquad \text{(VIII.14)}$$

whose cohomology will be denoted by $\{H_D^q(\Omega, \mathcal{D}') : q = 0, \ldots, n\}$.

The natural injections $C^\infty(\Omega, \underline{\Lambda}^q) \hookrightarrow \mathcal{D}'(\Omega, \underline{\Lambda}^q)$ commute with the operator D and then induce homomorphisms

$$H_D^q(\Omega) \longrightarrow H_D^q(\Omega, \mathcal{D}'). \qquad\qquad \text{(VIII.15)}$$

Finally we shall also consider the spaces

$$C_c^\infty(\Omega, \underline{\Lambda}^q) = \{f \in C^\infty(\Omega, \underline{\Lambda}^q) : \operatorname{supp} f \subset\subset \Omega\}; \qquad\qquad \text{(VIII.16)}$$

$$\mathcal{E}'(\Omega, \underline{\Lambda}^q) = \{f \in \mathcal{D}'(\Omega, \underline{\Lambda}^q) : \operatorname{supp} f \subset\subset \Omega\}. \qquad \text{(VIII.17)}$$

The natural pairing

$$C^\infty(\Omega, \underline{\Lambda}^q) \times C_c^\infty(\Omega, \underline{\Lambda}^{n-q}) \longrightarrow \mathbb{C}$$

defined by

$$(f, \psi) \longrightarrow \int f \wedge \mathrm{d}z \wedge \psi,$$

where $\mathrm{d}z = \mathrm{d}z_1 \wedge \cdots \wedge \mathrm{d}z_m$, extends to a bilinear form

$$\mathcal{D}'(\Omega, \underline{\Lambda}^q) \times C_c^\infty(\Omega, \underline{\Lambda}^{n-q}) \longrightarrow \mathbb{C}$$

which allows us to identify $\mathcal{D}'(\Omega, \underline{\Lambda}^q)$ with the topological dual of $C_c^\infty(\Omega, \underline{\Lambda}^{n-q})$, when the latter carries its natural structure of an inductive limit of Fréchet spaces. We shall use the standard notation of the de Rham theory: if $T \in \mathcal{D}'(\Omega, \underline{\Lambda}^q)$ and $\psi \in C_c^\infty(\Omega, \underline{\Lambda}^{n-q})$ we shall set

$$T[\psi] = (T \wedge \psi)[1] = \int T \wedge \mathrm{d}z \wedge \psi.$$

Likewise we have a natural identification between $\mathcal{E}'(\Omega, \underline{\Lambda}^q)$ and the topological dual of $C^\infty(\Omega, \underline{\Lambda}^{n-q})$, where now the latter carries its natural topology of a Fréchet space.

We shall always consider the weak topology in the spaces $\mathcal{D}'(\Omega, \underline{\Lambda}^q)$ and $\mathcal{E}'(\Omega, \underline{\Lambda}^q)$.

LEMMA VIII.3.2. *If $T \in \mathcal{D}'(\Omega, \underline{\Lambda}^q)$, $\psi \in C^\infty(\Omega, \underline{\Lambda}^{n-q-1})$ and one of them has compact support then*

$$\int \mathrm{D}_q T \wedge \mathrm{d}z \wedge \psi = (-1)^{q+m-1} \int T \wedge \mathrm{d}z \wedge \mathrm{D}_{n-q-1}\psi.$$

PROOF. Using the fact that $C^\infty(\Omega, \underline{\Lambda}^q) \subset \mathcal{D}'(\Omega, \underline{\Lambda}^q)$ as well as $C_c^\infty(\Omega, \underline{\Lambda}^q) \subset \mathcal{E}'(\Omega, \underline{\Lambda}^q)$ are dense inclusions we can assume that $T = f$ is smooth. We have

$$\begin{aligned}
\mathrm{d}_{2m+n'-1}(f \wedge \mathrm{d}z \wedge \psi) &= \mathrm{d}_q f \wedge \mathrm{d}z \wedge \psi + (-1)^q f \wedge \mathrm{d}_{m+n-q-1}(\mathrm{d}z \wedge \psi) \\
&= \mathrm{D}_q f \wedge \mathrm{d}z \wedge \psi + (-1)^{q+m} f \wedge \mathrm{d}z \wedge \mathrm{d}_{n-q-1}\psi \\
&= \mathrm{D}_q f \wedge \mathrm{d}z \wedge \psi + (-1)^{q+m} f \wedge \mathrm{d}z \wedge \mathrm{D}_{n-q-1}\psi.
\end{aligned}$$

Since

$$\int \mathrm{d}_{2m+n'-1}(f \wedge \mathrm{d}z \wedge \psi) = 0$$

we obtain the desired conclusion.   □

VIII.4 The Box operator associated with D

If $f, g \in C^\infty(\mathbb{C}^m \times \mathbb{R}^{n'}, \underline{\Lambda}^q)$ and one of them has compact support we set

$$(f, g)_q \doteq \sum_{|J|+|K|=q} \int_{\mathbb{C}^m \times \mathbb{R}^{n'}} f_{JK} \, \overline{g_{JK}} \, dx dy dt. \qquad \text{(VIII.18)}$$

The *formal adjoint* of the operator (VIII.13) is the differential operator

$$D_q^* : C^\infty(\mathbb{C}^m \times \mathbb{R}^{n'}, \underline{\Lambda}^{q+1}) \longrightarrow C^\infty(\mathbb{C}^m \times \mathbb{R}^{n'}, \underline{\Lambda}^q) \qquad \text{(VIII.19)}$$

defined by the expression

$$\left(D_q f, u \right)_{q+1} = \left(f, D_q^* u \right)_q, \qquad \text{(VIII.20)}$$

where $u \in C^\infty(\mathbb{C}^m \times \mathbb{R}^{n'}, \underline{\Lambda}^{q+1})$ and $f \in C^\infty(\mathbb{C}^m \times \mathbb{R}^{n'}, \underline{\Lambda}^q)$, the latter with compact support.

We then set $D_{-1} = D_{n+1} = 0$ and define

$$\Box_q = D_{q-1} D_{q-1}^* + D_q^* D_q. \qquad \text{(VIII.21)}$$

Notice that $\Box_q$ is a second-order differential operator acting on the space $C^\infty(\mathbb{C}^m \times \mathbb{R}^{n'}, \underline{\Lambda}^q)$. Actually an elementary but long computation shows that

$$\Box_q f = \sum_{|J|+|K|=q} (P f_{JK}) \, d\bar{z}_J \wedge dt_K, \qquad \text{(VIII.22)}$$

where f is as in (VIII.10) and

$$P = -\sum_{j=1}^m \frac{\partial^2}{\partial z_j \partial \bar{z}_j} - \sum_{k=1}^{n'} \frac{\partial^2}{\partial t_k^2}. \qquad \text{(VIII.23)}$$

The following crucial properties of the operators $\Box_q$, $q = 0, 1, \ldots, n$, will be used in the sequel:

$$D_q \Box_q = \Box_{q+1} D_q = D_q D_q^* D_q; \qquad \text{(VIII.24)}$$

$$D_q^* \Box_{q+1} = \Box_q D_q^* = D_q^* D_q D_q^*; \qquad \text{(VIII.25)}$$

$$\Box_0 \text{ is hypoelliptic in } \mathbb{C}^m \times \mathbb{R}^{n'}. \qquad \text{(VIII.26)}$$

Moreover, since any open subset of $\mathbb{C}^m \times \mathbb{R}^{n'}$ is P-convex for singular supports ([**H3**]), we also have

Given any open set $\Omega \subset \mathbb{C}^m \times \mathbb{R}^{n'}$ the maps $\qquad \text{(VIII.27)}$

$$\Box_q : \mathcal{D}'(\Omega, \underline{\Lambda}^q) \longrightarrow \mathcal{D}'(\Omega, \underline{\Lambda}^q) \text{ are surjective.}$$

PROPOSITION VIII.4.1. *For any open set* $\Omega \subset \mathbb{C}^m \times \mathbb{R}^{n'}$ *the maps (VIII.15) are isomorphims. More precisely:*

(i) *If* $u \in \mathcal{D}'(\Omega)$ *satisfies* $D_0 u = 0$ *then* $u \in C^\infty(\Omega)$.

(ii) *If* $q \geq 1$ *and if* $u \in \mathcal{D}'(\Omega, \underline{\Lambda}^{q-1})$ *is such that* $D_{q-1} u \in C^\infty(\Omega, \underline{\Lambda}^q)$ *then there is* $v \in C^\infty(\Omega, \underline{\Lambda}^{q-1})$ *such that* $D_{q-1} u = D_{q-1} v$.

(iii) *If* $q \geq 1$ *and if* $f \in \mathcal{D}'(\Omega, \underline{\Lambda}^q)$ *satisfies* $D_q f = 0$ *then there are* $g \in C^\infty(\Omega, \underline{\Lambda}^q)$ *and* $u \in \mathcal{D}'(\Omega, \underline{\Lambda}^{q-1})$ *such that* $f - g = D_{q-1} u$.

PROOF. (i) is a consequence of (VIII.26), since $\square_0 = D_0^* D_0$. Next take u as in (ii) and apply (VIII.27). We can solve

$$\square_{q-1} w = u$$

for some $w \in \mathcal{D}'(\Omega, \underline{\Lambda}^{q-1})$. Then, by (VIII.24),

$$D_{q-1} u = D_{q-1} \square_{q-1} w = \square_q D_{q-1} w.$$

If we apply (VIII.26) we conclude that $D_{q-1} w \in C^\infty(\Omega, \underline{\Lambda}^q)$ and consequently $v \doteq D_{q-1}^* D_{q-1} w \in C^\infty(\Omega, \underline{\Lambda}^{q-1})$. Since using (VIII.24) we also have

$$D_{q-1} u = D_{q-1} \square_{q-1} w = D_{q-1} D_{q-1}^* D_{q-1} w = D_{q-1} v$$

(ii) is proved.

Finally let f be as in (iii) and solve

$$\square_q U = D_{q-1} D_{q-1}^* U + D_q^* D_q U = f,$$

for some $U \in \mathcal{D}'(\Omega, \underline{\Lambda}^{q-1})$. We set $u \doteq D_{q-1}^* U$ and $g \doteq D_q^* D_q U$. In order to conclude the proof it remains to show that g is smooth. But (VIII.24) and (VIII.25) imply

$$\square_q g = \square_q D_q^* D_q U = D_q^* \square_{q+1} D_q U = D_q^* D_q \square_q U = D_q^* D_q f = 0.$$

By (VIII.26) g is smooth and we are done. $\qquad\square$

REMARK VIII.4.2. The preceding argument gives indeed the proof of a stronger statement than (iii): every cohomology class in $H_D^q(\Omega, \mathcal{D}')$ contains a representative which is in the kernel of $\square_q$ (and consequently it is real-analytic).

By a similar argument we have:

PROPOSITION VIII.4.3. *If* Ω *is any open set on* $\mathbb{C}^m \times \mathbb{R}^{n'}$ *then*

$$H_D^n(\Omega) = 0.$$

PROOF. Given $f \in C^\infty(\Omega, \underline{\Lambda}^n)$ we apply (VIII.26) and (VIII.27) in order to find $v \in C^\infty(\Omega, \underline{\Lambda}^n)$ solving

$$\square_n v = f \tag{VIII.28}$$

in Ω. Since moreover $\square_n = D_{n-1} D_{n-1}^*$ we then have $D_{n-1} u = f$, where $u = D_{n-1}^* v \in C^\infty(\Omega, \underline{\Lambda}^{n-1})$, thanks to (VIII.28). $\qquad\square$

Consider the function $E \in L_{\mathrm{loc}}^1(\mathbb{C}^m \times \mathbb{R}^{n'})$ defined by

$$E(z, t) = \begin{cases} \omega_{m,n}^{-1} \left\{|z|^2 + |t/2|^2\right\}^{-m-n'/2+1} & \text{if } m \geq 1, \\ -t/(2|t|) & \text{if } m = 0,\, n' = 1, \\ -(\log|t|)/2\pi & \text{if } m = 0,\, n' = 2, \\ \omega_{m,n}^{-1} \left\{|z|^2 + |t/2|^2\right\}^{-m-n'/2+1} & \text{if } m = 0 \text{ and } n' \geq 3, \end{cases}$$

where $\omega_{m,n} = 2^{n'-2}(2m+n'-2)|S^{2m+n'-1}|$. It is a well-known fact that E is a fundamental solution of P. If we then set, for $U \in \mathcal{E}'(\mathbb{C}^m \times \mathbb{R}^{n'}, \underline{\Lambda}^q)$,

$$E \star U \doteq \sum_{|J|+|K|=q} (E \star U_{JK}) \, d\bar{z}_J \wedge dt_K, \tag{VIII.29}$$

we obtain

$$\square_q(E \star U) = \square_q E \star U = U, \tag{VIII.30}$$

$$D_{q-1}\left[D_{q-1}^*(E \star U)\right] + D_q^*\left[E \star D_q U\right] = U. \tag{VIII.31}$$

We push the argument further. Let Ω be a regular, bounded open subset of $\mathbb{C}^m \times \mathbb{R}^{n'}$. If $f \in C^\infty(\overline{\Omega}, \underline{\Lambda}^q)$ we consider $f \mathcal{X}_\Omega \in \mathcal{E}'(\mathbb{C}^m \times \mathbb{R}^{n'}, \underline{\Lambda}^q)$, where $\mathcal{X}_\Omega$ denotes the characteristic function of Ω. We obtain, from (VIII.31),

$$f \mathcal{X}_\Omega = D_{q-1}\left[D_{q-1}^*(E \star f \mathcal{X}_\Omega)\right] + D_q^*\left[E \star (D_q f) \mathcal{X}_\Omega\right] + D_q^*\left[E \star (D_0 \mathcal{X}_\Omega \wedge f)\right]. \tag{VIII.32}$$

If we now introduce the operators

$$\mathfrak{G}_q : C^\infty(\overline{\Omega}, \underline{\Lambda}^q) \longrightarrow C^\infty(\Omega, \underline{\Lambda}^{q-1}), \quad \mathfrak{H}_q : C^\infty(\overline{\Omega}, \underline{\Lambda}^q) \longrightarrow C^\infty(\Omega, \underline{\Lambda}^q)$$

defined by the expressions

$$\mathfrak{G}_q(f) = D_{q-1}^*(E \star f \mathcal{X}_\Omega)|_\Omega, \quad \mathfrak{H}_q(f) = D_q^*[E \star (D_0 \mathcal{X}_\Omega \wedge f)]|_\Omega, \tag{VIII.33}$$

formula (VIII.32) then gives a natural extension of the so-called *Bochner–Martinelli formula*:

THEOREM VIII.4.4. *If Ω is a regular, bounded open subset of $\mathbb{C}^m \times \mathbb{R}^{n'}$ with a smooth boundary and if $f \in C^\infty(\overline{\Omega}, \underline{\Lambda}^q)$ then*

$$D_{q-1}\mathfrak{G}_q(f) + \mathfrak{G}_{q+1}(D_q f) + \mathfrak{H}_q(f) = f. \tag{VIII.34}$$

$\qquad\square$

Observe that E is real-analytic in the complement of the origin and that $\operatorname{supp} D_0 \mathcal{X}_\Omega \subset \partial\Omega$ and so there exists a neighborhood $\Omega^\bullet$ of Ω in the complexification of $\mathbb{C}^m \times \mathbb{R}^{n'}$ such that the following is true: for every $f \in \mathcal{C}^\infty(\overline{\Omega}, \underline{\Lambda}_q)$ the coefficients of $\mathfrak{H}_q(f)$ extend as holomorphic functions to $\Omega^\bullet$. This fact, in conjunction with Proposition VIII.2.1, provides another proof for the local solvability of the complex D.

VIII.5 The intersection number

We fix a pair (Ω, Ω') of open subsets of $\mathbb{C}^m \times \mathbb{R}^{n'}$, with $\Omega' \subset \Omega$, and an integer $q \geq 1$. The intersection number for the pair (Ω, Ω') in degree q is the $\mathbb{C}$-bilinear form defined on the product

$$\{f \in C^\infty(\Omega, \underline{\Lambda}^q) : D_q f = 0\} \times \{\Theta \in C_c^\infty(\Omega', \underline{\Lambda}^{n-q}) : D_{n-q}\Theta = 0\}$$

defined by

$$\mathcal{I}^q_{(\Omega,\Omega')}[f, \Theta] = \int f \wedge dz \wedge \Theta.$$

The intersection number for the pair (Ω, Ω') in degree 0 is the $\mathbb{C}$-bilinear form defined on the product

$$\{f \in C^\infty(\Omega) : D_0 f = 0\} \times \{\Theta \in C_c^\infty(\Omega', \underline{\Lambda}^n) : \int F dz \wedge \Theta = 0,$$

$$\forall F \in C^\infty(\mathbb{C}^m \times \mathbb{R}^{n'}), D_0 F = 0\}$$

defined by

$$\mathcal{I}^0_{(\Omega,\Omega')}[f, \Theta] = \int f dz \wedge \Theta.$$

We have the following result:

PROPOSITION VIII.5.1. *Let $q \geq 1$. The intersection number $\mathcal{I}^q_{(\Omega,\Omega')}$ vanishes identically if and only if for every $f \in C^\infty(\Omega, \underline{\Lambda}^q)$ satisfying $D_q f = 0$ its restriction to Ω' belongs to the closure of the image of the map*

$$D_{q-1} : C^\infty(\Omega', \underline{\Lambda}^{q-1}) \longrightarrow C^\infty(\Omega', \underline{\Lambda}^q). \tag{VIII.35}$$

PROOF. Let $f \in C^\infty(\Omega, \underline{\Lambda}^q)$ satisfy $D_q f = 0$. If

$$f|_{\Omega'} = \lim_{\nu \to \infty} D_{q-1} u_\nu \quad \text{in } C^\infty(\Omega', \underline{\Lambda}^q)$$

for some sequence (u_ν) in $C^\infty(\Omega', \underline{\Lambda}^{q-1})$, and if $\Theta \in C_c^\infty(\Omega', \underline{\Lambda}^{n-q})$ satisfies $D_{n-q}\Theta = 0$, we have

$$\mathfrak{I}_{(\Omega,\Omega')}^q [f, \Theta] = \lim_{\nu \to \infty} \int D_{q-1}u_\nu \wedge dz \wedge \Theta = \lim_{\nu \to \infty} (-1)^{q+m} \int u_\nu \wedge dz \wedge D_{n-q}\Theta = 0$$

thanks to Lemma VIII.3.2.

For the converse we reason by contradiction and apply the Hahn–Banach theorem. Thus we assume that there are $f_0 \in C^\infty(\Omega, \underline{\Lambda}^q)$ satisfying $D_q f_0 = 0$ and $T \in \mathcal{E}'(\Omega', \underline{\Lambda}^{n-q})$ such that

$$T[f_0] = 1, \qquad T[D_{q-1}u] = 0, \quad \forall u \in C^\infty(\Omega', \underline{\Lambda}^{q-1}).$$

In particular we have $D_{n-q}T = 0$.

Let now $\rho \in C_c^\infty(\mathbb{C}^m \times \mathbb{R}^{n'})$ be such that $\int \rho = 1$ and set, for $\epsilon > 0$,

$$\rho_\epsilon(z, t) = \frac{1}{\epsilon^{2m+n'}} \rho\left(\frac{z}{\epsilon}, \frac{t}{\epsilon}\right).$$

If we introduce the regularizations

$$\Theta_\epsilon = \rho_\epsilon \star T \in C_c^\infty(\mathbb{C}^m \times \mathbb{R}^{n'}, \underline{\Lambda}^{n-q})$$

then there is $\epsilon_0 > 0$ such that $\Theta_\epsilon \in C_c^\infty(\Omega', \underline{\Lambda}^{n-q})$ if $0 < \epsilon \leq \epsilon_0$. Moreover, $\Theta_\epsilon \to T$ in $\mathcal{E}'(\Omega', \underline{\Lambda}^{n-q})$ and $D_{n-q}\Theta_\epsilon = \rho_\epsilon \star D_{n-q}T = 0$ for every $\epsilon > 0$. Now

$$\int f_0 \wedge dz \wedge \Theta_\epsilon = \Theta_\epsilon[f_0] \longrightarrow T[f_0] = 1$$

and consequently there is $0 < \epsilon_1 \leq \epsilon_0$ such that

$$\mathfrak{I}_{(\Omega,\Omega')}^q [f_0, \Theta_{\epsilon_1}] = \int f \wedge dz \wedge \Theta_{\epsilon_1} \neq 0. \qquad \square$$

Next we turn to the case $q = 0$:

PROPOSITION VIII.5.2. *The intersection number $\mathfrak{I}_{(\Omega,\Omega')}^0$ vanishes identically if and only if the following holds:*

For every $f \in C^\infty(\Omega)$ satisfying $D_0 f = 0$ there is (VIII.36)

$\{F_\nu\} \subset C^\infty(\mathbb{C}^m \times \mathbb{R}^{n'})$ satisfying $D_0 F_\nu = 0$ such that

$$F_\nu|_{\Omega'} \longrightarrow f|_{\Omega'} \text{ in } C^\infty(\Omega').$$

PROOF. The proof that (VIII.36) implies the vanishing of $\mathfrak{I}_{(\Omega,\Omega')}^0$ is immediate. We then prove the converse and for this we argue by contradiction as in the proof of Proposition VIII.5.1. Thus we assume that there is $f_0 \in C^\infty(\Omega)$ satisfying $D_0 f = 0$ for which no sequence $\{F_\nu\} \subset C^\infty(\mathbb{C}^m \times \mathbb{R}^{n'})$ as stated exists and apply once more the Hahn–Banach theorem: there is $T \in \mathcal{E}'(\omega', \underline{\Lambda}^n)$ such that

$$T[f_0] = 1, \tag{VIII.37}$$

$$T[F] = 0, \quad \forall F \in C^\infty(\mathbb{C}^m \times \mathbb{R}^{n'}), \ D_0 F = 0. \tag{VIII.38}$$

We next observe that the vanishing of $H_D^1(\mathbb{C}^m \times \mathbb{R}^{n'})$ (Theorem VIII.3.1) implies, in particular, that the homomorphism of Fréchet spaces

$$D_0 : C^\infty(\mathbb{C}^m \times \mathbb{R}^{n'}) \longrightarrow C^\infty(\mathbb{C}^m \times \mathbb{R}^{n'}, \underline{\Lambda}^1)$$

has closed image. Consequently its transpose, which is the map

$$D_n : \mathcal{E}'(\mathbb{C}^m \times \mathbb{R}^{n'}, \underline{\Lambda}^{n-1}) \longrightarrow \mathcal{E}'(\mathbb{C}^m \times \mathbb{R}^{n'}, \underline{\Lambda}^n),$$

has a weakly closed image, that is, its image is precisely the orthogonal of the kernel of $D_0 : C^\infty(\mathbb{C}^m \times \mathbb{R}^{n'}) \to C^\infty(\mathbb{C}^m \times \mathbb{R}^{n'}, \underline{\Lambda}^1)$ in $\mathcal{E}'(\mathbb{C}^m \times \mathbb{R}^{n'}, \underline{\Lambda}^n)$.

Returning to our argument we conclude from (VIII.38) that there exists $S \in \mathcal{E}'(\mathbb{C}^m \times \mathbb{R}^{n'}, \underline{\Lambda}^{n-1})$ such that $D_{n-1} S = T$.

As in the proof of Proposition VIII.5.1 we introduce once more the regularizations

$$\Theta_\epsilon = \rho_\epsilon \star T \in C_c^\infty(\mathbb{C}^m \times \mathbb{R}^{n'}, \underline{\Lambda}^n).$$

There is $\epsilon_0 > 0$ such that $\Theta_\epsilon \in C_c^\infty(\Omega', \underline{\Lambda}^n)$ if $0 < \epsilon \le \epsilon_0$. Furthermore, if $F \in C^\infty(\mathbb{C}^m \times \mathbb{R}^{n'})$ satisfies $D_0 F = 0$ then

$$\int F \mathrm{d}z \wedge \Theta_\epsilon = \int F \mathrm{d}z \wedge (\rho_\epsilon \star D_{n-1} S) = \int F \mathrm{d}z \wedge D_{n-1}(\rho_\epsilon \star S)$$

$$= (-1)^{m-1} \int D_0 F \wedge \mathrm{d}z \wedge (\rho_\epsilon \star S) = 0$$

for $0 < \epsilon \le \epsilon_0$ and also

$$\mathfrak{I}^0_{(\Omega,\Omega')}[f_0, \Theta_\epsilon] = \int f_0 \wedge \mathrm{d}z \wedge \Theta_\epsilon \xrightarrow{\epsilon \to 0} 1,$$

thanks to (VIII.37), which leads to the desired contradiction. $\square$

REMARK VIII.5.3. It follows from the argument in the proof of Proposition VIII.5.2 that the space $\mathbb{C}^m \times \mathbb{R}^{n'}$ can be replaced in (VIII.36) by any open set containing Ω and of the form $U \times \Theta$, where U and Θ are as in Theorem VIII.3.1.

COROLLARY VIII.5.4. *Assume that $m = 0$ and let $\Omega' \subset \Omega$ be open subsets of $\mathbb{R}^{n'}$. Then, if $q \ge 1$, the vanishing of $\mathfrak{I}^q_{(\Omega,\Omega')}$ is equivalent to the vanishing of the natural map induced by restriction $H^q(\Omega, \mathbb{C}) \to H^q(\Omega', \mathbb{C})$. Also, the* *vanishing of* $\mathfrak{I}^0_{(\Omega,\Omega')}$ *is equivalent to the property that* Ω' *is contained in a single connected component of* Ω.

PROOF. Thanks to de Rham's theorem we can assert:

(a) The exterior derivative defines a map with closed image when defined on an arbitrary smooth manifold.
(b) The d-cohomology is isomorphic to the standard singular cohomology with complex coefficients for any smooth manifold.

These two properties in conjunction with Proposition VIII.5.1 prove the assertion for $q \geq 1$. Furthermore, since a scalar function is d-closed if and only if it is locally constant, the assertion for $q = 0$ is an immediate consequence of Proposition VIII.5.2. $\qquad\square$

We shall now draw an important corollary of Propositions VIII.5.1 and VIII.5.2. Let Ω be a regular, bounded open subset of $\mathbb{C}^m \times \mathbb{R}^{n'}$. Since we are dealing with an elliptic structure on $\mathbb{C}^m \times \mathbb{R}^{n'}$ it follows that $\partial\Omega$ is noncharacteristic and consequently we can apply the one-sided approximate Poincaré Lemma (Theorem VII.8.4) and obtain:

COROLLARY VIII.5.5. *Let* $p \in \partial\Omega$. *Given any neighborhood* W *of* p *in* $\mathbb{C}^m \times \mathbb{R}^{n'}$ *there is another such neighborhood* $W' \subset\subset W$ *such that* $\mathfrak{I}^q_{(W\cap\Omega, W'\cap\Omega)} = 0$ *for all* $q = 0, \ldots, m + n'$. $\qquad\square$

VIII.6 The intersection number under certain geometrical assumptions

In this section we shall give a special meaning to one of the complex variables. Thus we shall assume $m \geq 1$ and write the complex variables as $(z_1, \ldots, z_\nu, w)$, where now $m = \nu + 1$. If Ω is an open subset of $\mathbb{C}^m \times \mathbb{R}^{n'}$ we shall denote by $\mathcal{R}(\Omega)$ the space of all $u \in C^\infty(\Omega)$ which satisfy $\partial u/\partial\overline{w} = 0$. If Ω is an open subset of $\mathbb{C}^m \times \mathbb{R}^{n'}$ and if $w_0 \in \mathbb{C}$ we shall write

$$\Omega(w_0) = \{(z, w, t) \in \Omega : w = w_0\}.$$

In the sequel, when dealing with functions defined on $\Omega(w_0)$, we shall identify the latter with the open subset of $\mathbb{C}^\nu \times \mathbb{R}^{n'}$ given by $\{(z, t) \in \mathbb{C}^\nu \times \mathbb{R}^{n'} : (z, w_0, t) \in \Omega(w_0)\}$. We start with an important result:

PROPOSITION VIII.6.1. *Let* $\Omega \subset \mathbb{C}^m \times \mathbb{R}^{n'}$ *be open and* $\partial/\partial\overline{w}$*-convex, that is:*

The homomorphism $\partial/\partial\overline{w} : C^\infty(\Omega) \to C^\infty(\Omega)$ *is surjective.* (VIII.39)

Then given $w_0 \in \mathbb{C}$ the restriction map $\mathcal{R}(\Omega) \to C^\infty(\Omega(w_0))$ is surjective.

PROOF. There is a continuous function $\delta : \Omega(w_0) \to]0, \infty[$ such that the open set

$$\mathcal{U}_\delta = \{(z, w, t) : (z, w_0, t) \in \Omega(w_0), \ |w - w_0| < \delta(z, t)\}$$

is contained in Ω. Let $f = f(z, t) \in C^\infty(\Omega(w_0))$ and select $f^\bullet \in C^\infty(\Omega)$ such that $f^\bullet(z, w, t) = f(z, t)$ if $(z, w, t) \in \mathcal{U}_{\delta/2}$. In particular, $f^\bullet|_{\Omega(w_0)} = f$ and

$$\frac{\partial f^\bullet}{\partial \overline{w}} = 0 \quad \text{in} \quad \mathcal{U}_{\delta/2}. \tag{VIII.40}$$

We must find $g \in C^\infty(\Omega)$ such that

$$F(z, w, t) = f^\bullet(z, w, t) + (w - w_0)g(z, w, t)$$

belongs to $\mathcal{R}(\Omega)$. For this we must have

$$\frac{\partial f^\bullet}{\partial \overline{w}} + (w - w_0)\frac{\partial g^\bullet}{\partial \overline{w}} = 0,$$

which is possible to achieve, since by hypothesis and by (VIII.40) we can solve the equation

$$\frac{\partial g^\bullet}{\partial \overline{w}} = -(w - w_0)^{-1}\frac{\partial f^\bullet}{\partial \overline{w}}$$

in order to determine the desired g. $\qquad\square$

Denote by $C^\infty(\Omega, \underline{\underline{\Lambda}}^q)$, $q = 0, \ldots, n-1$, the space of all forms in $C^\infty(\Omega, \underline{\Lambda}^q)$ which do not involve $d\overline{w}$, that is, the space of all forms of the kind

$$f = \sum_{|J|+|K|=q} f_{JK} \, d\overline{z}_J \wedge dt_K, \tag{VIII.41}$$

with $f_{JK} \in C^\infty(\Omega)$. It is important to observe that if f is as in (VIII.41) and satisfies $D_q f = 0$ then *a fortiori* we have $f_{JK} \in \mathcal{R}(\Omega)$ for every J and K. Notice also that the pullback of an element in $C^\infty(\Omega, \underline{\underline{\Lambda}}^q)$ to any slice $\Omega(w_0)$ is simply obtained by setting $w = w_0$ in its coefficients.

PROPOSITION VIII.6.2. *Let $\Omega' \subset \Omega$ be open subsets of $\mathbb{C}^m \times \mathbb{R}^{n'}$, both satisfying (VIII.39). If for some $q \geq 1$ the homomorphism $H_D^q(\Omega) \to H_D^q(\Omega')$ is trivial then for every $w_0 \in \mathbb{C}$ and every $f \in C^\infty(\Omega(w_0), \underline{\underline{\Lambda}}^{q-1})$ satisfying $D_{q-1}f = 0$ there is $F \in C^\infty(\Omega', \underline{\underline{\Lambda}}^{q-1})$ satisfying $D_{q-1}F = 0$ and $F(z, w_0, t) = f(z, t)$ on $\Omega'(w_0)$.*

PROOF. Let $f = f(z, t) \in C^{\infty}(\Omega(w_0), \underline{\Lambda}^{q-1})$ satisfy $D_{q-1}f = 0$. We apply Proposition VIII.6.1 in order to get $f^{\bullet}(z, w, t) \in C^{\infty}(\Omega, \underline{\Lambda}^{q-1})$, with coefficients in $\mathcal{R}(\Omega)$, such that $f^{\bullet}(z, w_0, t) = f(z, t)$. We have

$$(D_{q-1}f^{\bullet})(z, w_0, t) = (D_{q-1}f)(z, t) = 0$$

and consequently we can write $D_{q-1}f^{\bullet} = (w - w_0)G$ for some $G \in C^{\infty}(\Omega, \underline{\Lambda}^q)$, also with coefficients in $\mathcal{R}(\Omega)$. It is clear that $D_q G = 0$ and thus by hypothesis there is $u \in C^{\infty}(\Omega', \underline{\Lambda}^{q-1})$ satisfying $D_{q-1}u = G$ in Ω'. Write

$$u = u_0 + u_1 \wedge d\overline{w},$$

with $u_j \in C^{\infty}(\Omega', \underline{\Lambda}^{q-j-1})$, $j = 0, 1$. We now use the fact that Ω' also satisfies (VIII.39) in order to solve

$$\frac{\partial v}{\partial \overline{w}} = (-1)^q u_1,$$

with $v \in C^{\infty}(\Omega', \underline{\Lambda}^{q-2})$ (we set $v = u_1 = 0$ if $q = 1$). A simple computation shows that $u - D_{q-2}v \in C^{\infty}(\Omega', \underline{\Lambda}^q)$ and consequently if we set

$$F \doteq f^{\bullet} - (w - w_0)\left(u - D_{q-2}v\right)$$

we obtain $F \in C^{\infty}(\Omega', \underline{\Lambda}^q)$, $F(z, w_0, t) = f(z, t)$, and

$$D_{q-1}F = D_{q-1}f^{\bullet} - (w - w_0)D_{q-1}u = (w - w_0)(G - D_{q-1}u) = 0. \qquad \square$$

We can now prove:

THEOREM VIII.6.3. *Let $\Omega'' \subset \Omega' \subset \Omega$ be open subsets of $\mathbb{C}^m \times \mathbb{R}^{n'}$, all of them satisfying (VIII.39). Let $q \geq 1$ and assume that $\mathfrak{I}^{q-1}_{(\Omega', \Omega'')} = 0$ and that $H^q_D(\Omega) \to H^q_D(\Omega')$ is the trivial map. Then $\mathfrak{I}^{q-1}_{(\Omega(w_0), \Omega''(w_0))} = 0$ for every $w_0 \in \mathbb{C}$.*

PROOF. First we observe that, after taking regularizations, the vanishing of $\mathfrak{I}^{q-1}_{(\Omega', \Omega'')} = 0$ allows one to assert that

$$\int F \wedge dz \wedge dw \wedge T = 0, \qquad \text{(VIII.42)}$$

for every $F \in C^{\infty}(\Omega', \underline{\Lambda}^{q-1})$ satisfying $D_{q-1}F = 0$ and every $T \in \mathcal{E}'(\Omega'', \underline{\Lambda}^{n-q+1})$ satisfying $D_{n-q+1}T = 0$ (when $q = 1$ this condition must be replaced by $T[G] = 0$ for every $G \in C^{\infty}(\mathbb{C}^m \times \mathbb{R}^{n'})$ satisfying $D_0 G = 0$).

Fix $w_0 \in \mathbb{C}$ and let $f \in C^{\infty}(\Omega(w_0), \underline{\Lambda}^{q-1})$, $\Theta \in C^{\infty}_c(\Omega''(w_0), \underline{\Lambda}^{n-q})$ be both D-closed (we assume $\Theta \in \{g \in C^{\infty}(\mathbb{C}^{\nu} \times \mathbb{R}^{n'}) : D_0 g = 0\}^{\perp}$ when $q = 1$). Thanks to our hypotheses we can apply Proposition VIII.6.2 in order to obtain $F \in C^{\infty}(\Omega', \underline{\Lambda}^{q-1})$ satisfying $D_{q-1}F = 0$ and $F|_{w=w_0} = f$.

On the other hand, if we write

$$\Theta = \sum_{|J|+|K|=n-q} \Theta_{JK}(z,t)\, d\overline{z}_J \wedge dt_K$$

and define $T_\Theta \in \mathcal{E}'(\Omega'', \underline{\Lambda}^{n-q+1})$ by the formula

$$T_\Theta \doteq \sum_{|J|+|K|=n-q} \Theta_{JK}(z,t) \otimes \delta(w-w_0)\, d\overline{z}_J \wedge d\overline{w} \wedge dt_K$$

we have $D_{n-q+1}T_\Theta = 0$ and also $T_\Theta \in \{G \in C^\infty(\mathbb{C}^m \times \mathbb{R}^{n'}) : D_0 G = 0\}^\perp$ when $q = 1$. Then (VIII.42) gives

$$\mathfrak{I}^{q-1}_{(\Omega(w_0),\Omega''(w_0))}(f,\Theta) = \int \left(F|_{w=w_0}\right) \wedge dz \wedge \Theta = \pm \int F \wedge dz \wedge dw \wedge T_\Theta = 0,$$

which concludes the proof. $\square$

VIII.7 A necessary condition for one-sided solvability

We keep the notation established in Section VIII.6 and consider now a regular open subset Ω of $\mathbb{C}^m \times \mathbb{R}^{n'}$. We fix a defining function ρ for $\partial\Omega$: thus ρ is a smooth, real-valued function such that $\partial\Omega$ is defined by the equation $\rho = 0$, with $d\rho \neq 0$ on $\partial\Omega$.

THEOREM VIII.7.1. *Let $p \in \partial\Omega$ be such that*

$$\frac{\partial\rho}{\partial w}(p) \neq 0. \tag{VIII.43}$$

Then if for some $q \geq 1$ D is solvable near p in degree q with respect to Ω it follows that the following property holds: given any open neighborhood U of p in $\mathbb{C}^m \times \mathbb{R}^{n'}$ there is another such neighborhood $V \subset U$ such that, for every $w_0 \in \mathbb{C}$, the intersection number $\mathfrak{I}^{q-1}_{(\Omega(w_0)\cap U,\Omega(w_0)\cap V)}$ vanishes identically.

This result is a direct consequence of Theorem VIII.6.3 in conjunction with Corollary VIII.5.5 and the following proposition:

PROPOSITION VIII.7.2. *Suppose that (VIII.43) is satisfied. Then there is an open neighborhood W of p in $\mathbb{C}^m \times \mathbb{R}^{n'}$ such that given any open convex set $D \subset W$ the set $D \cap \Omega$ is $\partial/\partial\overline{w}$-convex.*

PROOF. It suffices to prove the analogous statement for the operator

$$\Delta_w = 4\frac{\partial^2}{\partial w \partial\overline{w}},$$

since every open set which is Δ_w-convex is *a fortiori* $\partial/\partial\overline{w}$-convex.

We write $w = s + ir$, $p = (z_0, s_0 + ir_0, t_0)$, and assume, say, that $\partial\rho/\partial r \neq 0$ at p. By the implicit function theorem, there are an open neighborhood W of p and a smooth function $\psi : \mathbb{C}^\nu \times \mathbb{R} \times \mathbb{R}^{n'} \to \mathbb{R}$ such that $\psi(z_0, s_0, t_0) = r_0$ and

$$W \cap \Omega = \{(z, w, t) \in W : r < \psi(z, s, t)\}.$$

Now the set $\mathcal{U} = \{(z, w, t) \in \mathbb{C}^m \times \mathbb{R}^{n'} : r < \psi(z, s, t)\}$ is Δ_w-convex since Δ_w is real and any normal to $\partial\mathcal{U}$ is not a characteristic vector for Δ_w ([**H1**], theorem 3.7.4). Consequently, given any open convex set $D \subset \mathbb{C}^m \times \mathbb{R}^{n'}$, the set $D \cap \mathcal{U}$, being the intersection of Δ_w-convex open sets, is also Δ_w-convex. If we finally observe that if $D \subset W$ then $D \cap \Omega = D \cap \mathcal{U}$, the result follows at once. $\square$

REMARK VIII.7.3. As in Section VII.12, we introduce the spaces of germs:

$$C_\Omega^\infty(p, \underline{\Lambda}^q) = \lim_{U \to \{p\}} C^\infty(U \cap \Omega, \underline{\Lambda}^q);$$

$$C_{\overline{\Omega}}^\infty(p, \underline{\Lambda}^q) = \lim_{U \to \{p\}} C^\infty(U \cap \overline{\Omega}, \underline{\Lambda}^q).$$

It can be proved, via methods of hyperfunction theory, that if solvability for D near p in degree q with respect to Ω does not occur then there is $\underline{f} \in C_\Omega^\infty(p, \underline{\Lambda}^q)$ for which no $\mathbf{u} \in C_\Omega^\infty(p, \underline{\Lambda}^{q-1})$ satisfies $D\mathbf{u} = \underline{f}$. In particular, Corollary VIII.5.5 also gives a necessary condition for solvability for D near p in degree q with respect to $\overline{\Omega}$.

In the particular case when $m = 1$, Corollary VIII.5.4 allows us to state the necessary condition in terms of the de Rham cohomology. We give first a definition.

DEFINITION VIII.7.4. *Assume that $m = 1$ and let $p \in \Omega$. We shall say condition $(\star)_q$ $(q \geq 1)$ holds at p with respect to Ω if given any open neighborhood U of p in $\mathbb{C} \times \mathbb{R}^{n'}$ there is another such neighborhood $V \subset U$ such that, for all $w_0 \in \mathbb{C}$, the natural homomorphism $H^q(\Omega(w_0) \cap U, \mathbb{C}) \to H^q(\Omega(w_0) \cap V, \mathbb{C})$ is trivial. We further say that condition $(\star)_0$ holds at p with respect to Ω if given any open neighborhood U of p in $\mathbb{C} \times \mathbb{R}^{n'}$ there is another such neighborhood $V \subset U$ such that, for all $w_0 \in \mathbb{C}$, $\Omega(w_0) \cap V$ is contained in one of the connected components of $\Omega(w_0) \cap U$.*

COROLLARY VIII.7.5. *Suppose that $m = 1$ and that (VIII.43) is satisfied. Then if for some $q \geq 1$, D is solvable near $p \in \partial\Omega$ in degree q with respect to Ω it follows that condition $(\star)_{q-1}$ holds at p with respect to Ω.*

VIII.8 The sufficiency of condition $(\star)_0$

We shall now show the sufficiency, in a weak form, of condition $(\star)_0$ for solvability near $p \in \partial\Omega$ in degree 0 with respect to $\overline{\Omega}$ under the stronger assumption that

$$\text{The boundary of } \Omega \text{ is real-analytic.} \qquad \text{(VIII.44)}$$

In other words, we shall assume that Ω is defined by $\rho > 0$, where ρ is real-valued, real-analytic and such that $\partial\Omega$ is defined by $\rho = 0$, with $\partial\rho/\partial z \neq 0$ near p.

The next result is the key tool for the proof of the result. In all the arguments that follow we shall denote by $\pi : \mathbb{C} \times \mathbb{R}^{n'} \to \mathbb{C}$ the projection $\pi(z, t) = z$.

We assume that the central point in the analysis is $p = (z_0, t_0) \in \partial\Omega$ in $\mathbb{C} \times \mathbb{R}^{n'}$. By applying the implicit function theorem we can assume that

$$\rho(z, t) = y - \Phi(x, t), \quad z = x + iy, \qquad \text{(VIII.45)}$$

where Φ is real-analytic and $\Phi(x_0, t_0) = y_0$.

We shall also denote by $\mathfrak{V}(p)$ the set of all open sets D of the form $R \times \Theta$, where R (resp. Θ) is an open square in $\mathbb{C}$ with sides parallel to the coordinate axes (resp. open ball in $\mathbb{R}^{n'}$) centered at $z_0 \in \mathbb{C}$ (resp. $t_0 \in \mathbb{R}^{n'}$).

PROPOSITION VIII.8.1. *Assume that both (VIII.44) and condition $(\star)_0$ hold. Then given any $D \in \mathfrak{V}(p)$ there is $D_\bullet \subset\subset D$ also belonging to $\mathfrak{V}(p)$ and a constant $M > 0$ such that, for any $z \in \mathbb{C}$, any two points in $\Omega(z) \cap D_\bullet$ can be joined by a piecewise real-analytic curve contained in $\overline{\Omega}(z) \cap D$ and with length $\leq M$.*

PROOF. Given D as in the statement we take $D_1 \subset\subset D$ also in $\mathfrak{V}(p)$ and apply condition $(\star)_0$: there is $D_\bullet \subset D_1$, also in $\mathfrak{V}(p)$ such that, for any $z \in \mathbb{C}$, $\Omega(z) \cap D_\bullet$ is contained in a single component of $\Omega(z) \cap D_1$.

Next we observe that the set $K \doteq \overline{D_1 \cap \Omega}$ is compact and sub-analytic. We then apply a standard result on the theory of subanalytic sets which can be found in ([**Har**], section 8): there is $M > 0$ such that any two points in a component of $\pi^{-1}\{z\} \cap K$ may be joined by a piecewise analytic arc in $\pi^{-1}\{z\} \cap K$ of length $\leq M$.

Hence if t, s belong to $\Omega(z) \cap D_\bullet$ they belong to a component of $\Omega(z) \cap D_1$ and consequently to a component of $\pi^{-1}\{z\} \cap K$. Since $\pi^{-1}\{z\} \cap K \subset \overline{\Omega}(z) \cap D$ the result follows. $\qquad\square$

The key point in the argument is the following result:

PROPOSITION VIII.8.2. *Under the same hypotheses as in Proposition VIII.8.1, given $D \in \mathfrak{V}(p)$ there are $D_\star \in \mathfrak{V}(p)$, $D_\star \subset\subset D$ and a constant $C > 0$ such that the following is true: given $u \in \mathcal{C}^\infty(\overline{\Omega} \cap \overline{D}) \cap \mathcal{R}(\Omega \cap D)$ there is $v \in \mathcal{R}(\Omega \cap D_\star)$ such that $d_t v = d_t u$ and*

$$\sup_{(z,t) \in \Omega \cap D_\star} |v(z,t)(y - \Phi(x,t))| \le C \|d_t u\|_{L^\infty(\Omega \cap \overline{D})}. \qquad \text{(VIII.46)}$$

Before we embark on the (rather long) proof of this result, we will show how it can be used to derive our one-sided solvability result.

COROLLARY VIII.8.3. *Assume (VIII.44) and that condition $(\star)_0$ holds. Then given any $D_0 \in \mathfrak{V}(p)$ there is $D_\star \subset\subset D_0$ also belonging to $\mathfrak{V}(p)$ such that for every $f \in C^\infty(\overline{\Omega} \cap D_0, \underline{\Lambda}^1)$ satisfying $D_1 f = 0$ there is $v \in C^\infty(\Omega \cap D_\star)$ satisfying $D_0 v = f$ in $\Omega \cap D_\star$ and*

$$\sup_{(z,t) \in \Omega \cap D_\star} |v(z,t)(y - \Phi(x,t))| < \infty.$$

Notice that, in particular, (VIII.44) and condition $(\star)_0$ imply solvability for D near p in degree 1 with respect to Ω (*cf.* Remark VIII.7.3).

PROOF. Write

$$f = f_0 \, d\overline{z} + \sum_{j=1}^{n'} f_j \, dt_j.$$

If we extend f_0 to a smooth function on D_0 and then solve $(\partial v / \partial \overline{z}) = f_0$ in D_0 we obtain a new form $f - D_0 v \in C^\infty(\overline{\Omega} \cap D_0, \underline{\Lambda}^1)$ which has no $d\overline{z}$-component. In other words, we can start with $f \in C^\infty(\overline{\Omega} \cap D_0, \underline{\Lambda}^1)$ of the form

$$f = \sum_{j=1}^{n} f_j \, dt_j.$$

Notice that $D_1 f = 0$ means that $d_t f = 0$ and that each coefficient f_j is holomorphic in z, that is, $f_j \in \mathcal{R}(\Omega \cap D_0)$.

We apply the Approximate Poincaré Lemma: there is $D \in \mathfrak{V}(p)$, $D \subset D_0$ (which is independent of f) and a sequence $u_\nu \in C^\infty(\overline{\Omega} \cap \overline{D})$ such that $D_0 u_\nu \to f$ in $C^\infty(\overline{\Omega} \cap \overline{D}, \underline{\Lambda}^1)$. Notice that this means

$$d_t u_\nu \to f \quad \text{in} \quad C^\infty(\overline{\Omega} \cap \overline{D}, \underline{\Lambda}^1); \qquad \frac{\partial u_\nu}{\partial \overline{z}} \to 0 \quad \text{in} \quad C^\infty(\overline{\Omega} \cap \overline{D}).$$

Consider now a linear, continuous extension operator

$$\mathfrak{E} : C^\infty(\overline{\Omega} \cap \overline{D}) \longrightarrow C^\infty(\overline{D}).$$

and if $D = R \times \Theta$ let

$$A_\nu(z, t) = \frac{1}{\pi} \iint_R \mathfrak{E}\left(\frac{\partial u_\nu}{\partial \bar{z}}\right)(z', t) \frac{1}{z - z'} \, dx' dy'.$$

It is easily seen that $A_\nu \to 0$ as $\nu \to \infty$ in $C^\infty(\bar{D})$ and, clearly,

$$\frac{\partial u_\nu}{\partial \bar{z}} = \frac{\partial A_\nu}{\partial \bar{z}} \quad \text{in} \quad \bar{\Omega} \cap \bar{D}.$$

If we substitute $u_\nu - A_\nu$ for u_ν we then obtain a new sequence $u_\nu \in C^\infty(\bar{\Omega} \cap \bar{D})$ such that

$$d_t u_\nu \to f \text{ in } C^\infty(\bar{\Omega} \cap \bar{D}, \Lambda^1), \, u_\nu \text{ is holomorphic in } z. \qquad \text{(VIII.47)}$$

Finally we take $D_\bullet \subset\subset D$ as in Proposition VIII.8.2 and apply its conclusion to $u = u_\nu$: we can find $v_\nu \in \mathcal{R}(\Omega \cap D_\bullet)$ such that $d_t v_\nu = d_t u_\nu$ and, for some constant $C > 0$,

$$\sup_{(z, t) \in \Omega \cap D_\bullet} |v_\nu(z, t)(y - \Phi(x, t))| \leq C, \quad \forall \nu.$$

But then some subsequence v_{ν_k} converges weakly to a function v which satisfies the required properties. This concludes the proof of Corollary VIII.8.3. $\qquad \square$

VIII.9 Proof of Proposition VIII.8.2

We take $D_\bullet = R_\bullet \times \Theta_\bullet \subset\subset D$ as in Proposition VIII.8.1 and start by constructing a suitable covering of $\Omega \cap D_\bullet$. Set $\omega \doteq \pi(\Omega \cap D_\bullet)$ and for each $a \in \mathbb{R}^{n'}$ we set

$$W_a \doteq \{z \in \mathbb{C} : (z, a) \in \Omega \cap D_\bullet\}.$$

Notice that $\{W_a\}$ is an open covering of ω. We also set

$$U_a \doteq \pi^{-1}(W_a) \cap [\Omega \cap D_\bullet]. \qquad \text{(VIII.48)}$$

Then $\{U_a\}$ is an open covering of $\Omega \cap D_\bullet$ and $(z, t) \in U_a$ implies $(z, a) \in \Omega \cap D_\bullet$. Using the curves given in Proposition VIII.8.1 and the corresponding bound for their lengths we obtain

$$|u(z, t) - u(z, a)| \leq M \|d_t u\|_{L^\infty(\Omega \cap D)} \quad (z, t) \in U_a. \qquad \text{(VIII.49)}$$

The family $\{u(\cdot, a)\}$ defines a holomorphic one-cochain with respect to the open covering $\{W_a\}$ of ω which satisfies

$$|u(z, a) - u(z, b)| \leq M \|d_t u\|_{L^\infty(\Omega \cap D)} \quad z \in W_a \cap W_b. \qquad \text{(VIII.50)}$$

We shall now construct a new one-cochain $w_a \in \mathcal{O}(W_a)$ such that $w_a - w_b = u(\cdot, a) - u(\cdot, b)$ on $W_a \cap W_b$ and for which each term w_a can be estimated, on W_a, in terms of the right-hand side of (VIII.50).

Such a one-chain will be constructed through the following standard argument: start with a partition of unity $\{\psi_j\}$, $0 \leq \psi_j \leq 1$, subordinate to the covering $\{W_a\}$, that is for each j there corresponds a_j such that $\psi_j \in C_c^\infty(W_{a_j})$ and set

$$g_a(z) = \sum_k \psi_k(z) \left[u(z, a) - u(z, a_k) \right] .$$

Then $g_a \in C^\infty(W_a)$ and $g_a - g_b = u(\cdot, a) - u(\cdot, b)$ in $W_a \cap W_b$. Notice that this last equality implies $\partial g_a / \partial \bar{z} = \partial g_b / \partial \bar{z}$ in $W_a \cap W_b$ and consequently there is $G \in C^\infty(\omega)$, $G = \partial g_a / \partial \bar{z}$ in W_a for every a. Finally we solve

$$\frac{\partial F}{\partial \bar{z}} = G \tag{VIII.51}$$

in ω, with $F \in C^\infty(\omega)$, and set $w_a = g_a - F$.

Observe that such a solution F always exists (every open subset of $\mathbb{C}$ is a domain of holomorphy!) but in order to obtain (VIII.46) we will be forced to make a further contraction in the domain.

We have

$$|g_a(z)| \leq \sum_k \psi_k(z) |u(z, a) - u(z, a_k)| \leq M \|\mathrm{d}_t u\|_{L^\infty(\Omega \cap D)} \quad z \in W_a$$

for every a and thus the proof will be completed if we can show that, for some suitable choice of the partition of unity $\{\psi_j\}$, we can obtain a solution F to (VIII.51) on $R_* \cap \omega$, with $R_* \subset R_\bullet$ another square centered at z_0, satisfying

$$|F(z)(y - \Phi(z, t))| \leq M_1 \|\mathrm{d}_t u\|_{L^\infty(\Omega \cap D)}, \quad (z, t) \in \Omega \cap D_*, \tag{VIII.52}$$

where $D_* \doteq R_* \times \Theta_\bullet \in \mathfrak{V}(p)$.

Indeed $v \in \mathcal{R}(\Omega \cap D_*)$, defined on $U_a \cap D_*$ as

$$u - u(\cdot, a) - w_a = u - u(\cdot, a) - g_a + F,$$

satisfies $\mathrm{d}_t v = \mathrm{d}_t u$ and (VIII.46).

In order to achieve (VIII.52) we start by observing that

$$\left| \frac{\partial g_a}{\partial \bar{z}}(z) \right| \leq M \|\mathrm{d}_t u\|_{L^\infty(\Omega \cap D)} \sum_k \left| \frac{\partial \psi_k}{\partial \bar{z}}(z) \right|, \quad z \in W_a, \tag{VIII.53}$$

and take a closer look on the coverings $\{U_a\}$ and $\{W_a\}$. We have

$$\omega = \{ z \in R_\bullet : \exists t \in \Theta_\bullet, \, \rho(z, t) > 0 \},$$

$$W_a = \{z \in R_\bullet : \rho(z, a) > 0\}.$$

We set

$$\sigma(z) = \sup_{t \in \Theta_\bullet} \rho(z, t) = \max_{t \in \Theta_\bullet} \rho(z, t).$$

In particular, σ is Lipschitz continuous. Also

$$\omega = \{z \in R_\bullet : \sigma(z) > 0\}.$$

Set $\lambda(z) = \min\{\sigma(z), \text{dist}(z, \partial\omega)\}$ and observe that λ is also Lipschitz continuous. We then recall Lemma IV.3.11:

LEMMA VIII.9.1. *Let $\epsilon > 0$ be arbitrary. There is an open covering of ω by squares Q_j, with sides parallel to the coordinate axes, having the following properties:*

$$\text{diam } Q_j \leq \epsilon \inf_{Q_j} \lambda(z). \tag{VIII.54}$$

There are $\psi_j \in C_c^\infty(Q_j)$, $0 \leq \psi_j \leq 1$, such that $\sum \psi_j = 1$ and (VIII.55)

$$\sum_j \left| \frac{\partial \psi_j}{\partial \bar{z}}(z) \right| \leq C\lambda(z)^{-1}.$$

Next we claim that if we take $\epsilon < 1/(2K)$, where K is a Lipschitz constant for ρ, then for each j there is a_j such that $Q_j \subset W_{a_j}$. Indeed let $z_\flat \in Q_j$ and take $t_\flat \in \overline{\Theta_\bullet}$ such that $\sigma(z_\flat) = \rho(z_\flat, t_\flat)$. If $z \in \overline{Q_j}$ we have $|z - z_\flat| \leq \epsilon\sigma(z_\flat)$ and consequently

$$\begin{aligned} \rho(z, t_\flat) &= \rho(z, t_\flat) - \rho(z_\flat, t_\flat) + \rho(z_\flat, t_\flat) \\ &\geq \sigma(z_\flat) - K|z - z_\flat| \\ &\geq \sigma(z_\flat) - \epsilon K\sigma(z_\flat) \\ &\geq \frac{1}{2}\sigma(z_\flat) > 0, \end{aligned}$$

whence our assertion.

For this choice of partition of unity (VIII.53) gives

$$|G(z)| \leq 2MC\|d_t u\|_{L^\infty(\Omega \cap D)}\lambda(z)^{-1}, \quad z \in \omega. \tag{VIII.56}$$

Since $\rho(z_0, t_0) = 0$ we have $\sigma(z_0) \geq 0$. The case $\sigma(z_0) > 0$ is almost elementary, for we can take $R_\ast \subset \omega$ in such a way that $\lambda(z) \geq c > 0$ in $R_\ast$ and consequently the solution to (VIII.51), given by F defined via the formula

$$F(z) = \frac{1}{\pi} \iint_{R_\ast} \frac{G(z')}{z - z'} \, dx' \, dy'$$

satisfies

$$|F(z)| \leq M_1 \|\mathrm{d}_t u\|_{L^\infty(\Omega \cap D)}, \quad z \in R_\star.$$

Let us then assume that $\sigma(z_0) = 0$. We now take $R_\star$ as stated and such that

$$z \in R_\star \quad \Longrightarrow \quad \lambda(z) = \sigma(z). \tag{VIII.57}$$

Notice that, thanks to (VIII.45), we have

$$\sigma(z) = y - \Psi(x), \quad \Psi(x) = \inf_{t \in \Theta_\star} \Phi(x, t)$$

and then

$$\omega \cap R_\star = \{z \in R_\star : y > \Psi(x)\}.$$

We now apply the standard identity

$$\frac{1}{z - z'} = \frac{1}{z - \zeta} + \frac{1}{z - z'} \left(\frac{z' - \zeta}{z - \zeta} \right)$$

in order to obtain

$$\int_{\omega \cap R_\star} \frac{G(z')}{z - z'} \, \mathrm{d}x' \, \mathrm{d}y' = \iint_{\omega \cap R_\star} \frac{G(z')}{z - x' - i\Psi(x')} \, \mathrm{d}x' \, \mathrm{d}y'$$
$$- i \iint_{\omega \cap R_\star} \frac{G(z')(y' - \Psi(x'))}{(z - z')(z - x' - i\Psi(x'))^2} \, \mathrm{d}x' \, \mathrm{d}y'.$$

Since the first term on the right-hand side is holomorphic in $\omega \cap R_\star$ we can solve (VIII.51) by taking

$$F(z) = \frac{1}{\pi i} \iint_{\omega \cap R_\star} \frac{G(z')(y' - \Psi(x'))}{(z - z')(z - x' - i\Psi(x'))} \, \mathrm{d}x' \, \mathrm{d}y', \quad z \in \omega \cap R_\star. \tag{VIII.58}$$

It remains to verify (VIII.52). From (VIII.56) and (VIII.57) we obtain

$$(y - \Psi(x))|F(z)| \leq M_2 \|\mathrm{d}_t u\|_{L^\infty(\Omega \cap D)}$$
$$\times \iint_{\omega \cap R_\star} \frac{1}{|z - z'|} \cdot \frac{y - \Psi(x)}{|x - x'| + |y - \Psi(x')|} \mathrm{d}x' \mathrm{d}y', \quad z \in \omega \cap R_\star. \tag{VIII.59}$$

To conclude we just observe that, since Ψ is Lipschitz,

$$\frac{y - \Psi(x)}{|x - x'| + |y - \Psi(x')|} \leq \frac{|y - \Psi(x')| + M_3|x - x'|}{|x - x'| + |y - \Psi(x')|} \leq M_3 + 1$$

and hence (VIII.59) implies

$$(y - \Psi(x)))|F(z)| \leq M_4 \|d_t u\|_{L^\infty(\Omega \cap D)}.$$

We have thus proved (VIII.52) since

$$(z, t) \in \omega \cap D_\star \implies z \in R_\star, \, t \in \Theta_\bullet, \, y > \Phi(x, t),$$
$$\implies y - \Phi(x, t) \leq y - \Psi(x).$$

The proof of Proposition VIII.8.2 is now complete.

VIII.10 Solvability for corank one analytic structures

Since the solution v obtained in Corollary VIII.8.3 is holomorphic with respect to z and has tempered growth when $(z, t) \to \partial\Omega \cap D_\star$ the results in Chapter VI show that its boundary value is a well-defined distribution on $\partial\Omega \cap D_\star$ of order ≤ 2. If in addition we also assume the validity of condition $(\star)_0$ at p with respect to $\mathbb{C} \times \mathbb{R}^{n'} \setminus \Omega$, and if we denote by $\mathcal{V}_\bullet$ the bundle spanned by the vector fields $\partial/\partial\bar{z}$, $\partial/\partial t_j$, $j = 1, \ldots, n$ and by $L \doteq D_{\partial\Omega}$ the complex induced by the elliptic complex D on $\partial\Omega$, an almost immediate extension of (a) in Theorem VII.12.1 gives:

> *Given an open neighborhood U of p in $\partial\Omega$ there is another*   (VIII.60)
> *such neighborhood $V \subset U$ such that, given $f \in \mathfrak{U}^{\mathcal{V}_\bullet}(U)$*
> *satisfying $Lf = 0$ there is $u \in \mathcal{D}'_{(2)}(V)$ solving $Lu = f$ in V.*

Consider the complex vector fields

$$L_j^\bullet = \frac{\partial}{\partial t_j} - \left(\frac{\partial\rho}{\partial\bar{z}}\right)^{-1} \frac{\partial\rho}{\partial t_j} \frac{\partial}{\partial\bar{z}}, \quad j = 1, \ldots, n'.$$

Near p the vector fields $L_j^\bullet$ are tangent to $\partial\Omega$ and their restriction to $\partial\Omega$ span $\mathcal{V}_\bullet(\partial\Omega)$. As before we describe $\partial\Omega$ by the equation $y - \Phi(x, t) = 0$, with Φ real-analytic and take (x, y, t) as local coordinates in $\partial\Omega$. In these local coordinates the vector fields $L_j \doteq L_j^\bullet|_{\partial\Omega}$ are written as

$$L_j = \frac{\partial}{\partial t_j} - \frac{i\Phi_{t_j}}{1 + i\Phi_x} \frac{\partial}{\partial x}, \quad j = 1, \ldots, n'. \tag{VIII.61}$$

Hence $\mathcal{V}_\bullet(\partial\Omega)$ is exactly the locally integrable structure defined on a neighborhood of the point p in $\mathbb{R}^{n'+1}$ which is orthogonal to the sub-bundle of $\mathbb{C}T^*(\mathbb{R}^{n'+1})$ spanned by dZ, where $Z(x, t) = x + i\Phi(x, t)$.

The reverse argument is also true, that is, any smooth locally integrable structure $\mathcal{V}$ of corank one, say in a neighborhood of the origin in $\mathbb{R}^{n'+1}$, arises from the restriction of the elliptic structure $\mathcal{V}_\bullet$ on $\mathbb{C} \times \mathbb{R}^{n'}$ to a hypersurface Σ in $\mathbb{C} \times \mathbb{R}^{n'}$. Indeed if we choose local coordinates $(x, t) = (x, t_1, \ldots, t_{n'})$ in a neighborhood of the origin in $\mathbb{R}^{n'+1}$ in such a way that the orthogonal of $\mathcal{V}$ is generated by the differential of $Z(x, t) = x + i\Phi(x, t)$, with Φ smooth and real-valued, and if we denote by Σ the image of the imbedding $(x, t) \mapsto (Z(x, t), t)$, it follows easily that $\mathcal{V} = \mathcal{V}_\bullet(\Sigma)$.

Keeping this notation, and recalling Corollary I.10.2, we can (and will) even assume that $\Phi(0, 0) = \Phi_x(0, 0) = 0$. We emphasize that $\mathcal{V}$ is spanned by the pairwise commuting vector fields (VIII.61). We further take a small open neighborhood V of the origin in $\mathbb{C} \times \mathbb{R}^{n'}$ and set

$$V^+ = \{(z, t) \in V : z = x + iy, \, y > \Phi(x, t)\},$$

$$V^- = \{(z, t) \in V : z = x + iy, \, y < \Phi(x, t)\}.$$

DEFINITION VIII.10.1. *We shall say that condition* $(P)_0$ *holds at the origin for the locally integrable structure* $\mathcal{V}$ *if condition* $(\star)_0$ *holds at the origin in* $\mathbb{C} \times \mathbb{R}^{n'}$ *with respect to both* V^+ *and* V^-.

We shall then prove:

THEOREM VIII.10.2. *Let* $\mathcal{V}$ *be a corank one, real-analytic, locally integrable structure defined in an open neighborhood of the origin in* $\mathbb{R}^{n'+1}$ *and let* d' *be the associated differential complex. Then* d' *is solvable near the origin in degree one if and only if condition* $(P)_0$ *holds at the origin.*

PROOF. The necessity of condition $(P)_0$ follows from Theorem VII.12.1, Corollary VIII.8.3 and Remark VIII.7.3.

We now embark on the proof of the sufficiency. Let us denote by $\mathfrak{W}(0)$ the family of all open neighborhoods of the origin in $\mathbb{R}^{n'+1}$ of the form $U = I \times \Theta$, where I (resp. Θ) is an open interval (resp. ball) centered at the origin in $\mathbb{R}$ (resp. $\mathbb{R}^{n'}$). If $(p, q) \in \mathbb{R}^2$ and if $U \in \mathfrak{W}(0)$ we shall denote by $L^{2,r,s}_{loc}(U)$ the local Sobolev space of order r with respect to x and of order s with respect to t.

We recall that if we set $M = Z_x^{-1}(\partial/\partial x)$ then the vector fields $L_1, \ldots, L_{n'}, M$, (cf. (VIII.61)), are pairwise commuting, linearly independent (see (I.38)).

We now make use of (VIII.60). Then there is $(r_0, s_0) \in \mathbb{R}^2$ such that the following is true: given $U \in \mathfrak{W}(0)$ there is $U' = I' \times \Theta' \in \mathfrak{W}(0)$, $U' \subset\subset U$, such that given

$$f(x, t) = \sum_{j=1}^{n'} f_j(x, t) \, dt_j \in \mathfrak{U}^v(U), \quad Lf = 0 \tag{VIII.62}$$

there is $v \in L^{2,r_0,s_0}_{\text{loc}}(U')$ satisfying $Lv = f$ in U'.

We then fix f as in (VIII.62). Noticing that, for each $k \in \mathbb{N}$, $M^{2k}f$ is also L-closed (here M^{2k} acts componentwise on the one-form f), we can find $v_k \in L^{2,r_0,s_0}_{\text{loc}}(U')$ solving $Lv_k = M^{2k}f$ in U'. Next we solve, in U', $M^{2k}w_k = v_k$. Thus

$$M^{2k}[Lw_k - f] = 0$$

and consequently we can write

$$Lw_k - f = \sum_{j=0}^{2k-1} g_{jk}(t) Z(x,t)^j,$$

where g_{jk} are d_t-closed one-forms with distributional coefficients. We can find distributions $G_{jk} \in \mathcal{D}'(\Theta')$ such that $d_t G_{jk} = g_{jk}$ and hence we have

$$L\left[w_k + \sum_{j=0}^{2k-1} G_{jk}(t) Z(x,t)^j\right] = f. \tag{VIII.63}$$

Since $w_k \in L^{2,r_0+2k,s_0}_{\text{loc}}(U')$ it follows that

$$Lw_k - f \in L^{2,r_0+2k-1,s_0-1}_{\text{loc}}(U')$$

and hence $g_{jk} \in L^{2,s_0-1}_{\text{loc}}(\Theta')$. Consequently $G_{jk} \in L^{2,s_0}_{\text{loc}}(\Theta')$ and then, if we set

$$u_k \doteq w_k + \sum_{j=0}^{2k-1} G_{jk}(t) Z(x,t)^j$$

we have

$$Lu_k = f, \quad u_k \in L^{2,r_0+2k,s_0}_{\text{loc}}(U'). \tag{VIII.64}$$

Explicitly (VIII.64) means

$$\frac{\partial u_k}{\partial t_j} = \frac{i\Phi_{t_j}}{1+i\Phi_x} \frac{\partial u_k}{\partial x} + f_j, \quad j = 1, \dots, n'.$$

This expression implies that it is possible to trade differentiability with respect to x for differentiability with respect to t_j, $j = 1, \dots, n'$, that is, we also have $u_k \in L^{2,r_0+k,s_0+k}_{\text{loc}}(U')$.

Let $U_\bullet \in \mathfrak{W}(0)$, $U_\bullet \subset\subset U'$. By the Sobolev imbedding theorem it then follows that for each $\nu \in \mathbb{N}$ we can find a solution $u^\bullet_\nu \in C^\nu(\overline{U_\bullet})$ to the equation $Lu^\bullet_\nu = f$ in $U_\bullet$.

We finally apply, for each $\nu \in \mathbb{N}$, the C^ν-version of the Baouendi–Treves approximation formula (*cf.* Theorem II.1.1). There are $U_1 \in \mathfrak{W}(0)$, $U_1 \subset\subset U_\bullet$, depending only on $U_\bullet$ and on $\mathcal{V}$, and a sequence of holomorphic polynomials $\{p_\nu\} \subset \mathbb{C}[z]$ such that

$$\|u_{\nu+1}^\bullet - u_\nu^\bullet - p_\nu(Z)\|_{C^\nu(\overline{U_1})} \leq \frac{1}{2^\nu}. \tag{VIII.65}$$

If we set

$$u_{(1)} = u_1^\bullet, \quad u_{(\nu)} = u_\nu^\bullet - p_1(Z) - \ldots - p_{\nu-1}(Z), \quad \nu \geq 2,$$

then (VIII.65) gives

$$\|u_{\nu+1} - u_\nu\|_{C^\nu(\overline{U_1})} \leq \frac{1}{2^\nu}.$$

This shows that, for each $p \in \mathbb{N}$, the sequence $(u_\nu)_{\nu \geq p}$ converges to an element $u \in C^p(\overline{U_1})$, of course independent of p, and belonging to $\mathcal{C}^\infty(\overline{U_1})$. Since moreover $Lu_\nu = f$ in U_1 for every ν we also have $Lu = f$ in U_1.

The proof of Theorem VIII.10.2 is complete. $\qquad\square$

Notes

Until now, complete answers for local solvability in locally integrable structures, besides the cases $n = 1$ (a situation which has already been discussed in Chapter IV), $\mathcal{V}$ defines an essentially real structure (Section VIII.1) and when $\mathcal{V}$ defines an elliptic structure (Theorem VIII.3.1) are known in the following cases: (i) $\mathcal{V}$ defines a nondegenerate locally integrable CR structure of codimension one ([**AH2**]); (ii) $\mathcal{V}$ defines a nondegenerate real-analytic structure ([**T9**]); (iii) $m = 1$ ([**CorH3**]).

We also mention a necessary condition for nondegenerate CR structures of arbitrary codimension proved in [**AFN**], which was extended to general locally integrable structures with additional nondegeneracy conditions in [**T5**].

The notion of intersection number and the necessary condition given in Theorem VIII.11.4 is due to [**CorT1**].

As far as sufficiency is concerned, we point out the works by Kashiwara–Schapira ([**KaS**]) and Michel ([**Mi**]), which deal with locally integrable CR structures of codimension one and whose Levi form satisfies weaker nondegeneracy conditions.

Locally integrable structures with $m = 1$: for this class of locally integrable structures we have seen in Sections VIII.7 and VIII.8 that condition $(P)_0$ is necessary and (in the real-analytic category) sufficient for local solvability near the origin (*cf.* Corollary VIII.7.5 and Theorem VIII.10.2). This result can be generalized much more. Let us introduce, for each $q = 0, 1, \ldots, n-1$, the following property:

(P)$_q$ *Given any open neighborhood V of the origin there is another such neighborhood $V' \subset V$ such that, for every regular value $z_0 \in \mathbb{C}$ of the map Z, either $Z^{-1}\{z_0\} \cap V' = \emptyset$ or else the homomorphism*

$$\tilde{H}_q(Z^{-1}\{z_0\} \cap V') \longrightarrow \tilde{H}_q(Z^{-1}\{z_0\} \cap V)$$

induced by the inclusion map

$$Z^{-1}\{z_0\} \cap V' \subset Z^{-1}\{z_0\} \cap V$$

vanishes identically. Here $\tilde{H}_*$ denotes the reduced homology with complex coefficients.

In 1981 F. Treves proposed the following conjecture: *local solvability near the origin holds for $\mathcal{V}$ if and only if property* (P)$_{q-1}$ *is verified*. Several articles were published towards its verification; see [**MenT**], [**CorH1**], [**CorH2**], [**CorT3**], [**ChT**]. The complete proof of the conjecture was finally achieved in [**CorH3**]. The main point in the proof of Theorem VIII.10.2 that we presented is the use of the special covering (VIII.48), an idea inspired by the work [**H10**].

Solvability in top degree: one of the main questions in the theory is how to generalize condition (P)$_q$ in order to state a plausible conjecture for local solvability for general locally integrable systems. Observe that when $m \geq 2$ the sets $Z^{-1}\{z_0\}$ no longer carry enough information: for instance, in the CR case they are reduced to points.

There is one particular situation where a conjecture can be stated and at least verified in some particular but important cases: this is when $q = n$ (local solvability in maximum degree). Returning to the notation established at the beginning of this chapter, in particular to the vector fields (VIII.4), the equation under study is now

$$\sum_{j=1}^{n} L_j u_j = f, \tag{VIII.66}$$

where no compatibility condition occurs. This makes this case, in some sense, the closest to the single equation situation.

Before we introduce the solvability condition for (VIII.66) we introduce the following definition: a real-valued function F defined on a topological space X is said to *assume a local minimum over a compact set $K \subset X$* if there exist $a \in \mathbb{R}$ and $K \subset V \subset X$ open such that $F = a$ on K and $F > a$ on $V \backslash K$.

DEFINITION VIII.10.3. *We shall say that V satisfies condition $(\mathfrak{P})_{n-1}$ near the origin if there is an open neighborhood $U \subset \Omega$ of the origin such that given any open set $V \subset U$ and given any $h \in C^\infty(V)$ satisfying $L_j h = 0$, $j = 1, \ldots, n$, then $\Re h$ does not assume a local minimum over any nonempty compact subset of V.*

By using a classical device due to Lars Hörmander [**H6**], it was proved in [**CorH1**] that local solvability near the origin for (VIII.66) implies condition $(\mathfrak{P}_{n-1})$. This result would be of limited importance if no evidence that $(\mathfrak{P}_{n-1})$ is also a sufficient condition could be presented. This however is not the case, as the discussion that follows will show, and we can even conjecture at this point that *local solvability of (VIII.66) near the origin is equivalent to $(\mathfrak{P}_{n-1})$*.

When $n = 1$ condition $(\mathfrak{P}_0)$ is equivalent to the Nirenberg–Treves condition $(\mathcal{P})$: this result was proved in [**CorH1**] in the analytic category and in the general case in [**T3**]. When $m = 1$ condition $(\mathfrak{P}_{n-1})$ is equivalent to condition $(P)_{n-1}$ ([**CorH1**], [**T3**]). Thus, in these extreme cases, $(\mathfrak{P}_{n-1})$ unifies both known solvability conditions.

Let us pause here to discuss again the case when $m = 1$. The proof of the Treves conjecture in top degree as presented in [**CorH2**] is obtained by proving that $(P)_{n-1}$ implies, when $n \geq 2$, the following property: *there are an open neighborhood U of the origin and constants $C > 0$ and $k \in \mathbb{Z}_+$ such that*

$$\|\phi\|_\infty \leq C \sum_{j=1}^n \sum_{|\alpha| \leq k} \|D^\alpha L_j \phi\|_\infty, \quad \phi \in C_c^\infty(U). \tag{VIII.67}$$

Indeed k can be taken any integer $\geq [n/2] + 1$ and equal to zero when the structure is real-analytic ([**CorH1**]). The completion of the argument is quite standard, and holds whatever the value of m: by applying the Hahn–Banach theorem it is easily seen that (VIII.67) implies the existence of weak solutions to (VIII.66), and a general result due to [**T5**] proves the existence of smooth solutions.

For the tube structures it is not difficult to prove that property $(\mathfrak{P}_{n-1})$ implies (VIII.67) and consequently the preceding discussion shows that our conjecture is also satisfied for this particular class.

When V defines a CR structure of codimension one then condition $(\mathfrak{P}_{n-1})$ is equivalent to the existence of an open neighborhood U of the origin such that at every characteristic point over U the Levi form is not definite. In this case a partial answer was given in [**Mi**], where the existence of *hyperfunction* solutions is proved.

Finally we mention another general class of locally integrable structures that satisfy condition $(\mathfrak{P}_{n-1})$: these are the hypocomplex structures (*cf.* Definition VIII.5.4). For hypocomplex structures it is not still known if $(\mathfrak{P}_{n-1})$ implies the local solvability of (VIII.66). Neverthless, again in this case we can find hyperfunction solutions, as a consequence of more general results proved in [**CorTr**].

Epilogue

In this epilogue we describe briefly some results that are closely connected with the theory and tools developed in previous chapters and have been obtained in recent years but, in spite of their importance, could not be fully treated without increasing too much the size of this book.

1 The similarity principle and applications

In this section we will briefly discuss the first-order equation

$$Lu = Au + B\overline{u} \tag{1}$$

where L is a complex vector field in the plane and A and B are bounded, measurable functions. We will also present two applications of equation (1). The first application concerns uniqueness in the Cauchy problem for a class of semilinear equations. The second application will be to the theory of bending of surfaces.

Equation (1) generalizes the classical elliptic equation

$$\frac{\partial u}{\partial \overline{z}} = Au + B\overline{u} \tag{2}$$

which was investigated by numerous researchers (see for example [**Be**], [**CoHi**], [**Re**], and [**V**]). In the literature, solutions of (2) are called pseudo-analytic functions or generalized analytic functions. Such functions share many properties with holomorphic functions of one variable. These properties follow easily from the similarity principle according to which every continuous solution of (2) has the local form

$$u = \exp\{g\}\, h \tag{3}$$

where h is a holomorphic function and g is Hölder continuous. Thus, for example, the zero set of u is the same as that of h. The similarity principle holds for any elliptic vector field L (where the holomorphy of h is replaced by the condition $Lh = 0$) since any such vector field is a multiple of $\frac{\partial}{\partial \overline{z}}$ in appropriate coordinates. In [**Me2**] Meziani explored the validity of the similarity principle for the following three classes of vector

361

fields:

$$L_k = \frac{\partial}{\partial y} - iy^{2k}\frac{\partial}{\partial x}, \quad K_n = \frac{\partial}{\partial y} - ix^n\frac{\partial}{\partial x}, \quad M = \frac{\partial}{\partial y} - iy\frac{\partial}{\partial x}$$

where k and n are non-negative integers. It was proved in [**Me2**] that the similarity principle is valid for the L_k and K_n (under some vanishing assumption on $B(x, y)$ on the characteristic sets of the vector fields) in the following sense: if w is a continuous solution of

$$Lw = Aw + B\overline{w}$$

where $L \in \{L_k, K_n\}$, then w has the local form $w = \exp\{g\}\,h$ where $Lh = 0$ and g is Hölder continuous. It was also shown in [**Me2**] that this principle does not hold for M. The vector fields L_k and K_n are locally solvable while M is not. With this observation as a point of departure, it was shown in [**BHS**] and [**HdaS**] that a weaker version of the similarity principle is valid for all locally solvable vector fields L. In this weaker version, the functions g and h in the representation $w = \exp\{g\}\,h$ may no longer be continuous. However, this representation was still good enough to yield the uniqueness result mentioned below.

1.1 Application to uniqueness in the Cauchy problem

Let the vector field

$$L = \frac{\partial}{\partial t} + i\sum_{k=1}^{n} b_k(x, t)\frac{\partial}{\partial x_k}$$

satisfy condition $\mathcal{P}$ in some neighborhood $\Omega = \Omega_1 \times (-T, T)$ of the origin in $\mathbb{R}^{n+1}$. Here each b_k is real-valued, of class C^{1+r}, $0 < r < 1$. Let $f(x, t, \zeta, \overline{\zeta})$ be a bounded measurable complex-valued function defined for $(x, t) \in \Omega$, $\zeta \in \mathbb{C}$ satisfying the Lipschitz condition in ζ

$$|f(x, t, \zeta, \overline{\zeta}) - f(x, t, \zeta', \overline{\zeta'})| \le K\,|\zeta - \zeta'|.$$

If L and f are as above, the following result on uniqueness in the Cauchy problem was proved in [**HdaS**] (see also [**BHS**]):

THEOREM 1.1. *Suppose* $u(x, t)$, $w(x, t) \in L^p(\Omega)$, $p \ge 2$, *satisfy* $Lu = f(x, u, \overline{u})$, $Lw = f(x, w, \overline{w})$, *and* $u(x, 0) = w(x, 0)$. *Then* $u \equiv w$ *in a neighborhood of the origin.*

If the coefficients of L are smooth, Theorem 1.1 was proved in [**BHS**] under the weaker assumption that u and w belong to L^p, $p > 1$. These results were proved by applying the similarity principle to the difference $v = u - w$ which in view of the assumptions satisfies an equation of the form $Lv = Av + B\overline{v}$ with A and B bounded. The fact that L satisfies condition $\mathcal{P}$ is then used to reduce matters to a planar situation.

1.2 Application to infinitesimal bendings of surfaces

In a series of papers (see [**Me3**], [**Me4**], and the references therein) Meziani has demonstrated an intimate link between the study of the equation

$$Lu = Au + B\bar{u}$$

(L a planar vector field) and the study of infinitesimal deformations of surfaces with non-negative curvature. Here we will summarize some of the results in [**Me4**] to indicate this link.

Let S be a surface of class C^l, $l > 2$, embedded in $\mathbb{R}^3$ and given by parametric equations as

$$S = \{R(s, t) = (x(s, t), y(s, t), z(s, t)) \in \mathbb{R}^3, \quad (s, t) \in D \subset \mathbb{R}^2\} \tag{4}$$

with D an open subset of $\mathbb{R}^2$. An infinitesimal bending of S is a deformation

$$S_\epsilon = \{R_\epsilon(s, t) = R(s, t) + \epsilon U(s, t), \ (s, t) \in D\}, \quad -\delta < \epsilon < \delta \tag{5}$$

for some $\delta > 0$ and

$$U(s, t) = (\xi(s, t), \eta(s, t), \zeta(s, t)) \tag{6}$$

satisfying

$$dR(s, t) \cdot dU(s, t) = 0 \quad \forall (s, t) \in D. \tag{7}$$

This means that the first fundamental forms of S and S_ϵ satisfy

$$dR_\epsilon^2 = dR^2 + O(\epsilon^2).$$

Note that equation (7) is equivalent to the system of three equations

$$R_s \cdot U_s = 0, \quad R_t \cdot U_t = 0, \quad R_s \cdot U_t + R_t \cdot U_s = 0. \tag{8}$$

Recall that the coefficients of the first fundamental form of S are

$$E = R_s \cdot R_s, \qquad F = R_s \cdot R_t, \qquad G = R_t \cdot R_t \tag{9}$$

and those of the second fundamental form are

$$e = R_{ss} \cdot N, \qquad f = R_{st} \cdot N, \qquad g = R_{tt} \cdot N, \tag{10}$$

where

$$N = \frac{R_s \times R_t}{|R_s \times R_t|}$$

is the unit normal to S. The Gaussian curvature of S is

$$K = \frac{eg - f^2}{EG - F^2}.$$

We will assume that the curvature $K \geq 0$ everywhere on S. The (complex) asymptotic directions of S are given by the quadratic equation

$$\lambda^2 + 2f\lambda + eg = 0.$$

That is,

$$\lambda = -f + i\sqrt{eg - f^2}.$$

Let L be a vector field of asymptotic direction:

$$L = a(s, t)\left(g(s, t)\frac{\partial}{\partial s} + \lambda(s, t)\frac{\partial}{\partial t}\right), \tag{11}$$

where a is any function defined in D. Note that since $K \geq 0$, if $a \neq 0$, then L is an elliptic vector field that degenerates along the set where the curvature $K = 0$.

Let w be the $\mathbb{C}$-valued function defined by

$$w = LR \cdot U \tag{12}$$

where U is as given in (6). In **[Me4]**, the following theorem was proved.

THEOREM 1.2. *With w as in (12) and L as in (11), if $U(s,t)$ is a field of infinitesimal bending for the surface S, then the function w satisfies the equation*

$$CLw = Aw + B\overline{w}$$

where A, B, and C are invariants of the surface S.

1.3 Application to uniqueness in the Cauchy problem in elliptic structures

Let $(\mathcal{M}, \mathcal{V})$ define an elliptic structure. If $u \in L^1_{\mathrm{loc}}(\mathcal{M})$ we shall say that u is an *approximate solution* for the structure $\mathcal{V}$ if for any smooth section L of $\mathcal{V}$, Lu has coefficients belonging to $L^1_{\mathrm{loc}}(\mathcal{M})$ and given any point $p \in \mathcal{M}$, there is an open neighborhood U of p and a constant $M > 0$ such that

$$|Lu| \leq M|u| \quad \text{a.e. in } U.$$

In **[Cor2]** the author established a similarity principle for approximate solutions in the following sense: every approximate solution which belongs to $L^p_{\mathrm{loc}}(\mathcal{M})$ with $p > N = \dim \mathcal{M}$ can locally be written as $u = \exp\{S\}h$, where S is Hölder continuous and h is a solution.

This similarity principle was then used to show that every approximate solution that vanishes on a maximally real submanifold $\mathcal{X}$ vanishes identically in a neighborhood of $\mathcal{X}$.

2 Mizohata structures

The vector field in $\mathbb{R}^2$, where the coordinates are denoted (x, t), given by

$$M = \frac{\partial}{\partial t} - it\frac{\partial}{\partial x} \tag{13}$$

is called the (standard) Mizohata vector field (or operator) after the work of S. Mizohata ([**M**]) who studied the analytic hypoellipticity of a class of related operators of which M is the simplest example. A globally defined first integral of M is the function $Z(x, t) = x + it^2/2$. Notice that $t \mapsto t^2$ fails to be monotone in any neighborhood of a point $(x_0, 0)$, i.e., condition $(\mathcal{P})$ in not satisfied at any point of the x-axis and, as discussed in Chapter IV, fails to be locally solvable at those points. Thus, it is the simplest example of a nonlocally solvable operator and, in fact, its lack of local solvability at points of the x-axis can be proved by ad hoc elementary arguments, as shown by L. Nirenberg ([**N1**]). Off the x-axis, M is elliptic. In his Lectures Notes, Nirenberg constructed a perturbation of the Mizohata operator

$$L = \frac{\partial}{\partial t} - it(1 + \rho(x, t)) \frac{\partial}{\partial x} \tag{14}$$

with $\rho(x, t)$ real-valued and vanishing to infinite order at $t = 0$, which is not locally integrable in any neighborhood of the origin. As a matter of fact, any smooth function u that satisfies the homogeneous equation $Lu = 0$ in a connected open set U that contains the origin must be constant. In spite of the fact that the perturbed vector fields L and M behave differently with respect to local integrability, they have important geometric features in common. We have

(1) M and its conjugate $\overline{M}$ are linearly dependent precisely on the x-axis;
(2) M and $[M, \overline{M}]$ are linearly independent whenever M and $\overline{M}$ are linearly dependent.

These properties are shared by L in a neighborhood of the origin.

DEFINITION 2.1. *A vector field L defined on a connected 2-manifold Ω is called a Mizohata vector field if for a nonempty subset $\Sigma \subset \Omega$ the following holds:*

(1) *L and $\overline{L}$ are linearly dependent precisely on Σ;*
(2) *L and $[L, \overline{L}]$ are linearly independent on Σ.*

We also say that a Mizohata vector field L is of standard type at $p \in \Sigma$ if there exist local coordinates (x, t) in a neighborhood of p in terms of which Σ is given by $\{t = 0\}$ and $\mathcal{L}$ has the form (13). A Mizohata structure $\mathcal{L}$ on Ω is a structure which is locally generated in the neighborhood of every point by a Mizohata vector field.

Notice that (1) means that Σ is the image of the characteristic set $\{(p, \xi) \in T^*(\Omega) : \ell(p, \xi) = 0\}$, ℓ being the symbol of L, under the canonical projection $\Pi : T^*(\Omega) \longrightarrow \Omega$. With this terminology, the vector field (13) is a Mizohata vector field of standard type and (14) is also a Mizohata vector field but not of standard type. Indeed, (14) cannot be of standard type because it is not locally integrable.

Notice that a Mizohata vector field is elliptic on $\Omega \setminus \Sigma$, which is a relatively small set, since an application of the implicit function theorem shows that Σ is an embedded curve. The following question was considered by Treves [**T7**]: when is a Mizohata vector field L of standard type at a given point? Of course, since this is a local question, it is enough to study the case when L is defined in a neighborhood of the origin in $\mathbb{R}^2$. He showed that local coordinates can be found so that L becomes of the form (14) with $\rho(x, t)$ real-valued and vanishing to infinite order at $t = 0$, in other words, every Mizohata vector field has this form locally and it will be of standard type if we are able to take $\rho \equiv 0$. Furthermore, L is of standard type at the origin if and only if it is locally integrable. Then Sjöstrand ([**Sj2**]) took a closer look into the nonlocally integrable case. To describe his results, let us consider the problem of finding a smooth function $Z^+(x, t)$ satisfying $dZ(0, 0) \neq 0$ and $LZ^+ = 0$ on $U^+ = U \cap \{t \geq 0\}$, where U is a small disk centered at the origin. By the proof of Lemma I.13.4, to find Z^+ it is enough to find a smooth function u that satisfies $Lu = t\rho_x$ on U^+. This is, in fact, possible because L satisfies condition $(\mathcal{P})$ for $t > 0$ ([**BH6**]). Similarly, shrinking U if necessary, we can also find a smooth function $Z^-(x, t)$ satisfying $dZ^-(0, 0) \neq 0$ and $LZ^- = 0$ on $U^- = U \cap \{t \leq 0\}$. We can always choose Z^+ and Z^- satisfying $Z^\pm(0, 0) = 0$, $\Im Z_x^\pm(0, 0) = 0$, and $\Re Z_x^\pm(0, 0) > 0$ and we will do so. If we are so lucky that $Z^+(x, 0) = Z^-(x, 0)$, $(x, 0) \in U$, we may patch Z^+ and Z^- to get a single continuous solution Z of $LZ = 0$ on U and it is easy to see using the equation that Z is actually smooth. So the obstruction to the local integrability of L is related to the difficulty of finding a pair (Z^+, Z^-) such that $LZ^\pm = 0$ on $U^\pm$ and $Z^+ = Z^-$ on $U^+ \cap U^-$. Given such a pair, it can be shown that the range of $Z^\pm$ lies on one side of the smooth curve $\{Z^\pm(x, 0)\}$ (in fact, above the curve because $\Re Z_x^\pm(0, 0) > 0$), so let $H^\pm(z)$ be a smooth function defined on the range of $Z^\pm$ and holomorphic in its interior with $H^\pm(0) = 0$, $(H^\pm)'(0) = \Re(H^\pm)'(0) > 0$. Then, $\tilde{Z}^\pm = H^\pm \circ Z^\pm$ satisfies $d\tilde{Z}^\pm(0, 0) \neq 0$ and $L\tilde{Z}^\pm = 0$ on $U^\pm$. By the Riemann mapping theorem we may find H^+ and H^- so that the range of $\tilde{Z}^+$ and $\tilde{Z}^-$ is the upper half-plane. In other words, we may restrict ourselves to consider pairs (Z^+, Z^-) such that $Z^\pm(U^\pm) = \{\Im z \geq 0\}$ and $Z^\pm(U^+ \cap U^-) = \mathbb{R}$. Given such a pair and a smooth function H defined on $\Im z \geq 0$, holomorphic on $\Im z > 0$, real for z real and satisfying $H(0) = 0$, $H'(0) > 0$, a new pair $(Z^+, \tilde{Z}^-) = (Z^+, H \circ Z^-)$ may be considered and L will be locally integrable if $Z^+(x, 0) = \tilde{Z}^-(x, 0)$. It turns out that L is locally integrable if and only if there exists a pair (Z^+, Z^-) such that $H(z) = Z^+ \circ (Z^-)^{-1}(z)$ is holomorphic for $\Im z > 0$ and smooth up to $\Im z = 0$. Since $H(z)$ is real for z real, H has, by the reflection principle, an extension to a holomorphic function. By uniqueness, $H(x + iy)$ is determined by its trace $bH(x) = H(x + i0)$ so it is enough to look at the restrictions $bZ^\pm(x) = Z^\pm(x, 0)$ and check whether $\kappa \doteq bZ^+ \circ (bZ^-)^{-1} : \mathbb{R} \longrightarrow \mathbb{R}$ has a holomorphic extension to a neighborhood. Summing up, to each Mizohata vector field L we have associated an increasing diffeomorphism $\kappa : \mathbb{R} \longrightarrow \mathbb{R}$ such that L is locally integrable if an only if $\kappa = bH$ for some $H \in \mathcal{H}(\mathbb{C})$, i.e., κ has a holomorphic extension. More generally, we may consider the following question: given two Mizohata vector fields L_1, L_2, when are they equivalent in the sense that one can be locally transformed into a multiple of the other by a change of variables? The answer, due to Sjöstrand, can be stated as follows. Consider the associated diffeomorphisms $\kappa_1 = bZ_1^+ \circ (bZ_1^-)^{-1}$ and $\kappa_2 = bZ_2^+ \circ (bZ_2^-)^{-1}$, then L_1 and L_2 are equivalent if and only if there are holomorphic functions, $H_1(z)$, $H_2(z)$, real and increasing for z real, such that $\kappa_1(H_1(x)) = \kappa_2(H_2(x))$, $x \in \mathbb{R}$.

The local questions of standardness and equivalence for Mizohata vector fields have their global counterpart. For instance, it was established in [**BCH**] that a locally standard Mizohata planar vector field has a first integral globally defined in a tubular neighborhood of the characteristic set Σ. The standardness of a particular class of Mizohata structures on the sphere S^2 was proved in [**Ho4**] and Jacobowitz ([**J2**]) studied Mizohata structures on compact surfaces Ω, in particular, he proved that the existence of a first integral is equivalent to the fact that the genus is even. In the case of the sphere, he gave a classification of Mizohata structures in the spirit of Sjöstrand's result, proving in particular the existence of nonstandard Mizohata structures. These topics were developed further by Meziani in [**Me5**] and [**Me6**].

2.1 Mizohata structures in higher-dimensional manifolds

The questions discussed in the previous section admit natural generalization to higher dimension. A formally integrable structure $\mathcal{V}$ defined on a manifold Ω of dimension N is said to be a Mizohata structure if the following holds:

(1) $\mathcal{V}$ has rank $n = N - 1$;
(2) the characteristic set $T^0 = T' \cap T^*(\Omega)$ is not empty;
(3) the Levi form is nondegenerate at every point of $T^0 \backslash \{0\}$.

EXAMPLE 2.2. Denote by $t = (t_1, \ldots, t_n)$ the variables in $\mathbb{R}^n$, $n \geq 1$, and write $t = (t', t'')$, $t' = (t_1, \ldots, t_\nu)$, $t'' = (t_{\nu+1}, \ldots, t_n)$, for some $1 \leq \nu \leq n$. Consider the function $Z(x, t) = x + i(|t'|^2 - |t''|^2)/2$ defined on $\mathbb{R}_x \times \mathbb{R}_t$ and the locally integrable structure $\mathcal{V}$ determined by imposing that T' is spanned by $dZ(x, t)$. Then, $\mathcal{V}$ is spanned by the vector fields

$$M_j = \frac{\partial}{\partial t_j} - i\varepsilon_j t_j \frac{\partial}{\partial x}, \quad j = 1, \ldots, n, \tag{15}$$

with $\varepsilon_j = 1$ for $1 \leq j \leq \nu$ and $\varepsilon_j = -1$ for $\nu + 1 \leq j \leq n$. Then $\mathcal{V}$ is a Mizohata structure such that at every characteristic point its Levi form has ν eigenvalues with one sign and $n - \nu$ eigenvalues with the opposite sign and when this happens we say that $\mathcal{V}$ has type $\{\nu, n - \nu\}$. Thus, we have examples of Mizohata structures with all possible types. Notice that the projection of the characteristic set is the curve $\Sigma = \{t = 0\}$, i.e., the x-axis. A Mizohata structure with type $\{\nu, n - \nu\}$ is standard if for any point lying in the projection of the characteristic set we can choose local coordinates (x, t) so that the vector fields (15) span $\mathcal{V}$ in a neighborhood of that point. Let $\mathcal{V}$ be a Mizohata structure with type $\{\nu, n - \nu\}$. By analogy with the case $n = 1$, it turns out that for any $n \in \mathbb{N}$ and $1 \leq \nu \leq n$ ([**T5**]) it is possible to find local coordinates in a neighborhood U of a generic point p in the projection of Σ such that $x(p) = t(p) = 0$ and $\mathcal{V}$ is generated over U by the vector fields

$$L_j = \frac{\partial}{\partial t_j} - i\varepsilon_j t_j (1 + \rho_j(x, t)) \frac{\partial}{\partial x}, \quad j = 1, \ldots, n, \tag{16}$$

where the functions $\rho_j(x, t)$, $j = 1, \ldots, n$, vanish to infinite order at $t = 0$. In other words, every Mizohata structure has at a given point a contact of infinite order with a standard Mizohata structure of the same type. In particular, if we can take all the functions ρ_j identically zero $\mathcal{V}$ will have a first integral in U and will be standard in U. Conversely, if $\mathcal{V}$ has a first integral it is possible to choose the coordinates so that $\mathcal{V}$ is generated by the vector fields (15).

For the case $\nu = 1$, i.e., if the type is $\{1, n - 1\}$, Treves showed the existence of functions $\rho_j(x, t)$ vanishing to infinite order at $t = 0$ such that the structure $\mathcal{V}$ spanned by (16) is formally integrable (i.e., $[\mathcal{V}, \mathcal{V}] \subset \mathcal{V}$) and not locally integrable. On the other hand, Meziani proved in [**Me7**] that Mizohata structures of all other types $\{\nu, n - \nu\} \neq \{1, n - 1\}$ are always locally integrable. His proof is delicate and beyond the scope of this book: he first constructs first integrals on the connected components of $\{(x, t', t'') \in \mathbb{R}_x \times \mathbb{R}_t : |t'|^2 \neq |t''|^2\}$ which can be 2 (if $n > 2$ and $\nu < n-2$), 3 (if $n > 2$ and $\nu = n-2$), or 4 (if $n = 2$ and $\nu = 0$). When the components are 2 or 4, these first integrals can be patched together to yield a globally defined first integral of class C^1 which, by the hypoellipticity of the structure, is in fact smooth. The possibility of patching together these partially defined first integrals depends on a careful analysis of the holonomy of a certain foliation with leaves of dimension $n-1$ defined by the structure. For the case of type $\{1, n-1\}$ he gives a classification of Mizohata structures analogous to Sjöstrand's result for a single vector field. The local integrability for Mizohata structures of type $\{0, n\}$, $n \geq 3$, was first proved in [**HMa2**], by techniques akin to those used in the proof of Kuranishi's embedding theorem for CR structures ([**Ku1**], [**Ku2**], [**Ak**], [**W2**], [**W3**]), which also fall beyond the scope of this book. The restriction $n \geq 3$ comes from a technical fact: Kuranishi's approach depends on the existence of certain so-called homotopy formulas that do not exist when $n = 2$ ([**HMa3**]). However, the local integrability of Mizohata structures of type $\{0, n\}$ in $\mathbb{R}^{n+1}$, $n \geq 2$, can be proved by elementary methods. Consider a system of n commuting vector fields

$$L_j = \frac{\partial}{\partial t_j} - it_j(1 + \rho_j(x, t))\frac{\partial}{\partial x}, \quad j = 1, \ldots, n. \tag{17}$$

Here a generic point is described by coordinates $(x, t_1, \ldots, t_n)$ and the smooth functions $\rho_j(x, t)$ vanish to infinite order at $\Gamma = \{t = 0\} = \mathbb{R}_x \times \{0\}$. We regard the L_j's as perturbations of the Mizohata vector fields

$$M_j = \frac{\partial}{\partial t_j} - it_j\frac{\partial}{\partial x}, \quad j = 1, \ldots, n.$$

A simple computation using polar coordinates, $t = r\theta$, $r > 0$, $\theta \in S^{n-1}$ shows that the standard Mizohata structure spanned by the M_j's is also spanned on $\mathbb{R}^{n+1} \setminus \Gamma$ by

$$\begin{cases} M = \dfrac{\partial}{\partial r} - ir\dfrac{\partial}{\partial x} \\[2mm] \partial_k = \dfrac{\partial}{\partial \theta_k} \qquad k = 1, \ldots, n-1, \end{cases}$$

with $(\theta_1, \ldots, \theta_{n-1})$ angular variables in S^{n-1}. Then, the change of variables $s = r^2/2$ (x and θ are kept unchanged) takes M into a multiple of the Cauchy–Riemann operator

$$\partial_{\bar{z}} = \frac{1}{2}\left(\frac{\partial}{\partial x} + i\frac{\partial}{\partial s}\right), \quad z = x + is, \ s > 0,$$

and does not change ∂_k. If we perform the same operations on the perturbed system (17) we may find a set of generators of $\mathcal{V}$ in the variables $(x, s, \theta) \in \mathbb{R}_x \times \mathbb{R}_s^+ \times S^{n-1}$ of the form

$$\begin{cases} \tilde{L}_1 = \dfrac{\partial}{\partial \bar{z}} + \sigma_1 \dfrac{\partial}{\partial z} \\[3mm] \tilde{L}_k = \dfrac{\partial}{\partial \theta_{k-1}} + \sigma_k \dfrac{\partial}{\partial z} \quad k=2,\dots,n, \end{cases} \tag{18}$$

with smooth coefficients $\sigma_j(x, s, \theta)$, $j = 1, \dots, n$, that converge to zero when $s \searrow 0$ together with their derivatives of any order. Thus, we may smoothly extend the coefficients σ_j as zero for $\Im z = s \leq 0$ and obtain an elliptic system defined on $\mathbb{C} \times S^{n-1} \simeq \mathbb{R}_x \times \mathbb{R}_s \times S^{n-1}$ which for $\Im z < 0$ has the first integral $z = x + is$. The process that produced an elliptic system starting from a nonelliptic one was obtained by a combination of singular changes of variables (polar coordinates that are singular at the origin of $\mathbb{R}^n_t$ and $s = r^2/2$ which is singular at $r = 0$) and blows up the line $\mathbb{R}_x \times \{t = 0\}$ to the n-manifold $\mathbb{R}_x \times S^{n-1}$. Although we know from Theorem I.12.1 that elliptic structures are locally integrable, applying that result to (18) would only give us a first integral defined in a neighborhood of a point $s = 0$, $x = 0$, $\theta = \theta_0 \in S^{n-1}$ while only a first integral defined for all $\theta \in S^{n-1}$ can give us a first integral defined in a neighborhood of the origin of the original variables (x, t). Let's consider first the case $n = 2$, that is the system of two vector fields

$$\begin{cases} \tilde{L}_1 = \dfrac{\partial}{\partial \bar{z}} + \sigma_1 \dfrac{\partial}{\partial z}, \quad z = x + is \in \mathbb{C}, \\[3mm] \tilde{L}_2 = \dfrac{\partial}{\partial \theta} + \sigma_2 \dfrac{\partial}{\partial z}, \quad 0 \leq \theta \leq 2\pi, \end{cases} \tag{*}$$

defined in $\mathbb{C} \times S^1$, where the $\sigma_j(x, s, \theta)$, $j = 1, 2$, are C^∞ functions, 2π-periodic in θ, and vanish for $s = \Im z \leq 0$. Choose a smooth function $\beta = \beta(x, s, \theta)$ such that $X = \tilde{L}_2 + \beta \tilde{L}_1$ is a real vector. It is easy to check that this is possible if $|\Re \sigma_1| < 1$ (in particular for (x, s) close to the origin). Thus, X is a real generator of the structure $\tilde{\mathcal{V}}_2$ spanned by $\tilde{L}_1$ and $\tilde{L}_2$ for $|x| < 1$, $|s| < \varepsilon$ and $0 \leq \theta \leq 2\pi$. It is clear that $X = \partial/\partial\theta$ for $s \leq 0$, and that the orbits of X stemming from points $(x_0, s_0, 0)$, $s_0 \leq 0$, are the closed circles $\tau \to (x_0, s_0, \tau)$, $0 \leq \tau \leq 2\pi$. Notice also that the component of X along $\partial/\partial\theta$ is 1, i.e.,

$$X = \frac{\partial}{\partial\theta} + \rho_1 \frac{\partial}{\partial x} + \rho_2 \frac{\partial}{\partial s}$$

for some smooth functions ρ_1 and ρ_2 which are 2π-periodic in θ and vanish for $s \leq 0$. Since the commutator $[X, \tilde{L}_1] \in \tilde{\mathcal{V}}_2$ it must be a linear combination of $\tilde{L}_1$ and $\tilde{L}_2$; on the other hand, it does not contain derivations with respect to θ so it has to be proportional to $\tilde{L}_1$. This shows that there exists a smooth function $\lambda = \lambda(x, s, \theta)$ such that

$$[X, \tilde{L}_1] = \lambda \tilde{L}_1. \tag{19}$$

Now pick once and for all a local solution $W(x, s)$ of

$$\tilde{L}_{10} W \doteq \frac{\partial W}{\partial \bar{z}} + \sigma_1(x, s, 0) \frac{\partial W}{\partial z} = 0, \tag{20}$$

$$W_x(0,0) \neq 0.$$

We may assume that in a neighborhood of the origin any other solution $W^\flat(x, s)$ of $\tilde{L}_{10}W^\flat = 0$ is a holomorphic function of W, in fact, W is a local diffeomorphism that takes $\tilde{L}_{10}$ into a multiple of the Cauchy–Riemann operator. Let γ denote the closed orbit of X stemming from $(0, 0, 0)$, given by $\tau \to (0, 0, \tau)$, $0 \leq \tau \leq 2\pi$. We now solve the Cauchy problem

$$XV = 0, \tag{21}$$
$$V(x, s, 0) = W(x, s)$$

in a tubular neighborhood of γ made up of orbits of X. Let us set $U = \tilde{L}_1 V$ and observe that it follows from (19), (20) and (21) that U satisfies the Cauchy problem

$$XU - \lambda U = 0,$$
$$U(x, s, 0) = 0$$

so it must vanish identically in a tubular neighborhood of γ. This proves that dV is orthogonal to $\tilde{\mathcal{V}}_2$ because $\tilde{L}_1$ and X form a basis of $\tilde{\mathcal{V}}_2$. Differentiating (21) with respect to x and setting $s = x = 0$ it is easy to conclude that $V_{x\theta}(0, 0, \theta) = 0$, $0 \leq \theta \leq 2\pi$, so $V_x(0, 0, \theta) = W_x(0, 0)$ is constant, in particular it does not vanish in a neighborhood of γ. This already implies that dV is a generator of the orthogonal of $\tilde{\mathcal{V}}_2$, but we do not know yet that V is 2π-periodic in θ. Since the coefficients of $\tilde{L}_1$ are 2π-periodic we have that $V(x, s, 2\pi)$ satisfies $\tilde{L}_{10}V(x, s, 2\pi) = 0$ and therefore, there exists a holomorphic function G such that $V(x, s, 2\pi) = G \circ V(x, s, 0) = G \circ W(x, s)$ hold for (x, s) in a neighborhood of the origin. But $X = \partial/\partial\theta$ for $s \leq 0$, which implies that $V(x, s, 0) = V(x, s, 2\pi)$ for $s \geq 0$, and it turns out that $G(z) = z$. Thus, $V(x, s, 0) = V(x, s, 2\pi)$ in a neighborhood of $x = s = 0$. This proves that V is well-defined in $\mathbb{C} \times S^1$ and is a first integral globally defined in $\theta \in S^1$ of the system (∗). Furthermore, using $\theta' = \theta$, $x' = \Re V$ and $s' = \Im V$ as local coordinates in a neighborhood of the origin we see that $\partial_{x'} + i\partial_{s'}$ and $\partial_{\theta'}$ generate the same structure as $\tilde{L}_1$, $\tilde{L}_2$.

In the case of the system (18) with $n > 2$ the arguments above can be applied to the first two equations keeping the variables $\theta_2, \ldots, \theta_{n-1}$ as parameters. Thus, after a change of variables $(x, s) \mapsto (x', s')$, we may now assume that $\sigma_1 \equiv 0$ in (18). But then we have $\sigma_k \equiv 0$ for all values of k. Indeed, since $\tilde{L}_1$ commutes with $\tilde{L}_k$, it follows that σ_k, $k \geq 2$, depends holomorphically on z and then has to be identically zero because it vanishes for $\Im z \leq 0$. Thus, all the σ_k are identically zero in the new variables and $z = x + is$ is a first integral of the system. Returning to the original variables (x, s, θ) this shows the existence of a solution $V(x, s, \theta)$ of system (18) for $|x|$ and $|s|$ small and $\theta \in S^{n-1}$ that satisfies $V_x(0, 0, 0) = V_x(0, 0, \theta) \neq 0$. Finally, the function $(x, t) \mapsto V(x, |t|^2/2, \theta(t))$ is smooth in a neighborhood of the origin and its differential spans $\mathcal{V}$.

3 Hypoanalytic structures

Let Ω be a smooth manifold of dimension N. By a *hypoanalytic structure* on Ω (*cf.* [T5]) we mean a collection of pairs $\mathcal{A} = \{(U_\ell, Z_\ell)\}$, with U_ℓ an open subset of Ω and

$Z_\ell = (Z_{\ell,1}, \dots, Z_{\ell,m}) : U_\ell \to \mathbb{C}^m$ a smooth map, where $1 \le m \le N$ is independent of ℓ, such that the following conditions are satisfied:

$(H)_1$ $\{U_\ell\}$ is an open covering of Ω;
$(H)_2$ $dZ_{\ell,1}, \dots, dZ_{\ell,m}$ are $\mathbb{C}$-linearly independent at each point of U_ℓ;
$(H)_3$ if $\ell \ne \ell'$ and if $p \in U_\ell \cap U_{\ell'}$ there exists a biholomorphism $F_{\ell',p}^\ell$ of an open neighborhood of $Z_\ell(p)$ in $\mathbb{C}^m$ onto one of $Z_{\ell'}(p)$ such that $Z_{\ell'} = F_{\ell',p}^\ell \circ Z_\ell$ in a neighborhood of p in $U_\ell \cap U_{\ell'}$.

A complex-valued function f defined on an open subset U of Ω is called *hypoanalytic* if in a neighborhood of any point p of U we can write $f = h_\ell \circ Z_\ell$, where ℓ is such that $p \in U_\ell$ and h_ℓ is a holomorphic function in a neighborhood of $Z_\ell(p)$ in $\mathbb{C}^m$. By a *hypoanalytic chart* we shall mean a pair (U, Z) where $U \subset X$ is open, $Z = (Z_1, \dots, Z_m) : U \to \mathbb{C}^m$ has hypoanalytic components and $dZ_1 \wedge \dots \wedge dZ_m \ne 0$ in U.

If $\mathcal{A} = \{(U_\ell, Z_\ell)\}$ is a hypoanalytic structure on Ω and if $\Omega_\bullet \subset \Omega$ is open then we can induce a hypoanalytic structure $\mathcal{A}_{\Omega_\bullet}$ by the rule

$$\mathcal{A}_{\Omega_\bullet} = \{(U_\ell \cap \Omega_\bullet, Z_\ell|_{U_\ell \cap \Omega_\bullet})\}.$$

To each hypoanalytic structure $\mathcal{A} = \{(U_\ell, Z_\ell)\}$ on Ω we can canonically associate a locally integrable structure $\mathcal{V}$ on Ω in the following way: for each ℓ its orthogonal on U_ℓ is defined by

$$T'|_{U_\ell} = \text{span}\{dZ_{\ell,1}, \dots, dZ_{\ell,m}\}.$$

By properties $(H)_1$, $(H)_2$, and $(H)_3$ it follows that T' is indeed a subbundle of $\mathbb{C}T^*\Omega$ of rank m.

Notice however that two different hypoanalytic structures can define the same locally integrable structure. Indeed, to give an example it suffices to take $\Omega = \mathbb{R}$ and consider the hypoanalytic structure $\{(\mathbb{R}, \text{Id})\}$, where $\text{Id}(x) = x$, and the hypoanalytic structure $\{(\mathbb{R}, f)\}$, where $f : \mathbb{R} \to \mathbb{R}$ is smooth but *not* real-analytic and $f' \ne 0$ at each point.

By a *hypoanalytic manifold* we shall mean a pair $(\Omega, \mathcal{A})$, where Ω is a smooth manifold and $\mathcal{A}$ is a hypoanalytic structure on Ω. Notice that if $(\Omega, \mathcal{A})$ is a hypoanalytic manifold, endowed with the hypoanalytic structure $\mathcal{A} = \{(U_\ell, Z_\ell)\}$, if Ω' is another smooth manifold and if $f : \Omega' \to \Omega$ is a smooth submersion, then we can pull back the hypoanalytic structure $\mathcal{A}$ to a hypoanalytic structure $f^*\mathcal{A}$ on Ω' by defining

$$f^*\mathcal{A} = \{(f^{-1}(U_\ell), Z_\ell \circ f)\}.$$

Finally we shall say that two hypoanalytic manifolds $(\Omega' \mathcal{A}')$ and $(\Omega, \mathcal{A})$ are equivalent if there is a smooth diffeomorphism $f : \Omega' \to \Omega$ such that $f^*\mathcal{A} = \mathcal{A}'$.

4 The local model for a hypoanalytic manifold

Let $N \ge 1$ and write $N = m + n$. The variable in $\mathbb{C}^N = \mathbb{C}^m \times \mathbb{C}^n$ will be denoted by (z, z') with $z = (z_1, \dots, z_m)$, $z' = (z_1', \dots, z_n')$. In this space we consider the hypoanalytic structure defined by $\mathcal{A}^{\bullet} = \{(\mathbb{C}^N, (z_1, \ldots, z_m))\}$. The corresponding hypoanalytic functions are just the holomorphic functions of z that are locally independent of z'.

Let Ω and $\{(U_\ell, Z_\ell)\}$ be as in Section 3. An arbitrary point p of Ω has an open neighborhood U_p in which there are defined hypoanalytic functions $Z_1, \ldots, Z_m$ and a complementary number of C^∞ functions $Z'_1, \ldots, Z'_n$, with $m + n = N$, such that

$$dZ_1 \wedge \cdots \wedge dZ_m \wedge dZ'_1 \wedge \cdots \wedge dZ'_n \neq 0 \text{ at } p.$$

Possibly after contracting U_p about p we may assume that

$$\lambda : (Z, Z') \doteq (Z_1, \ldots, Z_m, Z'_1, \ldots, Z'_n)$$

is a smooth diffeomorphism of U_p onto a smooth, *maximally real* submanifold Σ_p of $\mathbb{C}^m \times \mathbb{C}^n$. We refer to the triplet (U_p, Z, Z') as an *extended hypoanalytic chart*.

The hypoanalytic $\mathcal{A}^{\bullet}$ induces a hypoanalytic structure $\mathcal{A}^{\#}$ on Σ_p, simply by setting

$$\mathcal{A}^{\#} = \{(\Sigma_p, (z_1|_{\Sigma_p}, \ldots, z_m|_{\Sigma_p}))\},$$

and it is easily seen that

$$\mathcal{A}_{U_p} = \lambda^* \mathcal{A}^{\#}. \tag{22}$$

This remark is crucial for what follows.

5 The sheaf of hyperfunction solutions on a hypoanalytic manifold

The sheaf of hyperfunctions can be introduced on any real-analytic manifold. This is a fundamental result, due to M. Sato ([Sa]). It is also possible to extend such a concept to hypoanalytic manifolds where no real-analyticity is required, but in order to obtain an invariant meaning, we must restrict ourselves to the hyperfunctions that are solutions in some sense. We give now a brief description of this theory.

It is a consequence of a result due to Harvey ([Ha]) that over any maximally real submanifold $\mathcal{M}$ of $\mathbb{C}^N$ it is also possible to define the sheaf of hyperfunctions $\mathcal{B}_{\mathcal{M}}$. Moreover, the following description is valid: given $q \in \mathcal{M}$ there is an open neighborhood V of q in $\mathcal{M}$ such that the following is true: if $W \subset\subset V$ is open then

$$\mathcal{B}_{\mathcal{M}}(W) = \mathcal{O}'(\overline{W})/\mathcal{O}'(\partial W). \tag{23}$$

Here the boundary of W is taken in $\mathcal{M}$ and for a compact subset K of $\mathbb{C}^N$ we are denoting by $\mathcal{O}'(K)$ the space of analytic functionals of $\mathbb{C}^N$ carried by K.

We return to the discussion of Section 4. We fix $p \in \mathcal{M}$ and Σ_p as described. Since the holomorphic derivatives act on $\mathcal{O}'(K)$ by transposition we can consider the space of hyperfunctions u on Σ_p which satisfy the system

$$\frac{\partial u}{\partial z'_j} = 0, \quad j = 1, \ldots, n. \tag{24}$$

The main result presented in the monograph [CorT2] states that the sheaf of these hyperfunctions on Σ_p, when pulled back to U_p, gives rise to a *well-defined sheaf* Sol_Ω

on Ω, *which is furthermore a hypoanalytic invariant.* The proof of this fundamental result relies on (22). We call Sol_Ω the sheaf of germs of *hyperfunction solutions* on Ω. This sheaf contains, as a subsheaf, the sheaf of germs of distribution solutions with respect to the associated locally integrable structure $\mathcal{V}$. Moreover, if Ω and the maps Z_ℓ are real-analytic then Sol_Ω equals the sheaf of hyperfunctions on Ω that are annihilated by the (real-analytic) sections of $\mathcal{V}$.

Many of the basic results that were proved in this book remain valid within this more general concept of solution, as for instance the propagation of the support of solutions by the orbits of the underlying structure and the uniqueness in the Cauchy problem ([**CorT2**]). Another important feature is that a certain class of infinite-order operators, which are local in the sense of Sato, act as endomorphisms of Sol_Ω ([**Cor1**]). It can then be proved that every hyperfunction solution can be obtained, locally, as the action of one such operator on a *smooth solution* and then, as a consequence, a version of the approximation formula for hyperfunction solutions can be derived (*cf.* [**Cor1**]).

Appendix A: Hardy space lemmas

A.1 Multipliers in h^1

We recall that ω is a modulus of continuity if $\omega : [0, \infty) \longrightarrow \mathbb{R}^+$ is continuous, increasing, $\omega(0) = 0$ and $\omega(2t) \leq C\omega(t), 0 < t < 1$. A modulus of continuity determines the Banach space $C_\omega(\mathbb{R})$ of bounded continuous functions $f : \mathbb{R} \longrightarrow \mathbb{C}$ such that

$$|f|_{C_\omega} \doteq \sup_{x \neq y} \frac{|f(y) - f(x)|}{\omega(|x - y|)} < \infty,$$

equipped with the norm $\|f\|_{C_\omega} = \|f\|_{L^\infty} + |f|_{C_\omega}$. Note that C_ω is only determined by the behavior of $\omega(t)$ for values of t close to 0. Consider a modulus of continuity $\omega(t)$ that satisfies

$$\frac{1}{h^n} \int_0^h \omega(t) t^{n-1} \, dt \leq K \left(1 + \log \frac{1}{h} \right)^{-1}, \quad 0 < h < 1, \tag{A.1}$$

and the corresponding space $C_\omega(\mathbb{R}^n)$.

LEMMA A.1.1. *Let* $b \in C_\omega(\mathbb{R}^n)$ *and* $f \in h^1(\mathbb{R}^n)$. *Then* $bf \in h^1(\mathbb{R}^n)$ *and there exists* $C > 0$ *such that*

$$\|bf\|_{h^1} \leq C\|b\|_{C_\omega}\|f\|_{h^1}, \quad b \in C_\omega(\mathbb{R}^n), f \in h^1(\mathbb{R}^n).$$

PROOF. Let $b(x) \in C_\omega$. It is enough to check that $\|bf\| \leq C\|b\|_{C_\omega}$ for every h^1-atom a with C an absolute constant. This fact is obvious for atoms supported in balls B with radius $\rho \geq 1$ without moment condition because b is bounded so $ba/\|b\|_{L^\infty}$ is again an atom without moment condition. If $B = B(x_0, \rho), \rho < 1$, we may write $a(x)b(x) = b(x_0)a(x) + (b(x) - b(x_0))a(x) = \beta_1(x) + \beta_2(x)$. Then $\beta_1(x)/\|b\|_{L^\infty}$ is again an atom while $\beta_2(x)$ is supported in B and satisfies

$$\|\beta_2\|_{L^\infty} \leq 2\|b\|_{L^\infty}\|a\|_{L^\infty} \leq \frac{C}{\rho^n},$$

$$\|\beta_2\|_{L^1} \leq C\|a\|_{L^\infty} \int_B \omega(|x - x_0|) \, dx \leq \frac{C'}{(1 + |\log \rho|)}.$$

374

We wish to conclude that $\|m_\Phi \beta_2\|_{L^1} < \infty$. Let $B^* = B(x_0, 2\rho)$. Since $m_\Phi \beta_2(x) \leq \|\beta_2\|_{L^\infty}$, we have

$$J_1 = \int_{B^*} m_\Phi \beta_2(x)\, dx \leq C|B^*|\rho^{-n} \leq C'.$$

It remains to estimate

$$J_2 = \int_{\mathbb{R} \setminus B^*} m_\Phi \beta_2(x)\, dx = \int_{2\rho \leq |x - x_0| \leq 2} m_\Phi \beta_2(x)\, dx \tag{A.2}$$

(observe that $m_\Phi \beta_2$ is supported in $B(x_0, 2)$ because supp $\Phi \subset B(0, 1)$). If $0 < \varepsilon < 1$ and $\Phi_\varepsilon * \beta_2(x) \neq 0$ for some $|x - x_0| \geq 2\rho$ it is easy to conclude that $\varepsilon \geq |x - x_0|/2$, which implies

$$|\Phi_\varepsilon * \beta_2(x)| \leq \left| \int \Phi_\varepsilon(y) \beta_2(x - y)\, dy \right| \leq \frac{C\|\beta_2\|_{L^1}}{\varepsilon^n} \leq \frac{C'|x - x_0|^{-n}}{(1 + |\log \rho|)}$$

so

$$m_\Phi \beta_2(x) \leq \frac{C'}{|x - x_0|^n (1 + |\log \rho|)} \qquad \text{for} \quad |x - x_0| \geq 2\rho. \tag{A.3}$$

It follows from (A.2) and (A.3) that

$$J_2 \leq \int_{2\rho \leq |x - x_0| \leq 2} \frac{C'}{|x - x_0|^n (1 + |\log \rho|)}\, dx \leq C''$$

which leads to

$$\|ba\|_{h^1} \leq \|\beta_1\|_{h^1} + \|\beta_2\|_{h^1} \leq C_1 + J_1 + J_2 \leq C_2.$$

Inspection of the proof shows that C_2 may be estimated by $C\|b\|_{C_\omega}$. $\qquad \square$

EXAMPLE A.1.2. Suppose that a modulus of continuity $\omega(t)$ satisfies:

$$\omega(t)/t^n \quad \text{is a decreasing function of } t \tag{A.4}$$

and

$$D = \int_0^1 \frac{\omega(t)}{t}\, dt < \infty. \tag{A.5}$$

A short and elegant argument shows (*cf.* [**Ta**], page 25) that under these conditions $h^1(\mathbb{R}^n)$ is stable under multiplication by elements of $C_\omega(\mathbb{R}^n)$. On the other hand, (A.5) alone already implies that

$$\omega(h) \log \frac{1}{h} = \int_h^1 \frac{\omega(h)}{t}\, dt \leq \int_h^1 \frac{\omega(t)}{t}\, dt \leq D, \qquad 0 < h < 1,$$

which keeping in mind the obvious estimate

$$\frac{1}{h^n} \int_0^h \omega(t) t^{n-1}\, dt \leq \frac{\omega(h)}{n}$$

shows that the modulus of continuity ω satisfies (A.1) and Lemma A.1.1 can be applied, proving the mentioned stability of $h^1(\mathbb{R}^n)$ under multiplication by elements of $C_\omega(\mathbb{R}^n)$.

Consider now a modulus of continuity $\omega(t)$ such that

$$\omega(t) = \frac{1 - n \log t}{\log^2 t}, \quad \text{for } 0 < t < 1/2.$$

Since $\omega(t) \geq |\log t|^{-1}$ it follows that $\int_0^{1/2} (\omega(t)/t)\,dt = \infty$ and the Dini condition (A.5) is not satisfied. On the other hand,

$$\frac{1}{h^n} \int_0^h \omega(t) t^{n-1}\,dt = \left(\log \frac{1}{h}\right)^{-1} \approx \left(1 + \log \frac{1}{h}\right)^{-1}, \quad \text{as } h \to 0,$$

so criterion (A.1) is satisfied. This shows that (A.5) is strictly more stringent than (A.1).

A.2 Commutators

We consider now a bounded smooth function $\psi(\xi)$, $\xi \in \mathbb{R}$, such that

$$\left|\frac{d^k}{dx^k} \psi(\xi)\right| \leq C_k \frac{1}{(1 + |\xi|)^k}, \qquad \xi \in \mathbb{R}, \quad k = 0, 1, 2, \ldots$$

Then $\psi(\xi)$ is a symbol of order zero and defines the pseudo-differential operator

$$\psi(D)u(x) = \frac{1}{2\pi} \int_{\mathbb{R}} e^{ix\xi} \psi(\xi) \widehat{u}(\xi)\,d\xi, \quad u \in \mathcal{S}(\mathbb{R}).$$

In particular, $\psi(D)$ is bounded in $h^1(\mathbb{R})$. The Schwartz kernel of $\psi(D)$ is the tempered distribution $k(x - y)$ defined by $\widehat{k}(\xi) = \psi(\xi)$ which is smooth outside the diagonal $x \neq y$. Moreover, $k(x - y)$ may be expressed as

$$k(x - y) = \lim_{\varepsilon \to 0} \frac{1}{2\pi} \int e^{i(x-y)\xi - \varepsilon|\xi|^2} \psi(\xi)\,d\xi = \lim_{\varepsilon \to 0} k_\varepsilon(x - y),$$

where the limit holds both in the sense of $\mathcal{S}'$ and pointwise for $x \neq y$. Furthermore, the approximating kernels $k_\varepsilon(x - y)$ satisfy uniformly in $0 < \varepsilon < 1$ the pointwise estimates

$$|k_\varepsilon(x - y)| \leq \frac{C_j}{|x - y|^j}, \quad j = 1, 2, \ldots \tag{A.6}$$

which of course also hold for $k(x - y)$ itself when $x \neq y$.

We consider a function $b(x)$ of class $C^{1+\sigma}$, $0 < \sigma < 1$, and wish to prove that the commutator $[\psi(D), b\partial_x]$ is bounded in $h^1(\mathbb{R})$. A simple standard computation that includes an integration by parts gives

$$[\psi(D), b\partial_x]u(x) = \int k'(x - y)(b(y) - b(x))u(y)\,dy - \psi(D)(b'u)$$

where the integral should be interpreted as the pairing

$$\langle < k'(x - \cdot)(b(\cdot) - b(x)), u(\cdot) \rangle$$

between a distribution depending on the parameter x and a test function u. Since multiplication by b' is bounded in $h^1(\mathbb{R})$ with norm controlled by $\|b'\|_{C^\sigma}$, we need only worry about the remaining integral term that can be rewritten as

$$Tu(x) = \int (y - x)k'(x - y)\frac{b(x) - b(y)}{x - y}u(y)\,dy$$

$$= \int k_1(x - y)\beta(x, y)u(y)\,dy \tag{A.7}$$

where

$$\beta(x, y) = \int_0^1 b'(\tau x + (1 - \tau)y)\,d\tau \quad \text{and} \quad k_1(x) = -xk'(x).$$

Observe that $\beta \in C^\sigma(\mathbb{R}^2)$.

LEMMA A.2.1. *Assume T is given by* (A.7) *with kernel*

$$K(x, y) = k_1(x - y)\beta(x, y).$$

Then T is bounded in $h^1(\mathbb{R})$.

PROOF. It follows that $\widehat{k_1}(\xi) = (\xi k(\xi))' = \psi(\xi) + \xi\psi'(\xi)$. In other words, $\widehat{k_1}(\xi) = \psi_1(\xi)$ is a symbol of order zero and T has kernel $k_1(x - y)\beta(x, y)$. We may write $\beta(x, y) = b'(x) + |x - y|^\sigma r(x, y)$ with $r(x, y) \in L^\infty(\mathbb{R}^2)$ so

$$Tu(x) = b'(x)\psi_1(D)u(x) + \int k_1(x - y)|x - y|^\sigma r(x, y)u(y)\,dy,$$

$$= T_1u(x) + T_2u(x).$$

The first operator T_1 is obviously bounded in h^1 because it is the composite of $\psi_1(D)$ with multiplication by a C^σ function. To analyze T_2 we check—writing $k_1 = \lim_{\varepsilon \to 0} k_{1,\varepsilon}$ and using (A.6) for $k_{1,\varepsilon}$—that its Schwartz kernel is a locally integrable distribution given by the integrable function $k_2(x, y) = k_1(x - y)|x - y|^\sigma r(x, y)$. Hence, $|k_2(x, y)| \le C_1|k_1(x - y)|\,|x - y|^\sigma = k_3(x - y)$. Observe that $k_3(x) \le C\min(|x|^{\sigma-1}, |x|^{-2})$ so $k_3 \in L^1(\mathbb{R})$. We will now show that

$$m_\Phi k_3(x) = \sup_{0 < \varepsilon < 1} |\Phi_\varepsilon * k_3(x)| \in L^1(\mathbb{R}),$$

where $\Phi \ge 0 \in C_c^\infty([-1/2, 1/2])$, $\int \Phi dz = 1$, $\Phi_\varepsilon(x) = \varepsilon^{-1}\Phi(x/\varepsilon)$. Since $m_\Phi k_3$ is pointwise majorized by the restricted Hardy–Littlewood maximal function

$$mk_3(x) = \sup_{0 < \varepsilon < 1}\frac{1}{2\varepsilon}\int_{x-\varepsilon}^{x+\varepsilon} k_3(t)\,dt$$

we start by observing that

$$\sup_{0 < \varepsilon < 1}\frac{1}{2\varepsilon}\int_{x-\varepsilon}^{x+\varepsilon} |t|^{\sigma-1}\,dt \le \frac{|x|^{\sigma-1}}{\sigma}. \tag{A.8}$$

In doing so we may assume that $x > 0$. If $0 < \varepsilon \le x$ we have

$$\frac{1}{2\varepsilon}\int_{x-\varepsilon}^{x+\varepsilon} |t|^{\sigma-1}\,dt = \frac{(x+\varepsilon)^\sigma - (x-\varepsilon)^\sigma}{2\varepsilon\sigma} \le \frac{(x+\varepsilon)^{\sigma-1}}{\sigma} \le \frac{x^{\sigma-1}}{\sigma}$$

where we have used the elementary inequality

$$\frac{b^\sigma - a^\sigma}{b - a} \leq b^{\sigma - 1}, \quad 0 \leq a < b, \quad 0 < \sigma < 1.$$

Similarly, if $0 < x < \varepsilon$,

$$\frac{1}{2\varepsilon} \int_{x-\varepsilon}^{x+\varepsilon} |t|^{\sigma-1} \, dt = \frac{(x+\varepsilon)^\sigma + (x-\varepsilon)^\sigma}{2\varepsilon\sigma} \leq \frac{(x+\varepsilon)^{\sigma-1}}{\sigma} \leq \frac{x^{\sigma-1}}{\sigma}.$$

This proves (A.8). Thus,

$$m_\Phi k_3(x) \leq C \, m k_3(x) \leq C' \, |x|^{\sigma-1}$$

which shows that $m_\Phi k_3$ is locally integrable. For large $|x|$ the inequality $k_3(x) \leq C|x|^{-2}$ easily implies $m_\Phi k_3(x) \leq C|x|^{-2}$ and we conclude that $m_\Phi k_3 \in L^1$. Finally, we see that

$$|\Phi_\varepsilon * T_2 u(x)| \leq \Phi_\varepsilon * k_3 * |u|(x) \leq m_\Phi k_3 * |u|(x)$$

so $m_\Phi T_2 u(x) \leq m_\Phi k_3 * |u|(x)$, which implies that $\|T_2 u\|_{h^1} \leq C\|u\|_{L^1} \leq C\|u\|_{h^1}$. This proves that $T = T_1 + T_2$ is bounded in $h^1(\mathbb{R})$. □

Summing up, we have proved:

PROPOSITION A.2.2. *If $\psi(\xi)$, $\xi \in \mathbb{R}$, is a smooth symbol of order 0 and $b(x) \in C^{1+\sigma}(\mathbb{R})$, $0 < \sigma < 1$, the commutator*

$$[\psi(D), b\partial_x]$$

is bounded in $h^1(\mathbb{R})$.

A.3 Change of variables

Consider a diffeomorphism $F : \mathbb{R}^n \to \mathbb{R}^n$ of class C^1, with Jacobian F' such that for some $K \geq 1$

$$K^{-1}|x - y| \leq |F(x) - F(y)| \leq K|x - y|, \quad x, y \in \mathbb{R}^n. \tag{A.9}$$

Write $H = F^{-1}$, denote by H' the Jacobian matrix of H, and assume that

$$\det H' \in C_\omega, \tag{A.10}$$

where the modulus of continuity $\omega(t)$ satisfies

$$\frac{1}{h^n} \int_0^h \omega(t) t^{n-1} \, dt \leq K \left(1 + \log \frac{1}{h} \right)^{-1}, \quad 0 < h < 1.$$

Notice that if F is a diffeomorphism of Hölder class $C^{1+\varepsilon}$, $\varepsilon > 0$, then (A.9) and (A.10) hold.

PROPOSITION A.3.1 (S. Chanillo, [Ch2]). *If F satisfies (A.9) and (A.10), the map $h^1(\mathbb{R}^n) \ni g \mapsto g \circ F$ is bounded in $h^1(\mathbb{R}^n)$.*

The main step in the proof of the proposition is

LEMMA A.3.2. *Let* $H : \mathbb{R}^n \to \mathbb{R}^n$ *be a homeomorphism such that for some* $K \geq 1$

$$K^{-1}|x-y| \leq |H(x) - H(y)| \leq K|x-y|, \quad x, y \in \mathbb{R}^n. \tag{A.11}$$

Let $\Phi \in C_c^\infty(B(0,1))$, $\Phi_t(x) = t^{-n}\Phi(x/t)$, $u \in H^1(\mathbb{R}^n)$ *and set*

$$U(x,t) = \int \Phi_t(H(x) - H(z))u(z)\,\mathrm{d}z, \quad 0 < t < 1,$$

$$U^*(x) = \sup_{0<t<1} |U(x,t)|.$$

Then there exists a constant $C > 0$ *depending on the dimension n, on K and on* Φ *but not on u such that*

$$\int U^*(x)\,\mathrm{d}x \leq C\|u\|_{h^1}. \tag{A.12}$$

PROOF. In view of the atomic decomposition it is enough to prove (A.12) when $u(x)$ is an atom, that we denote by $a(x)$. We must show that if $a(x)$ is an h^1-atom and

$$A(x,t) = \int \Phi_t(H(x) - H(z))a(z)\,\mathrm{d}z, \quad 0 < t < 1,$$

$$A^*(x) = \sup_{0<t<1} |A(x,t)|,$$

then $\|A^*\|_{L^1} \leq C$ with C independent of $a(x)$. Let $a(x)$ be an atom supported in ball $B = B(z_0, r)$ such that $\|a\|_{L^\infty} \leq |B|^{-1}$. Note that in view of (A.11) and the hypothesis on Φ

$$|x-z| \geq Kt \implies |H(x) - H(z)| \geq t \implies \Phi_t(H(x) - H(z)) = 0$$

for $0 < t < 1$ so

$$|A(x,t)| \leq \|a\|_{L^\infty}\|\Phi\|_{L^\infty} \int_{|z-x|<Kt} \frac{1}{t^n}\,\mathrm{d}z \leq \frac{C}{r^n},$$

showing that

$$|A^*(x)| \leq \frac{C}{r^n}. \tag{A.13}$$

If we write $B^* = B(z_0, 2r)$ we see right away that

$$\int_{B^*} A^*(x)\,\mathrm{d}x \leq C$$

and we need only concern ourselves with the integral

$$\int_{\mathbb{R}^n \setminus B^*} A^*(x)\,\mathrm{d}x.$$

We first consider the case $0 < r < 1$ so that $a(x)$ has vanishing mean $\int a(x)\,\mathrm{d}x = 0$. We will initially show that $A(x,t) = 0$ if $x \notin B^*$ and $2Kt \leq |x - z_0|$. Since $|x - z_0| \geq 2r$ and $|z - z_0| \leq r$ implies that $|z - z_0| \leq |x - z_0|/2$ we obtain from the triangular property that $|x - z| \geq |x - z_0|/2$ if $|x - z_0| \geq 2r$ and $|z - z_0| \leq r$. Thus, $2Kt \leq |x - z_0| \leq 2|x - z| \leq 2K|H(x) - H(z)|$. This implies that $|H(x) - H(z)|/t \geq 1$ so $\Phi_t(H(x) - H(z))a(z) = 0$.

Hence, $A(x, t) = 0$ if $|x - z_0| \geq 2r$ and $t \leq |x - z_0|/(2K)$ and when we estimate $A^*(x)$ on $\mathbb{R}^n \backslash B^*$ we may take the supremum of $|A(x, t)|$ for t in the range $|x - z_0|/(2K) \leq t < 1$. We may write

$$
\begin{aligned}
|A(x, t)| &= \left| \int \left(\Phi_t(H(x) - H(z)) - \Phi_t(H(x) - H(z_0)) \right) a(z) \, dz \right| \\
&\leq \frac{C \|a\|_{L^\infty}}{t^{n+1}} \int_{B(z_0, r)} |H(z) - H(z_0)| \, dz \\
&\leq \frac{Cr}{|x - z_0|^{n+1}}
\end{aligned}
$$

to conclude that

$$
A^*(x) \leq \frac{Cr}{|x - z_0|^{n+1}} \quad \text{for} \quad x \notin B^*
$$

and

$$
\int_{B^*} A^*(x) \, dx \leq C.
$$

Assume now that $r \geq 1$. Then, for $|z - z_0| \leq r$ and $|x - z_0| \geq (K + 1)r$ we have $|x - z| \geq (K + 1)r - r = Kr$ so

$$
|H(x) - H(z)| \geq r \geq 1 \quad \text{and} \quad \Phi_t(H(x) - H(z)) = 0.
$$

This shows that supp $A(x, t) \subset B(z_0, (K + 1)r)$ and also supp $A^* \subset B(z_0, (K + 1)r)$. Hence, we get

$$
\|A^*\|_{L^1} \leq \|A^*\|_{L^\infty} |\text{supp } A^*| \leq C,
$$

where we have used (A.13). $\qquad \square$

Proof of Proposition A.3.1. Let $g \in h^1(\mathbb{R}^n)$. Choose some test function $0 \leq \Phi \in C_c^\infty(B(0, 1))$ with $\int \Phi(x) \, dx = 1$ and set $v = g \circ F$. We must show that $v^*(x) = \sup_{0 < t < 1} |\Phi_t * [g \circ F](x)|$ satisfies $\|v^*\|_{L^1} \leq C \|g\|_{h^1}$. Since

$$
\int v^*(y) \, dy = \int v^* \circ H(x) |\det H'(x)| \, dx \leq C \|v^* \circ H\|_{L^1},
$$

it is enough to estimate

$$
\|v^* \circ H\|_{L^1} = \int \sup_{0 < t < 1} \left| \int \Phi_t(H(x) - z) g(F(z)) \, dz \right| \, dx
$$

which after the change of variables $z = H(y)$ may be written as

$$
I = \int \sup_{0 < t < 1} \left| \int \Phi_t(H(x) - H(y)) g(y) |\det H'(y)| \, dy \right| \, dx
$$

because $H = F^{-1}$. Notice that $u(y) = \pm g(y) \det H'(y) \in h^1(\mathbb{R}^n)$ by Lemma A.1.1 and (A.10); furthermore, $\|u\|_{h^1} \leq C \|g\|_{h^1}$. Using Lemma A.3.2 we get $I \leq C \|u\|_{h^1} \leq C' \|g\|_{h^1}$, as we wished to prove. $\qquad \square$

Bibliography

[Ak]   T. Akahori, *A new approach to the local embedding theorem of CR-structures for n ≥ 4*, Memoirs of the AMS **366** (1987), Providence, R.I.

[AFN]   A. Andreotti, G. Fredricks and M. Nacinovich, *On the absence of Poincaré lemma in tangential Cauchy–Riemann complexes,* Ann. Scuola Norm. Sup. Pisa **8** (1981), 365–404.

[AH1]   A. Andreotti and C. D. Hill, *Complex characteristic coordinates and tangential Cauchy–Riemann equations,* Ann. Scuola Norm. Sup. Pisa **26** (1972), 299–324.

[AH2]   A. Andreotti and C. D. Hill, *E. E. Levi convexity and the Hans Lewy problem, I and II*, Ann. Scuola Norm. Sup. Pisa **26** (1972), 325–363, 747–806.

[A]   C. Asano, *On the C^∞ wave-front set of solutions of first-order nonlinear pde's*, Proceedings of the AMS **123** (1995), 3009–3019.

[BE]   T. N. Bailey and M. G. Eastwood, *Zero-energy Fields on Real Projective Space,* Geometriae Dedicata **67** (1997), 245–258.

[BEGM]   T. N. Bailey, M. G. Eastwood, A. R. Gover and L. J. Mason, *The Funk Transform as a Penrose Transform,* Mathematical Proceedings of the Cambridge Philosophical Society **125** (1999), 67–81.

[BCT]   M. S. Baouendi, C. H. Chang and F. Treves, *Microlocal hypo-analyticity and extension of CR functions,* J. Differential Geom. **18** (1983), 331–391.

[BER]   M. S. Baouendi, P. Ebenfelt and L. P. Rothschild, *Real Submanifolds in Complex Space and their Mappings,* Princeton Math. Ser. **47** (1999), Princeton Univ. Press.

[BR]   M. S. Baouendi and L. P. Rothschild, *Cauchy–Riemann functions on manifolds of higher codimension in complex space,* Invent. Math. **101** (1990), 45–56.

[BRT]   M. S. Baouendi, L. P. Rothschild and F. Treves, *CR structures with group action and extendability of CR functions,* Invent. Math. **82** (1985), 359–396.

[BT1]   M. S. Baouendi and F. Treves, *A property of the functions and distributions annihilated by a locally integrable system of complex vector fields,* Ann. of Math. **113** (1981), 387–421.

[BT2]   M. S. Baouendi and F. Treves, *A local constancy principle for the solutions of certain overdetermined systems of first order linear partial differential equations,* Math. Analysis and Applications, Part A, Advances in Math. Supplementary Studies **7A** (1981), 245–262.

[BT3]   M. S. Baouendi and F. Treves, *About the holomorphic extension of CR func-*
        tions on real hypersurfaces in complex space, Duke Math. J. **51** (1984),
        77–107.

[BT4]   M. S. Baouendi and F. Treves, *Unique continuation in CR manifolds and in*
        hypo-analytic structures, Arkiv för Mat. **26** (1988), 21–40.

[BT5]   M. S. Baouendi and F. Treves, *A microlocal version of Bochner's tube*
        theorem, Indiana Math. Jour. **31** (1982), 885–895.

[BK]   K. Barbey and H. Konig, *Abstract analytic function theory and Hardy alge-*
        bras, Lecture Notes in Math. **593** (1977), Springer, Berlin.

[BCH]   A. P. Bergamasco, P. Cordaro and J. Hounie, *Global properties of a class of*
        vector fields in the plane, J. Diff. Equations **74** (1988), 179–199.

[BC]   S. Berhanu and S. Chanillo, *Unpublished notes.*

[BH1]   S. Berhanu and J. Hounie, *An F. and M. Riesz theorem for planar vector*
        fields, Math. Ann. **320** (2001), 463–485.

[BH2]   S. Berhanu and J. Hounie, *Uniqueness for locally integrable solutions of*
        overdetermined systems, Duke Math. J. **105** (2000), 387–410.

[BH3]   S. Berhanu and J. Hounie, *A strong uniqueness theorem for planar vector*
        fields, Bol. Soc. Brasil. Mat. **32:3** (2001), 359–376.

[BH4]   S. Berhanu and J. Hounie, *On boundary properties of solutions of complex*
        vector fields., J. Funct. Anal. **192** (2002), 446–490.

[BH5]   S. Berhanu and J. Hounie, *Traces and the F. and M. Riesz theorem for planar*
        vector fields, Annales de l'Institut Fourier **53** (2003), 1425–1460.

[BH6]   S. Berhanu and J. Hounie, *On boundary regularity for one-sided locally*
        solvable vector fields, Indiana Univ. Math. Jour. **52** (2003), 1447–1477.

[BH7]   S. Berhanu and J. Hounie, *An F. and M. Riesz theorem for a system of vector*
        fields, Inventiones Math. **162** (2005), 357–380.

[BH8]   S. Berhanu and J. Hounie, *On the F. and M. Riesz theorem on wedges with*
        edges of class $C^{1,\alpha}$, Math. Zeitschrift **255** (2007), 161–175.

[BHS]   S. Berhanu, J. Hounie and P. Santiago, *A similarity principle for complex*
        vector fields and applications, Trans. Amer. Math. Soc. **353** (2001), 1661–
        1675.

[BM]   S. Berhanu and G. A. Mendoza, *Orbits and global unique continuation for*
        systems of vector fields, J. Geom. Anal. **7** (1997), 173–194.

[BMe]   S. Berhanu and A. Meziani, *Global properties of a class of planar vector*
        fields of infinite type, Commun. in P. D. E. **22** (1997), 99–142.

[Be]   L. Bers, *An outline of the theory of pseudoanalytic functions*, Bull. Amer.
        Math. Soc. **62** (1956), 291–331.

[Bog]   A. Boggess, *CR Manifolds and the Tangential Cauchy–Riemann Complex,*
        Studies in Advanced Mathematics (1991), CRC Press.

[BP]   A. Boggess and J. Polking, *Holomorphic extension of CR functions,* Duke
        Math. J. **49** (1982), 757–784.

[Bo]   J. M. Bony, *Principe du maximum, inégalité de Harnack et unicité du*
        problème de Cauchy pour les operateurs elliptiques dégenerés, Ann. Inst.
        Fourier Grenoble **10** (1969), 277–304.

[Boo]   W. Boothby, *An introduction to differentiable manifolds and Riemannian*
        geometry, Academic Press, second edition (1986).

[Br]   H. Brézis, *On a characterization of flow-invariant sets,* Comm. Pure Appl. Math. **23** (1970), 261–263.

[B]   R. G. M. Brummelhuis, *A Microlocal F. and M. Riesz Theorem with applications,* Revista Matemática Iberoamericana **5** (1989), 21–36.

[C1]   A. P. Calderón, *Lebesgue spaces of differentiable functions and distributions,* Proc. of Symp. in Pure Math. **4** (1961), 33–49.

[C2]   A. P. Calderón, *Intermediate spaces and interpolation, the complex method,* Studia Math. **24** (1964), 113–190.

[CH]   F. Cardoso and J. Hounie, *First-order linear PDE's and uniqueness in the Cauchy problem,* J. of Diff. Equations **33** (1979), 239–248.

[Ch1]   S. Chanillo, L^p *estimates for multiplier transformations in* $\mathbb{R}^2$, PhD Thesis, Purdue University, 1980.

[Ch2]   Private communication.

[ChT]   S. Chanillo and F. Treves, *Local exactness in a class of differential complexes,* J. Amer. Math. Soc. **10(2)** (1997), 393–426.

[Che]   J. Y. Chemin, *Calcul paradifferentiel precise et applications a des equations aux derivées partielles non semilinéaires,* Duke Math. J. **56** (1988), 431–469.

[CR1]   E. M. Chirka and C. Rea, *Differentiable CR mappings and CR orbits,* Duke Math. J. **94(2)** (1998), 325–340.

[CR2]   E. M. Chirka and C. Rea, *F. and M. Riesz theorem for CR functions,* Math. Zeitschrift **250** (2005), 1–6.

[Co]   P. Cohen, *The non-uniqueness of the Cauchy problem,* O.N.R. Tech. Report **93**, Stanford (1960).

[Cor1] P. D. Cordaro, *Representation of hyperfunction solutions in a hypo-analytic structure*, Math. Zeitschrift **233** (2000), 633–654.

[Cor2] P. D. Cordaro, *Approximate solutions in locally integrable structures,* Differential equations and dynamical systems (Lisbon, 2000), 97–112, Fields Inst. Commun., 31, Amer. Math. Soc. (2002).

[CorH1] P. D. Cordaro and J. Hounie, *On local solvability of underdetermined systems of vector fields,* Amer. J. of Math. **112** (1990), 243–270.

[CorH2] P. D. Cordaro and J. Hounie, *Local solvability for top degree forms in a class of systems of vector fields,* Amer. J. of Math. **121** (1999), 487–495.

[CorH3] P. D. Cordaro and J. Hounie, *Local solvability for a class of differential complexes,* Acta Math. **187** (2001), 191–212.

[CorTr] P. D. Cordaro and J. M. Trépreau, *On the solvability of linear partial differential equations in spaces of hyperfunctions,* Arkiv Mat. **36** (1998), 41–71.

[CorT1] P. D. Cordaro and F. Treves, *Homology and cohomology in hypo-analytic structures of the hypersurface type,* J. Geometric Analysis **1** (1991), 39–70.

[CorT2] P. D. Cordaro and F. Treves, *Hyperfunctions in hypo-analytic manifolds,* Annals of Mathematics Studies **136** (1994), Princeton University Press.

[CorT3] P. D. Cordaro and F. Treves, *Necessary and sufficient conditions for the local solvability in hyperfunctions of a class of systems of complex vector fields,* Invent. Math. **120** (1995), 339–360.

[CoHi] R. Courant and D. Hilbert, *Methods of Mathematical Physics,* Vol. II, Wiley, New York (1962).

[D]   G. David, *Opérateurs intégraux singuliers sur certain courbes du plan complexe*, Ann. Scient. Éc. Norm. Sup. (4^0 série) **17** (1984), 157–189.

[DH]   J. J. Duistermaat and L. Hormander, *Fourier integral operators II,* Acta Math. **128** (1972), 183–269.

[Du]   P. Duren, *Theory of H^p spaces*, Academic Press (1970).

[EG1]   M. G. Eastwood and C. Robin Graham, *Edge of the wedge theory in hypo-analytic manifolds,* Comm. PDE **28** (2003), 2003–2028.

[EG2]   M. G. Eastwood and C. Robin Graham, *The Involutive Structure on the Blow-Up of $\mathbb{R}^n$ in $\mathbb{C}^n$*, Communications in Geometry and Analysis **7** (1999), 609–622.

[Fo]   G. H. Folland, *Introduction to Partial Differential Equations,* Mathematical Notes, second edition, Princeton University Press (1995).

[FS]   G. B. Folland and E. M. Stein, *Estimates for the $\bar{\partial}_b$ complex and Analysis on the Heisenberg Group,* Comm. Pure Appl. Math. **27** (1974), 429–522.

[F]   W. H. J. Fuchs, *Topics in the Theory of Functions of one complex variable,* Van Nostrand Mathematical Studies **12** (1967).

[Ga]   P. Garabedian, *An unsolvable equation,* Proc. Amer. Math. Soc. **25** (1970), 207–208.

[G]   D. Goldberg, *A local version of real Hardy spaces,* Duke Math. J. **46** (1979), 27–42.

[GG]   M. Golubitsky and V. Guillemin, *Stable mappings and their singularities,* Graduate Texts in Mathematics, **14**, Springer, New York (1973).

[Gr]   V. Grušin, *A certain example of an equation without solutions,* Mat. Zametki. **10** (1971), 125–128.

[Gu]   P. Guan, *Hölder regularity of subelliptic pseudo-differential operators,* PhD Thesis, Princeton (1989).

[HS]   N. Hanges and J. Sjöstrand, *Propagation of analyticity for a class of nonmicrocharacteristic operators,* Ann. of Math. **116** (1982), 559–577.

[HaT]   N. Hanges and F. Treves, *On the analyticity of solutions of first-order nonlinear pde's,* Transactions of the AMS **331** (1992), 627–638.

[Har]   R. M. Hardt, *Some analytic bounds for subanalytic sets,* Differential Geometric Control Theory (R. Brockett, R. Millman and H. Sussmann, editors), Progress in Mathematics **27**, Birkhäuser (1993), 259–267.

[Ha]   F. R. Harvey, *The theory of hyperfunctions on totally real subsets of a complex manifold with applications to extension problems,* Am. J. of Math. **91** (1969), 853–873.

[HaP]   R. Harvey and J. Polking, *Removable singularities of solutions of linear partial differential equations,* Acta Math. **125** (1970), 39–55.

[Hi]   A. Himonas, *Semirigid partial differential operators and microlocal analytic hypoellipticity,* Duke Math. J. **59** (1989), 265–287.

[HH]   G. Hoepfner and J. Hounie, *Locally solvable vector fields and Hardy spaces,* J. Funct. Anal. (2007), to appear.

[H1]   L. Hörmander, *Linear Partial Differential Operators,* Fourth Printing, Springer-Verlag (1976).

[H2]   L. Hörmander, *The analysis of linear partial differential operators I,* Second edition, Springer-Verlag (1990).

[H3]   L. Hörmander, *The analysis of linear partial differential operators II*, Springer-Verlag (1983).

[H4]   L. Hörmander, *The analysis of linear partial differential operators III*, Springer-Verlag (1985).

[H5]   L. Hörmander, *The analysis of linear partial differential operators IV*, Springer-Verlag (1990).

[H6]   L. Hörmander, *Differential equations without solutions,* Math. Ann. **140** (1960), 169–174.

[H7]   L. Hörmander, *Differential operators of principal type,* Math. Ann. **140** (1960), 124–146.

[H8]   L. Hörmander, *Uniqueness theorems and wave-front sets for solutions of linear differential equations with analytic coefficients*, Comm. Pure Appl. Math. **24** (1971), 671–704.

[H9]   L. Hörmander, *Propagation of singularities and semi-global existence theorems for (pseudo)-differential operators of principal type,* Ann. of Math. **108** (1978), 569–609.

[H10]   L. Hörmander, *Notions of convexity*, Progress in Mathematics **127**, Birkhäuser Boston, Inc., Boston, MA (1994).

[Ho1]   J. Hounie, *Globally hypoelliptic vector fields on compact surfaces*, Comm. PDE **7** (1982), 343–370.

[Ho2]   J. Hounie, *A note on global solvability of vector fields,* Proceedings of the AMS **94** (1985), 61–64.

[Ho3]   J. Hounie, *Local solvability of first order linear operators with Lipschitz coefficients,* Duke Math. J. **62** (1991), 467–477.

[Ho4]   J. Hounie, *The Mizohata structure on the sphere,* Trans. Amer. Math. Soc. **334** (1992), 641–649.

[HMa1] J. Hounie and P. Malagutti, *On the convergence of the Baouendi–Treves approximation formula,* Comm. P.D.E. **23** (1998), 1305–1347.

[HMa2] J. Hounie and P. Malagutti, *Local integrability of Mizohata structures,* Trans. Amer. Math. Soc. **338** (1993), 337–362.

[HMa3] J. Hounie and P. Malagutti, *Existence and nonexistence of homotopy formulas for the Mizohata complex,* Annali di Matematica Pura ed Applicata **173** (1997), 31–42.

[HM1]   J. Hounie and M. E. Moraes Melo, *Local solvability of first order linear operators with Lipschitz coefficients in two variables,* J. of Diff. Equations **121** (1995), 406–416.

[HM2]   J. Hounie and M. E. Moraes Melo, *Local a priori estimates in L^p for first order linear operators with non smooth coefficients,* Manuscripta Mathematica **94** (1997), 151–167.

[HP]   J. Hounie and E. Perdigão, *On local solvability in L^p of first-order equations,* J. of Math. An. and Appl. **197** (1996), 42–53.

[HT1]   J. Hounie and J. Tavares, *Radó's theorem for locally solvable vector fields,* Proc. Amer. Math. Soc. **119** (1993), 829–836.

[HT2]   J. Hounie and J. Tavares, *On removable singularities of locally solvable differential operators,* Inventiones Math. **126** (1996), 589–625.

[HT3]   J. Hounie and J. Tavares, *Removable singularities of vector fields and the Nirenberg–Treves property,* Contemporary Math. **205** (1997), 127–139.

[HT4]   J. Hounie and J. Tavares, *On BMO singularities of solutions of analytic complex vector fields,* Contemporary Math. **251** (2000), 295–308.

[HT5]   J. Hounie and J. Tavares, *The Hartogs property for tube structures,* Indag. Math. (N.S.) **1** (1990), 51–61.

[HdaS] J. Hounie and E. da Silva, *A similarity principle for locally solvable vector fields,* J. Math. Pure Appl. **81** (2002), 715–746.

[Hu]   L. R. Hunt, *The complex Frobenius Theorem and uniqueness of solutions to the tangential Cauchy–Riemann equations,* J. of Diff. Eq. **27** (1978), 214–233.

[HPS]   L. R. Hunt, J. C. Polking and M. J. Strauss, *Unique continuation for solutions to the induced Cauchy–Riemann equations,* J. of Diff. Eq. **23** (1977), 436–447.

[J1]   H. Jacobowitz, *A nonsolvable complex vector field with Hölder coefficients,* Proc. Amer. Math. Soc. **116** (1992), 787–795.

[J2]   H. Jacobowitz, *Global Mizohata structures,* J. Geom. Analysis **3** (1993), 153–193.

[JT1]   H. Jacobowitz and F. Treves, *Aberrant CR structures,* Hokkaido Math. J. **12** (1983), 276–292.

[JT2]   H. Jacobowitz and F. Treves, *Nowhere solvable homogeneous partial differential equations,* Bull. AMS **8** (1983), 467–469.

[Jo1]   B. Joricke, *Boundaries of singularity sets, removable singularities, and CR-invariant subsets of CR manifolds,* J. of Geom. Anal. **9** (1999), 257–300.

[Jo2]   B. Joricke, *Deformation of CR manifolds, minimal points and CR manifolds with the microlocal analytic extension property,* J. of Geom. Anal. **6** (1996), 555–611.

[KaS]   M. Kashiwara and P. Schapira, *A vanishing theorem for a class of systems with simple characteristics,* Invent. Math. **82** (1985), 579–592.

[K]   C. Kenig, *Progress on two problems posed by Rivière,* Contemp. Math. **107** (1990), 101–107.

[KT1]   C. Kenig and P. Tomas, *On conjectures of Rivière and Schtrichartz,* Bull. Amer. Math. Soc. **1** (1979), 694–697.

[KT2]   C. Kenig and P. Tomas, L^p *behavior of certain second order differential equations,* Trans. Amer. Math. Soc. **262** (1980), 521–531.

[K1]   S. Koshi, *Topics in complex analysis: Recent developments on the F. and M. Riesz theorem,* Banach Center Publ. **31** (1995), Polish Acad. Sci., Warsaw.

[K2]   S. Koshi, *The F. and M. Riesz theorem on locally compact abelian groups,* Infinite-dimen. Harmonic Analysis (Tubingen) (1995), 138–145.

[Ku1]   M. Kuranishi, *Strongly pseudo-convex CR structures over small balls, Part I,* Ann. of Math. **115** (1982), 451–500.

[Ku2]   M. Kuranishi, *Strongly pseudo-convex CR structures over small balls, Part II and III,* Ann. of Math. **116** (1982), 1–64 and 249–330.

[KR]   A. M. Kytmanov and C. Rea, *Elimination of* L^1 *singularities on Holder peak sets for CR functions,* Ann. Scuola Norm. Sup. Pisa Cl. Sci. **22** (1995), 211–226.

[L]   L. Lanzani, *Cauchy transform and Hardy spaces for rough planar domains.* Contemporary AMS **251** (2000), 409–428.

[L1]   H. Lewy, *On the local character of the solution of an atypical Differential Equation in three variables and a related problem for regular functions of two complex variables,* Ann. of Math. **64** (1956), 514–522.

[L2]   H. Lewy, *An example of a smooth linear partial differential equation without solution,* Ann. of Math. **66** (1957), 155–158.

[Li]   W. Littman, *The wave operator and L^p norms,* J. Math. Mech. **12** (1963), 55–68.

[LP]   N. Lusin and J. Priwaloff, *Sur l'unicité et la multiplicité des fonctions analytiques,* Ann. Sci. École Norm. Sup. **42** (1992), 143–194.

[Mal]   B. Malgrange, *Sur l'intégrabilité de structures presque-complexes,* Symposia Math., vol. II (INDAM, Rome, 1968), Academic Press, London (1969), 289–296.

[Ma]   M. E. Marson, *Wedge extendability for hypo-analytic structures,* Comm. in P. D. E. **17** (1992), 579–592.

[MenT] G. Mendoza and F. Treves *Local solvability in a class of overdetermined systems of linear PDE,* Duke Math. J. **63** (1991), 355–377.

[Mer1] J. Merker, *Global minimality of generic manifolds and holomorphic extendability of CR functions,* Internat. Math. Res. Notices **8** (1994), 329–343.

[Mer2] J. Merker, *On removable singularities for CR functions in higher codimension,* Internat. Math. Res. Notices **1** (1997), 21–56.

[MP1]   J. Merker and E. Porten, *Metrically thin singularities of integrable CR functions,* Internat. J. Math. **11** (2000), 857–872.

[MP2]   J. Merker and E. Porten, *On removable singularities for integrable CR functions,* Indiana Univ. Math. J. **48** (1999), 805–856.

[MP3]   J. Merker and E. Porten, *Holomorphic extension of CR functions, envelopes of homomorphy and removable singularities,* International Math. Research Surveys, to appear.

[Met]   G. Metivier, *Uniqueness and approximation of solutions of first order non linear equations,* Inventiones Math. **82** (1985), 263–282.

[Me1]   A. Meziani, *On real analytic planar vector fields near the characteristic set,* Contemporary Math. **251** (2000), 429–438.

[Me2]   A. Meziani, *On the similarity principle for planar vector fields: Application to second order pde,* J. Diff. Equations **157** (1999), 1–19.

[Me3]   A. Meziani, *Infinitesimal bending of homogeneous surfaces with nonnegative curvature,* Comm. Analysis and Geometry **11** (2003), 697–719.

[Me4]   A. Meziani, *Planar complex vector fields and infinitesimal bendings of surfaces with nonnegative curvature,* Contemporary Math. **400** (2006), to appear.

[Me5]   A. Meziani, *Mizohata structures on S^2: automorphisms and standardness,* Contemporary Math. **205** (1997), 235–246.

[Me6]   A. Meziani, *On the integrability of Mizohata structures on the sphere S^2,* J. Geom. Anal. **9** (1999), 301–315.

[Me7]   A. Meziani, *Classification of germs of Mizohata structures,* Commun. in P. D. E. **20** (1995), 499–539.

[Mi]   V. Michel, *Sur la regularité C^∞ du $\bar{\partial}$ au bord d'un domaine de $\mathbf{C}^n$ dont la forme de Levi a exactement s valeurs propres strictement négatives.* Math. Ann. **295** (1993), 135–161.

[M] S. Mizohata, *Solutions nulles et solutions non analytiques,* J. Math. Kyoto Univ. **1** (1962), 271–302.

[Mo] R. D. Moyer, *Local solvability in two dimensions: Necessary conditions for the principal type case,* Mimeographed manuscript, University of Kansas (1978).

[Na] T. Nagano, *Linear differential systems with singularities and an application to transitive Lie algebras,* J. Math. Soc. Japan **18** (1966), 398–404.

[N1] L. Nirenberg, *Lectures on Linear Partial Differential Equations,* A.M.S. Regional Conf. Series in Math. **17** (1973).

[N2] L. Nirenberg, *A complex Frobenius theorem,* Seminar on analytic functions I, Princeton (1957), 172–189.

[N3] L. Nirenberg, *On a question of Hans Lewy,* Russian Mathematical Surveys **29** (1974), 251–262.

[NT] L. Nirenberg and F. Treves, *Solvability of a first order linear partial differential equation,* Comm. Pure Appl. Math. **16** (1963), 331–351.

[NN] A. Newlander and L. Nirenberg, *Complex coordinates in almost complex manifolds,* Ann. of Math. **65** (1957), 391–404.

[Po] Ch. Pommerenke, *Boundary behaviour of Conformal maps,* Springer-Verlag, Wiley-Interscience (1992).

[Pr] J. Pradines, *How to define the graph of a singular foliation,* Cahiers de Top. et Geom. Diff. **26** (1985), 339–380.

[Re] H. Renelt, *Elliptic systems and quasiconformal mappings,* Pure and Applied Mathematics, Wiley-Interscience (1988).

[Ro] J. P. Rosay, *On the radial maximal function and the Hardy Littlewood maximal function in wedges,* Contemp. Math. **137** (1992), 383–398.

[RS] J-P. Rosay and E. Stout, *Radó's theorem for CR functions,* Proc. Amer. Math. Soc. **106** (1989), 1017–1026.

[Sa] M. Sato, *Theory of hyperfunctions,* J. Fac. Sci., Univ. Tokyo, Part I in **1** (1959), 139–193, Part II in **1** (1960), 387–437.

[daS] E. da Silva, *Estimativas a priori em espaços de Hardy para campos vetoriais complexos localmente resolúveis,* Tese de Doutorado, UFSCar, DM (2000).

[Sj1] J. Sjöstrand, *Singularites analytiques microlocales,* Asterisque **95** (1982), Soc. Math. France.

[Sj2] J. Sjöstrand, *Note on a paper of F. Treves concerning Mizohata type operators,* Duke Math. J. **47** (1980), 77–107.

[Sm] H. Smith, *An elementary proof of local solvability in two dimensions under condition* (Ψ), Ann. of Math. **136** (1992), 335–337.

[S1] E. M. Stein, *Singular integrals and differentiability properties of functions,* Princeton Univ. Press (1970).

[S2] E. M. Stein, *Harmonic Analysis: Real-variable methods, orthogonality, and oscillatory integrals,* Princeton Univ. Press (1993).

[SW] E. M. Stein and G. Weiss, *On the theory of harmonic functions of several variables I,* Acta Math. **103** (1960), 25–62.

[ST] M. J. Strauss and F. Treves, *First-order linear PDEs and uniqueness in the Cauchy problem,* J. of Diff. Eq. **15** (1974), 195–209.

[Su] H. J. Sussmann, *Orbits of families of vector fields and integrability of distributions,* Trans. Amer. Math. Soc. **180** (1973), 171–188.

[Sz] R. Szöke, *Involutive structures on the tangent bundle of symmetric spaces,* Math. Ann. **319** (2001), 319–348

[Ta] M. Taylor, *Tools for PDE,* Mathematical Surveys and Monographs **81**, AMS (2000).

[Tr] J. M. Trepreau, *Sur la propagation des singularités dans les variétés CR,* Bull. Soc. Math. Fr. **118** (1990), 403–450.

[T1] F. Treves, *Topological Vector Spaces, Kernels and Distributions,* Academic Press (1967).

[T2] F. Treves, *Local solvability in L^2 for first order linear PDEs,* Amer. J. of Math. **92** (1970), 369–380.

[T3] F. Treves, *On the local solvability for top-degree forms in hypo-analytic structures,* Amer. J. of Math. **112** (1999), 403–421.

[T4] F. Treves, *Approximation and representation of functions and distributions annihilated by a system of complex vector fields,* École Polytech., Centre de Math., Palaiseau, France (1981).

[T5] F. Treves, *Hypo-analytic structures,* Princeton University Press (1992).

[T6] F. Treves, *Integral representation of solutions of first-order linear partial differential equations I,* Ann. Scuola Norm. Sup. Pisa **3** (1976), 1–35.

[T7] F. Treves, *Remarks about certain first-order linear PDE in two variables,* Comm. PDE **5** (1980), 381–425.

[T9] F. Treves, *A remark on the Poincaré lemma in analytic complexes with nondegenerate Levi form,* Comm. in P.D.E. **7** (1982), 1467–1482.

[Tu1] A. E. Tumanov, *Extending CR functions on a manifold of finite type over a wedge,* Mat. Sbornik **136** (1988), 129–140.

[Tu2] A. E. Tumanov, *Connections and propagation of analyticity for CR functions,* Duke Math. J. **73** (1994), 1–24.

[U] A. Uchiyama, *A constructive proof of the Fefferman–Stein decomposition of $BMO(\mathbb{R}^n)$,* Acta Math. **148** (1982), 215–241.

[V] I. V. Vekua, *Generalized Analytic Functions,* Pergamon Press (1962).

[W1] S. M. Webster, *Analytic disks and the regularity of CR mappings of real submanifolds of $\mathbb{C}^n$,* Proc. of Symp. in Pure Math. **41** (1984), 199–208.

[W2] S. M. Webster, *A new proof of the Newlander–Nirenberg theorem,* Math. Z. **201**(1989), 303–316.

[W3] S. M. Webster, *On the local solution of the tangential Cauchy–Riemann equations,* Ann. Inst. Henri Poincaré, Anal. Non Linéaire **6** (1989), 167–182.

[Z] E. C. Zachmanoglou, *Propagation of the zeros and uniqueness in the Cauchy problem for first-order partial differential equations,* Arch Ratl. Mech. Anal. **38** (1970), 178–188.

[Zu] C. Zuily, *Uniqueness and non-uniqueness in the Cauchy Problem,* Birkhäuser, Boston-Basel-Stuttgart (1983).

[Zy] A. Zygmund, *Trigonometric Series,* Cambridge Univ. Press (1968).

Index